Gravitation and Inertia

PRINCETON SERIES IN PHYSICS

Edited by Philip W. Anderson, Arthur S. Wightman, and Sam B. Treiman (published since 1976)

Studies in Mathematical Physics: Essays in Honor of Valentine Bargmann *edited by Elliott H. Lieb, B. Simon, and A. S. Wightman*

Convexity in the Theory of Lattice Gases *by Robert B. Israel*

Works on the Foundations of Statistical Physics *by N. S. Krylov*

Surprises in Theoretical Physics *by Rudolf Peierls*

The Large-Scale Structure of the Universe *by P. J. E. Peebles*

Statistical Physics and the Atomic Theory of Matter, From Boyle and Newton to Landau and Onsager *by Stephen G. Brush*

Quantum Theory and Measurement *edited by John Archibald Wheeler and Wojciech Hubert Zurek*

Current Algebra and Anomalies *by Sam B. Treiman, Roman Jackiw, Bruno Zumino, and Edward Witten*

Quantum Fluctuations *by E. Nelson*

Spin Glasses and Other Frustrated Systems *by Debashish Chowdhury*
(*Spin Glasses and Other Frustrated Systems* is published in co-operation with World Scientific Publishing Co. Pte. Ltd., Singapore.)

Weak Interactions in Nuclei *by Barry R. Holstein*

Large-Scale Motions in the Universe: A Vatican Study Week *edited by Vera C. Rubin and George V. Coyne, S.J.*

Instabilities and Fronts in Extended Systems *by Pierre Collet and Jean-Pierre Eckmann*

More Surprises in Theoretical Physics *by Rudolf Peierls*

From Perturbative to Constructive Renormalization *by Vincent Rivasseau*

Supersymmetry and Supergravity (2nd ed.) *by Julius Wess and Jonathan Bagger*

Maxwell's Demon: Entropy, Information, Computing *edited by Harvey S. Leff and Andrew F. Rex*

Introduction to Algebraic and Constructive Quantum Field Theory *by John C. Baez, Irving E. Segal, and Zhengfang Zhou*

Principles of Physical Cosmology *by P. J. E. Peebles*

Scattering in Quantum Field Theories: The Axiomatic and Constructive Approaches *by Daniel Iagolnitzer*

QED and the Men Who Made It: Dyson, Feynman, Schwinger, and Tomonga *by Silvan S. Schweber*

The Interpretation of Quantum Mechanics *by Roland Omnès*

Gravitation and Inertia *by Ignazio Ciufolini and John Archibald Wheeler*

Gravitation and Inertia

IGNAZIO CIUFOLINI

AND

JOHN ARCHIBALD WHEELER

Princeton Series in Physics

PRINCETON UNIVERSITY PRESS · PRINCETON, NEW JERSEY

Published by Princeton University Press, 41 William Street,
Princeton, New Jersey 08540
In the United Kingdom: Princeton University Press,
Chichester, West Sussex

Library of Congress Cataloging-in-Publication Data

Ciufolini, Ignazio, 1951–
Gravitation and inertia / Ignazio Ciufolini and John Archibald
Wheeler.
p. cm. — (Princeton series in physics)
Includes bibliographical references and indexes.
ISBN 0–691–03323–4
1. Geometrodynamics. 2. General relativity (Physics)
3. Gravitation 4. Inertia (Mechanics) I. Wheeler, John
Archibald, 1911– . II. Title. III. Series.
QC173.59.G44C58 1995
530.1′1—dc20 94–29874 CIP

This book has been composed in Times Roman

Princeton University Press books are printed on acid-free paper and meet the
guidelines for permanence and durability of the Committee on Production
Guidelines for Book Longevity of the Council on Library Resources

Printed in the United States of America
by Princeton Academic Press

1 3 5 7 9 10 8 6 4 2

Contents

PREFACE ix

CHART OF MAIN TOPICS xii

1. A First Tour 1

1.1 Spacetime Curvature, Gravitation, and Inertia 1
1.2 Relation of Gravity to the Other Forces of Nature 9
References 10

2. Einstein Geometrodynamics 13

2.1 The Equivalence Principle 13
2.2 The Geometrical Structure 19
2.3 The Field Equation 21
2.4 Equations of Motion 27
2.5 The Geodesic Deviation Equation 31
2.6 Some Exact Solutions of the Field Equation 36
2.7 Conservation Laws 42
2.8 [The Boundary of the Boundary Principle and Geometrodynamics] 49
2.9 Black Holes and Singularities 61
2.10 Gravitational Waves 71
References 78

3. Tests of Einstein Geometrodynamics 87

3.1 Introduction 87
3.2 Tests of the Equivalence Principle 90
3.2.1 The Weak Equivalence Principle 91
3.2.2 Gravitational Redshift—Gravitational Time Dilation 97
3.2.3 The Equivalence Principle and Spacetime Location Invariance 109
3.2.4 The Equivalence Principle and Lorentz Invariance 111
3.2.5 The Very Strong Equivalence Principle 112

3.3 Active and Passive Gravitational Mass 115
3.4 Tests of the Geometrical Structure and of the Geodesic Equation of Motion 116
3.4.1 Gravitational Deflection of Electromagnetic Waves 117
3.4.2 Delay of Electromagnetic Waves 122
3.4.3 [de Sitter or Geodetic Effect and Lense-Thirring Effect] 128
3.4.4 Other Tests of Space Curvature 138
3.5 Other Tests of Einstein General Relativity, of the Equations of Motion, and of the Field Equations 139
3.5.1 Pericenter Advance 141
3.6 [Proposed Tests to Measure Gravitational Waves and Gravitomagnetic Field] 147
3.6.1 [Resonant Detectors] 148
3.6.2 [LIGO: Laser-Interferometer Gravitational-Wave Observatory] 154
3.6.3 [MIGO: Michelson Millimeter Wave Interferometer Gravitational- Wave Observatory] 157
3.6.4 [LISA or LAGOS: Laser Gravitational-Wave Observatory in Space] 158
3.6.5 [SAGITTARIUS-LINE: Laser Interferometer Near Earth] 160
3.6.6 [Interplanetary Doppler Tracking] 162
3.7 [Appendix: Metric Theories and PPN Formalism] 163
References 168

4. Cosmology, Standard Models, and Homogeneous Rotating Models 185

4.1 The Universe on a Large Scale 185
4.2 Homogeneity, Isotropy, and the Friedmann Cosmological Models 193
4.3 [Closure in Time Versus Spatial Compactness] 220
4.4 Spatially Homogeneous Models: The Bianchi Types 231
4.5 Expansion, Rotation, Shear, and the Raychaudhuri Equation 234
4.6 [The Gödel Model Universe] 240
4.7 [Bianchi IX Rotating Cosmological Models] 244
4.8 [Cosmology and Origin of Inertia] 249
References 256

5. The Initial-Value Problem in Einstein Geometrodynamics 269

5.1 [From the Initial-Value Problem to the Origin of Inertia in Einstein Geometrodynamics] 269
5.2 [The Initial-Value Problem and the Interpretation of the Origin of Inertia in Geometrodynamics] 271
5.3 [The Solution of the Initial-Value Equations] 289
5.4 [The Finally Adjusted Initial-Value Data for the Dynamics of Geometry] 293
5.5 [The Dynamics of Geometry] 294
5.6 [Further Perspectives on the Connection between Mass-Energy There and Inertia Here] 295
5.7 [Poor Man's Account of Inertial Frame] 298
5.8 [A Summary of Energy There Ruling Inertia Here] 300
5.9 [A Summary of the Initial-Value Problem and Dragging of Inertial Frames] 303
References 306

6. The Gravitomagnetic Field and Its Measurement 315

6.1 The Gravitomagnetic Field and the Magnetic Field 315
6.2 Gravitomagnetism and the Origin of Inertia in Einstein Geometrodynamics 324
6.3 The Gravitomagnetic Field in Astrophysics 327
6.4 The Pail, the Pirouette, and the Pendulum 328
6.5 Measurement of the Gravitomagnetic Field 330
6.6 [The Stanford Gyroscope: Gravity Probe-B (GP-B)] 332
6.7 [LAGEOS III] 334
6.8 [Error Sources and Error Budget of the LAGEOS III Experiment] 339
6.9 [Other Earth "Laboratory" Experiments] 347
6.10 The de Sitter or Geodetic Precession 351
6.11 [Gravitomagnetism, Dragging of Inertial Frames, Static Geometry, and Lorentz Invariance] 353
6.12 Appendix: Gyroscopes and Inertial Frames 361
References 374

7. Some Highlights of the Past and a Summary of Geometrodynamics and Inertia 384

7.1 Some Highlights of the Past 384
7.2 Geometrodynamics and Inertia 394
References 399

Mathematical Appendix 403

Symbols and Notations 437

Author Index 445

Subject Index of Mathematical Appendix 455

Subject Index 461

Fundamental and Astronomical Constants and Units 493

Preface

This book is on Einstein's theory of general relativity, or geometrodynamics. It may be used as an introduction to general relativity, as an introduction to the *foundations* and *tests* of gravitation and geometrodynamics, or as a monograph on the meaning and origin of inertia in Einstein theory.

The local equivalence of "gravitation" and "inertia," or the local "cancellation" of the gravitational field by local inertial frames, inspired Einstein to the theory of general relativity. This equivalence is realized through the geometrodynamical structure of spacetime. A gravitational field is affected by mass-energy distributions and currents, as are the local inertial frames. Gravitational field and local inertial frames are both characterized by the spacetime metric, which is determined by mass-energy distributions and currents.

The precise way by which the spacetime metric is determined by mass-energy and mass-energy currents is clarified by the initial-value formulation of general relativity. Central to the understanding of the origin of inertia in Einstein theory are: (a) the geometrodynamical formulation of the initial-value problem on a spacelike three-manifold and the Cauchy problem; (b) cosmological considerations on the compactness of space of some model universes and on hypothetical rotations of the cosmological fluid with respect to the local inertial observers, that is, with respect to the local gyroscopes; and (c) the theory and the measurement of the "gravitomagnetic field" and "dragging of inertial frames" by mass-energy currents. Some emphasis is given in the book to these topics.

Together with the great theoretical and experimental successes of Einstein standard geometrodynamics, come two main conceptual problems. First, the theory predicts the occurrence of spacetime singularities, events which are not part of a smooth spacetime manifold, where usually the curvature diverges and where the Einstein field equation and the known physical theories cease to be valid. Second, Einstein's theory of gravitation, unlike the other fundamental interactions, has not yet been "successfully" quantized. Therefore, it is of critical importance to test and to interpret correctly each prediction and each part of the foundations of geometrodynamics and of other gravitational theories.

In more than three-quarters of a century general relativity has achieved an experimental triumph. Nevertheless, all the impressive *direct* confirmations of Einstein geometrodynamics, from gravitational time dilation to deflection of path of photons, and from time delay of electromagnetic waves to Lunar Laser Ranging analyses and de Sitter or geodetic effect, are confirmations of weak-field corrections to the classical Galilei-Newton mechanics. However, so far, apart from the impressive astrophysical observations and *indirect* evidence from the orbital energy loss of the binary pulsar PSR 1913+16, in agreement with the calculated general relativistic emission of gravitational radiation, we have no direct detection and measurement of two fundamentally new phenomena predicted by general relativity but not by classical mechanics: gravitational waves, the gravitational "analogue" of electromagnetic waves, and gravitomagnetic field, generated by mass-energy currents, the gravitational "analogue" of a magnetic field. Furthermore, today general relativity is no longer a theory far removed from every practical application. Einstein theory is an active field of investigation in relation to space research and navigation in the solar system. Impressive examples are the need to include general relativistic corrections to be able to describe the motion of the Global Positioning Satellites to the needed level of accuracy, the need of general relativistic corrections in the techniques of Very Long Baseline Interferometry and Satellite Laser Ranging, and even the proposed hypothetical propulsion of a future spacecraft using the gravitational lensing effect of the Sun. Therefore, it is crucial to test and to measure accurately each prediction of general relativity. A special emphasis is thus given in this book to the status of past, present, and proposed future tests of gravitational interaction, metric theories, and general relativity.

In chapter 1 we briefly introduce the reader to some illuminating ideas on gravitation, inertia, and cosmology in geometrodynamics.

In chapter 2 we describe the foundations and the main features of Einstein general relativity.

In chapter 3 we present the impressive confirmations of the foundations of geometrodynamics and some proposed experiments to test some of its basic predictions never directly measured. In particular, we present the main laboratory and space experiments proposed to directly measure gravitational waves.

In chapter 4 we describe the standard Friedmann models of the universe and the relations between spatial compactness and time recollapse of some model universes. We briefly report on some beautiful and impressive observations made by COBE (Cosmic Background Explorer) and by the Hubble Space Telescope. We then introduce the spatially homogeneous Bianchi cosmological models, and we discuss the Gödel rotating models and other spatially homogeneous rotating models.

In chapter 5 we describe the formulation of the initial-value problem in the case of a model universe admitting a foliation in compact spacelike hypersurfaces.

In chapter 6 we introduce the concept of gravitomagnetic field generated by mass currents, in partial analogy with electrodynamics, and the main experiments proposed to detect it. Its measurement, through the measurement of the dragging of inertial frames, will constitute direct *experimental* evidence against an absolute inertial frame of reference and will experimentally display the basic role in nature of the local inertial frames.

Finally, in chapter 7 we briefly report on some of the historical discussions on inertia, inertial frames, and gravitation. These arguments go back to a Leibniz-Newton debate, to Huygens, Berkeley, Mach, Einstein, and others. We also give a brief summary of gravitation and inertia in geometrodynamics.

In the mathematical appendix, with a related analytical index, we briefly introduce the main elements of tensor calculus and differential geometry used in geometrodynamics. The mathematical appendix may be useful to a reader not familiar with differential geometry. However, a complete introduction to these topics requires a separate book. References to some introductory books on differential geometry are given in the text. In the mathematical appendix, definitions and treatment are given both in coordinate-independent form and in components in a coordinate system. However, in order to make the book easier to understand for a reader not familiar with the coordinate-independent notation of differential geometry, and since in physics most of the calculations are simpler in a particular coordinate system, also to gain some physical insight, most of the equations in the book are in components. Furthermore, to make the use of the book easier for different readers, following the style of *Gravitation* by Misner, Thorne, and Wheeler, we have supplied two tracks. Track 1 topics may be used as an introduction to general relativity, while track 2, indicated by section headings in square brackets, covers more specialized topics on gravitation, inertia, foundations, and tests of Einstein geometrodynamics. To help in selecting different topics and arguments, at the beginning of the book we have supplied a chart of the main topics. A table of notations and symbols used in the book is also given after the mathematical appendix. Since references come at the end of each chapter, we have listed at the end of the book, after the table of notations and symbols, all authors cited in the references. Finally, after the mathematical analytical index and the general analytical index, we have included a table with some useful physical quantities.

It is a pleasure for us to thank, for all their invaluable suggestions and comments, our colleagues: Allen Anderson, John Anderson, Carlo Bernardini, Bruno Bertotti, Carl Brans, Gaetano Chionchio, James Condon, George Ellis, Richard Ellis, Frank Estabrook, Ephraim Fischbach, Luciano Guerriero, Ronald Hellings, Tom Hutchings, James Isenberg, Arkadi Kheyfets, Vladimir Lukash, Charles Misner, Changbom Park, James Peebles, Wolfgang Rindler, Wolfgang Schleich, Dennis Sciama, Massimo Visco, and Stefano Vitale, and *in particular* Peter Bender, Emily Bennett, Manlio Bordoni, Milvia Capalbi, Francesco De Paolis, Harry King, Richard Matzner, Kenneth Nordtvedt, Mike Reisenberger, Lawrence Shepley, James York, and Luciano Vanzo.

Finally, we express our appreciation to *Brunella* and *Loreta Ciufolini* and *Janette Wheeler*, for their patience and encouragement in the completion of this work.

December 1993

Main Topics

- Some Highlights of the Past on Inertia and Gravitation — 7.1
- **Summary of Gravitation and Inertia** — 7.2

[GI]

- **Tests of Geometrodynamics** — 3.1
- Equivalence Principle — 3.2, 3
- Spacetime Curvature — 3.4
- Equations of Motion — 3.4, 5
- Field Equations — 3.5

* de Sitter Effect and Lense-Thirring Effect — 3.4.3
* **Gravitational Waves** — 3.6
* Gravitational Metric Theories — 3.7

[I] [ET] [GI]

- Basic Standard **Cosmology** — 4.1, 2
- Homogeneous Cosmological Models, Bianchi Models — 4.4
- Expansion, Rotation, Shear, Raychaudhuri Equation — 4.5

* Time "Closure" of a Model Universe — 4.3
* Space "Closure" of a Model Universe — 4.3
* "Closure" in Time Implies "Closure" in Space? And Vice Versa? — 4.3
* Gödel Models — 4.6
* Rotating Model Universes — 4.6, 7
* Cosmology and Inertia — 4.7, 8

[I] [GI]

---

* **Initial-Value Formulation of Geometrodynamics** Chapter 5
* Initial-Value Formulation and Interpretation of the Origin of Inertia (Local Inertial Frames) in Geometrodynamics 5.1, 8, 9

[GI]

- **Gravitomagnetic Field** — 6.1, 2, 3, 4, 5
- Gravitomagnetism, Dragging of Inertial Frames and Origin of Inertia — 6.2, 4
- de Sitter Effect, Papapetrou Equation — 6.10
- Gyroscopes — 6.12

* Measurement of the Gravitomagnetic Field — 6.5, 6, 7, 8, 9
* Gravitomagnetism, Dragging of Inertial Frames, and de Sitter Effect — 6.11

[ET] [GI]

---

t ↑

- **Introduction to Einstein's Theory of General Relativity, or Geometrodynamics:** Equivalence Principle, Geometrical Structure, Field Equation, Equations of Motion, Geodesic Deviation Equation, Exact Solutions, Conservation Laws, Black Holes, Singularities, and Gravitational Waves — 2.1, 2, 3, 4, 5, 6, 7, 9, 10

* The Boundary of the Boundary Principle and Geometrodynamics — 2.8

[I]

- **Introduction to Gravitation and Inertia** — 1.1, 2

[I]

- **Track 1 Reader**
* **Track 2 Reader**

I = Introductory Material
ET = Experiments and Tests
GI = Gravitation and Inertia

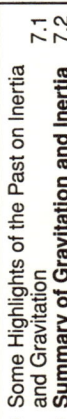

Gravitation and Inertia

Inertia here arises from mass there

1

A First Tour

In this chapter we introduce some basic concepts and ideas of Einstein's General Theory of Relativity which are developed in the book.

1.1 SPACETIME CURVATURE, GRAVITATION, AND INERTIA

"**Gravity** is not a foreign and physical force transmitted through space and time. It is a manifestation of the *curvature of spacetime*." That, in a nutshell, is **Einstein's theory**.[1,2]

What this theory is and what it means, we grasp more fully by looking at its intellectual antecedents. First, there was the idea of Riemann[3] that space, telling mass how to move, must itself—by the principle of action and reaction—be affected by mass. It cannot be an ideal Euclidean perfection, standing in high mightiness above the battles of matter and energy. Space geometry must be a participant in the world of physics. Second, there was the contention of Ernst Mach[4] that the "acceleration relative to absolute space" of Newton is only properly understood when it is viewed as acceleration relative to the sole significant mass there really is, the distant stars. According to this "Mach principle," *inertia here arises from mass there*. Third was that great insight of Einstein that we summarize in the phrase "free fall is free float": the **equivalence principle**, one of the best-tested principles in physics, from the inclined tables of Galilei and the pendulum experiments of Galilei, Huygens, and Newton to the highly accurate torsion balance measurements of the twentieth century, and the Lunar Laser Ranging experiment (see chaps. 2 and 3). With those three clues vibrating in his head, the magic of the mind opened to Einstein what remains one of mankind's most precious insights: gravity is manifestation of **spacetime curvature**.

Euclid's (active around 300 B.C.) fifth postulate states that, given any straight line and any point not on it, we can draw through that point one and only one straight line parallel to the given line, that is, a line that will never meet the given one (this alternative formulation of the fifth postulate is essentially due to Proclos). This is the parallel postulate. In the early 1800s the discussion grew lively about whether the properties of parallel lines as presupposed in Euclidean geometry could be derived from the other postulates and axioms, or whether the parallel postulate had to be assumed independently. More than two thousand years after Euclid, Karl Friedrich Gauss, János Bolyai, and Nikolai Ivanovich Lobačevskij discovered pencil-and-paper geometric systems that satisfy all the axioms and postulates of Euclidean geometry except the parallel postulate. These geometries showed not only that the parallel postulate must be assumed in order to obtain Euclidean geometry but, more important, that non-Euclidean geometries as mathematical abstractions can and do exist (see § 2.2, 3.4, and 4.3).

Consider the two-dimensional surface of a sphere, itself embedded in the three-dimensional space geometry of everyday existence. Euclid's system accurately describes the geometry of ordinary three-dimensional space, but not the geometry on the surface of a sphere. Let us consider two lines locally parallel on the surface of a sphere (fig. 1.1). They propagate on the surface as straight as any lines could possibly be, they bend in their courses one whit neither to left or right. Yet they meet and cross. Clearly, geodesic lines (on a surface, a geodesic is the shortest line between two nearby points) on the curved surface of a sphere do not obey Euclid's parallel postulate.

The thoughts of the great mathematician Karl Friedrich Gauss about curvature stemmed not from theoretical spheres drawn on paper but from concrete, down-to-Earth measurements. Commissioned by the government in 1827 to make a survey map of the region for miles around Göttingen, he found that the sum of the angles in his largest survey triangle was different from 180°. The deviation from 180° observed by Gauss—almost 15 seconds of arc—was both inescapable evidence for and a measure of the curvature of the surface of Earth.

To recognize that straight and initially parallel lines on the surface of a sphere can meet was the first step in exploring the idea of a curved space. Second came the discovery of Gauss that we do not need to consider a sphere or other two-dimensional surface to be embedded in a three-dimensional space to define its geometry. It is enough to consider measurements made entirely within that two-dimensional geometry, such as, would be made by an ant forever restricted to live on that surface. The ant would know that the surface is curved by measuring that the sum of the internal angles of a large triangle differs from 180°, or by measuring that the ratio between a large circumference and its radius R differs from 2π.

Gauss did not limit himself to thinking of a curved two-dimensional surface floating in a flat three-dimensional universe. In an 1824 letter to Ferdinand Karl Schweikart, he dared to conceive that space itself is curved: "Indeed I have therefore from time to time in jest expressed the desire that Euclidean geometry would not be correct." He also wrote: "Although geometers have given much attention to general investigations of curved surfaces and their results cover a significant portion of the domain of higher geometry, this subject is still so far from being exhausted, that it can well be said that, up to this time, but a small portion of an exceedingly fruitful field has been cultivated" (Royal Society of

(a)

(b)

(c)

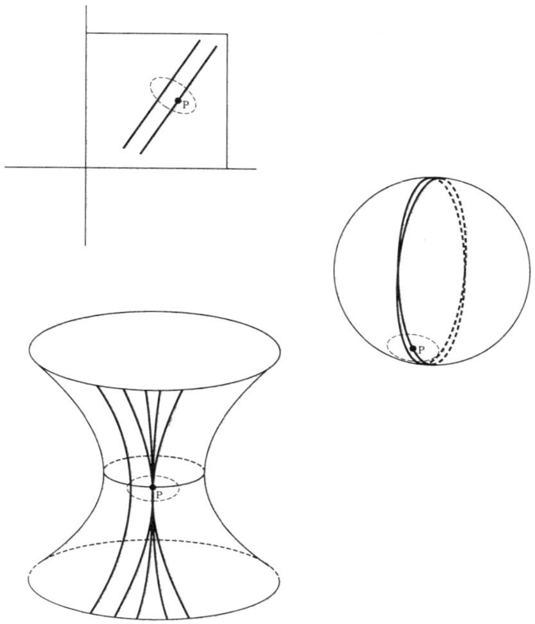

FIGURE 1.1. Surfaces with different curvature.

(a) Zero curvature surface: through any point P not lying on a given straight line (geodesic line), there is one, and only one, straight line parallel to the given line.

(b) Positive curvature: there are no geodesic lines parallel (in the sense that they will never intersect) to a given geodesic (every geodesic line through P will intersect the given geodesic).

(c) Negative curvature: there are infinitely many geodesic lines parallel to a given geodesic (in the sense that infinitely many geodesic lines through P will never intersect the given geodesic).

Göttingen, 8 October 1827). The inspiration of these thoughts, dreams, and hopes passed from Gauss to his student, Bernhard Riemann.

Bernhard Riemann went on to generalize the ideas of Gauss so that they could be used to describe curved spaces of three or more dimensions. Gauss had found that the curvature in the neighborhood of a given point of a specified two-dimensional space geometry is given by a single number: the Gaussian curvature. Riemann found that six numbers are needed to describe the curvature of a three-dimensional space at a given point, and that 20 numbers at each point are required for a four-dimensional geometry: the 20 independent components of the so-called Riemann curvature tensor.

In a famous lecture he gave 10 June 1854, entitled *On the Hypotheses That Lie at the Foundations of Geometry*,[3] Riemann emphasized that the truth about space is to be discovered not from perusal of the 2000-year-old books of Euclid but from physical experience. He pointed out that space could be highly irregular at very small distances and yet appear smooth at everyday distances. At very great distances, he also noted, large-scale curvature of space might show up, perhaps even bending the universe into a closed system like a gigantic ball. He wrote: "Space [in the large] if one ascribes to it a constant curvature, is necessarily finite, provided only that this curvature has a positive value, however small. . . . It is quite conceivable that the geometry of space in the very small does not satisfy the axioms of [Euclidean] geometry. . . . The curvature in the three directions can have arbitrary values if only the entire curvature for every sizeable region of space does not differ greatly from zero. . . . The properties which distinguish space from other conceivable triply-extended magnitudes are only to be deduced from experience."

But as Einstein was later to remark, "Physicists were still far removed from such a way of thinking: space was still, for them, a rigid, homogeneous something, susceptible of no change or conditions. Only the genius of Riemann, solitary and uncomprehended, had already won its way by the middle of the last century to a new conception of space, in which space was deprived of its rigidity, and in which its power to take part in physical events was recognized as possible."

Even as the 39-year-old Riemann lay dying of tuberculosis at Selasca on Lake Maggiore in the summer of 1866, having already achieved his great mathematical description of space curvature, he was working on a unified description of electromagnetism and gravitation. Why then did he not, half a century before Einstein, arrive at a geometric account of gravity? No obstacle in his way was greater than this: he thought only of space and the curvature of space, whereas Einstein discovered that he had to deal with **spacetime** and the *spacetime curvature*. Einstein could not thank Riemann, who ought to have been still alive. A letter of warm thanks he did, however, write to Mach.[5] In it he explained how mass there does indeed influence inertia here, through its influence on the enveloping spacetime geometry. Einstein's geometrodynamics had transmuted Mach's bit of philosophy into a bit of physics, susceptible to calculation, prediction, and test.

Let us bring out the main idea in what we may call poor man's language. **Inertia here**, in the sense of **local inertial frames**, that is the grip of spacetime here on mass here, is fully defined by the geometry, the curvature, the structure of spacetime here. The geometry here, however, has to fit smoothly to the geometry of the immediate surroundings; those domains, onto their surroundings; and so on, all the way around the great curve of space. Moreover, the geometry in each local region responds in its curvature to the mass in that region. Therefore every bit of momentum-energy, wherever located, makes its influence felt on the geometry of space throughout the whole universe—and felt, thus, on inertia right here.

The bumpy surface of a potato is easy to picture. It is the two-dimensional analogue of a bumpy three-sphere, the space geometry of a universe loaded irregularly here and there with concentrations and distributions of momentum-energy. If the spacetime has a Cauchy surface, that three-geometry once known—mathematical solution as it is of the so-called initial-value problem of geometrodynamics (see chap. 5)—the future evolution of that geometry follows straightforwardly and deterministically.

In other words, inertia (local inertial frames) everywhere and at all times is totally fixed, specified, determined, by the initial distribution of momentum-energy, of mass and mass-in-motion. The mathematics cries out with all the force at its command that **mass there** does determine inertia here.

We will enter into the mathematics of this initial-value problem—so thoroughly investigated in our day—in chapter 5. Of all the contributions made to inertia here by all the mass and mass-in-motion in the whole universe, the fractional contribution made by a particular mass at a particular distance is of the order of magnitude[6]

$$\left[\begin{array}{c}\text{fractional contribution}\\ \text{by a given mass, } \textit{there}\\ \text{to the determination}\\ \text{of the direction of}\\ \text{axes of the local gyroscopes,}\\ \text{the compass of inertia, } \textit{here}\end{array}\right] \text{ is of order of } \frac{(\text{mass, } \textit{there})}{(\text{distance, } \textit{there} \text{ to } \textit{here})} . \tag{1.1.1}$$

In this rough measure of the voting power, the "inertia-contributing power" of any object or any concentration of energy, its mass is understood to be expressed in the same geometric units as the distance.

Does this whole idea of voting rights and inertia-contributing power make sense? It surely does so if the total voting power of all the mass there is in the whole universe adds up to 100%. But does it? Let us run a check on the closed Friedmann model universe (chap. 4).[7] There the total amount of mass is of the order of $\sim 6 \times 10^{56}$ g (see § 4.2). This amount, translated into geometric units by way of the conversion factor 0.742×10^{-28} cm/g, is $\sim 4.5 \times 10^{28}$ cm of mass. It is much harder to assign an effective distance at which that mass lies from us, and for two reasons. First, distances are changing with time. So at what time is it that we think of the distance as being measured?

If we got into all the subtleties of that question we would be right back at the full mathematics of the so-called initial-value problem (see chap. 5), mathematics sufficiently complicated to have persuaded us to short circuit it in this introduction by our quick and rough poor man's account of inertia. This problem of "when" to measure distances let us therefore resolve by taking them at the instant of maximum expansion, when they are not changing.

What is the other problem about defining an effective distance to all the mass in the universe? Simple! Some lies at one distance, some at another. What separation does it make sense to adopt as a suitable average measure of the distance between inertia here and mass there? It is surely too much to accept as an average the mileage all the way around to the opposite side of the three-sphere, the most remote bit of mass in the whole universe! As a start let us nevertheless look at that maximum figure a minute. It is half of the circumference of the three-sphere. In other words, it is π times the radius of the three-sphere at maximum expansion, $\sim$ 10 to 20×10^9 yr $\times 3 \times 10^7$ s/yr $\times 3 \times 10^{10}$ cm/s $\approx 2 \times 10^{28}$ cm, or $\sim 6 \times 10^{28}$ cm. It is too big, we know, but surely not tenfold too big, not fivefold too big, conceivably not even twice as great as a reasonable figure for an effective average distance from here to mass there. Then, we say, why not adopt for that fuzzy average distance figure the very amount of mass figure that we have for the three-sphere model universe, $\sim 4.5 \times 10^{28}$ cm? This figure is surely of the right order of magnitude, and in our rough and ready way of figuring inertia, gives the satisfying 100% signal of "all votes in" when all the mass is counted. This is the poor man's version of the origin of inertia!

Now for inertia-determination in action! Mount a gyroscope on frictionless gimbals or, better, float it weightless in space to eliminate the gravity force that here on Earth grinds surface against surface. Picture our ideal gyro as sitting on a platform at the North Pole with the weather so cloudy that it has not one peek at the distant stars. Pointing initially to the flag and flagpole at a corner of the support platform, will the gyro continue to point that way as a 24 hour candle gradually burns away? No. The clouds do not deceive it. It does not see the star to which its spin axis points, but to that star it nevertheless continues to point as the day wears on. Earth turns beneath the heedless gyro, one rotation and many another as the days go by. That is the inertia-determining power of the mass spread throughout space, as that voting power is seen in its action on the gyro. Wait! As conscientious workers against election fraud, let us inspect the ballot that each mass casts. Where is the one signed "Earth"? Surely it was entitled to vote on what would be the free-float frame at the location of the gyro. The vote tabulator sheepishly confesses. "I know that Earth voted for having the frame of reference turn with it. However, its vote had so little inertia-determining power compared to all the other masses in the universe that I left it out." We inspectors reply, "you did wrong. Let us see how much damage you have done." Yes, the voting power of Earth at the location of the gyro is small, but it is not zero. It is of the order of magnitude of

$$\left[\frac{\text{mass of Earth}}{\text{radius of Earth}}\right] \cong \frac{0.44\ \text{cm}}{6.4 \times 10^8\ \text{cm}} \cong 0.69 \times 10^{-9} \tag{1.1.2}$$

roughly only one-billionth as much influence as all the rest of the universe together. "It was not only that its voice was so weak that I threw out its vote," the vote tabulator interrupts to say. "The free-float frame of reference that Earth wanted the gyro axis to adhere to was so little different from the frame demanded for the gyro by the faraway stars that I could not believe that Earth really took its own wishes seriously. It wants the gyroscope axis to creep slowly around in a twenty-four hour day rather than keep pointing at one star? Who cares about that peanut difference?" "Again you are wrong," we tell him. "You do not know our friends the astronomers. For a long time there was so little that they could measure about a faraway star except its direction that they developed the art of measuring angles to a fantastic discrimination. With Very Long Baseline Interferometry (VLBI; see § 3.4.1) one can measure the angular separation of the distant quasars down to less than a tenth of millisecond of arc and the rate of turning of Earth down to less than 1 milliarcsec per year.[8] Content to wait a day to see the effect of Earth on the direction of the spin axis of the gyro? No. Figure our friends as watching it for a whole 365-day year. Do you know how many milliseconds of arc the axis of the gyro would turn through in the course of a whole year, relative to the distant stars, if it followed totally and exclusively the urging of Earth? Four-hundred and seventy-three billion milliarcseconds! And that is the turning effect which you considered so miniscule that you threw away Earth's ballot! We are here to see justice done."

The corrected turning rate of the gyro, relative to the distant stars, is of the order of magnitude of

$$\begin{bmatrix}\text{voting power}\\ \text{of}\\ \text{Earth}\end{bmatrix} \times \begin{bmatrix}\text{rate of turn}\\ \text{desired}\\ \text{by Earth}\end{bmatrix} = \begin{bmatrix}\text{0.698 billionth}\\ \text{of total}\\ \text{voting power}\end{bmatrix} \times \begin{bmatrix}\text{473 billion}\\ \text{milliarcsec}\\ \text{per year}\end{bmatrix}$$

$$= \begin{bmatrix}\text{330 milliarcsec}\\ \text{per year}\end{bmatrix}. \qquad ((1.1.3))$$

Rough and ready though the reasoning that leads to this estimate, it gives the right order of magnitude. However, nobody has figured out how to operate on Earth's surface a gyroscope sufficiently close to friction-free that it can detect the predicted effect.

What to call this still undetected phenomenon? "Dragging," Einstein called it in his 1913 letter of thanks to Mach,[5] already two years before he arrived at the final formulation of his geometric theory of gravity; and *frame-dragging* the effect is often still called today following the lead of Jeffrey Cohen (see below, ref. 24). Others often call it the *Lense-Thirring effect*, after the 1918 paper of Hans Thirring and J. Lense,[9] that gave a quantitative analysis of it (see also W. de Sitter 1916[10] and §§ 3.4.3 and 6.10 below). Today still other words are used for that force at work on the gyro which causes the slow turning of its axis: *gyrogravitation* or **gravitomagnetism**[11] (see fig. 1.2). Such a new name is justified for a new force. This force differs as much from everyday gravity as a magnetic force differs from an electric force. Magnetism known since Greek times was analyzed (William Gilbert, *De Magnete*, 1600) long after electricity.

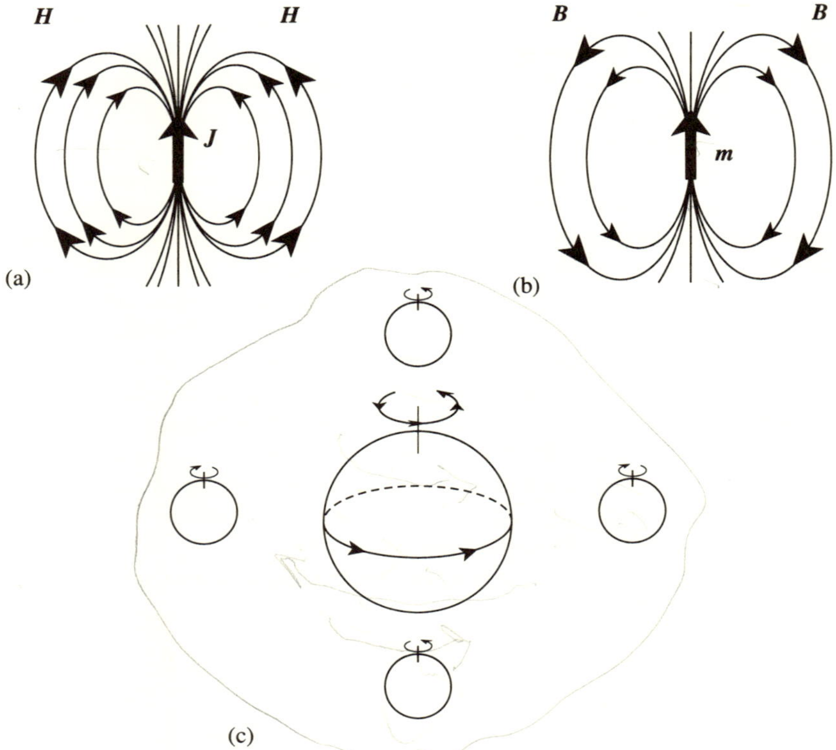

FIGURE 1.2. The directionalities of gravitomagnetism, magnetism, and fluid drag compared and contrasted.

(a) The gravitomagnetic field $\boldsymbol{H}$ in the weak field approximation.[11] $\boldsymbol{J}$ is the angular momentum of the central body.

(b) The magnetic induction $\boldsymbol{B}$ in the neighborhood of a magnetic dipole moment $\boldsymbol{m}$.

(c) Rotation of sphere induces by viscous drag a motion in the surrounding fluid that imparts to an immersed ball a rotation in the same sense as gravitomagnetism[24–26] drags the inertial frame defined by gyroscopes.

It took even longer to recognize that an electric charge going round and round in a circuit produces magnetism (H. C. Oersted, 1820). Gravitomagnetism or gyrogravitation, predicted in 1896–1916[27,28,5,10,9] (see § 6.5) but still not brought to the light of day, is produced by the motion of mass around and around in a circle. How amazing that a new force of nature should be enveloping us all and yet still stand there undetected!

Active work is in progress to detect gravitomagnetism. The detecting gyroscopes will be orbiting Earth. The one gyroscope is a quartz sphere smaller than a hand, carried along with an electrostatic suspension and sensitive readout devices, the whole under development by the Stanford group of C. W. F. Everitt and his associates for more than 20 years.[12–15] (see § 6.6).

The other gyroscope is huge, roughly four Earth radii in diameter. It consists of a sphere of 30 cm of radius covered with reflecting mirrors, going round and round in Earth orbit every 3.758 hours. An axis perpendicular to that orbit we identify with the gyroscope axis (see §§ 6.7 and 6.8). Such a satellite, LAGEOS,[16,17] already exists. It was launched in 1976. Its position is regularly read with a precision better than that of any other object in the sky, or an uncertainty of about a centimeter. For it the predicted gravitomagnetic precession is 31 milliarcsec per year,[18] quite enough to be detected if there were not enormously larger precessions at work that drown it out, about 126° per year.[19] This devastating competition arises from the nonsphericity of the Earth. Fortunately the possibility exists to launch another LAGEOS satellite going around at another inclination, so chosen that the anomalies in Earth's shape give a reversed orbit turning but the Earth's angular momentum gives the same Lense-Thirring frame-dragging (Ciufolini 1984).[18,19] In other words the two LAGEOS satellites, with their supplementary inclinations, will define an enormous gyroscope unaffected by the multipole mass moments of the Earth but only affected by the Lense-Thirring drag (see chap. 6). If this is done the small "difference" between the "large" turning rates will yield a direct measure of a force of nature new to man. Still more important, we will have direct evidence that mass there governs inertia here!

1.2 RELATION OF GRAVITY TO THE OTHER FORCES OF NATURE

In this book we will trace out the meaning and consequences of Einstein's great 1915 battle-tested and still standard picture of gravity as manifestation of the curvature of four-dimensional spacetime.

> I weigh all that is.
> Nowhere in the universe
> Is there anything over which
> I do not have dominion.
> As spacetime,
> As curved all-pervading spacetime,
> I reach everywhere.
> My name is gravity.[29]

What of the other forces of nature?

Every other force—the electric force that rules the motion of the atomic electron, the weak nuclear force that governs the emission of electrons and neutrinos from radioactive nuclei, and the strong nuclear force that holds together the constituents of particles heavier than the electron—demands for its understanding, today's researches argue, a geometry of more than four dimensions, perhaps as many as ten. The extra six dimensions are envisaged as curled up into an ultrasmall cavity, with one such cavity located at each point in spacetime. This tiny world admits of many a different organ-pipe resonance, many a different vibration frequency, each with its own characteristic quantum energy. Each of these modes of vibration of the cavity shows itself to us of the larger world as a particle of a particular mass, with its own special properties.

Such ideas are pursued by dozens of able investigators in many leading centers today, under the name of "grand unified field theories (GUTs)" or "string theories."[20,21] They open up exciting prospects for deeper understanding. However, this whole domain of research is in such a turmoil that no one can report with any confidence any overarching final view, least of all any final conclusion as to how gravity fits into the grand pattern.

The theories of the unification of forces with greatest promise today all have this striking feature that they, like the battle-tested, but simpler and older Einstein gravitation theory, build themselves on the **boundary of a boundary** principle, though in a higher dimensional version; for example, "the eight-dimensional boundary of the nine-dimensional boundary of a ten-dimensional region is automatically zero." Hidden to the uninstructed student of modern field theory is this vital organizing power of the boundary principle (see § 2.8). Hidden to the casual observer of a beautiful modern building is the steel framework that alone supports it; and hidden in Einstein's great account of geometric gravity, hidden until Élie Cartan's[22] penetrating insight brought it to light, is the unfolding of it all, from the grip of spacetime on mass to the grip of mass on spacetime, and from the automatic conservation of momentum-energy—without benefit of gears and pinions or any sophisticated manual to instruct nature what to do—to all the rich garden of manifestations of gravity in nature, the unfolding of all this from "the one-dimensional boundary of the two-dimensional boundary of a three-dimensional region is zero" and "the two-dimensional boundary of the three-dimensional boundary of a four-dimensional region is zero."[23]

Nowhere more decisively than in the fantastic austerity of this organizing idea—the boundary principle—does nature instruct man that it is at heart utterly simple—and that some day we will see it so.

REFERENCES CHAPTER 1

1. A. Einstein, Die Grundlage der allgemeinen Relativitätstheorie, *Ann. Physik* 49:769–822 (1916). Translation in *The Principle of Relativity* (Methuen, 1923; reprinted by Dover, 1952); see also ref. 2.

2. A. Einstein, *The Meaning of Relativity*, (5th ed., Princeton, Princeton University Press, 1955).
3. G.F.B. Riemann, Bei die Hypothesen, welche der Geometrie zu Grunde liegen, in *Gesammelte Mathematische Werke* (1866); reprint of 2d ed., ed. H. Weber (Dover, New York, 1953); see also the translation by W.K. Clifford, *Nature* 8:14–17 (1873).
4. E. Mach, *Die Mechanik in ihrer Entwicklung historisch-kritisch dargestellt* (Brockhaus, Leipzig, 1912); trans. T.J. McCormack with an introduction by Karl Menger as *The Science of Mechanics* (Open Court, La Salle, IL, 1960).
5. A. Einstein, letter to E. Mach, Zurich, 25 June 1913. See, e.g., ref. 7, p. 544.
6. J. Isenberg and J.A. Wheeler, Inertia here is fixed by mass-energy there in every W model universe, in *Relativity, Quanta, and Cosmology in the Development of the Scientific Thought of Albert Einstein*, ed. M. Pantaleo and F. de Finis (Johnson Reprint Corp., New York, 1979), vol. 1, 267–93.
7. See, e.g., C.W. Misner, K.S. Thorne, and J.A. Wheeler, *Gravitation* (Freeman, San Francisco, 1973).
8. D.S. Robertson and W.E. Carter, Earth orientation determinations from VLBI observations, in *Proc. Int. Conf. on Earth Rotation and the Terrestrial Reference Frame*, Columbus, OH, ed. I. Mueller (Department of Geodetic Science and Survey, Ohio State University, 1985), 296–306.
9. J. Lense and H. Thirring, Über den Einfluss der Eigenrotation der Zentralkörper auf die Bewegung der Planeten und Monde nach der Einsteinschen Gravitationstheorie, *Phys. Z.* 19:156–63 (1918); trans. by B. Mashhoon et al., *Gen. Relativ. Gravit.* 16:711–750 (1984).
10. W. de Sitter, Einstein's theory of gravitation and its astronomical consequences, *Mon. Not. Roy. Astron. Soc.* 76:699–728 (1916); in this paper the author gives a quantitative analysis of the precession of the Mercury perihelion due to the Sun's angular momentum.
11. See, e.g., K.S. Thorne, R.H. Price, and D.A. Macdonald, eds., *Black Holes: The Membrane Paradigm* (Yale University Press, New Haven, 1986), 72.
12. G.E. Pugh, Proposal for a satellite test of the Coriolis prediction of general relativity, WSEG Research Memorandum No. 11, Weapons Systems Evaluation Group, The Pentagon, Washington, DC, 12 November 1959.
13. L.I. Schiff, Motion of a gyroscope according to Einstein's theory of gravitation, *Proc. Natl. Acad. Sci.* (*USA*) 46:871–82 (1960). The author, using the Lense–Thirring potential, discusses in this paper the effect that the Stanford group will try to measure.
14. C.W.F. Everitt, The gyroscope experiment. I. General description and analysis of gyroscope performance, in *Experimental Gravitation*, ed. B. Bertotti (Academic Press, New York, 1974), 331–60.
15. J.A. Lipa, W.M. Fairbank, and C.W.F. Everitt, The gyroscope experiment. II. Development of the London-moment gyroscope and of cryogenic technology for space, in *Experimental Gravitation*, ed. B. Bertotti (Academic Press, New York, 1974), 361–80.
16. D.E. Smith and P.J. Dunn, Long term evolution of the LAGEOS orbit, *J. Geophys. Res. Lett.* 7:437–40 (1980).

17. C.F. Yoder, J.G. Williams, J.O. Dickey, B.E. Schutz, R.J. Eanes, and B.D. Tapley, Secular variation of Earth's gravitational harmonic J_2 coefficient from LAGEOS and nontidal acceleration of Earth rotation, *Nature* 303:757–62 (1983).

18. I. Ciufolini, Measurement of the Lense-Thirring drag on high-altitude laser-ranged artificial satellites, *Phys. Rev. Lett.* 56:278–81 (1986).

19. I. Ciufolini, A comprehensive introduction to the LAGEOS gravitomagnetic experiment: From the importance of the gravitomagnetic field in physics to preliminary error analysis and error budget, *Int. J. Mod. Phys. A* 4:3083–145 (1989).

20. J.H. Schwarz, *Superstrings, The First Fifteen Years of Superstring Theory* (World Scientific, Singapore, 1985); see also: *String Theory and Quantum Gravity '92*, ed. J. Harey, R. Lengo, K.S. Narain, S. Randjbar-Daemi, and H. Verlinde (World Scientific, Singapore, 1993).

21. M.B. Green, J.H. Schwarz, and E. Witten, *Superstring Theory* (Cambridge University Press, New York, 1987).

22. É. Cartan, *Leçon sur la géométrie des espaces de Riemann*, 2d ed. (Gauthier-Villars, Paris, 1963).

23. C.W. Misner and J.A. Wheeler, Conservation laws and the boundary of a boundary, in *Gravitatsiya: Problemi i Perspektivi: pamyati Alekseya Zinovievicha Petrova posvashaetsya*, ed. V.P. Shelest (Naukova Dumka, Kiev, 1972), 338–51.

24. J.M. Cohen, Dragging of inertial frames by rotating masses, in *1965 Lectures in Applied Mathematics 8, Relativity Theory and Astrophysics*, ed. J. Ehlers (American Mathematical Society, Providence, RI, 1965), 200–202.

25. K.S. Thorne, Relativistic stars, black holes, and gravitational waves, in *General Relativity and Cosmology, Proc. Course 47 of Int. School of Physics Enrico Fermi*, ed. R.K. Sachs (Academic Press, New York, 1971), 238–83.

26. J.M. Cohen, Gravitational collapse of rotating bodies, *Phys. Rev.* 173:1258–63 (1968).

27. B. and I. Friedländer, *Absolute und relative Bewegung*? (Simion-Verlag, Berlin, 1896).

28. A. Föppl, Übereinen Kreiselversuch zur Messung der Umdrehungsgeschwindigkeit der Erde *Sitzber. Bayer. Akad. Wiss.* 34:5–28 (1904); also in *Phys. Z.* 5:416–425 (1904). See also A. Föppl, Über absolute und relative Bewegung, *Sitzber Bayer. Akad. Wiss.* 34:383–95 (1904).

29. J.A. Wheeler, *A Journey into Gravity and Spacetime* (Freeman, New York, 1990).

2

Einstein Geometrodynamics

If Einstein gave us a geometric account of motion and gravity, if according to his 1915 and still-standard geometrodynamics spacetime tells mass how to move and mass tells spacetime how to curve, then his message requires mathematical tools to describe position and motion, curvature and the action of mass on curvature. The tools (see the mathematical appendix) will open the doorways to the basic ideas—equivalence principle, geometric structure, field equation, equation of motion, equation of geodesic deviation—and these ideas will open the doorways to more mathematical tools—exact solutions of Einstein's geometrodynamics field equation, equations of conservation of source, and the principle that the boundary of a boundary is zero. The final topics in this chapter—black holes, singularities, and gravitational waves—round out the interplay of mathematics and physics that is such a central feature of Einstein's geometrodynamics.

2.1 THE EQUIVALENCE PRINCIPLE

At the foundations of Einstein[1–10] geometrodynamics[11–21] and of its geometrical structure is one of the best-tested principles in the whole field of physics (see chap. 3): the equivalence principle.

Among the various formulations of the **equivalence principle**[16,21] (see § 3.2), we give here three most important versions: the **weak form**, also known as the *uniqueness of free fall* or the *Galilei equivalence principle* at the base of most known viable theories of gravity; the **medium strong form**, at the base of metric theories of gravity; and the **very strong form**, a cornerstone of Einstein geometrodynamics.

Galilei in his *Dialogues Concerning Two New Sciences*[22] writes: "The variation of speed in air between balls of gold, lead, copper, porphyry, and other heavy materials is so slight that in a fall of 100 cubits a ball of gold would surely not outstrip one of copper by as much as four fingers. Having observed this, I came to the conclusion that in a medium totally void of resistance all bodies would fall with the same speed."

We therefore formulate the **weak equivalence principle**, or *Galilei equivalence principle*[22,23] in the following way: *the motion of any freely falling test* *particle is independent of its composition and structure.* A test particle is defined to be electrically neutral, to have negligible gravitational binding energy compared to its rest mass, to have negligible angular momentum, and to be small enough that inhomogeneities of the gravitational field within its volume have negligible effect on its motion.

The weak equivalence principle—that all test particles fall with the same acceleration—is based on the principle[24] that the ratio of the inertial mass to the gravitational—passive—mass is the same for all bodies (see chap. 3). The principle can be reformulated by saying that in every local, nonrotating, freely falling frame the line followed by a freely falling test particle is a straight line, in agreement with special relativity.

Einstein generalized[10] the weak equivalence principle to all the laws of special relativity. He hypothesized that in no local freely falling frame can we detect the existence of a gravitational field, either from the motion of test particles, as in the weak equivalence principle, or from any other special relativistic physical phenomenon. We therefore state the **medium strong form of the equivalence principle**, also called the *Einstein equivalence principle*, in the following way: *for every pointlike event of spacetime, there exists a sufficiently small neighborhood such that in every local, freely falling frame in that neighborhood, all the nongravitational laws of physics obey the laws of special relativity.* As already remarked, the medium strong form of the equivalence principle is satisfied by Einstein geometrodynamics and by the so-called metric theories of gravity, for example, Jordan-Brans-Dicke theory, etc. (see chap. 3).

If we replace[18] *all the nongravitational laws of physics* with *all the laws of physics* we get the **very strong equivalence principle**, which is at the base of Einstein geometrodynamics.

The medium strong and the very strong form of the equivalence principle differ: the former applies to all phenomena except gravitation itself whereas the latter applies to all phenomena of nature. This means that according to the medium strong form, the existence of a gravitational field might be detected in a freely falling frame by the influence of the gravitational field on local gravitational phenomena. For example, the gravitational binding energy of a body might be imagined to contribute differently to the inertial mass and to the passive gravitational mass, and therefore we might have, for different objects, different ratios of inertial mass to gravitational mass, as in the Jordan-Brans-Dicke theory. This phenomenon is called the Nordtvedt effect[26,27] (see chap. 3). If the very strong equivalence principle were violated, then Earth and Moon, with different gravitational binding energies, would have different ratios of inertial mass to passive gravitational mass and therefore would have different accelerations toward the Sun; this would lead to some polarization of the Moon orbit around Earth. However, the Lunar Laser Ranging[28] experiment has put strong limits on the existence of any such violation of the very strong equivalence principle.

The equivalence principle, in the medium strong form, is at the foundations of Einstein geometrodynamics and of the other metric theories of gravity, with a "locally Minkowskian" spacetime. Nevertheless, it has been the subject of many discussions and also criticisms over the years.[13,25,29,30]

First, the equivalence between a gravitational field and an accelerated frame in the absence of gravity, and the equivalence between a flat region of spacetime and a freely falling frame in a gravity field, has to be considered valid only locally and not globally.[29] However, the content of the strong equivalence principle has been criticized even "locally." It has been argued that if one puts a spherical drop of liquid in a gravity field, after some time one would observe a tidal deformation from sphericity of the drop. Of course, this deformation does not arise in a flat region of spacetime. Furthermore, let us consider a freely falling frame in a small neighborhood of a point in a gravity field, such as the cabin of a spacecraft freely falling in the field of Earth. Inside the cabin, according to the equivalence principle, we are in a local inertial frame, without any observable effect of gravity. However, let us take a gradiometer, that is, an instrument which measures the gradient of the gravity field between two nearby points with great accuracy (present room temperature gradiometers may reach a sensitivity of about 10^{-11} (cm/s^2)/cm per Hz$^{-1/2}$ $\equiv$ 10^{-2} Eötvös per Hz$^{-1/2}$ between two points separated by a few tens of cm; future superconducting gradiometers may reach about 10^{-5} Eötvös Hz$^{-1/2}$ at certain frequencies, see §§ 3.2 and 6.9). No matter if we are freely falling or not, the gradiometer will eventually detect the gravity field and thus will allow us to distinguish between the freely falling cabin of a spacecraft in the gravity field of a central mass and the cabin of a spacecraft away from any mass, in a region of spacetime essentially flat. Then, may we still consider the strong equivalence principle to be valid?

From a mathematical point of view, at any point P of a pseudo-Riemannian, Lorentzian, manifold (see § 2.2 and mathematical appendix), one can find coordinate systems such that, at P, the metric tensor $g_{\alpha\beta}$ (§ 2.2) is the Minkowski metric $\eta_{\alpha\beta} = \mathrm{diag}(-1, +1, +1, +1)$ and the first derivatives of $g_{\alpha\beta}$, with respect to the chosen coordinates, are zero. However, one cannot in general eliminate certain combinations of second derivatives of $g_{\alpha\beta}$ which form a tensor called the Riemann curvature tensor: $R^{\alpha}{}_{\beta\gamma\delta}$ (see § 2.2 and mathematical appendix). The Riemann curvature tensor represents, at each point, the intrinsic curvature of the manifold, and, since it is a tensor, one cannot transform it to zero in one coordinate system if it is nonzero in some other coordinate system. For example, at any point P on the surface of a sphere one can find coordinate systems such that, at P, the metric is $g_{11}(P) = g_{22}(P) = 1$. However, the Gaussian curvature of the sphere (see mathematical appendix), that is, the R_{1212} component of the Riemann tensor, is, at each point, an intrinsic (independent of coordinates) property of the surface and therefore cannot be eliminated with a coordinate transformation. The metric tensor can indeed be written using the Riemann tensor, in a neighborhood of a spacetime event, in a freely falling, nonrotating, local inertial frame, to second order in the separation, δx^α, from the origin:

$$\begin{aligned} g_{00} &= -1 - R_{0i0j}\delta x^i \delta x^j \\ g_{0k} &= -\frac{2}{3} R_{0ikj}\delta x^i \delta x^j \\ g_{kl} &= \delta_{kl} - \frac{1}{3} R_{kilj}\delta x^i \delta x^j . \end{aligned} \tag{2.1.1}$$

These coordinates are called *Fermi Normal Coordinates*.

In section 2.5 we shall see that in general relativity, and other metric theories of gravity, there is an important equation, the *geodesic deviation equation*, which connects the physical effects of gravity gradients just described with the mathematical structure of a manifold, that is, which connects the physical quantities measurable, for example with a gradiometer, with the mathematical object representing the curvature: the Riemann curvature tensor. We shall see via the geodesic deviation equation that the relative, covariant, acceleration between two freely falling test particles is proportional to the Riemann curvature tensor, that is, $\ddot{\delta x}^\alpha \sim R^\alpha{}_{\beta\mu\nu}\delta x^\mu$, where δx^α is the "small" spacetime separation between the two test particles. On a two-surface, this equation is known as the Jacobi equation for the second derivative of the distance between two geodesics on the surface as a function of the Gaussian curvature.

The Riemann curvature tensor, however, cannot be eliminated with a coordinate transformation. Therefore, the relative, covariant, acceleration cannot be eliminated with a change of frame of reference. In other words, by the measurement of the second rate of change of the relative distance between two test particles, we can detect, in every frame, the gravitational field, and indeed, at least in principle, we can measure all the components of the Riemann curvature tensor and therefore completely determine the gravitational field. Furthermore, the motion of one test particle in a local freely falling frame can be described by considering the origin of the local frame to be comoving with another nearby freely falling test particle. The motion of the test particle in the local frame, described by the separation between the origin and the test particle, is then given by the geodesic deviation equation of section 2.5. This equation gives also a rigorous description of a falling drop of water and of a freely falling gradiometer, simply by considering two test particles connected by a spring, that is, by including a force term in the geodesic deviation equation (see § 3.6.1).

From these examples and arguments, one might think that the strong equivalence principle does not have the content and meaning of a fundamental principle of nature. Therefore, one might think to restrict to interpreting the equivalence principle simply as the equivalence between inertial mass M_i and gravitational mass M_g. However, $M_i = M_g$ is only a part of the medium (and very strong) equivalence principle whose complete formulation is at the basis of the locally Minkowskian spacetime structure.

In general relativity, the content and meaning of the strong equivalence principle is that *in a sufficiently small neighborhood of any spacetime event, in a local freely falling frame, no gravitational effects are observable*. Here, neighborhood means neighborhood in *space* and *time*. Therefore, one might formulate the medium strong equivalence principle, or Einstein equivalence principle, in the following form: for every spacetime event (then excluding singularities), for any experimental apparatus, with some limiting accuracy, there exists a neighborhood, in space and time, of the event, and infinitely many local freely falling frames, such that for every nongravitational phenomenon the difference between the measurements performed (assuming that the smallness of the spacetime neighborhood does not affect the experimental accuracy) and the theoretical results predicted by special relativity (including the Minkowskian character of the geometry) is less than the limiting accuracy and therefore undetectable in the neighborhood. In other words, in the spacetime neighborhood considered, in a freely falling frame all the nongravitational laws of physics agree with the laws of special relativity (including the Minkowskian character of spacetime), apart from a small difference due to the gravitational field that is; however, unmeasurable with the given experimental accuracy. We might formulate the very strong equivalence principle in a similar way.

For a test particle in orbit around a mass M, the geodesic deviation equation gives

$$\delta\ddot{x}^{\alpha} \sim R^{\alpha}{}_{0\beta 0}\delta x^{\beta} \sim \omega_0^2 \delta x^{\alpha} \tag{2.1.2}$$

where ω_0 is the orbital frequency. Thus, one would sample large regions of the spacetime if one waited for even one period of this "oscillator." We must limit the dimensions in space and time of the domain of observation to values small compared to one period if we are to uphold the equivalence principle.

A liquid drop which has a surface tension, and which resists distortions from sphericity, supplies an additional example of how to interpret the equivalence principle. In order to detect a gravitational field, the *measurable* quantity—the *observable*—is the tidal deformation $\boldsymbol{\delta x}$ of the drop. If a gravity field acts on the droplet and if we choose a small enough drop, we will not detect any deformation because the tidal deformations from sphericity are proportional to the size D of the small drop, and even for a self-gravitating drop of liquid in some external gravitational field, the tidal deformations $\boldsymbol{\delta x}$ are proportional to its size D. This can be easily seen from the geodesic deviation equation with a springlike force term (§ 3.6.1), in equilibrium: $\frac{k}{m}\,\delta x \sim R^{i}{}_{0j0}D \sim \frac{M}{R^3}\,D$, where M is the mass of an external body and $R^{i}{}_{0j0} \sim \frac{M}{R^3}$ are the leading components of the Riemann tensor generated by the external mass M at a distance R. Thus, in a spacetime neighborhood, with a given experimental accuracy, the deformation $\boldsymbol{\delta x}$, is unmeasurable for sufficiently small drops.

We overthrow yet a third attempt to challenge the equivalence principle—this time by use of a modern gravity gradiometer—by suitably limiting the scale or time of action of the gradiometer. Thus either one needs large distances over which to measure the gradient of the gravity field, or one needs to wait a period of time long enough to increase, up to a detectable value, the amplitude of the oscillations measured by the gradiometer. Similarly, with gravitational-wave detectors (resonant detectors, laser interferometers, etc.; see § 3.6), measuring the time variations of the gravity field between two points, one may be able to detect very small changes of the gravity field (present relative sensitivity to a metric perturbation or fractional change in physical dimensions $\sim 10^{-18}$ to 10^{-19}, "near" future sensitivity $\sim 10^{-21}$ to 10^{-22}; see § 3.6) during a small interval of time (for example a burst of gravitational radiation of duration $\sim 10^{-3}$ s). However, all these detectors basically obey the geodesic deviation equation, with or without a force term, and in fact their sensitivity to a metric perturbation decreases with their dimensions or time of action (see § 3.6).

In a final attempt to challenge the equivalence principle one may try to measure the *local* deviations from geodesic motion of a spinning particle, given by the Papapetrou equation described in section 6.10. In agreement with the geodesic deviation equation, these deviations are of type $\delta\ddot{x}^i \sim R^i{}_{0\mu\nu} J^{\mu\nu}$, where $J^{\mu\nu}$ is the spin tensor of the particle and $u^0 \cong 1$, defined in section 6.10. However, general relativity is a classical—nonquantized—theory. Therefore, in the formulation of the strong equivalence principle one has to consider only *classical* angular momentum of finite size particles. However, the classical angular momentum of a particle goes to zero as its size goes to zero, and we thus have a case analogous to the previous ones: sufficiently limited in space and time, no observations of motion will reveal any violation of the equivalence principle.

Of course, the local "eliminability" of gravitational effects is valid for gravity only. Two particles with arbitrary electric charge to mass ratios, $\frac{q_1}{m_1} \neq \frac{q_2}{m_2}$, for example $q_1 = 0$ and $\frac{q_2}{m_2} \gg 1$ (in geometrized units), placed in an external electric field, will undergo a relative acceleration that can be very large independently from their separation going to zero.

In summary, since the gravitational field is represented by the Riemann curvature tensor it cannot be transformed to zero in some frame if it is different from zero in some other frame; however, the measurable effects of the gravitational field, that is, of the spacetime curvature, between two nearby events, go to zero as the separation in space and time between the two events, or equivalently as the separation between the space and time origin of a freely falling frame and another local event. Thus, *any effect of the gravitational field is unmeasurable, in a sufficiently small spacetime neighborhood in a local freely falling frame of reference.*

2.2 THE GEOMETRICAL STRUCTURE

In 1827 Carl Friedrich Gauss (1777–1855) published what is thought to be the single most important work in the history of differential geometry: *Disquisitiones generales circa superficies curvas* (General Investigations of Curved Surfaces).[31] In this work he defines the curvature of two-dimensional surfaces, the Gaussian curvature, from the intrinsic properties of a surface. He concludes that all the properties that can be studied within a surface, without reference to the enveloping space, are independent from deformations, without stretching, of the surface—*theorema egregium*—and constitute the intrinsic geometry of the surface. The distance between two points, measured along the shortest line between the points within the surface, is unchanged for deformations, without stretching, of the surface.

The study of non-Euclidean geometries really began with the ideas and works of Gauss, Nikolai Ivanovich Lobačevskij (1792–1856),[32] and János Bolyai (1802–1860). In non-Euclidean geometries, Euclid's 5th postulate on straight lines is not satified (that through any point not lying on a given straight line, there is one, and only one, straight line parallel to the given line; see § 1.1).

In 1854 Georg Friedrich Bernhard Riemann (1826–1866) delivered his qualifying doctoral lecture (published in 1866): *Über die Hypothesen, welche der Geometrie zu Grunde liegen* (On the Hypotheses Which Lie at the Foundations of Geometry).[33] This work is the other cornerstone of differential geometry; it extends the ideas of Gauss from two-dimensional surfaces to higher dimensions, introducing the notions of what we call today Riemannian manifolds, Riemannian metrics, and the Riemannian curvature of manifolds, a curvature that reduces to the Gaussian curvature for ordinary two-surfaces. He also discusses the possibilities of a curvature of the universe and suggests that space geometry may be related to physical forces (see § 1.1).

The absolute differential calculus is also known as tensor calculus or Ricci calculus. Its development was mainly due to Gregorio Ricci Curbastro (1853–1925) who elaborated the theory during the ten years 1887–1896.[34,35] Riemann's ideas and a formula (1869) of Christoffel[36] were at the basis of the tensor calculus. In 1901 Ricci and his student Tullio Levi-Civita (1873–1941) published the fundamental memoir: *Méthods de calcul différential absolu et leurs applications* (Methods of Absolute Differential Calculus and their Applications),[35] a detailed description of the tensor calculus; that is, the generalization, on a Riemannian manifold, of the ordinary differential calculus. At the center of attention are geometrical objects whose existence is independent of any particular coordinate system.

From the medium strong equivalence principle, it follows that spacetime must be at an event, in suitable coordinates, Minkowskian; furthermore, it may be possible to show some theoretical evidence for the existence of a curvature of the spacetime.[37] The **Lorentzian, pseudo-Riemannian**[38–43] character of spacetime is the basic ingredient of general relativity and other metric theories of gravity; we therefore assume the **spacetime** to be a **Lorentzian manifold**: that is, a four-dimensional pseudo-Riemannian manifold, with signature +2 (or −2, depending on convention); that is, a smooth manifold M^4 with a continuous two-index tensor field $\boldsymbol{g}$, the **metric tensor**, such that $\boldsymbol{g}$ is covariant (see the mathematical appendix), symmetric, and nondegenerate or, simply, at each point of M, in components:

$$\begin{aligned} &g_{\beta\alpha} = g_{\alpha\beta} \\ &\det(g_{\alpha\beta}) \neq 0; \qquad \text{and signature}(g_{\alpha\beta}) = +2 \text{ (or } -2). \end{aligned} \tag{2.2.1}$$

The metric $g_{\alpha\beta}(\boldsymbol{x})$ determines the spacetime squared "distance" ds^2 between two nearby events with coordinates x^α and $x^\alpha + dx^\alpha$: $ds^2 \equiv g_{\alpha\beta}dx^\alpha dx^\beta$. On a pseudo-Riemannian manifold (the spacetime), for a given vector $\boldsymbol{v}_P$ in P, the squared norm $g_{\alpha\beta}v^\alpha_P v^\beta_P$ can be positive, negative, or null, the vector is then respectively called spacelike, timelike, or null. The metric tensor with both indices up, that is, *contravariant*, $g^{\alpha\beta}$, is obtained from the *covariant* components, $g_{\alpha\beta}$, by $g^{\alpha\beta}g_{\beta\gamma} \equiv \delta^\alpha{}_\gamma$, where $\delta^\alpha{}_\gamma$ is the Kronecker tensor, 0 for $\alpha \neq \gamma$ and 1 for $\alpha = \gamma$.

Let us briefly recall the definition of a few basic quantities of tensor calculus on a Riemannian manifold;[38–43] for a more extensive description see the mathematical appendix. We shall mainly use quantities written in components and referred to a coordinate basis on an n-dimensional Riemannian manifold.

A ***p*-covariant tensor** $T_{\alpha_1\cdots\alpha_p}$, or $\boldsymbol{T}$, is a mathematical object made of n^p quantities that under a coordinate transformation, $x'^\alpha = x'^\alpha(x^\alpha)$, change according to the transformation law $T'_{\alpha_1\cdots\alpha_p} = \partial^{\beta_1\cdots\beta_p}_{\alpha'_1\cdots\alpha'_p}T_{\beta_1\cdots\beta_p}$, where $\partial^{\beta_1\cdots\beta_p}_{\alpha'_1\cdots\alpha'_p} \equiv \frac{\partial x^{\beta_1}}{\partial x'^{\alpha_1}} \cdots \frac{\partial x^{\beta_p}}{\partial x'^{\alpha_p}}$ denotes the partial derivatives of the old coordinates x^α with respect to the new coordinates x'^α : $\partial^\beta_{\alpha'} \equiv \frac{\partial x^\beta}{\partial x'^\alpha}$.

A ***q*-contravariant tensor** $T^{\alpha_1\cdots\alpha_q}$ is a mathematical object made of n^q quantities that transform according to the rule $T'^{\alpha_1\cdots\alpha_q} = \partial^{\alpha'_1\cdots\alpha'_q}_{\beta_1\cdots\beta_q}T^{\beta_1\cdots\beta_q}$ where $\partial^{\alpha'_1\cdots\alpha'_q}_{\beta_1\cdots\beta_q} \equiv \frac{\partial x'^{\alpha_1}}{\partial x^{\beta_1}} \cdots \frac{\partial x'^{\alpha_q}}{\partial x^{\beta_q}}$. The covariant and contravariant components of a tensor are obtained from each other by lowering and raising the indices with $g_{\alpha\beta}$ and $g^{\alpha\beta}$.

The **covariant derivative** ∇_γ of a tensor $T^{\alpha\cdots}{}_{\beta\cdots}$, written here with a semicolon "; γ" is a tensorial generalization to curved manifolds of the standard partial derivative of Euclidean geometry. Applied to an n-covariant, m-contravariant tensor $T^{\alpha\cdots}{}_{\beta\cdots}$ it forms a $(n+1)$-covariant, m-contravariant tensor $T^{\alpha\cdots}{}_{\beta\cdots;\gamma}$ defined as

$$T^{\alpha\cdots}{}_{\beta\cdots;\gamma} \equiv T^{\alpha\cdots}{}_{\beta\cdots,\gamma} + \Gamma^\alpha_{\sigma\gamma}T^{\sigma\cdots}{}_{\beta\cdots} - \Gamma^\sigma_{\beta\gamma}T^{\alpha\cdots}{}_{\sigma\cdots} \tag{2.2.2}$$

where the $\Gamma^{\alpha}_{\beta\gamma}$ are the **connection coefficients**. They can be constructed, on a Riemannian manifold, from the first derivatives of the metric tensor:

$$\Gamma^{\alpha}_{\gamma\beta} = \Gamma^{\alpha}_{\beta\gamma} = \frac{1}{2} g^{\alpha\sigma} (g_{\sigma\beta,\gamma} + g_{\sigma\gamma,\beta} - g_{\beta\gamma,\sigma}) \equiv \left\{ \begin{matrix} \alpha \\ \beta\gamma \end{matrix} \right\}. \qquad (2.2.3)$$

On a Riemannian manifold, in a **coordinate basis** (**holonomic basis**), the connection coefficients have the above form, $\left\{ \begin{smallmatrix} \alpha \\ \beta\gamma \end{smallmatrix} \right\}$, as a function of the metric and of its first derivatives, and are usually called Christoffel symbols (see § 2.8 and mathematical appendix). The **Christoffel symbols** $\Gamma^{\alpha}_{\beta\gamma}$ are not tensors, but transform according to the rule $\Gamma'^{\alpha}_{\beta\gamma} = \partial^{\alpha'}_{\sigma} \partial^{\mu}_{\beta'} \partial^{\nu}_{\gamma'} \Gamma^{\sigma}_{\mu\nu} + \partial^{\alpha'}_{\delta} \partial^{\delta}_{\beta'\gamma'}$ where $\partial^{\delta}_{\beta'\gamma'} \equiv \frac{\partial^2 x^{\delta}}{\partial x'^{\beta} \partial x'^{\gamma}}$.

The **Riemann curvature tensor** $R^{\alpha}{}_{\beta\gamma\delta}$ is the generalization to n-dimensional manifolds of the Gaussian curvature K of a two-dimensional surface; it is defined as the commutator of the covariant derivatives of a vector field $\boldsymbol{A}$,

$$A^{\alpha}{}_{;\beta\gamma} - A^{\alpha}{}_{;\gamma\beta} = R^{\alpha}{}_{\sigma\gamma\beta} A^{\sigma}. \qquad (2.2.4)$$

In terms of the Christoffel symbols (2.2.3) the curvature is given by

$$R^{\alpha}{}_{\beta\gamma\delta} = \Gamma^{\alpha}_{\beta\delta,\gamma} - \Gamma^{\alpha}_{\beta\gamma,\delta} + \Gamma^{\alpha}_{\sigma\gamma} \Gamma^{\sigma}_{\beta\delta} - \Gamma^{\alpha}_{\sigma\delta} \Gamma^{\sigma}_{\beta\gamma}. \qquad (2.2.5)$$

The various symmetry properties of the Riemann curvature tensor are given in the mathematical appendix.

2.3 THE FIELD EQUATION

In electromagnetism[44] the four components of the electromagnetic vector potential A^{α} are connected with the density of charge ρ and with the three components of the density of current, $j^i = \rho v^i$, by the Maxwell equation

$$F^{\alpha\beta}{}_{,\beta} \equiv (A^{\beta,\alpha} - A^{\alpha,\beta})_{,\beta} = 4\pi j^{\alpha} \equiv 4\pi \rho u^{\alpha} \qquad (2.3.1)$$

in flat spacetime. Here $F^{\alpha\beta} \equiv A^{\beta,\alpha} - A^{\alpha,\beta}$ is the electromagnetic field tensor, $j^{\alpha} \equiv \rho u^{\alpha}$ is the charge current density four-vector, and $u^{\alpha} \equiv \frac{dx^{\alpha}}{ds}$ is the four-velocity of the charge distribution. The comma ", β" means partial derivative with respect to x^{β} : $\frac{\partial A^{\alpha}}{\partial x^{\beta}} \equiv A^{\alpha}{}_{,\beta}$.

We search now for a field equation that will connect the gravitational tensor potential $g_{\alpha\beta}$ with the density of mass-energy and its currents. Let us follow David Hilbert[45] (1915) to derive this Einstein field equation[6] from a variational principle, or principle of least action. We are motivated by Richard Feynman's later insight that classical action for a system reveals and follows the phase of the quantum mechanical wave function of that system (see below, refs. 128 and 129). We write the total action over an arbitrary spacetime region Ω as

$$I = \int_{\Omega} (\mathcal{L}_G + \mathcal{L}_M) d^4x \qquad (2.3.2)$$

where $d^4x = dx^1 \cdot dx^2 \cdot dx^3 \cdot dx^0$ and $\mathcal{L}_G$ and $\mathcal{L}_M$ are the Lagrangian densities for the geometry and for matter and fields, respectively, $\mathcal{L}_G \equiv L_G\sqrt{-g}$ and $\mathcal{L}_M \equiv L_M\sqrt{-g}$, and g is the determinant of the metric $g_{\alpha\beta} : g = \det(g_{\alpha\beta})$. The field variables describing the geometry, that is, the gravitational field, are the ten components of the metric tensor $g_{\alpha\beta}$. In order to have a tensorial field equation for $g_{\alpha\beta}$, we search for a $(\mathcal{L}_G + \mathcal{L}_M)$ that is a scalar density, that is, we search for an action I that is a scalar quantity. By analogy with electromagnetism we then search for a field equation of the second order in the field variables $g_{\alpha\beta}$, which, to be consistent with the observations, in the weak field and slow motion limit, must reduce to the classical Poisson equation. Therefore, the **Lagrangian density for the geometry** should contain the field variables $g_{\alpha\beta}$ and their first derivatives $g_{\alpha\beta,\gamma}$ only. In agreement with these requirements we assume

$$\mathcal{L}_G = \frac{1}{2\chi}\sqrt{-g} \cdot R. \tag{2.3.3}$$

Here $\frac{1}{2\chi}$ is a constant to be determined by requiring that we recover classical gravity theory in the weak field and slow motion limit, $R \equiv R^\alpha{}_\alpha \equiv g^{\alpha\beta}R_{\alpha\beta}$ is the **Ricci** or **curvature scalar**, and $R_{\alpha\beta}$ is the **Ricci tensor** constructed by contraction from the Riemann curvature tensor, $R_{\alpha\beta} = R^\sigma{}_{\alpha\sigma\beta}$. The curvature scalar R has a part linear in the second derivatives of the metric; however, it turns out that the variation of this part does not contribute to the field equation (see below).

Before evaluating the variation of the action I, we need to introduce a few identities and theorems, valid on a Riemannian manifold, that we shall prove at the end of this section.

1. The covariant derivative (defined by the Riemannian connection, see § 2.8) of the metric tensor $g^{\alpha\beta}$ is zero (Ricci theorem):

$$g^{\alpha\beta}{}_{;\gamma} = 0. \tag{2.3.4}$$

2. The variation, δg, with respect to $g_{\alpha\beta}$, of the determinant of the metric g is given by

$$\delta g = g \cdot g^{\alpha\beta} \cdot \delta g_{\alpha\beta} = -g \cdot g_{\alpha\beta} \cdot \delta g^{\alpha\beta}. \tag{2.3.5}$$

3. For a vector field v^α, we have the useful formula

$$v^\alpha{}_{;\alpha} = \left(\sqrt{-g}v^\alpha\right)_{,\alpha}\frac{1}{\sqrt{-g}}, \tag{2.3.6}$$

and similarly for a tensor field $T^{\alpha\beta}$

$$T^{\alpha\beta}{}_{;\beta} = \left(\sqrt{-g}T^{\alpha\beta}\right)_{,\beta}\frac{1}{\sqrt{-g}} + \Gamma^\alpha_{\sigma\beta}T^{\sigma\beta}. \tag{2.3.7}$$

4. Even though the Christoffel symbols $\Gamma^\alpha_{\beta\gamma}$ are not tensors and transform according to the rule that follows expression (2.2.3), $\Gamma'^\alpha_{\beta\gamma} = \partial^{\alpha'}_\sigma\partial^\mu_{\beta'}\partial^\nu_{\gamma'}\Gamma^\sigma_{\mu\nu} +$

$\partial^{\alpha'}_{\delta}\partial^{\delta}_{\beta'\gamma'}$, the difference between two sets of Christoffel symbols on the manifold M, $\delta\Gamma^{\alpha}_{\beta\gamma}(x) \equiv \Gamma^{*\alpha}_{\beta\gamma}(x) - \Gamma^{\alpha}_{\beta\gamma}(x)$, is a tensor. This immediately follows from the transformation rule for the $\Gamma^{\alpha}_{\beta\gamma}(x)$. The two sets of Christoffel symbols on M, $\Gamma^{*\alpha}_{\beta\gamma}(x)$ and $\Gamma^{\alpha}_{\beta\gamma}(x)$, may, for example, be related to two tensor fields, $g^{*}_{\alpha\beta}(x)$ and $g_{\alpha\beta}(x)$, on M.

5. The variation $\delta R_{\alpha\beta}$ of the Ricci tensor $R_{\alpha\beta}$ is given by

$$\delta R_{\alpha\beta} = \left(\delta\Gamma^{\sigma}_{\alpha\beta}\right)_{;\sigma} - \left(\delta\Gamma^{\sigma}_{\alpha\sigma}\right)_{;\beta}. \tag{2.3.8}$$

6. The generalization of the **Stokes divergence theorem**, to a **four-dimensional** manifold M, is

$$\int_{\Omega} v^{\sigma}{}_{;\sigma}\sqrt{-g}\,d^4x = \int_{\Omega}\left(v^{\sigma}\sqrt{-g}\right)_{,\sigma}d^4x = \int_{\partial\Omega}\sqrt{-g}\,v^{\sigma}\,d^3\Sigma_{\sigma}. \tag{2.3.9}$$

Here v^{σ} is a vector field, Ω is a four-dimensional spacetime region, $d^4x = dx^0\,dx^1\,dx^2\,dx^3$ its four-dimensional integration element, $\partial\Omega$ is the three-dimensional boundary (with the induced orientation; see § 2.8 and mathematical appendix) of the four-dimensional region Ω, and $d\Sigma_{\sigma}$ the three-dimensional integration element of $\partial\Omega$ (see § 2.8).

We now require the action I to be stationary for arbitrary variations $\delta g^{\alpha\beta}$ of the field variables $g^{\alpha\beta}$, with certain derivatives of $g^{\alpha\beta}$ fixed on the boundary of Ω: $\delta I = 0$. By using expression (2.3.5) we then find that

$$\begin{aligned}\delta I = {}& \frac{1}{2\chi}\int_{\Omega}\left(R_{\alpha\beta} - \frac{1}{2}g_{\alpha\beta}R\right)\sqrt{-g}\,\delta g^{\alpha\beta}d^4x + \frac{1}{2\chi}\int_{\Omega}g^{\alpha\beta}\sqrt{-g}\,\delta R_{\alpha\beta}d^4x \\ &+ \int_{\Omega}\frac{\delta\mathcal{L}_M}{\delta g^{\alpha\beta}}\,\delta g^{\alpha\beta}d^4x = 0.\end{aligned} \tag{2.3.10}$$

The second term of this equation can be written

$$\begin{aligned}&\frac{1}{2\chi}\int_{\Omega}g^{\alpha\beta}\sqrt{-g}\,\delta R_{\alpha\beta}d^4x \\ &\quad = \frac{1}{2\chi}\int_{\Omega}g^{\alpha\beta}\sqrt{-g}\left[\left(\delta\Gamma^{\sigma}_{\alpha\beta}\right)_{;\sigma} - \left(\delta\Gamma^{\sigma}_{\alpha\sigma}\right)_{;\beta}\right]d^4x \\ &\quad = \frac{1}{2\chi}\int_{\Omega}\sqrt{-g}\left[\left(g^{\alpha\beta}\delta\Gamma^{\sigma}_{\alpha\beta}\right)_{;\sigma} - \left(g^{\alpha\beta}\delta\Gamma^{\sigma}_{\alpha\sigma}\right)_{;\beta}\right]d^4x \\ &\quad = \frac{1}{2\chi}\int_{\Omega}\left[\left(\sqrt{-g}\,g^{\alpha\beta}\delta\Gamma^{\sigma}_{\alpha\beta}\right) - \left(\sqrt{-g}\,g^{\alpha\sigma}\delta\Gamma^{\rho}_{\alpha\rho}\right)\right]_{,\sigma}d^4x.\end{aligned} \tag{2.3.11}$$

where $\delta\Gamma^{\alpha}_{\beta\gamma} = \frac{1}{2}g^{\alpha\sigma}[(\delta g_{\beta\sigma})_{;\gamma} + (\delta g_{\sigma\gamma})_{;\beta} - (\delta g_{\gamma\beta})_{;\sigma}]$. This is an integral of a divergence and by the four-dimensional Gauss theorem can be transformed into an integral over the boundary $\partial\Omega$ of Ω, where it vanishes if certain derivatives of $g_{\alpha\beta}$ are fixed on the boundary $\partial\Omega$ of Ω. Then, this term gives no contribution to the field equation. Indeed, the integral over the boundary $\partial\Omega = \sum_I S_I$ of Ω can be rewritten (see York 1986)[46] as

$$\sum_I \frac{\varepsilon_I}{2\chi} \int_{S_I} \gamma_{\alpha\beta} \delta N^{\alpha\beta} d^3x \tag{2.3.12}$$

where $\varepsilon_I \equiv \boldsymbol{n}_I \cdot \boldsymbol{n}_I = \pm 1$ and $\boldsymbol{n}_I$ is the unit vector field normal to the hypersurface S_I, $\gamma_{\alpha\beta} = g_{\alpha\beta} - \varepsilon_I n_\alpha n_\beta$ is the three-metric on each hypersurface S_I of the boundary $\partial\Omega$ of Ω, and

$$N^{\alpha\beta} \equiv \sqrt{|\gamma|}(K\gamma^{\alpha\beta} - K^{\alpha\beta}) = -\frac{1}{2} g\gamma^{\alpha\mu}\gamma^{\beta\nu} £_{\boldsymbol{n}}(g^{-1}\gamma_{\mu\nu}) \tag{2.3.13}$$

where γ is the three-dimensional determinant of $\gamma_{\alpha\beta}$, $K_{\alpha\beta} = -\frac{1}{2} £_{\boldsymbol{n}} \gamma_{\alpha\beta}$ is the so-called second fundamental form or "extrinsic curvature" of each S_I (see § 5.2.2 and mathematical appendix), $K \equiv \gamma^{\alpha\beta} K_{\alpha\beta}$, and $£_{\boldsymbol{n}}$ is the Lie derivative (see § 4.2 and mathematical appendix) along the normal $\boldsymbol{n}$ to the boundary $\partial\Omega$ of Ω. Therefore, if the quantities $N^{\alpha\beta}$ are fixed on the boundary $\partial\Omega$, for an arbitrary variation $\delta g^{\alpha\beta}$, from the first and last integrals of (2.3.10), we have the **field equation**

$$\boxed{G_{\alpha\beta} = \chi T_{\alpha\beta}} \tag{2.3.14}$$

where $G_{\alpha\beta} \equiv R_{\alpha\beta} - \frac{1}{2} R g_{\alpha\beta}$ is the **Einstein tensor**, and—following the last integral of 2.3.10—we have defined the **energy-momentum tensor** of matter and fields $T_{\alpha\beta}$ (see below) as:

$$T_{\alpha\beta} \equiv -2 \frac{\delta L_M}{\delta g^{\alpha\beta}} + L_M g_{\alpha\beta}. \tag{2.3.15}$$

Let us now determine the constant χ by comparison with the classical, weak field, Poisson equation, $\Delta U = -4\pi\rho$, where U is the standard Newtonian gravitational potential. We first observe that in any metric theory of gravity (see chap. 3), without any assumption on the field equations, in the weak field and slow motion limit (see § 3.7), the metric $\boldsymbol{g}$ can be written at the lowest order in U, $g_{00} \cong -1 + 2U$, $g_{ik} \cong \delta_{ik}$, and $g_{i0} \cong 0$ and the energy-momentum tensor, at the lowest order, $T_{00} \cong -T \cong \rho$. From the definition of Ricci tensor $R_{\alpha\beta}$, it then follows that $R_{00} \cong -\Delta U$. From the field equation (2.3.14) we also have

$$R^\alpha{}_\alpha - \frac{1}{2} R\delta^\alpha{}_\alpha = -R = \chi T^\alpha{}_\alpha \equiv \chi T \tag{2.3.16}$$

where $T \equiv T^\alpha{}_\alpha$ is the trace of $T^{\alpha\beta}$. Therefore, the field equation can be rewritten in the alternative form

$$R_{\alpha\beta} = \chi\left(T_{\alpha\beta} - \frac{1}{2} T g_{\alpha\beta}\right). \tag{2.3.17}$$

From the 00 component of this equation, in the weak field and slow motion limit, we have

$$R_{00} \cong \chi\left(T_{00} + \frac{1}{2}T\right), \tag{2.3.18}$$

and therefore

$$\Delta U \cong -\frac{\chi}{2}\rho. \tag{2.3.19}$$

Requiring the agreement of the very weak field limit of general relativity with the classical Newtonian theory and comparing this equation (2.3.19) with the classical Poisson equation, we finally get $\chi = 8\pi$.

An alternative method of variation—the **Palatini method**[47]—is to take as independent field variables not only the ten components $g^{\alpha\beta}$ but also the forty components of the affine connection $\Gamma^{\alpha}_{\beta\gamma}$, assuming, a priori, no dependence of the $\Gamma^{\alpha}_{\beta\gamma}$ from the $g^{\alpha\beta}$ and their derivatives. Taking the variation with respect to the $\Gamma^{\alpha}_{\beta\gamma}$ and the $g^{\alpha\beta}$, and assuming L_M to be independent from any derivative of $g^{\alpha\beta}$, we thus have

$$\begin{aligned}
&\frac{1}{2\chi}\int_{\Omega}\left(R_{\alpha\beta} - \frac{1}{2}g_{\alpha\beta}R\right)\delta g^{\alpha\beta}\sqrt{-g}\,d^4x \\
&+ \frac{1}{2\chi}\int_{\Omega} g^{\alpha\beta}\left(\delta\Gamma^{\sigma}_{\alpha\beta;\sigma} - \delta\Gamma^{\sigma}_{\alpha\sigma;\beta}\right)\sqrt{-g}\,d^4x \\
&+ \int_{\Omega}\left(\frac{\delta L_M}{\delta g^{\alpha\beta}} - \frac{1}{2}g_{\alpha\beta}L_M\right)\delta g^{\alpha\beta}\sqrt{-g}\,d^4x = 0.
\end{aligned} \tag{2.3.20}$$

From the second integral, after some calculations,[11] one then gets

$$g_{\alpha\beta;\gamma} = g_{\alpha\beta,\gamma} - g_{\alpha\sigma}\Gamma^{\sigma}_{\beta\gamma} - g_{\sigma\beta}\Gamma^{\sigma}_{\alpha\gamma} = 0, \tag{2.3.21}$$

and therefore, by calculating from expression (2.3.21): $g^{\alpha\sigma}(g_{\beta\sigma,\gamma} + g_{\sigma\gamma,\beta} - g_{\beta\gamma,\sigma})$, on a Riemannian manifold, one gets the expression of the affine connection as a function of the $g_{\alpha\beta}$, that is, the Christoffel symbols $\left\{ {\alpha \atop \beta\gamma} \right\}$

$$\Gamma^{\alpha}_{\beta\gamma} = \frac{1}{2}g^{\alpha\sigma}\left(g_{\beta\sigma,\gamma} + g_{\sigma\gamma,\beta} - g_{\beta\gamma,\sigma}\right) \equiv \left\{ {\alpha \atop \beta\gamma} \right\}. \tag{2.3.22}$$

From the first and third integral in expression (2.3.20), we finally have the field equation (2.3.14).

Let us give the expression of the energy-momentum tensor in two cases: an electromagnetic field and a matter fluid.

In special relativity the energy-momentum tensor for an electromagnetic field[44] is $T^{\alpha\beta} = \frac{1}{4\pi}(F^{\alpha}{}_{\sigma}F^{\beta\sigma} - \frac{1}{4}\eta^{\alpha\beta}F_{\gamma\delta}F^{\gamma\delta})$, where $F^{\alpha\beta}$ is the electromagnetic field tensor. Moreover the *energy-momentum tensor*[131,132] of a matter fluid can be written $T^{\alpha\beta} = (\varepsilon + p)u^{\alpha}u^{\beta} + (q^{\alpha}u^{\beta} + u^{\alpha}q^{\beta}) + p\eta^{\alpha\beta} + \pi^{\alpha\beta}$, where ε is the *total energy density* of the fluid, u^α its *four-velocity*, q^α the *energy flux* relative to u^α (*heat flow*), p the *isotropic pressure*, and $\pi^{\alpha\beta}$ the tensor representing *viscous stresses* in the fluid. Therefore, by replacing $\eta_{\alpha\beta}$ with $g_{\alpha\beta}$ (in agreement with the equivalence principle), we define in Einstein geometrodynamics:

$$T^{\alpha\beta} = \frac{1}{4\pi}\left(F^{\alpha}{}_{\sigma}F^{\beta\sigma} - \frac{1}{4}g^{\alpha\beta}F_{\gamma\delta}F^{\gamma\delta}\right) \tag{2.3.23}$$

for an electromagnetic field, and

$$T^{\alpha\beta} = (\varepsilon + p)u^\alpha u^\beta + (q^\alpha u^\beta + u^\alpha q^\beta) + pg^{\alpha\beta} + \pi^{\alpha\beta} \tag{2.3.24}$$

for a matter fluid, where $\pi^{\alpha\beta}$ may be written:[11] $\pi^{\alpha\beta} = -2\eta\sigma^{\alpha\beta} - \zeta\Theta(g^{\alpha\beta} + u^\alpha u^\beta)$, where η is the *coefficient of shear viscosity*, ζ the *coefficient of bulk viscosity*, and $\sigma^{\alpha\beta}$ and Θ are the *shear tensor* and the *expansion scalar* of the fluid (see § 4.5).

In the case of a perfect fluid, defined by $\pi_{\alpha\beta} = q_\alpha = 0$, we then have

$$T^{\alpha\beta} = (\varepsilon + p)u^\alpha u^\beta + pg^{\alpha\beta}. \tag{2.3.25}$$

The general relativity expressions (2.3.23) and (2.3.24), for the energy-momentum tensor of an electromagnetic field and for a matter fluid, agree with the previous definition (2.3.15) of energy-momentum tensor, with a proper choice of the matter and fields Lagrangian density $\mathcal{L}_M$.

Let us finally prove the identities used in this section.

1. From the definition of covariant derivative and Christoffel symbols, we have

$$\begin{aligned} g^{\alpha\beta}{}_{;\gamma} &= g^{\alpha\beta}{}_{,\gamma} + \frac{1}{2}g^{\alpha\mu}g^{\beta\nu}(g_{\gamma\nu,\mu} + g_{\nu\mu,\gamma} - g_{\mu\gamma,\nu}) \\ &\quad + \frac{1}{2}g^{\mu\beta}g^{\alpha\nu}(g_{\mu\nu,\gamma} + g_{\nu\gamma,\mu} - g_{\gamma\mu,\nu}) \\ &= g^{\alpha\beta}{}_{,\gamma} + g^{\alpha\mu}g^{\beta\nu}g_{\nu\mu,\gamma} \\ &= g^{\alpha\beta}{}_{,\gamma} + g^{\alpha\beta}{}_{,\gamma} - g^{\beta\nu}{}_{,\gamma}g^{\alpha\mu}g_{\nu\mu} - g^{\alpha\mu}{}_{,\gamma}g^{\beta\nu}g_{\nu\mu} = 0. \end{aligned} \tag{2.3.26}$$

2. By using the symbol $\delta^{\alpha\beta\gamma\lambda}_{\mu\nu\rho\sigma}$, defined to be equal to $+1$ if $\alpha\beta\gamma\lambda$ is an even permutation of $\mu\nu\rho\sigma$, equal to -1 if $\alpha\beta\gamma\lambda$ is an odd permutation of $\mu\nu\rho\sigma$, and 0 otherwise (see § 2.8), we can write the determinant of a 4×4 tensor, $g_{\alpha\beta}$, in the form

$$g \equiv \det g_{\alpha\beta} = \delta^{\alpha\beta\gamma\lambda}_{0\,1\,2\,3}g_{\alpha 0}\,g_{\beta 1}\,g_{\gamma 2}\,g_{\lambda 3}. \tag{2.3.27}$$

By taking the variation of g we then have

$$\delta g = \delta g_{\alpha\beta} \cdot (g^{\alpha\beta} \cdot g) \tag{2.3.28}$$

and therefore

$$\delta\sqrt{-g} = \frac{1}{2}\sqrt{-g}\, g^{\alpha\beta}\delta g_{\alpha\beta}. \tag{2.3.29}$$

Moreover, from $\delta(g^{\alpha\beta}g_{\alpha\beta}) = 0$, we also have

$$\delta g_{\alpha\beta} \cdot g^{\alpha\beta} = -\delta g^{\alpha\beta} \cdot g_{\alpha\beta}. \tag{2.3.30}$$

3. From the definition (2.3.22) of Christoffel symbols, we have

$$\Gamma^{\sigma}_{\sigma\alpha} = \frac{1}{2} g_{\mu\nu,\alpha} g^{\mu\nu}, \tag{2.3.31}$$

and therefore, from the rule for differentiation of a determinant, $g_{,\alpha} = g g^{\mu\nu} g_{\mu\nu,\alpha}$, we get

$$\left(\ln\sqrt{-g}\right)_{,\alpha} = \Gamma^{\sigma}_{\sigma\alpha} \tag{2.3.32}$$

and finally

$$v^{\alpha}{}_{;\alpha} = v^{\alpha}{}_{,\alpha} + v^{\sigma}\Gamma^{\alpha}_{\alpha\sigma} = \left(\sqrt{-g}\, v^{\alpha}\right)_{,\alpha} \frac{1}{\sqrt{-g}}. \tag{2.3.33}$$

4. From the rule for transformation of the connection coefficients, it immediately follows that the difference between two sets of connection coefficients is a tensor.
5. At any event of the spacetime Lorentzian manifold, we can find infinitely many local inertial frames of reference where $\overset{(i)}{g}_{\alpha\beta} = \eta_{\alpha\beta}$, $\overset{(i)}{g}_{\alpha\beta,\gamma} = 0$ and therefore $\overset{(i)}{\Gamma}{}^{\alpha}_{\mu\nu} = 0$. From the definition of Ricci tensor (contraction of the Riemann tensor $R^{\alpha}{}_{\beta\gamma\delta}$ of expression (2.2.5) on the two indices α and γ) we then have at the event in any such local inertial frame

$$\delta \overset{(i)}{R}_{\alpha\beta} = \left(\delta \overset{(i)}{\Gamma}{}^{\sigma}_{\alpha\beta}\right)_{,\sigma} - \left(\delta \overset{(i)}{\Gamma}{}^{\sigma}_{\alpha\sigma}\right)_{,\beta}, \tag{2.3.34}$$

or equivalently

$$\delta \overset{(i)}{R}_{\alpha\beta} = \left(\delta \overset{(i)}{\Gamma}{}^{\sigma}_{\alpha\beta}\right)_{;\sigma} - \left(\delta \overset{(i)}{\Gamma}{}^{\sigma}_{\alpha\sigma}\right)_{;\beta}, \tag{2.3.35}$$

and since this is a tensorial equation, it is valid in any coordinate system

$$\delta R_{\alpha\beta} = \left(\delta \Gamma^{\sigma}_{\alpha\beta}\right)_{;\sigma} - \left(\delta \Gamma^{\sigma}_{\alpha\sigma}\right)_{;\beta}. \tag{2.3.36}$$

2.4 EQUATIONS OF MOTION

According to the *field equation*, $G^{\alpha\beta} = \chi T^{\alpha\beta}$, mass-energy $T^{\alpha\beta}$ "tells" geometry $g^{\alpha\beta}$ how to "curve"; furthermore, from the field equation itself, geometry "tells" mass-energy how to move. The key to the proof is **Bianchi's second**

identity[48,49] (for the "boundary of a boundary interpretation" of which see § 2.8):

$$R^{\alpha}{}_{\beta\gamma\nu;\mu} + R^{\alpha}{}_{\beta\nu\mu;\gamma} + R^{\alpha}{}_{\beta\mu\gamma;\nu} = 0.$$

Raising the indices β and ν and summing over α and γ, and over β and μ, we get the **contracted Bianchi's identity**:

$$G^{\nu\beta}{}_{;\beta} = \left(R^{\nu\beta} - \frac{1}{2}Rg^{\nu\beta}\right)_{;\beta} = 0. \tag{2.4.1}$$

By taking the covariant divergence of both sides of the field equation (2.3.14), we get

$$\boxed{T^{\nu\beta}{}_{;\beta} = 0.} \tag{2.4.2}$$

This statement summarizes the dynamical equations for matter and fields described by the energy-momentum tensor $T^{\alpha\beta}$. Therefore, as a consequence of the field equation, we have obtained the **dynamical equations** for matter and fields.

There exists an alternative approach to get the contracted Bianchi's identity. Consider an infinitesimal coordinate transformation:

$$x'^{\alpha} = x^{\alpha} - \xi^{\alpha}. \tag{2.4.3}$$

Under this transformation the metric tensor changes to (see § 4.2)

$$g'_{\alpha\beta} = g_{\alpha\beta} + \delta g_{\alpha\beta} = g_{\alpha\beta} + \xi_{\alpha;\beta} + \xi_{\beta;\alpha}. \tag{2.4.4}$$

This coordinate change bringing with it no real change in the geometry or the physics, we know that the action cannot change with this alteration. In other words, from the variational principle, $\delta \int \mathcal{L}_G \, d^4x = 0$, corresponding to the variation $\delta g^{\alpha\beta} = \xi^{\alpha;\beta} + \xi^{\beta;\alpha}$, we have

$$\delta I_G = \frac{1}{2\chi} \int G_{\alpha\beta}(\xi^{\alpha;\beta} + \xi^{\beta;\alpha})\sqrt{-g}d^4x = 0. \tag{2.4.5}$$

We translate

$$G_{\alpha\beta}\xi^{\alpha;\beta} = -G_{\alpha\beta}{}^{;\beta}\xi^{\alpha} + (G_{\alpha\beta}\xi^{\alpha})^{;\beta} = -G_{\alpha\beta}{}^{;\beta}\xi^{\alpha} + \frac{1}{\sqrt{-g}}\left(\sqrt{-g}G_{\alpha}{}^{\beta}\xi^{\alpha}\right)_{,\beta}$$

and use the four-dimensional divergence theorem (2.3.9), to get

$$\delta I_G = -\frac{1}{\chi} \int G_{\alpha\beta}{}^{;\beta}\xi^{\alpha}\sqrt{-g}d^4x = 0. \tag{2.4.6}$$

Since I_G is a scalar its value is independent of coordinate transformations; therefore this expression must be zero for every infinitesimal ξ^{α}, whence the contracted Bianchi identities (2.4.1).

For a pressureless perfect fluid, $p = 0$, that is, for dust particles, from expression (2.3.25) we have

$$T^{\alpha\beta} = \varepsilon u^{\alpha} u^{\beta}, \tag{2.4.7}$$

and from the equation of motion $T^{\alpha\beta}{}_{;\beta} = 0$,

$$T^{\alpha\beta}{}_{;\beta} = (\varepsilon u^{\alpha} u^{\beta})_{;\beta} = u^{\alpha}{}_{;\beta}\varepsilon u^{\beta} + \left(\varepsilon u^{\beta}\right)_{;\beta} u^{\alpha} = 0. \tag{2.4.8}$$

Multiplying this equation by u_{α} (and summing over α), recognizing $u^{\alpha}u_{\alpha} = -1$, and $(u^{\alpha}u_{\alpha})_{;\beta} = 0$ or $u^{\alpha}{}_{;\beta}u_{\alpha} = 0$, we get $(\varepsilon u^{\beta})_{;\beta} = 0$. Then, on substituting this result back into equation (2.4.8) we obtain the **geodesic equation**

$$u^{\alpha}{}_{;\beta} u^{\beta} = 0. \tag{2.4.9}$$

Therefore, each particle of dust follows a geodesic,[50,51] in agreement with the equivalence principle and with the equation of motion of special relativity, $\frac{du^{\alpha}}{ds} = u^{\alpha}{}_{,\beta}u^{\beta} = 0$. In a local inertial frame, from expression (2.4.8), we get to lowest order the classical equation of continuity, $\rho_{,0} + (\rho v^{i})_{,i} = 0$, and also the Euler equations for fluid motion, $\rho(v^{i})_{,0} + \rho(v^{i})_{,k}v^{k} = 0$, where ρ is the fluid mass density.

In general, we assume that the equation of motion of any test particle is a geodesic, where we define[39] a **geodesic** as the **extremal curve**, or history, $x^{\alpha}(t)$ that extremizes the integral of half of the squared interval E^{b}_{a} between two events $a = x(t_a)$ and $b = x(t_b)$:

$$E^{b}_{a}(x(t)) \equiv \frac{1}{2}\int_{t_a}^{t_b} g_{\alpha\beta}\left(x(t)\right)\frac{dx^{\alpha}}{dt}\frac{dx^{\beta}}{dt}\,dt. \tag{2.4.10}$$

In this sense a geodesic counts as a critical point in the space of all histories. We demand that any first-order small change $\delta x^{\alpha}(t)$ of the history, that keeps the end point fixed $\delta x^{\alpha}(t_a) = \delta x^{\alpha}(t_b) = 0$, shall cause a change in the integral $E^{b}_{a}(x(t))$ that is of higher order. The first-order change is required to vanish: $\delta E^{b}_{a}(x^{\alpha}(t)) = 0$. It is the integral of the product of $\delta x^{\alpha}(t)$ with the Lagrange expression:

$$\frac{\partial L}{\partial x^{\alpha}} - \frac{d}{dt}\frac{\partial L}{\partial\left(\frac{dx^{\alpha}}{dt}\right)} = 0, \tag{2.4.11}$$

where $L = \frac{1}{2} g_{\alpha\beta}(x(t))\frac{dx^{\alpha}}{dt}\frac{dx^{\beta}}{dt}$, and we have

$$g_{\alpha\beta}\frac{d^2x^{\beta}}{dt^2} + g_{\alpha\beta,\gamma}\frac{dx^{\beta}}{dt}\frac{dx^{\gamma}}{dt} - \frac{1}{2}g_{\beta\gamma,\alpha}\frac{dx^{\beta}}{dt}\frac{dx^{\gamma}}{dt} = 0. \tag{2.4.12}$$

This **equation for a geodesic** translates into the language (2.3.22) of the Christoffel symbols:

$$\boxed{\frac{d^2x^\alpha}{dt^2} + \Gamma^\alpha_{\beta\gamma}\frac{dx^\beta}{dt}\frac{dx^\gamma}{dt} = 0.} \tag{2.4.13}$$

The geodesic equation keeps the standard form (2.4.13) for every transformation of the parameter t of the type $s = ct + d$, where $c \neq 0$ and d are two constants; when the geodesic equation has the standard form (2.4.13), t is called **affine parameter**. A special choice of parameter p is the **arc-length** itself $s(p)$ along the curve $s(p) = L^p_a(x) = \int_a^p \sqrt{\pm g_{\alpha\beta}(x(p'))\frac{dx^\alpha}{dp'}\frac{dx^\beta}{dp'}}\,dp'$ (+ sign for spacelike geodesics and − sign for timelike geodesics), where p is a parameter along the curve. When $p = s$, the geodesic is said to be parametrized by arc-length. For a timelike geodesic, $s \equiv \tau$ is the **proper time** measured by a clock comoving with the test particle ("wrist-watch time").

On a proper Riemannian manifold there is a variational principle that gives the geodesic equation parametrized with any parameter. This principle defines a geodesic[39] as the **extremal curve for the length** $L^a_b(x(p))$:

$$L^b_a\big(x(p)\big) = \int_{p_a}^{p_b}\sqrt{g_{\alpha\beta}\big(x(p)\big)\frac{dx^\alpha}{dp}\frac{dx^\beta}{dp}}\,dp. \tag{2.4.14}$$

From

$$\delta L^b_a\big(x(p)\big) = 0 \tag{2.4.15}$$

for any variation $\delta x^\alpha(p)$ of the curve $x^\alpha(p)$, such that $\delta x^\alpha(p_a) = \delta x^\alpha(p_b) = 0$, taking the variation of $L^b_a(x(p))$, from the Lagrange equations, we thus find

$$\frac{d^2x^\alpha}{dp^2} + \Gamma^\alpha_{\beta\gamma}\frac{dx^\beta}{dp}\frac{dx^\gamma}{dp} - \frac{dx^\alpha}{dp}\left(\frac{d^2s/dp^2}{ds/dp}\right) = 0 \tag{2.4.16}$$

where $s(p)$ is the arc-length.

Extremal curve for the quantity E^b_a and extremal curve for the length L^b_a? When are the two the same on a proper Riemannian manifold? When and only when the two equations (2.4.13) and (2.4.16) are both satisfied: that is, when the quantity $\frac{d^2s}{dp^2}$ vanishes—that is, when the parameter p grows linearly with arc-length. Therefore, an extremal curve for the quantity E^b_a is also an extremal curve for the length, L^b_a, and vice versa; it is always possible[39] to reparametrize a curve that on a proper Riemannian manifold is an extremal curve for the length and with $\frac{dx^\alpha}{dp} \neq 0$ everywhere, to give an extremal curve for the quantity E^b_a.

For a test particle with proper mass different from zero, the geodesic equation of motion is the curve that extremizes the proper time $\tau = \int d\tau = \int\sqrt{-g_{\alpha\beta}dx^\alpha dx^\beta}$ along the world line of the particle. For a photon, the equation of motion follows from the variational principle for E^b_a, (2.4.10), and is a null geodesic (with $ds^2 = 0$), in agreement with special relativity and with the equivalence principle. On a timelike geodesic, we can write

$$\frac{D}{d\tau} u^\alpha = 0 \qquad (2.4.17)$$

where τ is the proper time measured by a clock moving on the geodesic, $u^\alpha \equiv \frac{dx^\alpha}{d\tau}$ its four-velocity, and $u^\alpha u_\alpha = -1$.

Parallel transport of a vector v^α along a curve $x^\alpha(t)$, with tangent vector $n^\alpha(t) \equiv \frac{dx^\alpha}{dt}(t)$, is defined by requiring $\boldsymbol{n} \cdot \boldsymbol{v}$ to be covariantly constant along the curve:

$$\frac{D}{dt}(n^\alpha v_\alpha) = (n^\alpha v_\alpha)_{;\beta} n^\beta = 0. \qquad (2.4.18)$$

Therefore, for a geodesic, from equation (2.4.13), we have that $v^\alpha{}_{;\beta} n^\beta = 0$.

In particular, *a geodesic is a curve with tangent vector*, n^α, *transported parallel to itself all along the curve*: $n^\alpha{}_{;\beta} n^\beta = 0$.

Finally, from the definition (2.2.5) of Riemann tensor, one can derive[39] the formula for the change of a vector v^α parallel transported around an infinitesimal closed curve determined by the infinitesimal displacements δx^α and $\widetilde{\delta x}^\alpha$ (infinitesimal "quadrilateral" which is closed apart from higher order infinitesimals in $\delta x \cdot \widetilde{\delta x}$):

$$\delta v^\alpha = -R^\alpha{}_{\beta\mu\nu} v^\beta \delta x^\mu \widetilde{\delta x}^\nu. \qquad (2.4.19)$$

This equation shows that, on a curved manifold, the vector obtained by parallel transport along a curve depends on the path chosen and on the curvature (and on the initial vector; see fig. 2.1).

2.5 THE GEODESIC DEVIATION EQUATION

A fundamental equation of Einstein geometrodynamics and other metric theories of gravity is the **equation of geodesic deviation.**[38,52] It connects the spacetime curvature described by the Riemann tensor with a measurable physical quantity: the relative "acceleration" between two nearby test particles.

The equation of geodesic deviation, published in 1925 by Levi-Civita,[38,52] gives the second covariant derivative of the distance between two infinitesimally close geodesics, on an arbitrary n-dimensional Riemannian manifold:

$$\frac{D^2(\delta x^\alpha)}{ds^2} = -R^\alpha{}_{\beta\mu\nu} u^\beta \delta x^\mu u^\nu. \qquad (2.5.1)$$

Here, δx^α is the infinitesimal vector that connects the geodesics, $u^\mu = \frac{dx^\mu[s]}{ds}$ is the tangent vector to the base geodesic, and $R^\alpha{}_{\mu\nu\delta}$ is the Riemann curvature tensor. This equation generalizes the classical **Jacobi equation** for the distance

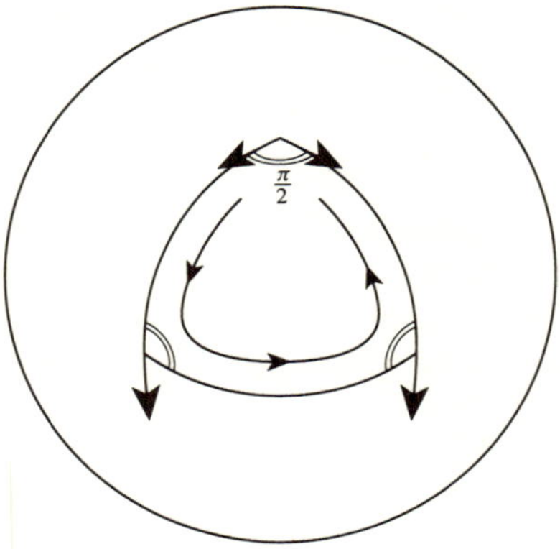

FIGURE 2.1. A vector transported parallel to itself around the indicated circuit, on the surface of a sphere of radius R, comes back to its starting point rotated through an angle of $\frac{\pi}{2}$. The curvature of the surface is given by

$$(\text{curvature}) = \frac{(\text{angle of rotation})}{(\text{area circumnavigated})} = \frac{\frac{\pi}{2}}{\frac{1}{8}(4\pi R^2)} = \frac{1}{R^2}.$$

y between two geodesics on a two-dimensional surface:

$$\frac{d^2 y}{d\sigma^2} + K y = 0 \tag{2.5.2}$$

where σ is the arc of the base geodesic and $K[\sigma]$ is the *Gaussian curvature* of the surface.[31,39]

The equation of geodesic deviation can be derived from the second variation of the quantity $E_a^b(x(t))$, defined by expression (2.4.10), set equal to zero. However, we follow here a more intuitive approach.

In order to derive the geodesic deviation equation (2.5.1) let us assume the following:

1. The two curves are geodesics:

$$\frac{Du_1^\alpha}{d\tau} = 0 \qquad \text{and} \qquad \frac{Du_2^\alpha}{d\sigma} = 0 \tag{2.5.3}$$

where τ, σ are affine parameters.

2. The law of correspondence between the points of the two geodesics—that is, the definition of the connecting vector $\delta x^\alpha[\tau]$—is such that, if $d\tau$ is an infinitesimal arc on geodesic 1 and $d\sigma$ the arc on geodesic 2 corresponding to the connecting vectors $\delta x^\alpha[\tau]$ and $\delta x^\alpha[\tau + d\tau]$, we have[38]

$$\frac{d\sigma}{d\tau} = 1 + \lambda, \qquad \text{where} \quad \frac{d\lambda}{d\tau} = 0 \tag{2.5.4}$$

3. The geodesics are infinitesimally close in a neighborhood U:

$$x_2^\alpha[\sigma] = x_1^\alpha[\tau] + \delta x^\alpha[\tau] \tag{2.5.5}$$

where $x_2^\alpha \epsilon U$ and $x_1^\alpha \epsilon U$, and where the relative change in the curvature is small:

$$\left| \frac{\mathcal{R}_{,\alpha}\delta x^\alpha}{\mathcal{R}} \right| \ll 1, \tag{2.5.6}$$

and $\mathcal{R}^{-2}$ is approximately the typical magnitude of the components of the Riemann tensor.

4. The difference between the tangent vectors to the two geodesics is infinitesimally small in the neighborhood U:

$$\left| \frac{\|\delta u^\alpha\|}{\|u^\alpha\|} \right| \ll 1 \tag{2.5.7}$$

where

$$\delta u^\alpha \equiv u_2^\alpha[\sigma] - u_1^\alpha[\tau]. \tag{2.5.8}$$

5. Equation (2.5.1) is derived neglecting terms higher than the first-order, ϵ^1, in δx^α and in δu^α. Furthermore, for simplicity, we define the connecting vector δx^α as connecting points of equal arc-lengths s on the two geodesics,* then, $\delta\tau = \delta\sigma = ds$ and s satisfies

$$u_1^\alpha[s]u_{1\alpha}[s] = -1, \qquad \text{where} \quad u_1^\alpha[s] \equiv \frac{dx_1^\alpha[s]}{ds} \tag{2.5.9}$$

and

$$u_2^\alpha[s]u_{2\alpha}[s] = -1, \qquad \text{where} \quad u_2^\alpha[s] \equiv \frac{dx_2^\alpha[s]}{ds}. \tag{2.5.10}$$

Physically s is the proper time measured by two observers comoving with two test particles following the two geodesics.

The equation of geodesic 1 is

$$\frac{Du_1^\alpha}{ds} = \frac{du_1^\alpha}{ds} + \Gamma^\alpha_{\mu\nu}[x_1]u_1^\mu u_1^\nu = 0, \tag{2.5.11}$$

and the equation of geodesic (2) is

$$\begin{aligned} \frac{Du_2^\alpha}{ds} &= \frac{du_2^\alpha}{ds} + \Gamma^\alpha_{\mu\nu}[x_1 + \delta x]u_2^\mu u_2^\nu = \frac{d^2}{ds^2}\left(x_1^\alpha + \delta x^\alpha\right) \\ &+ \Gamma^\alpha_{\mu\nu}[x_1 + \delta x]\frac{d}{ds}\left(x_1^\mu + \delta x^\mu\right)\frac{d}{ds}\left(x_1^\nu + \delta x^\nu\right) = 0. \end{aligned} \tag{2.5.12}$$

*For simplicity, in this derivation we do not consider null geodesics.

We also have

$$\frac{d}{ds}\left(\delta x^{\mu}[s]\right) \equiv \frac{d}{ds}\left(x_2^{\mu}[s] - x_1^{\mu}[s]\right) = u_2^{\mu}[s] - u_1^{\mu}[s] \equiv \delta u^{\mu}[s] \quad (2.5.13)$$

with this notation, and writing $u_1^{\mu} \equiv u^{\mu}$, we can rewrite equation (2.5.12), with a Taylor expansion to first order in δx^{α} and δu^{α}, as

$$\frac{d^2}{ds^2}\left(x_1^{\alpha}\right) + \frac{d^2}{ds^2}\left(\delta x^{\alpha}\right) + \left(\Gamma^{\alpha}_{\mu\nu} + \Gamma^{\alpha}_{\mu\nu,\rho}\delta x^{\rho}\right)\left(u^{\mu}u^{\nu} + 2u^{\mu}\delta u^{\nu}\right) = 0. \quad (2.5.14)$$

Taking the difference between equations (2.5.14) and (2.5.11) we find, to first order,

$$\frac{d^2(\delta x^{\alpha})}{ds^2} + \Gamma^{\alpha}_{\mu\nu,\rho}\delta x^{\rho}u^{\mu}u^{\nu} + 2\Gamma^{\alpha}_{\mu\nu}u^{\mu}\delta u^{\nu} = 0, \quad (2.5.15)$$

and using the definition $\frac{Dv^{\alpha}}{ds} = \frac{dv^{\alpha}}{ds} + \Gamma^{\alpha}_{\mu\nu}u^{\mu}v^{\nu}$ and the expression (2.2.5) of the Riemann tensor in terms of the Christoffel symbols and their derivatives, we have, to first order, the law of change of the geodesic separation,

$$\frac{D^2(\delta x^{\alpha})}{ds^2} = -R^{\alpha}{}_{\beta\mu\nu}u^{\beta}\delta x^{\mu}u^{\nu}. \quad (2.5.16)$$

In electromagnetism,[44] one can determine all the six independent components of the antisymmetric electromagnetic field tensor $F^{\alpha\beta}$, by measuring the accelerations of test charges in the field, and by using the Lorentz force equation

$$\frac{d^2x^{\alpha}}{ds^2} = \frac{e}{m}F^{\alpha}{}_{\beta}u^{\beta} \quad (2.5.17)$$

where e, m, and u^{β} are charge, mass, and four-velocity of the test particles. In electromagnetism, it turns out that the minimum number of test particles, with proper initial conditions, necessary to fully measure $F^{\alpha\beta}$ is two.[11]

Similarly, on a Lorentzian n-dimensional manifold, in any metric theory of gravity (thus with geodesic motion for test particles), one can determine all the $\frac{n^2(n^2-1)}{12}$ independent components of the Riemann tensor, by measuring the relative accelerations of a sufficiently large number of test particles and by using the equation of geodesic deviation (2.5.1).

However, which is the *minimum number of test particles necessary to determine the spacetime curvature fully*? As we observed, in a four-dimensional spacetime the Riemann tensor has twenty independent components. However, when the metric of the spacetime is subject to the Einstein equation in vacuum, $R_{\alpha\beta} = R^{\sigma}{}_{\alpha\sigma\beta} = 0$, the number of independent components of the Riemann tensor is reduced to ten, and they form the **Weyl tensor**[11] which is in general defined by

$$C_{\alpha\beta\gamma\delta} = R_{\alpha\beta\gamma\delta} + g_{\alpha[\delta}R_{\gamma]\beta} + g_{\beta[\gamma}R_{\delta]\alpha} + \frac{1}{3}Rg_{\alpha[\gamma}g_{\delta]\beta} \quad (2.5.18)$$

where $R = R_{\alpha\beta} g^{\alpha\beta}$.

Synge in his classic book on the general theory of relativity[13] describes a method of measuring the independent components of the Riemann tensor. Synge calls his device a five-point curvature detector. The five-point curvature detector consists of a light source and four mirrors. By performing measurements of the distance between the source and the mirrors and between the mirrors, one can determine the curvature of the spacetime. However, in order to measure all the independent components of the Riemann tensor with Synge's method, the experiment must be repeated several times with different orientations of the detector; equivalently—and when the spacetime is not, stationary—it is necessary to use several curvature detectors at the same time.

Instead, one can measure the relative accelerations of test particles moving on infinitesimally close geodesics and use equation (2.5.1). However, in order to minimize the number of test particles necessary to determine all the independent components of the Riemann tensor at one event, it turns out that one has to use nearby test particles, moving with arbitrarily different four-velocities.

It is then possible to derive a generalized geodesic deviation equation,[53] valid for any two geodesics, with arbitrary tangent vectors, not necessarily parallel, and describing the relative acceleration of two test particles moving with any four-velocity on neighboring geodesics. This generalized equation can be derived by dropping the previous condition (4): $\left| \frac{\|\delta u^\alpha\|}{\|u^\alpha\|} \right| \ll 1$, and by retaining the conditions (1), (2), (3), and (5) only,[53] and it is valid in any neighborhood in which the change of the curvature is small (condition 3). Of course, when the two geodesics are locally parallel one recovers the classical geodesic deviation equation. Physically, one would measure the relative acceleration of two test particles moving with arbitrary four-velocities (their difference $(u_2^\alpha - u_1^\alpha)$ need not necessarily be small) in an arbitrary gravitational field (in an arbitrary Riemannian manifold), in a region where the relative change of the gravitational field is small. The spacetime need not necessarily satisfy the Einstein field equation so long as the test particles follow geodesic motion (metric theories). It turns out[54] that the minimum number of test particles can be drastically reduced by using the generalized geodesic deviation equation instead of the standard geodesic deviation equation (2.5.1). This number is reduced either (1) under the hypothesis of an arbitrary four-dimensional Lorentzian manifold or (2) when we have an empty region of the spacetime satisfying the Einstein equations, $R_{\alpha\beta} = 0$ (the measurement of the Riemann tensor reduces then to the measurement of the Weyl tensor $C^\alpha{}_{\beta\mu\delta}$).

It turns out[54] that to fully determine the curvature of the spacetime in vacuum, in general relativity, it is *sufficient* to use four test particles, and in general spacetimes (twenty independent components of the Riemann tensor) it is sufficient to use six test particles. It is easy to show that in a vacuum, to fully determine the curvature, it is also *necessary* to use at least four test particles. With four test particles we have three independent geodesic deviation equations leading to twelve relations between the ten independent components of the Riemann tensor and the relative accelerations of the test particles. In general spacetimes it is necessary to use at least six test particles. Of course, it is possible to determine the curvature of the spacetime using test particles having approximately equal four-velocities and using the standard geodesic deviation equation. However, it turns out then that the minimum number of test particles which is required in general relativity increases to thirteen in general spacetimes and to six in vacuum.

2.6 SOME EXACT SOLUTIONS OF THE FIELD EQUATION

A Rigorous Derivation of a Spherically Symmetric Metric

Given a **three-dimensional Riemannian manifold** M^3, one may define M^3 to be **spherically symmetric**[20,38,41] about one point O (for the definition based on the isometry group see § 4.2), if, in some coordinate system, x^i, every rotation about O, of the type $x'^i = O^{i'}_k x^k$ where $\delta_{ij} = O^{m'}_i O^{n'}_j \delta_{m'n'}$, and $\det O^{i'}_k = +1$, is an **isometry** for the metric $\boldsymbol{g}$ of M^3. In other words, the metric $\boldsymbol{g}$ in M^3 is defined spherically symmetric if it is **formally invariant** *for rotations*; that is, the new components of $\boldsymbol{g}$ are the same functions of the new coordinates x'^α as the old components of $\boldsymbol{g}$ were of the old coordinates x^α for rotations

$$g_{\alpha\beta}(y^\alpha \equiv x^\alpha) = g'_{\alpha\beta}(y^\alpha \equiv x'^\alpha). \tag{2.6.1}$$

A **Lorentzian manifold** M^4 may then be defined **spherically symmetric** about one point O, if, in some coordinate system, the metric $\boldsymbol{g}$ is formally invariant for three-dimensional spatial rotations about $O : x'^i = O^{i'}_k x^k$ (as defined above), that is, three-dimensional spatial rotations are isometries for $\boldsymbol{g}: g_{\alpha\beta}(x^0, x^i) = g'_{\alpha\beta}(x^0, x^i)$. (In general, on a Lorentzian manifold a geometrical quantity $G(x^0, x^i)$ may be defined to be spherically symmetric if G is formally invariant for three-dimensional spatial rotations: $G(x^0, x^i) = G'(x^0, x^i)$.)

Formal invariance of the metric $\boldsymbol{g}$ under the infinitesimal coordinate transformation $x'^\alpha = x^\alpha + \varepsilon\xi^\alpha$, where $|\varepsilon| \ll 1$, is equivalent to the requirement that the **Lie derivative**[55,56] (see § 4.2 and mathematical appendix) of the metric tensor $\boldsymbol{g}$, with respect to $\boldsymbol{\xi}$, be zero:

$$\mathcal{L}_{\xi} g_{\alpha\beta} \equiv g_{\alpha\beta,\sigma}\xi^\sigma + g_{\sigma\beta}\xi^\sigma{}_{,\alpha} + g_{\alpha\sigma}\xi^\sigma{}_{,\beta} = 0. \tag{2.6.2}$$

This requirement follows from the definition (2.6.1) of formal invariance under the infinitesimal coordinate transformation $x'^\alpha = x^\alpha + \varepsilon\xi^\alpha$, thus

$$\begin{aligned} 0 &= g_{\alpha\beta}(x'^\gamma) - g'_{\alpha\beta}(x'^\gamma) \\ &= g_{\alpha\beta}(x^\gamma) + g_{\alpha\beta,\sigma}\varepsilon\xi^\sigma - \partial^{\sigma}_{\alpha'}{}^{\rho}_{\beta'} g_{\sigma\rho}(x^\gamma) \\ &= g_{\alpha\beta,\sigma}\varepsilon\xi^\sigma + \varepsilon\xi^\sigma{}_{,\alpha} g_{\sigma\beta} + \varepsilon\xi^\rho{}_{,\beta} g_{\alpha\rho}. \end{aligned} \tag{2.6.3}$$

As follows from the definition (2.3.22) of the Christoffel symbols that enter into covariant derivatives, this condition on the metric is equivalent (see § 4.2) to the **Killing equation**:

$$\xi_{\alpha;\beta} + \xi_{\beta;\alpha} = 0. \tag{2.6.4}$$

Therefore, the **Killing vector** $\boldsymbol{\xi}$ describes the symmetries of the metric tensor $\boldsymbol{g}$ by defining the isometric mappings of the metric onto itself, that is, the isometries.[57] We have just defined a metric $\boldsymbol{g}$ to be spherically symmetric if it is formally invariant under three-dimensional spatial rotations, therefore a metric is spherically symmetric if it satisfies the Killing equation for every Killing vector $\boldsymbol{\xi}_{ss}$ that represents a three-dimensional spatial rotation. The Killing vector representing **spherical symmetry**, in "generalized Cartesian coordinates," is

$$\xi^0_{ss} = 0, \qquad \xi^i_{ss} = c^{ij} x^j \tag{2.6.5}$$

where $c^{ik} = -c^{ki}$ are three constants. In other words, spherical symmetry about the point O is equivalent to axial symmetry around each of the three-axes Ox^a, represented by the Killing vector:

$$\xi^0 = \xi^a = 0; \qquad \xi^b = x^c; \qquad \xi^c = -x^b \tag{2.6.6}$$

where (a, b, c) is some permutation of $(1, 2, 3)$. In particular, using generalized Cartesian coordinates, we have

$$\begin{aligned} \xi_1'^{\alpha} &= (0, 0, z, -y) \\ \xi_2'^{\alpha} &= (0, -z, 0, x) \\ \xi_3'^{\alpha} &= (0, y, -x, 0) \end{aligned} \tag{2.6.7}$$

or using "generalized polar coordinates," defined by the usual transformation $x = r \sin\theta \cos\phi$, $y = r \sin\theta \sin\phi$ and $z = r \cos\theta$, we have

$$\begin{aligned} \xi_1^{\alpha} &= (0, 0, \sin\phi, \cot\theta \cos\phi) \\ \xi_2^{\alpha} &= (0, 0, -\cos\phi, \cot\theta \sin\phi) \\ \xi_3^{\alpha} &= (0, 0, 0, -1). \end{aligned} \tag{2.6.8}$$

From the Killing equation (2.6.2), using the Killing vector $\boldsymbol{\xi}_3$, we get[58]

$$g_{\alpha\beta,\phi} = 0, \tag{2.6.9}$$

and using the Killing vectors $\boldsymbol{\xi}_1$ and $\boldsymbol{\xi}_2$ in equation (2.6.2), we then get

$$g_{11,\theta} = 0, \qquad g_{00,\theta} = 0, \qquad g_{10,\theta} = 0, \tag{2.6.10}$$

and by applying equation (2.6.2) to ξ_1:

$$g_{22,\theta}\sin\phi = 2g_{23}\frac{\cos\phi}{\sin^2\theta}$$
$$(g_{33,\theta} - 2g_{33}\cot\theta)\sin\phi = -2g_{23}\cos\phi$$
$$g_{12,\theta}\sin\phi = g_{13}\frac{\cos\phi}{\sin^2\theta}$$
$$(g_{13,\theta} - g_{13}\cot\theta)\sin\phi = -g_{12}\cos\phi \tag{2.6.11}$$
$$(g_{23,\theta} - g_{23}\cot\theta)\sin\phi = \left(-g_{22} + g_{33}\frac{1}{\sin^2\theta}\right)\cos\phi$$
$$g_{20,\theta}\sin\phi = g_{30}\frac{\cos\phi}{\sin^2\theta}$$
$$(g_{30,\theta} - g_{30}\cot\theta)\sin\phi = -g_{20}\cos\phi,$$

plus the seven similar equations for ξ_2 obtained by replacing both $\sin\phi$ with $-\cos\phi$ and $\cos\phi$ with $\sin\phi$ in the equations (2.6.11). From equations (2.6.9), (2.6.10), and (2.6.11) and the seven similes we get

$$\begin{aligned} &g_{00} = g_{00}(r,t), \qquad g_{11} = g_{11}(r,t), \qquad g_{22} = g_{22}(r,t), \\ &g_{33} = g_{22}(r,t)\sin^2\theta \qquad \text{and} \qquad g_{01} = g_{01}(r,t), \end{aligned} \tag{2.6.12}$$

that is, g_{00}, g_{11}, g_{22}, $g_{33}/\sin^2\theta$, and g_{01} are functions of r and t only; all the other components of $\boldsymbol{g}$ are identically equal to zero.

The general form of a **four-dimensional spherically symmetric metric** is then

$$\begin{aligned} ds^2 = {} & A(r,t)dt^2 + B(r,t)dr^2 + C(r,t)drdt \\ & + D(r,t)(d\theta^2 + \sin^2\theta d\phi^2). \end{aligned} \tag{2.6.13}$$

This we simplify by performing the coordinate transformation

$$t' = t \qquad \text{and} \qquad r'^2 = D(r,t) \tag{2.6.14}$$

where we assume $D(r,t) \neq$ constant. We then get (dropping the prime in t' and r')

$$ds^2 = E(r,t)dt^2 + F(r,t)dr^2 + G(r,t)drdt + r^2(d\theta^2 + \sin^2\theta d\phi^2). \tag{2.6.15}$$

With the further coordinate transformation

$$t' = H(r,t) \qquad \text{and} \qquad r' = r \tag{2.6.16}$$

where we assume $H_{,t} \neq 0$, we have

$$g_{01} = \partial_0^{0'}\partial_1^{1'}g'_{01} + \partial_0^{0'}\partial_1^{0'}g'_{00} = H_{,t}g'_{01} + H_{,t}H_{,r}g'_{00} \tag{2.6.17}$$

and

$$g_{00} = \partial_0^{0'} \partial_0^{0'} g'_{00} = (H_{,t})^2 g'_{00}; \tag{2.6.18}$$

to simplify the metric in its new form, we impose the condition

$$g'_{01} = \frac{H_{,t} \cdot G}{2(H_{,t})^2} - \frac{H_{,r} \cdot E}{(H_{,t})^2} \equiv 0. \tag{2.6.19}$$

This condition can always be satisfied, for any function G and $E \neq 0$, by finding a solution to the differential equation:

$$\frac{1}{2} H_{,t} \cdot G - H_{,r} \cdot E = 0. \tag{2.6.20}$$

Therefore, we finally have (dropping the prime in t' and r')

$$ds^2 = -e^{m(r,t)} dt^2 + e^{n(r,t)} dr^2 + r^2(d\theta^2 + \sin^2\theta d\phi^2) \tag{2.6.21}$$

as **metric of a spherically symmetric spacetime** in a *particular* coordinate system. The signs were determined according to the Lorentzian character of the Riemannian manifold, in agreement with the equivalence principle: $\overset{(i)}{g}_{\alpha\beta} \to \eta_{\alpha\beta}$.

Let us now find the expression of a spherically symmetric metric satisfying the vacuum Einstein field equation (2.3.14), with $T^{\alpha\beta} = 0$:

$$G^{\alpha\beta} = 0 \qquad \text{or, equivalently,} \qquad R^{\alpha\beta} = 0. \tag{2.6.22}$$

From the definition of Ricci tensor, that we symbolically write here

$$R^{\sigma}{}_{\alpha\sigma\beta} = \begin{vmatrix} {}_{,\sigma} & {}_{,\beta} \\ \Gamma^{\sigma}_{\alpha\sigma} & \Gamma^{\sigma}_{\alpha\beta} \end{vmatrix} + \begin{vmatrix} \Gamma^{\sigma}_{\rho\sigma} & \Gamma^{\sigma}_{\rho\beta} \\ \Gamma^{\rho}_{\alpha\sigma} & \Gamma^{\rho}_{\alpha\beta} \end{vmatrix} \equiv \Gamma^{\sigma}_{\alpha\beta,\sigma} - \Gamma^{\sigma}_{\alpha\sigma,\beta} + \cdots, \tag{2.6.23}$$

and from the definition (2.3.22) of Christoffel symbols, we then get

$$\begin{aligned} R_{00} = &-e^{m-n}\left(\frac{1}{2}m_{,rr} - \frac{1}{4}m_{,r}n_{,r} + \frac{1}{4}m_{,r}{}^2 + \frac{m_{,r}}{r}\right) \\ &+ \frac{1}{2}n_{,tt} + \frac{1}{4}n_{,t}{}^2 - \frac{1}{4}m_{,t}n_{,t} = 0 \end{aligned} \tag{2.6.24}$$

$$\begin{aligned} R_{11} = &\frac{1}{2}m_{,rr} - \frac{1}{4}m_{,r}n_{,r} + \frac{1}{4}m_{,r}{}^2 - \frac{n_{,r}}{r} \\ &- e^{n-m}\left(\frac{1}{2}n_{,tt} + \frac{1}{4}n_{,t}{}^2 - \frac{1}{4}m_{,t}n_{,t}\right) = 0 \end{aligned} \tag{2.6.25}$$

$$R_{22} = -1 + e^{-n} + \frac{1}{2}e^{-n}r(m_{,r} - n_{,r}) = 0 \tag{2.6.26}$$

$$R_{33} = R_{22}\sin^2\theta = 0 \tag{2.6.27}$$

and

$$R_{01} = -\frac{n_{,t}}{r} = 0 \tag{2.6.28}$$

with all the other nondiagonal components of $R_{\alpha\beta}$ identically zero. From the 00 and 11 components we then have

$$(m+n)_{,r} = 0, \tag{2.6.29}$$

and from the 01 component (2.6.28) $\frac{\partial n}{\partial t} = 0$; therefore,

$$m + n(r) = f(t) \qquad \text{or} \qquad e^{m} = e^{f(t)} e^{-n(r)}. \tag{2.6.30}$$

The time dependence $f(t)$ can be absorbed in the definition of t with a coordinate transformation of the type $t' = \int e^{\frac{1}{2} f(t)} dt$. Therefore, in the new coordinates (dropping the prime in n' and m'), we have the result

$$\frac{\partial n}{\partial t} = \frac{\partial m}{\partial t} = 0 \qquad \text{and} \qquad e^{m(r)} = e^{-n(r)}. \tag{2.6.31}$$

Therefore, a spherically symmetric spacetime satisfying the vacuum Einstein field equation (2.6.22) is static, that is, there is a coordinate system in which the metric is time independent, $g_{\alpha\beta,0} = 0$, and in which $g_{0i} = 0$.

We recall that a **spacetime** is called **stationary** if it admits a timelike Killing vector field, $\boldsymbol{\xi}_t$. For it, there exists some coordinate system in which $\boldsymbol{\xi}_t$ can be written $\boldsymbol{\xi}_t = (1, 0, 0, 0)$. In this system, from the Killing equation (2.6.2), the metric $\boldsymbol{g}$ is then time independent, $g_{\alpha\beta,0} = 0$. A **spacetime** is called **static** if it is stationary and the timelike Killing vector field $\boldsymbol{\xi}_t$ is orthogonal to a foliation (§ 5.2.2) of spacelike hypersurfaces. Therefore, there exists some coordinate system, called adapted to $\boldsymbol{\xi}_t$, in which the metric $\boldsymbol{g}$ satisfies both $g_{\alpha\beta,0} = 0$ and $g_{0i} = 0$.

From the 22, or the 33, component of the vacuum field equation, plus equation (2.6.29), we then have

$$-1 + e^{-n} - re^{-n} n_{,r} = 0 \tag{2.6.32}$$

and therefore

$$(re^{-n})_{,r} = 1 \tag{2.6.33}$$

with the solution

$$e^{-n} = 1 + \frac{C}{r}. \tag{2.6.34}$$

By writing the constant $C \equiv -2M$, we finally have

$$ds^2 = -\left(1 - \frac{2M}{r}\right) dt^2 + \left(1 - \frac{2M}{r}\right)^{-1} dr^2 + r^2\left(d\theta^2 + \sin^2\theta \, d\phi^2\right). \tag{2.6.35}$$

This is the **Schwarzschild** (1916) **solution.**[59] In conclusion, any spherically symmetric solution of the vacuum Einstein field equation must be static and in some coordinate system must have the Schwarzschild form (**Birkhoff Theorem**).[60] By assuming that the spacetime geometry generated by a spherically symmetric object is itself spherically symmetric, and by requiring that we recover the classical gravity theory, for large r, in the weak field region, we find that M is the mass of the central body (see § 3.7).

However, inside a hollow, static, spherically symmetric distribution of matter, for $r \to 0$, to avoid $g_{00} \to \infty$ and $g_{11} \to 0$, we get $C \equiv 0$. Therefore, the solution internal to a nonrotating, empty, spherically symmetric shell is the Minkowski metric $\eta_{\alpha\beta}$ (for the weak field, slow motion solution inside a rotating shell, see § 6.1 and expression 6.1.37).

Other One-Body Solutions

A solution of the field equation with no matter but with an electromagnetic field, with three parameters M, Q, and J that in the weak field limit are identified with the mass M, the charge Q, and the angular momentum J of a central body, is the **Kerr-Newman solution**,[61,62] that in the t, r, θ, ϕ Boyer-Lindquist coordinates[63] can be written

$$\begin{aligned} ds^2 = & -\left(1 - \frac{(2Mr - Q^2)}{\rho^2}\right)dt^2 \\ & - \left(\frac{(4Mr - 2Q^2)a \sin^2\theta}{\rho^2}\right) dt d\phi + \frac{\rho^2}{\Delta} dr^2 + \rho^2 d\theta^2 \\ & + \left(r^2 + a^2 + \frac{(2Mr - Q^2)a^2 \sin^2\theta}{\rho^2}\right) \sin^2\theta d\phi^2 \end{aligned} \tag{2.6.36}$$

where

$$\begin{aligned} \Delta &\equiv r^2 - 2Mr + a^2 + Q^2 \\ \rho^2 &\equiv r^2 + a^2 \cos^2\theta \end{aligned} \tag{2.6.36'}$$

and $a \equiv \frac{J}{M}$ = angular momentum per unit mass.

In the case $Q = J = 0$ and $M \neq 0$ we have the Schwarzschild metric (2.6.35); when $J = 0$, $M \neq 0$ and $Q \neq 0$, we have the **Reissner-Nordstrøm metric**:[64,65]

$$\begin{aligned} ds^2 = & -\left(1 - \frac{2M}{r} + \frac{Q^2}{r^2}\right)dt^2 + \left(1 - \frac{2M}{r} + \frac{Q^2}{r^2}\right)^{-1} dr^2 \\ & + r^2\left(d\theta^2 + \sin^2\theta d\phi^2\right). \end{aligned} \tag{2.6.37}$$

This solution describes a spherically symmetric spacetime satisfying the Einstein field equation in a region with no matter, but with a radial electric field to be included in the energy-momentum tensor $T_{\alpha\beta}$ (see § 2.3),

$$\boldsymbol{E} = \frac{Q}{r^2}\boldsymbol{e}_r \qquad \boldsymbol{B} = \boldsymbol{0} \tag{2.6.38}$$

where $\boldsymbol{e}_r$ is the radial unit vector of a static orthonormal tetrad. In the weak field region, M and Q are identified with the mass and the charge of the central object.

Finally, when $Q = 0$ and $M \neq 0$, $J \neq 0$ we have the Kerr solution.[61] In the **weak field** and slow motion limit,[66–69] $M/r \ll 1$, $(J/M)/r \ll 1$, in Boyer-Lindquist coordinates, the **Kerr metric** (2.6.36) can be written

$$ds^2 \cong -\left(1 - \frac{2M}{r}\right)dt^2 + \left(1 + \frac{2M}{r}\right)dr^2 + r^2\left(d\theta^2 + \sin^2\theta d\phi^2\right) - \frac{4J}{r}\sin^2\theta d\phi dt. \tag{2.6.39}$$

This is the weak field metric generated by a central body with mass M and angular momentum J; we shall return to this important solution in chapter 6.

2.7 CONSERVATION LAWS

In classical electrodynamics[44] one defines the total charge on a three-dimensional spacelike hypersurface Σ, corresponding to t = constant: $Q = \int_\Sigma j^0 d^3\Sigma_0$. From the Maxwell equations with source $F^{\alpha\beta}{}_{,\beta} = 4\pi j^\alpha$ and from the antisymmetry of the electromagnetic tensor $F^{\alpha\beta}$, one has the differential conservation law of charge $j^\alpha{}_{,\alpha} = 0$. Therefore, by using the four-dimensional divergence theorem (2.3.9), we verify that Q is conserved:

$$0 = \int_\Omega j^\alpha{}_{,\alpha}\, d^4\Omega = \int_{\partial\Omega} j^\alpha d^3\Sigma_\alpha \tag{2.7.1}$$

where Ω is a spacetime region and $\partial\Omega$ its three-dimensional boundary, and where $d^4\Omega$ and $d^3\Sigma_\alpha$ are respectively the four-dimensional and the three-dimensional integration elements defined by expressions (2.8.21) and (2.8.20) below. By choosing $\partial\Omega$ composed of two spacelike hypersurfaces Σ and Σ', corresponding to the times t = constant and t' = constant$'$, plus an embracing hypersurface Λ, away from the source, on which j^α vanishes (see fig. 2.2), we then have

$$Q = \int_\Sigma j^0 d^3\Sigma_0 = \int_{\Sigma'} j'^0 d^3\Sigma'_0 = Q', \tag{2.7.2}$$

that is, the total charge Q = constant, or $\frac{dQ}{dt} = 0$.

Similarly, in special relativity, one defines the total four-momentum of a fluid described by energy momentum tensor $T^{\alpha\beta}$ (see § 2.3), on a spacelike hypersurface Σ, as

$$P^\alpha = \int_\Sigma T^{\alpha\beta} d^3\Sigma_\beta \tag{2.7.3}$$

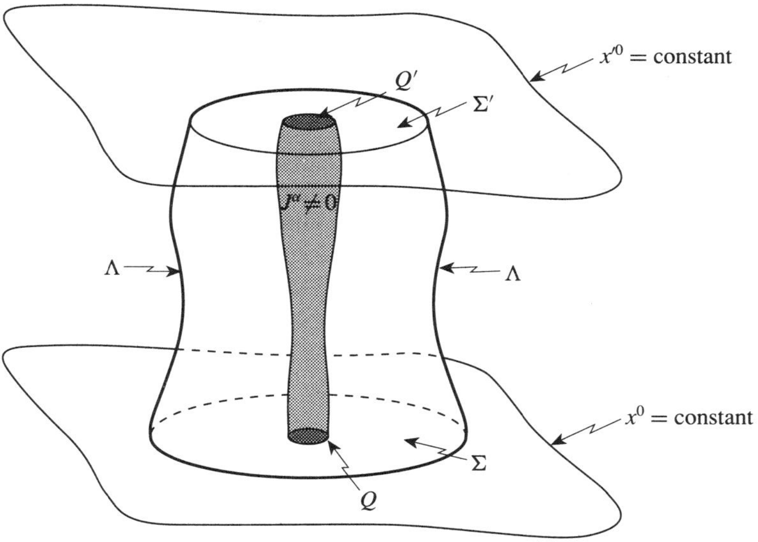

FIGURE 2.2. The hypersurface of integration $\partial\Omega^{(3)}$, boundary of $\Omega^{(4)}$ (see equation (2.7.2)).

where $E \equiv P^0 = \int T^{0\beta} d^3\Sigma_\beta$ is the energy, and the angular momentum of the fluid is defined (see also § 6.10) on a spacelike hypersurface Σ:

$$J^{\alpha\beta} = \int_\Sigma (x^\alpha T^{\beta\mu} - x^\beta T^{\alpha\mu}) d^3\Sigma_\mu . \tag{2.7.4}$$

From the special relativistic, differential conservation laws $T^{\alpha\beta}{}_{,\beta} = 0$, it then follows that these quantities are conserved:

$$0 = \int_\Omega T^{\alpha\beta}{}_{,\beta} d^4\Omega = \int_{\partial\Omega} T^{\alpha\beta} d^3\Sigma_\beta \tag{2.7.5}$$

and

$$P^\alpha = \int_\Sigma T^{\alpha 0} d^3\Sigma_0 = \int_{\Sigma'} T'^{\alpha 0} d^3\Sigma'_0 = P'^\alpha \tag{2.7.6}$$

(zero total outflow of energy and momentum), or

$$\frac{dP^\alpha}{dt} = 0, \tag{2.7.7}$$

and similarly, for the angular momentum:

$$\frac{dJ^{\alpha\beta}}{dt} = 0 \tag{2.7.8}$$

where, in formula (2.7.6), we have chosen the hypersurface $\partial\Omega$ as shown in figure 2.2, with Λ away from the source where $T^{\alpha\beta}$ vanishes, and Σ and Σ' corresponding to t = constant and t' = constant$'$.

In this section we generalize these Minkowski-space definitions to geometrodynamics, to get conserved quantities in curved spacetime. In geometrodynamics, the special relativistic dynamical equation generalize to the tensorial equation, $T^{\alpha\beta}{}_{;\beta} = 0$, consequence of the field equation and of the Bianchi identities—that is, of the fundamental principle that the boundary of the boundary of a region is zero (§ 2.8). However, the divergence theorem does not apply to the covariant divergence of a tensor, therefore the geometrodynamical conserved quantities cannot involve only the energy-momentum tensor $T^{\alpha\beta}$.

Before describing the mathematical details of the definition of the conserved quantities in general relativity, let us first discuss what one would expect from the fundamental analogies and differences between electrodynamics and geometrodynamics. First, the gravitational field $g_{\alpha\beta}$ has energy and momentum associated with it. We know that, in general relativity, gravitational waves carry energy[133–135] and momentum (see § 2.10); this has been experimentally indirectly confirmed with the observations of the decrease of the orbital period of the binary pulsar PSR 1913+1916, explained by the emission of gravitational waves, in agreement with the general relativistic formulae (§ 3.5.1). Two gravitons may create matter, an electron and a positron, by the standard Ivanenko process;[70] therefore, for the conservation of energy, gravitons and gravitational waves must carry energy. We also know that the gravitational geon,[71] made of gravitational waves (see § 2.10), carries energy and momentum. Therefore, since gravitational waves are curvature perturbations of the spacetime, the spacetime geometry must have energy and momentum associated with it. In general relativity the geometry $g_{\alpha\beta}$, where the various physical phenomena take place, is generated by the energy and the energy-currents in the universe, through the field equation. Since the gravity field $g_{\alpha\beta}$ has energy and momentum, the gravitational energy contributes itself, in a loop, to the spacetime geometry $g_{\alpha\beta}$. However, in special relativistic electrodynamics the spacetime geometry $\eta_{\alpha\beta}$ where the electromagnetic phenomena take place, is completely unaffected by these phenomena. Indeed, the fundamental difference between electrodynamics and geometrodynamics is the equivalence principle: locally, in a suitable spacetime neighborhood, it is possible to eliminate every *observable* effect of the gravitational field (see § 2.1). This is true for gravity only.

Therefore, what should one expect from this picture, before one defines the conserved quantities in geometrodynamics? First, one should not expect the conserved quantities to involve only the energy and momentum of matter and nongravitational fields, described by the energy-momentum tensor $T^{\alpha\beta}$ (see expressions 2.3.24 and 2.3.23 for the energy-momentum tensor of a fluid and of an electromagnetic field). Indeed, since the gravitational field $g_{\alpha\beta}$ itself carries energy and momentum, it must, somehow, be included in the definition of energy, momentum, and angular momentum. However, because of the equivalence principle, we should not expect any definition of the energy of the gravitational field to have any local validity; in general relativity, gravitational energy and momentum should only have nonlocal (or quasi-local)[74] validity. Indeed, the gravity field can be locally eliminated, in every freely falling frame, in the sense of eliminating the first derivatives of the metric $g_{\alpha\beta}$ and have $\overset{(i)}{g}_{\alpha\beta} \longrightarrow \eta_{\alpha\beta}$ at a pointlike event; and in the sense of locally (in a spacetime neighborhood of the event) eliminating any measurable effect of gravity, this should also apply to the gravitational energy.

Let us now define the general relativistic conserved quantities. In special relativity, one defines quantities that can be shown to be conserved by using the four-dimensional divergence theorem applied to the differential conservation laws $j^{\alpha}{}_{,\alpha} = 0$ and $T^{\alpha\beta}{}_{,\beta} = 0$. On a curved manifold, from the covariant divergence of the charge current density we can still define conserved quantities by using formula (2.3.6):

$$\begin{aligned} &\int j^{\alpha}{}_{;\alpha}\sqrt{-g}\,d^4\Omega = \\ &\int \left(j^{\alpha}\sqrt{-g}\right)_{,\alpha} d^4\Omega = \\ &\int j^{\alpha}\sqrt{-g}\,d^3\Sigma_{\alpha}. \end{aligned} \tag{2.7.9}$$

However, the four-dimensional divergence theorem is valid for standard divergences but not for the vanishing covariant divergence of the tensor $T^{\alpha\beta}$ in geometrodynamics, $T^{\alpha\beta}{}_{;\beta} = 0$; for a tensor field $T^{\alpha\beta}$, expression (2.3.7) holds, and we cannot directly apply the divergence theorem.

Therefore, we should define quantities $t^{\alpha\beta}$, representing the energy and momentum of the gravitational field, such that the sum of these quantities and of the energy-momentum tensor $T^{\alpha\beta}$

$$T^{\alpha\beta} + t^{\alpha\beta} \equiv T^{\alpha\beta}_{\text{eff}} \tag{2.7.10}$$

will satisfy an equation of the type $T^{\alpha\beta}_{\text{eff},\beta} = 0$. We could then apply the four-dimensional divergence theorem. Of course, on the basis of what we have just observed, we should not expect these quantities $t^{\alpha\beta}$ to form a tensor, since locally the gravity field and its energy should be eliminable.

There are several possible choices for $t^{\alpha\beta}$. We follow here the useful convention of Landau-Lifshitz.[17] By our making zero the first derivatives of the metric tensor at a pointlike event, the gravity field can be "eliminated" in a local inertial frame. Therefore, the quantities $t^{\alpha\beta}$ representing energy and momentum of the gravity field should go to zero in every local inertial frame, and should then be a function of the first derivatives of $g_{\alpha\beta}$. Indeed, at any event, in a local inertial frame, one can reduce the differential conservation laws to $T^{\alpha\beta}{}_{,\beta} = 0$.

Therefore, in order to define the pseudotensor, $t^{\alpha\beta}$, for the gravity field, we first write the field equation at an event, in a local inertial frame, where the first derivatives of the metric are zero. At this event the field equation will involve only the metric and its second derivatives. After some rearrangements, the field equation can then be written

$$\overset{(i)}{\Lambda}{}^{\alpha\beta\mu\nu}{}_{,\nu\mu} = (-\overset{(i)}{g})\overset{(i)}{T}{}^{\alpha\beta} \tag{2.7.11}$$

where

$$\overset{(i)}{g}_{\alpha\beta,\mu} = 0 \tag{2.7.12}$$

and

$$\Lambda^{\alpha\beta\mu\nu} \equiv \frac{1}{16\pi}(-g)\left(g^{\alpha\beta}g^{\mu\nu} - g^{\alpha\mu}g^{\beta\nu}\right). \tag{2.7.13}$$

We may now rewrite the field equation in a general coordinate system, where the first derivatives of $g_{\alpha\beta}$ are in general different from zero, by defining a quantity $(-g)t^{\alpha\beta}$ that represents the difference between the field equation written in the two systems (2.7.11 and 2.3.14), depending on the first derivatives of the metric:

$$(-g)t^{\alpha\beta} \equiv \Lambda^{\alpha\beta\mu\nu}{}_{,\nu\mu} - (-g)T^{\alpha\beta}. \tag{2.7.14}$$

Then this Einstein field equation (2.7.14) lets itself be translated into the language of the effective energy-momentum pseudotensor of expression (2.7.10); that is,

$$(-g)T^{\alpha\beta}_{\text{eff}} \equiv (-g)\left(T^{\alpha\beta} + t^{\alpha\beta}\right) = \Lambda^{\alpha\beta\mu\nu}{}_{,\nu\mu}. \tag{2.7.15}$$

From expression (2.7.13) we know that $\Lambda^{\alpha\beta\mu\nu}$ is antisymmetric with respect to β and μ. Hence the quantity $\Lambda^{\alpha\beta\mu\nu}{}_{,\nu\mu\beta}$ is zero, and therefore from the field equation we have

$$\left((-g)T^{\alpha\beta}_{\text{eff}}\right)_{,\beta} = \Lambda^{\alpha\beta\mu\nu}{}_{,\nu\mu\beta} = 0. \tag{2.7.16}$$

The explicit expression of the pseudotensor $t^{\alpha\beta}$ can be found after some cumbersome calculations. $t^{\alpha\beta}$ can be symbolically written in the form

$$\begin{pmatrix}\textbf{energy-momentum}\\ \textbf{pseudotensor for the}\\ \textbf{gravity field}\end{pmatrix} = t^{\alpha\beta} \sim \sum\left(g\cdot g\cdot\Gamma\cdot\Gamma\right), \tag{2.7.17}$$

that is, $t^{\alpha\beta}$ is the sum of various terms, each quadratic in both $g^{\alpha\beta}$ and $\Gamma^{\alpha}_{\mu\nu}$. The precise expression of $t^{\alpha\beta}$ is (see Landau-Lifshitz)[17]

$$
\begin{aligned}
t^{\alpha\beta} = \frac{1}{16\pi}\Big[& \left(g^{\alpha\mu}g^{\beta\nu} - g^{\alpha\beta}g^{\mu\nu}\right)\left(2\Gamma^{\sigma}_{\mu\nu}\Gamma^{\rho}_{\sigma\rho} - \Gamma^{\sigma}_{\mu\rho}\Gamma^{\rho}_{\nu\sigma} - \Gamma^{\sigma}_{\mu\sigma}\Gamma^{\rho}_{\nu\rho}\right) \\
& + g^{\alpha\mu}g^{\nu\sigma}\left(\Gamma^{\beta}_{\mu\rho}\Gamma^{\rho}_{\nu\sigma} + \Gamma^{\beta}_{\nu\sigma}\Gamma^{\rho}_{\mu\rho} - \Gamma^{\beta}_{\sigma\rho}\Gamma^{\rho}_{\mu\nu} - \Gamma^{\beta}_{\mu\nu}\Gamma^{\rho}_{\sigma\rho}\right) \\
& + g^{\beta\mu}g^{\nu\sigma}\left(\Gamma^{\alpha}_{\mu\rho}\Gamma^{\rho}_{\nu\sigma} + \Gamma^{\alpha}_{\nu\sigma}\Gamma^{\rho}_{\mu\rho} - \Gamma^{\alpha}_{\sigma\rho}\Gamma^{\rho}_{\mu\nu} - \Gamma^{\alpha}_{\mu\nu}\Gamma^{\rho}_{\sigma\rho}\right) \\
& + g^{\mu\nu}g^{\sigma\rho}\left(\Gamma^{\alpha}_{\mu\sigma}\Gamma^{\beta}_{\nu\rho} - \Gamma^{\alpha}_{\mu\nu}\Gamma^{\beta}_{\sigma\rho}\right)\Big].
\end{aligned}
\tag{2.7.18}
$$

Using the effective **energy-momentum pseudotensor for** matter, fields and **gravity field**, in analogy with special relativity and electromagnetism, we finally define the conserved quantities on an asymptotically flat spacelike hypersurface Σ (see below):

$$P^{\alpha} \equiv \int_{\Sigma}\left(T^{\alpha\beta} + t^{\alpha\beta}\right)(-g)d^3\Sigma_{\beta}: \quad \textbf{four-momentum} \tag{2.7.19}$$

$$E \equiv P^0: \quad \textbf{energy} \tag{2.7.20}$$

$$J^{\alpha\beta} \equiv \int_{\Sigma}\left(x^{\alpha}T^{\beta\mu}_{\text{eff}} - x^{\beta}T^{\alpha\mu}_{\text{eff}}\right)(-g)d^3\Sigma_{\mu}: \quad \textbf{angular momentum}. \tag{2.7.21}$$

From equations (2.7.16), as in special relativity, we then have that E, P^{α}, and $J^{\alpha\beta}$ are conserved.

Of course $t^{\alpha\beta}$ (and therefore $T^{\alpha\beta}_{\text{eff}}$) is not a tensor; however, it transforms as a tensor for linear coordinate transformations, as is clear from its expression (2.7.18). Even if the spacetime curvature is different from zero, the pseudotensor for the gravity field $t^{\alpha\beta}$ can be set equal to zero at an event. Vice versa, even in a flat spacetime, $t^{\alpha\beta}$ can be made different from zero with some simple nonlinear coordinate transformation, not even a physical change of frame of reference, but just a mathematical transformation of the spatial coordinates, for example, a simple spatial transformation from Cartesian to polar coordinates. However, the fact that $t^{\alpha\beta}$ can be made different from zero in a flat spacetime, and that it can be made zero, at an event, in a spacetime with curvature, is what we expected, even before defining $t^{\alpha\beta}$, on the basis of the equivalence principle, that is, on the basis that, locally, we can eliminate the observable effects of the gravity field, and therefore, locally, we should not be able to define an energy associated with the gravity field.

However, the situation is different nonlocally; for example, one can define the effective energy carried by a gravitational wave by integrating over a region large compared to a wavelength (see next section). In fact, the energy, momentum, and angular momentum, $E \equiv P^0$, P^{α}, and $J^{\alpha\beta}$, as defined by expressions (2.7.20), (2.7.19), and (2.7.21), have the fundamental property that in an asymptotically flat spacetime, if evaluated on a large region extending far from the source, have a value independent from the coordinate system chosen near the source, and behave as special relativistic four-tensors for any transformation that far from the source is a Lorentz transformation. This happy feature appears when the integrals are transformed to two-surface integrals evaluated far from the source. We have, in fact,

$$P^{\alpha} = \int_{\Sigma} T^{\alpha\beta}_{\text{eff}}(-g)d^3\Sigma_{\beta} = \int_{\Sigma} \Lambda^{\alpha\beta\mu\nu}{}_{,\nu\mu}d^3\Sigma_{\beta}. \tag{2.7.22}$$

By choosing a hypersurface x^0 = constant, with volume element $d^3\Sigma_0$, and by using the divergence theorem, we find

$$P^{\alpha} = \int_{\Sigma} \Lambda^{\alpha 0 i\nu}{}_{,\nu i}d^3\Sigma_0 = \int_{\partial\Sigma\equiv S} \Lambda^{\alpha 0 i\nu}{}_{,\nu}d^2S_i, \tag{2.7.23}$$

and similarly for $J^{\alpha\beta}$, where $d^2S_i \equiv ({}^*dS)_{0i}$ is defined by expression (2.8.19) below. Therefore, P^{α} is invariant for any coordinate transformation near the source, that far from the source, and thus on $\partial\Sigma$, leaves the metric unchanged. Then, since $t^{\alpha\beta}$ behaves as a tensor for linear coordinate transformations (see expression 2.7.18) and P^{α} and $J^{\alpha\beta}$ have a value independent from the coordinates chosen near the source, P^{α} and $J^{\alpha\beta}$ behave as special relativistic four-tensors for any transformation that far from the source is a Lorentz transformation.

In an asymptotically flat manifold, in the weak field region far from the source, where $g_{\alpha\beta} = \eta_{\alpha\beta} + h_{\alpha\beta}$, and $|h_{\alpha\beta}| \ll 1$, from expression (2.7.23), we have the **ADM** formula for the **total energy**:[72]

$$E \equiv P^0 = \frac{1}{16\pi} \int_S \left(g_{ij,j} - g_{jj,i}\right) d^2S_i. \tag{2.7.24}$$

In a spacetime that in the weak field region matches the Schwarzschild (or the Kerr) solution, one then gets, from the post-Newtonian expression (3.4.17) of chapter 3, in asymptotically Minkowskian coordinates, $E = M$, where M is the observed (Keplerian) mass of the central object.

If the interior of the hypersurface of integration Σ contains singularities with apparent horizons or wormholes, one can still prove[73] the gauge invariance and the conservation of P^{α}, without the use of the divergence theorem.

Penrose[74] has given an interesting *quasi-local definition* of energy-momentum and angular momentum, using twistors (a type of spinor field), valid, unlike the ADM formula,[72] even if the integration is done over a finite spacelike two-surface on a manifold *not necessarily asymptotically flat*.

One may now ask an important question. In general, when dealing with arbitrarily strong gravitational fields at the source and with arbitrary matter distributions as sources, is the total energy E of an isolated system positive in general relativity? The solution of this problem is given by the so-called Positive Energy Theorem.

The **Positive Energy Theorem** of Schoen and Yau[75–79] (see also Choquet-Bruhat, Deser, Teitelboim, Witten, York, etc.)[73,80–83] states that given a spacelike, asymptotically Euclidean, hypersurface Σ, and assuming the so-called dominance of energy condition, that is, $\varepsilon \geq (j^i j_i)^{\frac{1}{2}}$, where ε is the energy density on Σ and j^i is the momentum-density on Σ (the dominance of energy condition implies also the weak energy condition $\varepsilon \geq 0$; see § 2.9), and the validity of the Einstein field equation (2.3.14), then: (1) $|E| \equiv |P^0| > |\boldsymbol{P}|$, that is, *the ADM four-momentum is timelike*, and (2) *future-pointing*, $E > 0$, unless $P^\alpha = 0$ (occurring only for Minkowskian manifolds).

2.8 [THE BOUNDARY OF THE BOUNDARY PRINCIPLE AND GEOMETRODYNAMICS]

Einstein's "general relativity," or geometric theory of gravitation, or "geometrodynamics," has two central ideas: (1) Spacetime geometry "tells" mass-energy how to move; and (2) mass-energy "tells" spacetime geometry how to curve.

We have just seen that the way spacetime tells mass-energy how to move is automatically obtained from the Einstein field equation (2.3.14) by using the identity of Riemannian geometry, known as the Bianchi identity, which tells us that the covariant divergence of the Einstein tensor is zero.

According to an idea of extreme simplicity of the laws at the foundations of physics, what one of us has called "the principle of austerity" or "law without law at the basis of physics,"[84] in geometrodynamics it is possible to derive[85,11] the dynamical equations for matter and fields from an extremely simple but central identity of algebraic topology:[86,39] *the principle that the* **boundary of the boundary of a manifold is zero**. Before exploring the consequences of this principle in physics, we have to introduce some concepts and define some quantities of topology and differential geometry.[39–43,86,87]

An n-dimensional **manifold**, M, **with boundary** is a topological space, each of whose points has a neighborhood homeomorphic (two topological spaces are homeomorphic if there exists a mapping between them that is bijective and bicontinuous, called a homeomorphism; see mathematical appendix), that is, topologically equivalent, to an open set in half $\Re^n$, that is to the subspace H^n of all the points $(x^1, x^2, \cdots, x^n)$ of $\Re^n$ such that $x^n \geq 0$. The boundary ∂M of this manifold M is the $(n-1)$-dimensional manifold of all points of M whose images under one of these homeomorphisms lie on the submanifold of H^n corresponding to the points $x^n = 0$. An **orientable manifold** is a manifold that can be covered by a family of charts or coordinate systems $(x^1, \cdots, x^n)$, $(\overline{x}^1, \cdots, \overline{x}^n), \ldots$, such that in the intersections between the charts, the Jacobian, that is, the determinant $\left| \frac{\partial x^i}{\partial \overline{x}^j} \right| \equiv \det\left(\frac{\partial x^i}{\partial \overline{x}^j} \right)$ of the derivatives of the coordinates, is positive. Examples of nonorientable manifolds are the Möbius strip and the

FIGURE 2.3. Two examples of nonorientable manifolds: the Klein bottle or twisted torus and the Möbius strip.

Klein bottle or twisted torus (see fig. 2.3). In the theory of integration[39] on a manifold M which is smooth (that is, differentiable, or which is covered by a family of charts, such that in their intersections the $\frac{\partial x^i}{\partial \bar{x}^j}$ are C^∞ functions) and orientable, one defines a singular n-cube (see fig. 2.4) as a smooth map in the manifold M of an n-cube in the Euclidean $\Re^n$; singular means that the correspondence between a standard n-cube of $\Re^n$ and its image in the manifold M is not necessarily one to one. Then, n-chains c of n-cubes are formally defined as finite sums of n-cubes (multiplied by integers).[39] On these n-chains one defines integration. The boundary ∂c (see figs. 2.4, 2.5, and 2.6) of an n-chain c of n-cubes is the sum of all the properly oriented singular $(n-1)$-cubes which are the boundary of each singular n-cube of the n-chain c. One can then define an operator ∂ that gives the boundary, with a definite orientation, of an n-cube or of an n-chain. It is in general possible to prove[39] that the boundary of the boundary of any n-chain c is zero (see figs. 2.5 and 2.6), that is,

$$\partial(\partial c) = 0 \qquad \text{or formally} \qquad \partial^2 = 0. \tag{2.8.1}$$

Next, let us consider a **differential *n*-form θ** that is, a completely antisymmetric covariant n-tensor, in components $\theta_{\alpha_1 \cdots \beta\gamma \cdots \alpha_n} = -\theta_{\alpha_1 \cdots \gamma\beta \cdots \alpha_n}$, against exchange of any pair of nearby indices such as β, γ; n is the degree of the form. Similarly one can consider a completely antisymmetric contravariant n-tensor called ***n*-polyvector**. The operation of **antisymmetrization** of an n-tensor $T_{\alpha_1 \cdots \alpha_n}$, that we shall denote by writing the indices of the tensor within square brackets, is defined as

$$T_{[\alpha_1 \cdots \alpha_n]} = \frac{1}{n!} \sum_{\substack{\text{all} \\ \text{permutations, } p}} \epsilon_p T_{\alpha_1 \cdots \alpha_n} \tag{2.8.2}$$

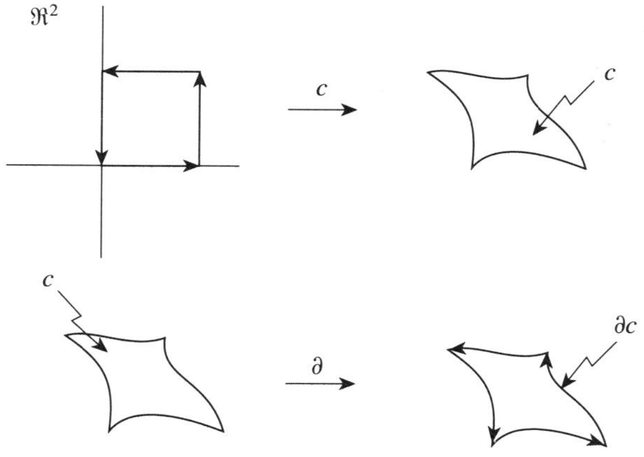

FIGURE 2.4. A standard two-cube c and its $(2-1)$-dimensional boundary ∂c.

where the sum is extended to all the permutations of $\alpha_1 \cdots \alpha_n$, with a plus sign for even permutations, $\epsilon_{p\,\text{even}} \equiv +1$, and minus sign for odd permutations, $\epsilon_{p\,\text{odd}} \equiv -1$. An n-form $\boldsymbol{\theta}$ can then be defined in components as

$$\theta_{\alpha_1 \cdots \alpha_n} = \theta_{[\alpha_1 \cdots \alpha_n]}. \tag{2.8.3}$$

From a p-form $\theta_{\alpha_1 \cdots \alpha_p}$ and from a q-form $\omega_{\alpha_1 \cdots \alpha_q}$, one can construct a $(p+q)$-form, by defining the **wedge product** or **exterior product** $\wedge$ between the two forms, in components

$$(\boldsymbol{\theta} \wedge \boldsymbol{\omega})_{\alpha_1 \cdots \alpha_{p+q}} = \frac{(p+q)!}{p!\,q!}\theta_{[\alpha_1 \cdots \alpha_p}\omega_{\alpha_{p+1} \cdots \alpha_{p+q}]} \tag{2.8.4}$$

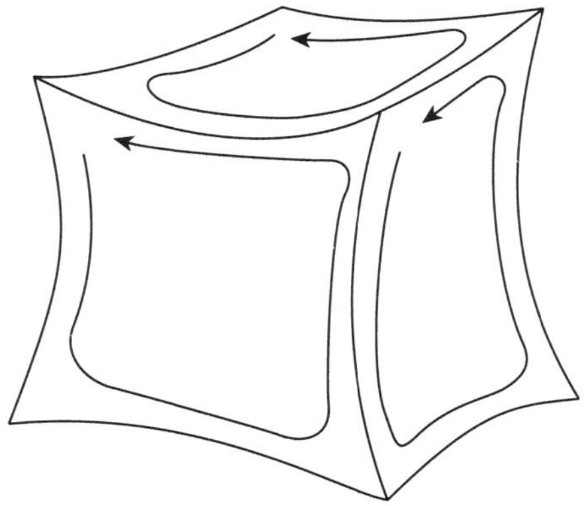

FIGURE 2.5. The oriented one-dimensional boundary of the two-dimensional boundary of a three-cube is zero.

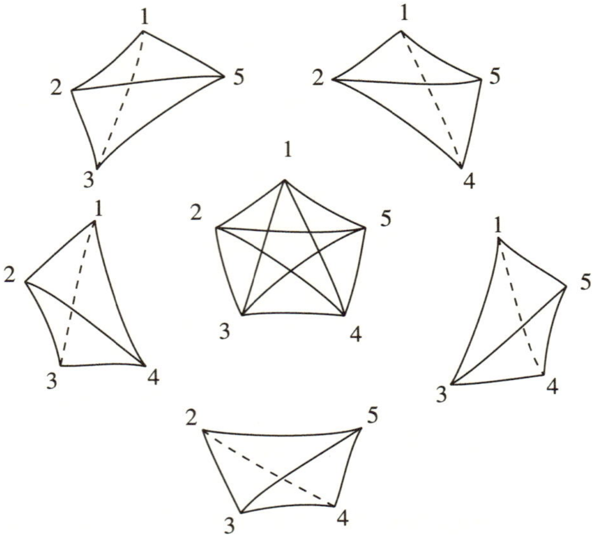

FIGURE 2.6. The two-dimensional boundary of the three-dimensional boundary of a four-dimensional singular four-cube, here a four-simplex, is zero. A two-dimensional projection of the four-simplex is shown in the center. A four-simplex has five vertices, ten edges, ten triangles, and five tetrahedrons. The three-dimensional boundary of the four-simplex is made out of the five tetrahedrons shown in the figure. Each of the ten, two-dimensional, triangles is counted twice with opposite orientations. Therefore, the two-dimensional boundary of the three-dimensional boundary of the four-simplex is zero (adapted from W. Miller 1988).[88]

where $[\alpha_1 \cdots \alpha_{p+q}]$ means antisymmetrization (2.8.2), with respect to the indices within square brackets. The wedge product satisfies the properties

$$
\begin{aligned}
(\boldsymbol{\theta}_1 \wedge \boldsymbol{\theta}_2) \wedge \boldsymbol{\theta}_3 &= \boldsymbol{\theta}_1 \wedge (\boldsymbol{\theta}_2 \wedge \boldsymbol{\theta}_3) \\
(\boldsymbol{\theta}_1 + \boldsymbol{\theta}_2) \wedge \boldsymbol{\omega} &= \boldsymbol{\theta}_1 \wedge \boldsymbol{\omega} + \boldsymbol{\theta}_2 \wedge \boldsymbol{\omega} \\
\boldsymbol{\theta} \wedge (\boldsymbol{\omega}_1 + \boldsymbol{\omega}_2) &= \boldsymbol{\theta} \wedge \boldsymbol{\omega}_1 + \boldsymbol{\theta} \wedge \boldsymbol{\omega}_2 \\
\boldsymbol{\theta} \wedge \boldsymbol{\omega} &= (-1)^{pq} \boldsymbol{\omega} \wedge \boldsymbol{\theta}.
\end{aligned}
\tag{2.8.5}
$$

Then, from an n-form $\theta_{\alpha_1 \cdots \alpha_n} = \theta_{[\alpha_1 \cdots \alpha_n]}$, one can construct an $(n+1)$-form, by defining the **exterior derivative** $\boldsymbol{d\theta}$ of $\boldsymbol{\theta}$, that is the exterior product of $\frac{\partial}{\partial x^\alpha}$

with $\theta_{\alpha_1\cdots\alpha_n}$, in components

$$d\theta_{\alpha_1\cdots\alpha_{n+1}} = (n+1)\frac{\partial}{\partial x^{[\alpha_1}}\theta_{\alpha_2\cdots\alpha_{n+1}]}$$
$$= \frac{1}{n!}\sum_{\substack{\text{all}\\ \text{permutations, } p}} \epsilon_p \frac{\partial}{\partial x^{\alpha_1}}\theta_{\alpha_2\cdots\alpha_{n+1}}. \tag{2.8.6}$$

The exterior derivative of the exterior product (where $\boldsymbol{\theta}$ is a p-form) satisfies the property

$$\boldsymbol{d}(\boldsymbol{\theta}\wedge\boldsymbol{\omega}) = \boldsymbol{d\theta}\wedge\boldsymbol{\omega} + (-1)^p\boldsymbol{\theta}\wedge\boldsymbol{d\omega}. \tag{2.8.7}$$

We introduce the **Levi-Civita pseudotensor**, $\epsilon_{\alpha\beta\gamma\lambda} \equiv \sqrt{-g}[\alpha\beta\gamma\lambda]$, where $\sqrt{-g}$ is the square root of minus the determinant of the metric (equal to one when $g_{\alpha\beta} = \eta_{\alpha\beta} = \text{diag}(-1, +1, +1, +1) =$ Minkowski tensor), and the symbol $[\alpha\beta\gamma\lambda]$ is equal to $+1$ for even permutations of (0, 1, 2, 3), -1 for odd permutations of (0, 1, 2, 3), and 0 when any indices are repeated. We then have $\epsilon^{\alpha\beta\gamma\lambda} = -\frac{1}{\sqrt{-g}}[\alpha\beta\gamma\lambda]$, and the Levi-Civita pseudotensor satisfies the following relations:

$$\epsilon^{\alpha\beta\gamma\lambda}\epsilon_{\alpha\beta\gamma\lambda} = -4! \tag{2.8.8}$$

$$\epsilon^{\rho\sigma\tau\alpha}\epsilon_{\rho\sigma\tau\beta} = -3!\,\delta^{\alpha}{}_{\beta} \tag{2.8.9}$$

$$\begin{aligned}\epsilon^{\rho\sigma\alpha\beta}\epsilon_{\rho\sigma\gamma\lambda} &= -2!\left(\delta^{\alpha}{}_{\gamma}\delta^{\beta}{}_{\lambda} - \delta^{\alpha}{}_{\lambda}\delta^{\beta}{}_{\gamma}\right)\\ &= -2!\left(2!\delta^{\alpha}{}_{[\gamma}\delta^{\beta}{}_{\lambda]}\right)\\ &\equiv -2!\delta^{\alpha\beta}{}_{\gamma\lambda}\end{aligned} \tag{2.8.10}$$

$$\epsilon^{\alpha\beta\gamma\sigma}\epsilon_{\lambda\mu\nu\sigma} = -3!\delta^{\alpha}{}_{[\lambda}\delta^{\beta}{}_{\mu}\delta^{\gamma}{}_{\nu]} \equiv -\delta^{\alpha\beta\gamma}{}_{\lambda\mu\nu} \tag{2.8.11}$$

and

$$\epsilon^{\alpha\beta\gamma\lambda}\epsilon_{\mu\nu\rho\sigma} = -4!\delta^{\alpha}{}_{[\mu}\delta^{\beta}{}_{\nu}\delta^{\gamma}{}_{\rho}\delta^{\lambda}{}_{\sigma]} \equiv -\delta^{\alpha\beta\gamma\lambda}{}_{\mu\nu\rho\sigma} \tag{2.8.12}$$

where $\delta^{\alpha_1\cdots\alpha_n}{}_{\beta_1\cdots\beta_n}$ is equal to $+1$ if $\alpha_1\cdots\alpha_n$ is an even permutation of $\beta_1\cdots\beta_n$ with no repeated indices ($1 \le n \le 4$), equal to -1 if it an odd permutation, and 0 otherwise. The δ-tensors satisfy

$$\begin{aligned}&\delta^{\alpha\beta\gamma\sigma}{}_{\lambda\mu\nu\sigma} = \delta^{\alpha\beta\gamma}{}_{\lambda\mu\nu}; \qquad \delta^{\alpha\beta\sigma}{}_{\mu\nu\sigma} = 2\delta^{\alpha\beta}{}_{\mu\nu};\\ &\delta^{\alpha\sigma}{}_{\beta\sigma} = 3\delta^{\alpha}{}_{\beta} \qquad \text{and} \qquad \delta^{\alpha}{}_{\alpha} = 4.\end{aligned} \tag{2.8.13}$$

They can be used to antisymmetrize a tensor

$$T_{[\alpha_1\cdots\alpha_n]} = \frac{1}{n!}T_{\beta_1\cdots\beta_n}\delta^{\beta_1\cdots\beta_n}{}_{\alpha_1\cdots\alpha_n} \tag{2.8.14}$$

(where in a four-manifold: $1 \leq n \leq 4$) and to write the determinant of a tensor $T^{\alpha}{}_{\beta}$

$$\det(T^{\alpha}{}_{\beta}) = \frac{1}{4!}\delta^{\alpha\beta\gamma\lambda}{}_{\mu\nu\rho\sigma}T^{\mu}{}_{\alpha}T^{\nu}{}_{\beta}T^{\rho}{}_{\gamma}T^{\sigma}{}_{\lambda} = [\mu\nu\rho\sigma]T^{\mu}{}_{0}T^{\nu}{}_{1}T^{\rho}{}_{2}T^{\sigma}{}_{3}. \tag{2.8.15}$$

Finally, by using the δ-tensors, one can compactly rewrite a **two-dimensional surface element** $dS^{\alpha\beta}$, a **three-dimensional hypersurface element** $d\Sigma^{\alpha\beta\gamma}$, and a **four-dimensional volume element** $d\Omega^{\alpha\beta\gamma\lambda}$, respectively built on two, three, and four infinitesimal displacements $dx^{\alpha}_{(\rho)}$:

$$dS^{\alpha\beta} \equiv \delta^{\alpha\beta}{}_{\mu\nu}dx^{\mu}_{(1)}dx^{\nu}_{(2)} = \begin{vmatrix} dx^{\alpha}_{(1)} & dx^{\alpha}_{(2)} \\ dx^{\beta}_{(1)} & dx^{\beta}_{(2)} \end{vmatrix} \tag{2.8.16}$$

$$d\Sigma^{\alpha\beta\gamma} \equiv \delta^{\alpha\beta\gamma}{}_{\mu\nu\rho}dx^{\mu}_{(1)}dx^{\nu}_{(2)}dx^{\rho}_{(3)} \tag{2.8.17}$$

$$d\Omega^{\alpha\beta\gamma\lambda} \equiv \delta^{\alpha\beta\gamma\lambda}{}_{\mu\nu\rho\sigma}dx^{\mu}_{(0)}dx^{\nu}_{(1)}dx^{\rho}_{(2)}dx^{\sigma}_{(3)}. \tag{2.8.18}$$

The **duals** of these elements, for $\sqrt{-g} = 1$, are defined as

$$({}^{*}dS)_{\alpha\beta} \equiv \frac{1}{2}[\rho\sigma\alpha\beta]dS^{\rho\sigma} \tag{2.8.19}$$

$$d^3\Sigma_{\alpha} \equiv \frac{1}{3!}[\alpha\mu\nu\rho]d\Sigma^{\mu\nu\rho} \tag{2.8.20}$$

$$d^4\Omega \equiv \frac{1}{4!}[\mu\nu\rho\sigma]d\Omega^{\mu\nu\rho\sigma}. \tag{2.8.21}$$

In particular, for the four infinitesimal coordinate displacements, $dx^{\alpha}_{(\rho)} = \delta^{\alpha}{}_{\rho}dx^{\alpha}$ (no summation over α), with $\rho\epsilon(0, 1, 2, 3)$, we have

$$d^4\Omega \equiv d^4x = dx^0dx^1dx^2dx^3, \tag{2.8.22}$$

and corresponding to a hypersurface $x^0 = \text{constant}$:

$$d^3\Sigma_0 \equiv d^3V = dx^1dx^2dx^3. \tag{2.8.23}$$

On an n-dimensional manifold, we can then define the $(n-p)$-polyvector ${}^{*}\boldsymbol{\theta}$ **dual** to the ***p*-form** $\boldsymbol{\theta}$ in components

$$({}^{*}\boldsymbol{\theta})^{\alpha_1\cdots\alpha_{n-p}} = \frac{1}{p!}\epsilon^{\beta_1\cdots\beta_p\alpha_1\cdots\alpha_{n-p}}\theta_{\beta_1\cdots\beta_p} \tag{2.8.24}$$

with a similar definition for the $(n-p)$-form, ${}^{*}\boldsymbol{v}$ dual of a p-polyvector $\boldsymbol{v}$.

Now, on an n-dimensional manifold M, we have the beautiful and fundamental **Stokes theorem** (for the mathematical details see Spivak 1979, vol. 2)[39]

$$\int_c d\boldsymbol{\theta} = \int_{\partial c}\boldsymbol{\theta} \qquad \textbf{Stokes theorem} \tag{2.8.25}$$

where c is an n-chain on the manifold M, ∂c the $(n-1)$-chain oriented boundary of c, $\boldsymbol{\theta}$ a $(n-1)$-form on M, and $\boldsymbol{d\theta}$ the n-form exterior derivative of $\boldsymbol{\theta}$. For an oriented, n-dimensional manifold M with boundary ∂M (with the induced orientation)[39] and for an $(n-1)$-form $\boldsymbol{\theta}$ on M, with compact support (i.e., the smallest closed set containing the region of M where $\boldsymbol{\theta}$ is nonzero is compact), we then have

$$\int_M \boldsymbol{d\theta} = \int_{\partial M} \boldsymbol{\theta} \qquad \textbf{Stokes theorem}. \tag{2.8.26}$$

Furthermore, as a consequence of the boundary of the boundary principle (2.8.1), for every $(n-2)$-form $\boldsymbol{\theta}$ on an n-dimensional, differentiable, oriented manifold M, we have

$$\int_{\partial\partial M} \boldsymbol{\theta} = 0. \tag{2.8.27}$$

Therefore, from the boundary of the boundary principle (2.8.1) and from Stokes theorem:

$$\int_c \boldsymbol{dd\theta} = \int_{\partial c} \boldsymbol{d\theta} = \int_{\partial\partial c} \boldsymbol{\theta} = 0. \tag{2.8.28}$$

By applying this result to an arbitrary neighborhood of an arbitrary point, one has then, automatically,

$$\boldsymbol{dd\theta} = \mathbf{0}, \qquad \text{or formally} \qquad d^2 = 0. \tag{2.8.29}$$

The exterior derivative of the exterior derivative of any form is zero. In other words, the exterior derivative of any exact form is zero, where **exact** is any ***n*-form** that can be written as $\boldsymbol{d\theta}$ and $\boldsymbol{\theta}$ is an $(n-1)$-form. Therefore, any exact form is **closed**, that is, with null exterior derivative (as one can also directly calculate from the definition of $\boldsymbol{d}$). For a vector field $\boldsymbol{W}$ in the three-dimensional Euclidean space $\Re^3$, from Stokes theorem we get two well-known corollaries, the so-called *divergence theorem* (*Ostrogradzky-Green formula* or *Gauss theorem*):

$$\int_V \boldsymbol{\nabla} \cdot \boldsymbol{W} d^3V = \int_{\partial V = S} \boldsymbol{W} \cdot \boldsymbol{n}\, d^2S \tag{2.8.30}$$

and the *Riemann-Ampère-Stokes formula*:

$$\int_S (\boldsymbol{\nabla} \times \boldsymbol{W}) \cdot \boldsymbol{n}\, d^2S = \int_{\partial S = l} \boldsymbol{W} \cdot d^1\boldsymbol{l} \tag{2.8.31}$$

where d^3V, d^2S and $d^1\boldsymbol{l}$ are the standard Euclidean volume, surface, and line elements, and $\boldsymbol{n}$ is the normal to the surface S.

We are now ready to investigate on some physical consequences[89,90] of the boundary of the boundary principle.

In electrodynamics, one defines (see § 2.3) the electromagnetic field tensor $\boldsymbol{F}$ as the 2-form:

$$\boldsymbol{F} = \boldsymbol{dA} \tag{2.8.32}$$

or in components, $F_{\alpha\beta} = A_{\beta,\alpha} - A_{\alpha,\beta}$, where $\boldsymbol{A}$ is the four-potential 1-form, with components A_α.

From the boundary of the boundary principle, in the form $d^2 = 0$, we *automatically* get the **sourceless Maxwell equations** for $\boldsymbol{F}$:

$$\boldsymbol{dF} = \boldsymbol{ddA} = \boldsymbol{0} \tag{2.8.33}$$

in components

$$F_{[\alpha\beta,\gamma]} = 0. \tag{2.8.34}$$

The **Maxwell equations with source** are

$$F^{\alpha\beta}{}_{,\beta} = 4\pi j^\alpha \tag{2.8.35}$$

where $j^\alpha = \rho u^\alpha$ is the charge current density four-vector. This equation can be rewritten by defining the dual form, $^\star\boldsymbol{F}$, of the form $\boldsymbol{F}$ and the dual form, $^\star\boldsymbol{j}$, of the charge current density 1-form $\boldsymbol{j}$ (see expression 2.8.63 for the general definition of $^\star(\cdots)$):

$$(^\star\boldsymbol{F})_{\mu\nu} \equiv \frac{1}{2}\epsilon_{\alpha\beta\mu\nu}F^{\alpha\beta} \tag{2.8.36}$$

$$(^\star\boldsymbol{j})_{\beta\mu\nu} \equiv \epsilon_{\alpha\beta\mu\nu}j^\alpha; \tag{2.8.37}$$

therefore

$$\begin{aligned}(\boldsymbol{d}^\star\boldsymbol{F})_{\alpha\beta\gamma} &= \frac{3}{2}\epsilon_{\mu\nu[\alpha\beta}F^{\mu\nu}{}_{,\gamma]} \\ &= 4\pi\epsilon_{\sigma\alpha\beta\gamma}j^\sigma\end{aligned} \tag{2.8.38}$$

or

$$\boldsymbol{d}^\star\boldsymbol{F} = 4\pi\,^\star\boldsymbol{j}. \tag{2.8.39}$$

From the boundary of the boundary principle, in the form $d^2 = 0$, we then *automatically* get the **dynamical equations for $\boldsymbol{j}$**:

$$4\pi\boldsymbol{d}^\star\boldsymbol{j} = \boldsymbol{dd}^\star\boldsymbol{F} = \boldsymbol{0} \tag{2.8.40}$$

in components

$$\left(j^\alpha\epsilon_{\alpha[\beta\mu\nu}\right)_{,\gamma]} = 0, \tag{2.8.41}$$

that is, multiplying by $\epsilon^{\beta\mu\nu\gamma}$ and summing over all its indices,

$$j^\alpha{}_{,\alpha} = 0. \tag{2.8.42}$$

Summarizing, in electrodynamics we have

$$F \equiv dA \begin{pmatrix}\text{definition} \\ \text{of } F\end{pmatrix} \overset{d^2=0}{\Longrightarrow} \left[dF = 0 \begin{pmatrix}\text{sourceless} \\ \text{Maxwell} \\ \text{equations}\end{pmatrix}\right. \tag{2.8.43}$$

and

$$\left\lfloor d^\star F = 4\pi {}^\star j \begin{pmatrix}\text{Maxwell} \\ \text{equations} \\ \text{with source}\end{pmatrix} \right. \overset{d^2=0}{\Longrightarrow} d^\star j = 0 \begin{pmatrix}\text{dynamical} \\ \text{equations} \\ \text{for } j\end{pmatrix} \tag{2.8.44}$$

In geometrodynamics, the Riemann curvature tensor satisfies the so-called **first Bianchi identity**:

$$R^\alpha{}_{[\beta\gamma\delta]} = 0, \tag{2.8.45}$$

and the **second Bianchi identity** (§ 2.4)

$$R^\alpha{}_{\beta[\gamma\delta;\mu]} = 0. \tag{2.8.46}$$

Consequently the Einstein tensor $G_{\alpha\beta}$ satisfies the contracted second Bianchi identities

$$G^\sigma{}_{\alpha;\sigma} \equiv \left(R^\sigma{}_\alpha - \frac{1}{2} R \delta^\sigma{}_\alpha \right)_{;\sigma} = 0. \tag{2.8.47}$$

As in electrodynamics, these identities can be derived from the boundary of the boundary principle, $\partial^2 = 0$, directly from its consequence that the second exterior derivative of any form is zero, $d^2 = 0$.

Let us first consider,[91,39,43] on an n-dimensional manifold, n linearly independent vector fields $X_1, \ldots, X_n$, called a *moving frame* (the *Cartan's Repère Mobile*). We can then consider the 1-forms θ^α which define the dual basis (different concept from the dual of a form (2.8.24) or the dual of a polyvector), that is, the forms θ^α such that $\theta^\alpha{}_\sigma X^\sigma{}_\beta = \delta^\alpha{}_\beta$. Furthermore, by using the exterior product (2.8.4), on a Riemannian manifold M with metric $g_{\alpha\beta}$, one can construct the **connection 1-forms** $\omega^\alpha{}_\beta = \Gamma^\alpha_{\beta\gamma}\theta^\gamma$, defined by

$$dg_{\alpha\beta} = g_{\alpha\sigma}\omega^\sigma{}_\beta + g_{\sigma\beta}\omega^\sigma{}_\alpha \tag{2.8.48}$$

where $dg_{\alpha\beta} = X_\rho(g_{\alpha\beta})\theta^\rho$, and in a coordinate basis $dg_{\alpha\beta} = g_{\alpha\beta,\rho}dx^\rho$, and by

$$\Theta^\alpha = 0 \tag{2.8.49}$$

where $\Theta^\alpha = d\theta^\alpha + \omega^\alpha{}_\sigma \wedge \theta^\sigma$ (first Cartan structure equation), and Θ^α are the **torsion 2-forms** (see below).

Using the connection 1-forms $\omega^\alpha{}_\beta$, the exterior derivative (2.8.6) and the exterior product (2.8.4), one can then construct the **curvature 2-forms** $\Omega^\alpha{}_\beta$, for the moving frame X^α:

$$\Omega^\alpha{}_\beta = d\omega^\alpha{}_\beta + \omega^\alpha{}_\sigma \wedge \omega^\sigma{}_\beta \qquad \text{(second Cartan structure equation).} \tag{2.8.50}$$

By taking the exterior derivative of expression (2.8.49), from the boundary of the boundary principle, in the form $d^2 = 0$, we get the **first Bianchi identity**:

$$\begin{aligned} \mathbf{0} = \boldsymbol{dd\theta}^\alpha &= \boldsymbol{d}\left(-\boldsymbol{\omega}^\alpha{}_\sigma \wedge \boldsymbol{\theta}^\sigma\right) = -\left(\boldsymbol{\Omega}^\alpha{}_\sigma - \boldsymbol{\omega}^\alpha{}_\rho \wedge \boldsymbol{\omega}^\rho{}_\sigma\right) \wedge \boldsymbol{\theta}^\sigma \\ &- \boldsymbol{\omega}^\alpha{}_\sigma \wedge \boldsymbol{\omega}^\sigma{}_\rho \wedge \boldsymbol{\theta}^\rho = -\boldsymbol{\Omega}^\alpha{}_\sigma \wedge \boldsymbol{\theta}^\sigma \end{aligned} \tag{2.8.51}$$

and by taking the exterior derivative of expression (2.8.50), from $d^2 = 0$, we get

$$\boldsymbol{d\Omega}^\alpha{}_\beta = \boldsymbol{d\omega}^\alpha{}_\sigma \wedge \boldsymbol{\omega}^\sigma{}_\beta - \boldsymbol{\omega}^\alpha{}_\sigma \wedge \boldsymbol{d\omega}^\sigma{}_\beta. \tag{2.8.52}$$

By substituting $\boldsymbol{d\omega}^\alpha{}_\beta = \boldsymbol{\Omega}^\alpha{}_\beta - \boldsymbol{\omega}^\alpha{}_\sigma \wedge \boldsymbol{\omega}^\sigma{}_\beta$, we then have

$$\boldsymbol{d\Omega}^\alpha{}_\beta + \boldsymbol{\omega}^\alpha{}_\sigma \wedge \boldsymbol{\Omega}^\sigma{}_\beta - \boldsymbol{\Omega}^\alpha{}_\sigma \wedge \boldsymbol{\omega}^\sigma{}_\beta = \mathbf{0}. \tag{2.8.53}$$

This is the **second Bianchi identity**. Finally, by defining the **exterior covariant derivative**, D: $\boldsymbol{D\Omega}^\alpha{}_\beta \equiv \boldsymbol{d\Omega}^\alpha{}_\beta + \boldsymbol{\omega}^\alpha{}_\sigma \wedge \boldsymbol{\Omega}^\sigma{}_\beta - \boldsymbol{\Omega}^\alpha{}_\sigma \wedge \boldsymbol{\omega}^\sigma{}_\beta$, which maps a tensor-valued p-form (a p-form with tensor indices) into a tensor-valued $(p+1)$-form, we can rewrite the second Bianchi identity as:

$$\boldsymbol{D\Omega}^\alpha{}_\beta = \mathbf{0}. \tag{2.8.54}$$

Equation (2.8.49) expresses that the torsion $\boldsymbol{\Theta}^\alpha$ is zero, and equation (2.8.48) that the connection is metric-compatible, that is, the covariant derivative of the metric is zero. It follows that the connection is uniquely[39] determined to be the standard Riemannian connection. Using the **natural coordinate basis**, $\{\overset{(c)}{\boldsymbol{X}}_\alpha\} = \{\frac{\partial}{\partial x^\alpha}\}$ (a coordinate basis is also called **holonomic,** and a noncoordinate basis **anholonomic**), on a Riemannian manifold, one has then

$$\left(\overset{(c)}{\boldsymbol{\omega}}{}^\alpha{}_\beta\right)_\gamma = \Gamma^\alpha_{\beta\gamma} = \text{Christoffel symbols (expression 2.2.3)} \tag{2.8.55}$$

(for the expression of $\Gamma^\alpha_{\beta\gamma}$ in a general basis see the mathematical appendix),

$$\left(\overset{(c)}{\boldsymbol{\Theta}}{}^\alpha\right)_{\beta\gamma} = \Gamma^\alpha_{\gamma\beta} - \Gamma^\alpha_{\beta\gamma} \equiv T^\alpha_{[\gamma\beta]} = 0, \qquad \text{i.e., no torsion,} \tag{2.8.56}$$

and

$$\left(\overset{(c)}{\boldsymbol{\Omega}}{}^\alpha{}_\beta\right)_{\gamma\delta} = R^\alpha{}_{\beta\gamma\delta} = \text{Riemann curvature tensor (expression 2.2.5)} \tag{2.8.57}$$

and we can rewrite equations (2.8.51) and (2.8.54), in components, as

$$R^\alpha{}_{[\beta\gamma\delta]} = 0, \qquad \text{(eq. 2.8.45)} \tag{2.8.58}$$

and

$$R^\alpha{}_{\beta[\gamma\delta;\mu]} = 0, \qquad \text{(eq. 2.8.46).} \tag{2.8.59}$$

Equations (2.8.50) and (2.8.57) define the curvature tensor $R^{\alpha}{}_{\beta\gamma\delta}$ without the use of the covariant derivatives as in the standard definition (2.2.4).

In geometrodynamics the contracted second Bianchi identity, consequence of $d^2 = 0$, is especially important. In fact, the dynamical equations for matter and fields automatically follow from this identity plus the Einstein field equation (2.3.14). To derive the dynamical equations from the boundary of the boundary principle we first construct[11] the double dual of the Riemann tensor:

$$\left({}^*\boldsymbol{R}^*\right)_{\alpha\beta}{}^{\gamma\delta} \equiv \frac{1}{4}\epsilon_{\alpha\beta\mu\nu}R^{\mu\nu}{}_{\rho\sigma}\epsilon^{\rho\sigma\gamma\delta}. \tag{2.8.60}$$

We can then rewrite the Einstein tensor, $G^{\alpha}{}_{\beta}$, as

$$G^{\alpha}{}_{\beta} = \left({}^*\boldsymbol{R}^*\right)^{\alpha\sigma}{}_{\beta\sigma}. \tag{2.8.61}$$

We have, in fact,

$$\begin{aligned}\left({}^*\boldsymbol{R}^*\right)^{\alpha\sigma}{}_{\beta\sigma} &= \frac{1}{4}\epsilon^{\alpha\sigma\gamma\lambda}\epsilon_{\mu\nu\beta\sigma}R^{\mu\nu}{}_{\gamma\lambda} = -\frac{1}{4}\delta^{\alpha\gamma\lambda}{}_{\mu\nu\beta}R^{\mu\nu}{}_{\gamma\lambda}\\ &= -\frac{1}{4}\left(2\delta^{\alpha}{}_{\beta}R - 2R^{\alpha\gamma}{}_{\beta\gamma} - 2R^{\alpha\lambda}{}_{\beta\lambda}\right) = G^{\alpha}{}_{\beta}\end{aligned} \tag{2.8.62}$$

where we have used the relation (2.8.11). We now define the *star operator* ${}^\star(\cdots)$, with a star on the *left*, a duality operator which acts only on m-forms (with $m \leq n$ on an n-dimensional manifold) and gives $(n - m)$-forms, that is, a duality operator which acts only on the m $(0 \leq m \leq n)$ antisymmetric covariant indices of a tensor and generates $n - m$ antisymmetric covariant indices. In other words, the ${}^\star(\cdots)$ operator acts only on the antisymmetric covariant indices of a tensor $T^{\alpha\beta\cdots}{}_{\gamma\delta\ldots}$, by first raising each covariant index with $g^{\mu\nu}$ and then by taking the dual, with $\epsilon_{\alpha\beta\cdots\mu}$, of these raised indices:

$$({}^\star\boldsymbol{T})^{\alpha\beta\cdots}{}_{\cdots\mu} = \frac{1}{m!}\epsilon_{\sigma\rho\cdots\mu}T^{\alpha\beta\cdots}{}_{\gamma\delta\cdots}g^{\sigma\gamma}g^{\rho\delta\cdots}. \tag{2.8.63}$$

Similarly, we define the *star operator* $(\cdots)^\star$, with a star on the *right*, as a duality operator which acts only on m-polyvectors (antisymmetric m-contravariant tensors) and gives $(n - m)$-polyvectors, that is, a duality operator which acts only on the m $(0 \leq m \leq n)$ antisymmetric contravariant indices of a tensor and generates $n - m$ antisymmetric contravariant indices:

$$(\boldsymbol{T}^\star)^{\cdots\mu}{}_{\gamma\delta\cdots} = \frac{1}{m!}\epsilon^{\sigma\rho\cdots\mu}T^{\alpha\beta\cdots}{}_{\gamma\delta\cdots}g_{\alpha\sigma}g_{\beta\rho\cdots}. \tag{2.8.64}$$

We then introduce the vector-valued (a form with a vector index) 1-form, $(\boldsymbol{dP})^{\alpha}{}_{\beta} = \delta^{\alpha}{}_{\beta}$, sometimes called the Cartan unit tensor. By taking the star dual ${}^\star(\cdots)$ of both sides of the Einstein field equation $G^{\alpha}{}_{\beta} = \chi T^{\alpha}{}_{\beta}$,

$${}^\star\boldsymbol{G} = \chi\,{}^\star\boldsymbol{T}, \tag{2.8.65}$$

we have in components

$$\epsilon_{\sigma\beta\gamma\delta}G^{\alpha\sigma} = \chi\epsilon_{\sigma\beta\gamma\delta}T^{\alpha\sigma}. \tag{2.8.66}$$

By defining

$$[\boldsymbol{dP} \wedge \boldsymbol{R}]^{\alpha\mu\nu}_{\beta\rho\sigma} \equiv \frac{3!}{2!}\frac{3!}{2!}\delta^{[\alpha}_{[\beta}R^{\mu\nu]}{}_{\rho\sigma]} \tag{2.8.67}$$

here by $[\boldsymbol{S} \wedge \boldsymbol{T}]$ we mean exterior product of *both* the covariant and the contravariant parts of the antisymmetric tensors $\boldsymbol{S}$ and $\boldsymbol{T}$, that is we mean both antisymmetrization of the covariant indices of the product of $\boldsymbol{S}$ with $\boldsymbol{T}$ times a factor $\frac{(p+q)!}{p!q!}$ and antisymmetrization of the *contravariant* indices times a factor $\frac{(n+m)!}{n!\,m!}$. We can rewrite the left-hand side of the star dual, $^{\star}\boldsymbol{G}$, of the Einstein tensor, $\boldsymbol{G}$,

$$[\boldsymbol{dP} \wedge \boldsymbol{R}]^{\star} = {}^{\star}\boldsymbol{G} = \chi\,{}^{\star}\boldsymbol{T} \tag{2.8.68}$$

Indeed, we have, in components, using expressions (2.8.14) and (2.8.62):

$$\begin{aligned} &\frac{3}{2}g^{\gamma\tau}\epsilon_{\alpha\mu\nu\tau}\delta^{[\alpha}_{[\beta}R^{\mu\nu]}{}_{\rho\sigma]} = \frac{1}{4}g^{\gamma\tau}\epsilon_{\alpha\mu\nu\tau}\delta^{\alpha\mu\nu}{}_{\lambda\theta\varphi}\delta^{\lambda}_{[\beta}R^{\theta\varphi}{}_{\rho\sigma]} = \\ &-\frac{3}{2}g^{\gamma\tau}\epsilon_{\mu\nu\tau[\beta}R^{\mu\nu}{}_{\rho\sigma]} = -\frac{1}{4}g^{\gamma\tau}\epsilon_{\mu\nu\tau\lambda}\delta_{\beta\rho\sigma}{}^{\lambda\theta\varphi}R^{\mu\nu}{}_{\theta\varphi} = \\ &-\frac{1}{4}g^{\gamma\tau}\epsilon_{\mu\nu\tau\lambda}\epsilon_{\beta\rho\sigma\alpha}\epsilon^{\alpha\lambda\theta\varphi}R^{\mu\nu}{}_{\theta\varphi} = \epsilon_{\alpha\beta\rho\sigma}G^{\alpha\gamma}. \end{aligned} \tag{2.8.69}$$

By taking the exterior covariant derivative of equation (2.8.68) we then have

$$\boldsymbol{D}[\boldsymbol{dP} \wedge \boldsymbol{R}]^{\star} = \big(\boldsymbol{D}[\boldsymbol{dP} \wedge \boldsymbol{R}]\big)^{\star} = \big([\boldsymbol{DdP} \wedge \boldsymbol{R}] - [\boldsymbol{dP} \wedge \boldsymbol{DR}]\big)^{\star} = \boldsymbol{0} \tag{2.8.70}$$

where $\boldsymbol{DdP} = \boldsymbol{0}$, that is, there is no torsion, and $\boldsymbol{DR} = \boldsymbol{0}$ is the second Bianchi identity (2.8.54) as a consequence of $d^2 = 0$, that is, as a consequence of the boundary of the boundary principle. Finally, from the Einstein field equation (2.8.68), we have

$$\boldsymbol{D}^{\star}\boldsymbol{G} = \boldsymbol{D}^{\star}\boldsymbol{T} = \boldsymbol{0}, \tag{2.8.71}$$

that is, in components, using (2.3.7),

$$T^{\alpha\beta}{}_{;\beta} = 0. \tag{2.8.72}$$

The quantity $[\boldsymbol{dP} \wedge \boldsymbol{R}]^{\star}$ has a geometrical interpretation.[91,84,11] It may be thought of as the star dual of the moment of rotation, of a vector, associated with a three-cube and induced by the Riemann curvature (see fig. 2.7). The Einstein field equation may then be geometrically interpreted as identifying the star dual of the moment of rotation associated with a three-cube with the amount of

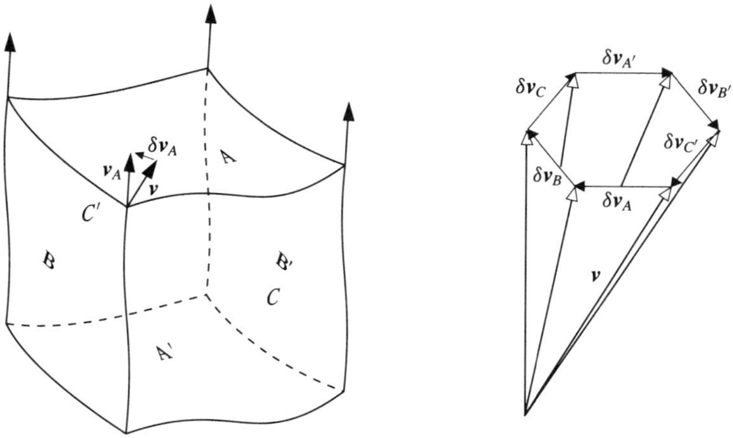

FIGURE 2.7. The rotation of a vector $\boldsymbol{v}$ associated with each face of a three-cube and induced by the Riemann curvature tensor, and the one-boundary of the two-boundary of a three-cube. *Left*: the rotation of a vector transported $\boldsymbol{v}_t$ parallel to itself around the indicated circuit, this rotation measures some components of the spacetime curvature (see eq. 2.4.19). *Right*: the rotations associated with all six faces together add up to zero; the diagram closes. It closes because each edge of the cube is traversed twice, and in opposite directions, in the circumnavigation of the two abutting faces of the cube: $\partial\partial = 0$.

energy-momentum of matter and fields contained in that three-cube:

$$\begin{pmatrix} \text{dual of} \\ \text{moment of} \\ \text{rotation} \\ \text{associated with} \\ \text{a three-cube} \end{pmatrix} = 8\pi \begin{pmatrix} \text{amount of} \\ \text{energy-momentum} \\ \text{in that} \\ \text{three-cube} \end{pmatrix}. \tag{2.8.73}$$

This is the geometrical content of the Einstein equation. Then, by applying to the Einstein field equation the simple but central topological 2-3-4 (in two-three-four dimensions) **boundary of the boundary principle**, $\partial^2 = 0$, one gets the *dynamical equations* for matter and fields.

2.9 BLACK HOLES AND SINGULARITIES

Black Holes and Gravitational Collapse

Collapse of a spherically symmetric star to a dense configuration[92–96] can on occasion put enough mass M inside a spherical surface of circumference $2\pi r$

as to make the terms $-(1 - \frac{2M}{r})dt^2$ and $(1 - \frac{2M}{r})^{-1}dr^2$ in the metric (2.6.35) reverse sign inside this surface. By analyzing the radial light cones (θ and ϕ constant), as calculated from $ds^2 = 0$ in the Schwarzschild coordinates of expression (2.6.35), we find that $\frac{dr}{dt} = \pm(1 - \frac{2M}{r})$ tends to zero as it approaches the region $r = 2M$, and inside this region, where $r < 2M$, the future light cones point inward, toward $r = 0$ (fig. 2.8). Since particles, or photons, propagate within, or on, the light cones, no photon can escape from such a region, nor any particle that follows classical physics. It is no wonder that such a collapsed star received[92] the name "**black hole**"[19,97–101,144] as early as 1967. This strange behavior of the Schwarzschild spacetime geometry extends over the region where r is less than the so-called Schwarzschild radius, $r_s = 2M$. A black hole with Earth mass has a Schwarzschild radius of about 0.88 cm and one of Sun's mass $M_\odot$ of about 3 km.

The X-ray telescope UHURU floating above the atmosphere discovered in 1971 (see ref. 130) the first compelling evidence for a black hole, Cygnus X-1. Its mass is today estimated as of the order of 10 $M_\odot$ (since then, other black hole candidates have been found in X-ray binary systems, for example in nova V404 Cygni[142] and in Nova Muscae[143]). Recently, H. Ford et al., using the Faint Object Spectrograph of the *Hubble Space Telescope*, have observed gas orbiting at high velocity near the nucleus of the elliptical galaxy M87.[147] This observation provides a decisive experimental evidence for a supermassive black hole, source of the strong gravitational field that keeps the gas orbiting (see picture 4.5, p. 204). A star collapses by contraction,[93–96] after the end of the nuclear reactions that kept the star in equilibrium, if the mass of the star is

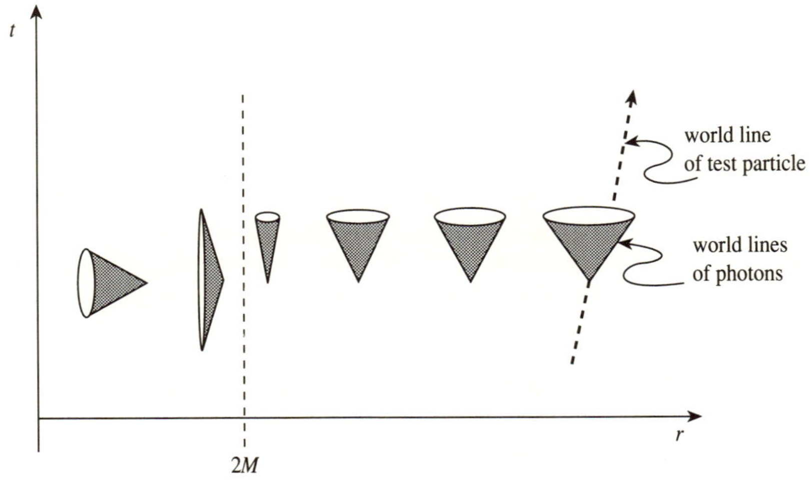

FIGURE 2.8. Future light cones in Schwarzschild coordinates outside, near, and inside the region $r = 2M$.

larger than a critical value, the *critical mass* (in general relativity, for a neutron star and depending from the equation of state used, at most $\sim$ 2–3$M_\odot$; the *Chandrasekhar limit* for the mass of a white dwarf is about 1.2 $M_\odot$).

The first detailed treatment of gravitational collapse within the framework of Einstein geometrodynamics was given in 1939 by Oppenheimer and Snyder.[93] For simplicity they treated the collapsing system as a collection of dust particles ($p = 0$), so that all the problems of pressure and temperature could be overlooked. Each particle would then move freely under the gravitational attraction of the others. More realistic equations of state have been later used,[96,99] without avoiding the collapse.

However, do we know enough about matter to be sure that it cannot successfully oppose collapse? We understand electromagnetic radiation better than we understand the behavior of matter at high density. Then why not consider a star containing no matter at all, an object built exclusively out of light, a "gravitational electromagnetic entity" or "geon," described in section 2.10, deriving its mass solely from photons, and these photons held in orbit solely by the gravitational attraction of that very mass?[71] It turns out that a geon has the stability—and the instability—of a pencil standing on its tip.[95] The geon does not let its individual photons escape any more than the pencil lets its individual atoms escape. But that swarm of photons, collectively, like the assembly of atoms that make up the pencil, collectively, can fall one way or the other. Starting slowly at first, it can expand outward more and more rapidly and explode into its individual photons. Equally easily, it can fall the other way slowly at first, then more and more rapidly to complete gravitational collapse. Thus it does not save one from having to worry about gravitational collapse to turn from matter to "pure" radiation.

A closer look at matter itself shows that "the harder it resists, the harder it falls": pressure itself has weight, and that weight creates more pressure, a "regenerative cycle" out of which again the only escape is collapse (see § 4.5).[96]

Gravitational collapse will have quite a different appearance according as it is studied by a faraway observer or a traveler falling in with, and at the outskirts of, the cloud of dust. The traveler will arrive in a very short time at a condition of infinite gravitational stress. If he sends out a radio "beep" every second of his existence, he will get off only a limited number of messages before the collapse terminates. In contrast, the faraway observer will receive these beeps at greater and greater time intervals; and, wait as long as he will, he will never receive any of the signals given out by the traveler after his crossing of the intangible horizon, $r_s = 2M$. Moreover, the cloud of dust will appear to the faraway observer, not to be falling ever faster, but to slow up and approach asymptotically a limiting sphere with the dimensions of the horizon. As it freezes down to this standard size it grows redder and fainter by the instant, and quickly becomes invisible. In other words, the observer on the surface of the collapsing star will pass through the horizon in a finite amount of his proper time, measured by his clocks. In contrast, an observer far from the collapsing star will see the collapse slow down and only asymptotically reach the horizon. However, since the intensity of the light he receives will exponentially decrease as the surface of the star approaches the horizon, after a short time he essentially will not receive any more light emitted from the collapsing star (however, see the Hawking radiation below). This phenomenon of different speed of the collapse is due to the gravitational time dilation of clocks, explained in section 3.2.2, and experimentally observed in a variety of experiments in weak fields (§ 3.2.2).[21] From the metric (2.6.35), we have

$$\begin{aligned}\Delta\tau|_{r'\cong 2M} &\equiv \begin{pmatrix}\text{interval of proper time}\\ \text{measured by an external}\\ \text{observer, at } r'\text{, near the}\\ \text{horizon } r_s = 2M\end{pmatrix}\\ &= \left(1-\frac{2M}{r'}\right)^{1/2}\Delta t = \left(1-\frac{2M}{r'}\right)^{1/2}\Delta\tau_\infty\\ &= \varepsilon\times\begin{pmatrix}\text{interval of proper time}\\ \text{measured by an asymptotic}\\ \text{observer}\end{pmatrix}\end{aligned} \tag{2.9.1}$$

where $\varepsilon \equiv (1-\frac{2M}{r'})^{1/2} \ll 1$. This is the sense in which time goes slower near a black hole. Put an atomic clock on the surface of a planet. Let it send signals to a higher point. The interval from pulse to pulse of this clock is seen to be greater than the interval between pulse and pulse of an identical clock located at the higher point. In this sense the clock closer to the planet's surface goes slower than the clock further away. Likewise a clock somehow suspended close above a black hole, measuring proper time: $\Delta\tau_{BH} = (1-\frac{2M}{r'})^{1/2}\Delta t = \varepsilon\Delta t$, will send signals to a faraway observer, equipped with an identical clock, measuring proper time: $\Delta\tau_\infty \cong \Delta t = \Delta\tau_{BH}/\varepsilon$. Therefore, the spacing between ticks of the clock just above the black hole is seen to be much larger than the spacing between ticks of the clock of the faraway observer.

Features of a Black Hole

Not even light signals or radio messages will escape from inside the horizon of the collapsed object. The only feature of the black hole that will be observed is its gravitational attraction[97–101,19] (however, see the Hawking radiation below). What falls into a black hole carries in mass and angular momentum, and it can also carry in electric charge. These are the only attributes that a black hole conserves out of the matter that falls into it. All other particularities, all other details, all other physical properties of "matter" are extinguished. The resulting stationary black hole, according to all available theoretical evidence, is completely characterized by its mass, its charge, and its angular momentum, and by nothing more. Jokingly put, "*a black hole has no hair*."[11]

Of the number of particles that went in not a trace is left, if present physics is safe as our guide. Not the slightest possibility is evident, even in principle, to distinguish between three black holes of the same mass, charge, and angular momentum, the first made from particles, the second made from antiparticles, and the third made primarily from pure radiation. This circumstance deprives us of all possibility to count or even define the number of particles at the end and compare it with the starting count. In this sense the laws of conservation of particle number are not violated; they are transcended.

The typical black hole is spinning and has angular momentum. This is a very strange kind of spin. One cannot "touch one's finger to the flywheel" to find it. The flywheel, the black hole, is so "immaterial," so purely geometrical, so untouchable, that no such direct evidence for its spin is available. Evidence for the spin of the black hole is obtainable by indirect means. For this purpose it is enough to put a gyroscopic compass in polar orbit around the black hole. The gyroscopic compass, pointed originally at a distant star, will slowly sweep about the circuit of the heavens, in sympathy with the rotation of the black hole, but at a far slower rate. At work on the gyro, in addition to the normal direct pull of gravity, is a new feature of geometry predicted by Einstein's theory. This "gravitomagnetic force" is as different from standard gravity as magnetism is different from electricity. An electric charge circling in orbit creates magnetism. A spinning mass creates gravitomagnetism.

We are far from being able today to observe gravitomagnetism of a spinning black hole. However, space experiments are in active development (GP-B and LAGEOS III; chap. 6) to measure the gravitomagnetic effects, on an orbiting gyroscope, due to the slow rotation of Earth.

The Event Horizon

Using the Schwarzschild coordinates of expression (2.6.35), at the Schwarzschild horizon, $r_S = 2M$, we have $g_{11} = -g_{00}^{-1} \xrightarrow{r=2M} \infty$. However, the **Schwarzschild horizon** is not a true singularity but just a **coordinate singularity**.

The quantities that have an intrinsic geometrical meaning, independent from the particular coordinates that are used, are the scalar invariants[15] constructed using the Riemann curvature tensor and the metric tensor. No invariant,[19] built with the curvature and metric tensors, diverges on the horizon, $r_s = 2M$. The Schwarzschild horizon is just a pathology of the coordinates of expression (2.6.35), but not a true geometrical singularity (see below). Indeed, with a coordinate transformation, for example to Eddington-Finkelstein[102,103] coordinates, or to Kruskal-Szekeres[104,105] coordinates, one can extend the solution (2.6.35) to a solution covering the whole Schwarzschild geometry with nonsingular

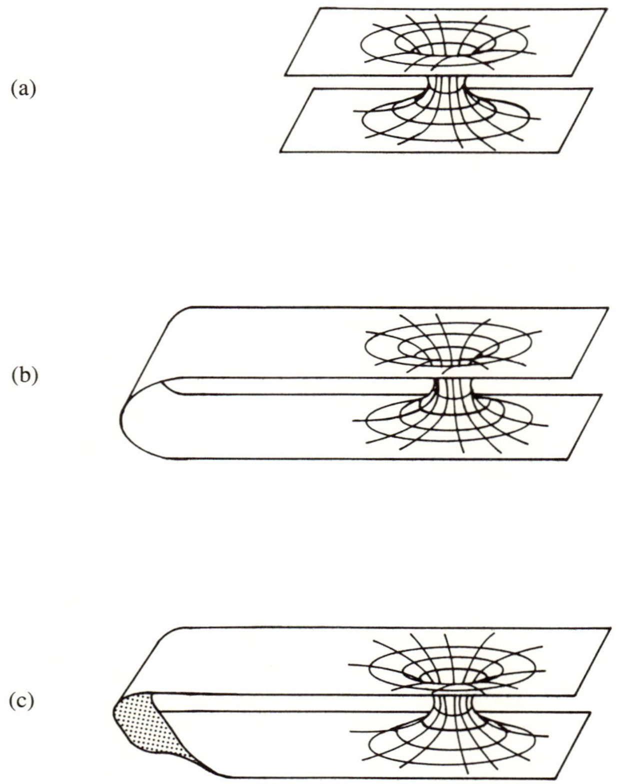

FIGURE 2.9. Alternative interpretations of the three-dimensional "maximally extended Schwarzschild metric" of Kruskal at time $t' = 0$. (a) A connection in the sense of Einstein and Rosen (Einstein-Rosen bridge)[106] between two otherwise Euclidean spaces. (b) and (c) A **wormhole** connecting two regions in one Euclidean space, in (c) not orientable with the topology of a Möbius strip (in the case where these regions are extremely far apart compared to the dimensions of the throat of the wormhole). Case (a) has the same curvature but different topology from cases (b) and (c). For a discussion on causality in a case of type (b) or (c) see refs. 107–109 and 138–141.

metric components at $r_s = 2M$ (see fig. 2.9). With the transformation to **Kruskal-Szekeres coordinates:**[11,19]

$$\begin{bmatrix} x' = \left(\dfrac{r}{2M} - 1\right)^{\frac{1}{2}} e^{r/4M} \cosh\left(\dfrac{t}{4M}\right) \\ t' = \left(\dfrac{r}{2M} - 1\right)^{\frac{1}{2}} e^{r/4M} \sinh\left(\dfrac{t}{4M}\right) \end{bmatrix} : \quad \text{for} \quad r > 2M$$

$$\begin{bmatrix} x' = \left(1 - \dfrac{r}{2M}\right)^{\frac{1}{2}} e^{r/4M} \sinh\left(\dfrac{t}{4M}\right) \\ t' = \left(1 - \dfrac{r}{2M}\right)^{\frac{1}{2}} e^{r/4M} \cosh\left(\dfrac{t}{4M}\right) \end{bmatrix} : \quad \text{for} \quad r < 2M, \tag{2.9.2}$$

one thus gets

$$\begin{aligned} ds^2 = & \left(\frac{32M^3}{r}\right) e^{-r/2M} \left(-dt'^2 + dx'^2\right) \\ & + r^2(t', x') \left(d\theta^2 + \sin^2\theta d\phi^2\right) \end{aligned} \tag{2.9.3}$$

where r is a function of x' and t' implicitly determined, from expression (2.9.2), by

$$\left(\frac{r}{2M} - 1\right) e^{r/2M} = x'^2 - t'^2. \tag{2.9.4}$$

The metric (2.9.3), in Kruskal-Szekeres coordinates, explicitly shows that the Schwarzschild geometry is well-behaved at $r_s = 2M$ and that is possible to extend analytically the Schwarzschild solution (2.6.35) to cover the whole Schwarzschild geometry (see fig. 2.9).

Black Hole Evaporation

In 1975 Hawking[110] discovered the so-called process of *black hole evaporation* (fig. 2.10). Quantum theory allows a process to happen at the horizon analogous to the Penrose process.[111] In the Penrose process two already existing particles trade energy in a region outside the horizon of a spinning black hole (see 2.6.36) called the ergosphere, the only domain where macroscopic masses of positive energy and of negative energy can coexist. Because the ergosphere shrinks to extinction when a black hole is deprived of all spin, the Penrose process applies only to a rotating, or "live," black hole. In contrast, the Hawking process takes place at the horizon itself and thus operates as effectively for a nonrotating black

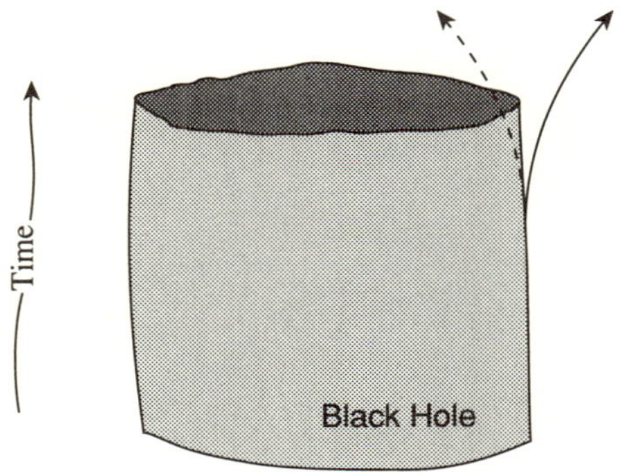

FIGURE 2.10. The Hawking[110] evaporation process capitalizes on the fact that space is nowhere free of so-called quantum vacuum fluctuations, evidence that everywhere latent particles await only opportunity—and energy—to be born. Associated with such fluctuations at the surface of a black hole, a might-have-been pair of particles or photons can be caught by gravity and transformed into a real-life particle or photon (solid arrow) that evaporates out into the surroundings and an antiparticle or counterphoton (dashed arrow) that "goes down" the black hole.

hole as for a rotating one. Furthermore, unlike the Penrose process, it involves a pair of newly created microscopic particles.

According to the uncertainty principle for the energy, $\Delta E \Delta t \gtrsim \hbar$, that is, space—pure, empty, energy-free space—all the time and everywhere experiences so-called quantum vacuum fluctuations at a very small scale of time, of the order of 10^{-44} s and less. During these quantum fluctuations, pairs of particles appear for an instant from the emptiness of space—perhaps an electron and an antielectron pair or a proton and an antiproton pair. Particle-antiparticle pairs are in effect all the time and everywhere being created and destroyed. Their destruction is so rapid that the particles never come into evidence at any everyday scale of observation. For this reason, the pairs of particles everywhere being born and dying are called virtual pairs. Under the conditions at the horizon, a virtual pair may become a real pair.

In the Hawking process, two newly created particles exchange energy, one acquiring negative energy $-E$ and the other positive energy E. Slightly outside the horizon of a black hole, the negative energy photon has enough time Δt to cross the horizon. Therefore, the negative energy particle flies inward from the horizon; the positive energy particle flies off to a distance. The energy it carries with it comes in the last analysis from the black hole itself. The massive object is no longer quite so massive because it has had to pay the debt of energy brought in by the negative energy member of the newly created pair of particles.

Radiation of light or particles from any black hole of a solar mass or more proceeds at an absolutely negligible rate—the bigger the mass the cooler the surface and the slower the rate of radiation. The calculated Bekenstein-Hawking temperature of a black hole of 3 $M_{\odot}$ is only 2×10^{-8} degrees above the absolute zero of temperature. The total thermal radiation calculated to come from its 986 square kilometers of surface is only about 1.6×10^{-29} watt, therefore this evaporation process would not be able to affect in any important way black holes of about one solar mass or more. A black hole of any substantial mass is thus deader than any planet, deader even than any dead moon—when it stands in isolation.

Singularities

The $r = 2M$ region of the Schwarzschild metric (2.6.35) is a mere coordinate singularity; however, the $r = 0$ region, where $g_{00} = -g_{11}^{-1} \xrightarrow{r=0} \infty$, is a true geometrical singularity,[19] where, as for the big bang and big crunch singularities of some cosmological models (see chap. 4), some curvature invariants diverge; for example the Kretschmann invariant for the Schwarzschild metric is $R_{\alpha\beta\gamma\delta} R^{\alpha\beta\gamma\delta} \sim \frac{m^2}{r^6} \xrightarrow{r=0} \infty$ (see § 6.11).

Indeed, besides coordinate singularities, or pathologies of a coordinate system removable with a coordinate transformation, there are various types of true geometrical singularities.[112–115]

Usually, in a physically realistic solution, a singularity is characterized by diverging curvature.[19] However, on a curved manifold the individual components of the Riemann tensor depend on the coordinates used. Therefore, one defines the true curvature singularities using the invariants built by contracting the Riemann tensor $R^{\alpha}{}_{\beta\mu\nu}$, with $g_{\alpha\beta}$ and with $\epsilon_{\alpha\beta\mu\nu}$. The regions where these invariants diverge are called **scalar polynomial curvature singularities**. One may also measure the components of the Riemann tensor with respect to a local basis parallel transported along a curve. In this case the corresponding curvature singularities are called **parallelly propagated curvature singularities**.

It is usual to assume that spacetime is a differentiable manifold (i.e., a manifold that is covered by a family of charts, such that in the intersections between the charts, the coordinates x^{α} of a chart as a function of the coordinates $\overline{x}^{\alpha}$ of another chart, $x^{\alpha} = x^{\alpha}(\overline{x}^{\alpha})$, are continuous and with continuous derivatives, C^{∞}), where space and time intervals and other physical quantities can be measured, and standard equations of physics hold in a neighborhood of every event. Then a curvature singularity is not part of the differentiable manifold called spacetime. Therefore, in such manifolds with singularities cut out, there will exist curves incomplete in the sense that they cannot be extended.

To distinguish between different types of incompleteness of a manifold, various definitions have been given. First, a manifold is called inextendible if it includes all the nonsingular spacetime points.[19] The definition of geodesic completeness is useful to characterize an incomplete manifold. A manifold is called geodesically complete if every geodesic can be extended to any value of its affine parameter (§ 2.4). In particular a manifold is **not** timelike or null **geodesically complete**, if it has incomplete timelike or null geodesics. In this case the history of a freely moving observer (or a photon), on one of these incomplete geodesics, cannot be extended after (or before) a finite amount of proper time. However, this definition does not include the type of singularity that a nonfreely falling observer, moving with rockets on a nongeodesic curve, may encounter in some manifolds. To describe these types of singularities on nongeodesic curves, one can give the definition of bundle-completeness or b-completeness. One first constructs on any continuous curve, with continuous first derivatives, a generalized affine parameter that in the case of a geodesic reduces to an affine parameter. An inextendible manifold (with all nonsingular points) is called **bundle-complete**, or *b-complete*, if for every curve of finite length, measured by the generalized affine parameter from a point p, there is an endpoint of the curve in the manifold. Bundle-completeness implies geodesic completeness, but not vice versa. Usually, in physically realistic solutions, a spacetime which is bundle-incomplete has curvature singularities on the b-incomplete curves (however, see the Hawking-Ellis discussion[19] of the Taub-NUT space).

In 1965 Roger Penrose proved a theorem about the existence of singularities,[112] of the type corresponding to null geodesic incompleteness, without using any particular assumption of exact symmetry.

Incomplete null geodesics exist on a manifold if:

1. The **null convergence condition** is satisfied: $R_{\alpha\beta}k^{\alpha}k^{\beta} \geq 0$, for every null vector k^{α}.
2. In the manifold there exists a noncompact Cauchy surface, that is, a noncompact spacelike hypersurface such that every causal path without endpoint intersects it once and only once (see chap. 5).
3. In the manifold there exists a **closed trapped surface.** A closed trapped surface is a closed (compact, without boundary) spacelike two-surface such that both the ingoing and the outgoing light rays moving on the null geodesics orthogonal to the surface converge toward each other.

Such a closed trapped surface is due to a very strong gravitational field that attracts back and causes the convergence even of the outgoing light rays. An example of closed trapped surface is a two-dimensional spherical surface inside the Schwarzschild horizon. Even the outgoing photons emitted from this surface are attracted back and converge due to the very strong gravitational field. Since not even the outgoing orthogonal light rays can escape from the closed trapped surface, all the matter, with velocity less than c, is also trapped and cannot escape from this surface. Closed trapped surfaces occur if a star collapses below its Schwarzschild radius. As we have previously observed, this should happen if a cold star, white dwarf, or neutron star, or white dwarf or neutron star core of a larger star, after the end of the nuclear reactions that kept the star in equilibrium, has a mass above a critical value of a few solar masses (in general relativity, for a neutron star, depending from the equation of state used, at most $\sim 3\ M_{\odot}$). Therefore, any such star or star core should collapse within the horizon and generate closed trapped surfaces and singularities, according to various singularity theorems[112–115] and in particular according to the 1965 Penrose theorem[112] and to the 1970 Hawking-Penrose theorem.[115]

Singularities of the type of incomplete timelike and null geodesics occur in a manifold, if:

1. $R_{\alpha\beta}u^{\alpha}u^{\beta} \geq 0$ for every nonspacelike vector u^{α}.
2. Every nonspacelike geodesic has at least a point where:

$$u_{[\alpha}R_{\beta]\gamma\delta[\mu}u_{\nu]}u^{\gamma}u^{\delta} \neq 0,$$

   where u^{α} is the tangent vector to the geodesics (the manifold is not too highly symmetric): this is the so-called **generic condition**.
3. There are no closed timelike curves; this causality condition is called **chronology condition** (see the 1949 Gödel model universe, discussed in § 4.6, as an example of solution violating the chronology condition).
4. There exists a closed trapped surface (or some equivalent mathematical condition is satisfied; see Hawking and Ellis).[19]

We note that the null convergence condition (1) of the Penrose theorem is a consequence of the **weak energy condition**, $T_{\alpha\beta}u^{\alpha}u^{\beta} \geq 0$, for every timelike vector u^{α}, plus the Einstein field equation (2.3.14) (even including a cosmological term), $R_{\alpha\beta} - \frac{1}{2}Rg_{\alpha\beta} + \Lambda g_{\alpha\beta} = \chi T_{\alpha\beta}$. The **timelike convergence condition**, $R_{\alpha\beta}u^{\alpha}u^{\beta} \geq 0$, for every timelike vector u^{α}, is a consequence of the Einstein field equation plus the condition $T_{\alpha\beta}u^{\alpha}u^{\beta} \geq u^{\alpha}u_{\alpha}(\frac{1}{2}T - \frac{1}{8\pi}\Lambda)$, for every timelike vector u^{α}; for $\Lambda = 0$ this is called the **strong energy condition** for the energy-momentum tensor.

We conclude this brief introduction to spacetime singularities by observing that, probably, the problem of the occurrence of the singularities in classical geometrodynamics might finally be understood[95] only when a consistent and complete quantum theory of gravity[116,145] is available. Question: Does a proper quantum theory of gravity rule out the formation of such singularities?

2.10 GRAVITATIONAL WAVES

As in electromagnetism in which there are electromagnetic perturbations propagating with speed c in a vacuum—electromagnetic waves—Einstein geometrodynamics predicts curvature perturbations propagating in the spacetime—**gravitational waves.**[117–121]

In this section we derive a simple, weak field, wave solution of the field equation (2.3.14). Let us first consider a perturbation of the flat Minkowski metric $\eta_{\alpha\beta}$:

$$g_{\alpha\beta} = \eta_{\alpha\beta} + h_{\alpha\beta} \tag{2.10.1}$$

where $h_{\alpha\beta}$ is a small perturbation of $\eta_{\alpha\beta}$: $|h_{\alpha\beta}| \ll 1$. We then define

$$\begin{aligned} h^{\alpha}{}_{\beta} &\equiv \eta^{\alpha\sigma} h_{\sigma\beta} \\ h^{\alpha\beta} &\equiv \eta^{\alpha\sigma} \eta^{\beta\rho} h_{\sigma\rho} \\ h &\equiv h^{\alpha}{}_{\alpha} = \eta^{\sigma\rho} h_{\sigma\rho}. \end{aligned} \tag{2.10.2}$$

Therefore, to first order in $|h_{\alpha\beta}|$, we have

$$g^{\alpha\beta} = \eta^{\alpha\beta} - h^{\alpha\beta}. \tag{2.10.3}$$

From the definition of Ricci tensor (§ 2.3), we then have up to first order

$$R^{(1)}_{\alpha\beta} = \Gamma^{\sigma}_{\alpha\beta,\sigma} - \Gamma^{\sigma}_{\alpha\sigma,\beta} = \frac{1}{2}\left(-\Box h_{\alpha\beta} + h^{\sigma}{}_{\beta,\sigma\alpha} + h^{\sigma}{}_{\alpha,\sigma\beta} - h_{,\alpha\beta}\right) \tag{2.10.4}$$

where $\Box = \eta^{\alpha\beta} \frac{\partial^2}{\partial x^\alpha \partial x^\beta}$ is the d'Alambertian operator. Therefore the Einstein field equation, in the alternative form (2.3.17), can be written

$$-\Box h_{\alpha\beta} + h^{\sigma}{}_{\beta,\sigma\alpha} + h^{\sigma}{}_{\alpha,\sigma\beta} - h_{,\alpha\beta} = 16\pi\left(T_{\alpha\beta} - \frac{1}{2}\eta_{\alpha\beta}T\right) \tag{2.10.5}$$

where $T = \eta^{\sigma\rho}T_{\sigma\rho} = -\frac{1}{8\pi}R$. With an infinitesimal coordinate, or gauge, transformation, $x'^{\alpha} = x^{\alpha} + \xi^{\alpha}$ (see § 2.6), we then have

$$h'_{\alpha\beta} = h_{\alpha\beta} - \xi_{\alpha,\beta} - \xi_{\beta,\alpha} \tag{2.10.6}$$

where, of course, $h'_{\alpha\beta}$ is still a solution of the field equation (gauge invariance of the field equation). Therefore, if for ξ_α we choose a solution of the differential equation

$$\Box\xi_{\alpha} = h^{\sigma}{}_{\alpha,\sigma} - \frac{1}{2}h^{\sigma}{}_{\sigma,\alpha}\,, \tag{2.10.7}$$

we have

$$h'^{\sigma}{}_{\alpha,\sigma} - \frac{1}{2}h'^{\sigma}{}_{\sigma,\alpha} = h^{\sigma}{}_{\alpha,\sigma} - \frac{1}{2}h^{\sigma}{}_{\sigma,\alpha} - \xi_{\alpha,\sigma}{}^{\sigma} = 0. \tag{2.10.8}$$

In this gauge, $(h'^{\sigma}{}_{\alpha} - \frac{1}{2}\delta^{\sigma}{}_{\alpha}h')_{,\sigma} = 0$, sometimes called the *Lorentz gauge,* the field equation becomes (dropping the prime in $h'_{\alpha\beta}$)

$$\Box h_{\alpha\beta} = -16\pi\left(T_{\alpha\beta} - \frac{1}{2}\eta_{\alpha\beta}T\right). \tag{2.10.9}$$

As in electromagnetism,[44] a solution to this equation is the retarded potential:

$$h_{\alpha\beta}(\boldsymbol{x}, t) = 4 \int \left\{ \frac{\left[T_{\alpha\beta} - \frac{1}{2}\eta_{\alpha\beta}T\right](\boldsymbol{x}', t - |\boldsymbol{x} - \boldsymbol{x}'|)}{|\boldsymbol{x} - \boldsymbol{x}'|} \right\} d^3x'. \qquad (2.10.10)$$

This solution represents a gravitational perturbation propagating at the speed of light, $c \equiv 1$. When $T_{\alpha\beta} = 0$, we then have, in the Lorentz gauge,

$$\Box h_{\alpha\beta} = 0. \qquad (2.10.11)$$

This is the wave equation for $h_{\alpha\beta}$. We recall that in electromagnetism, in the Lorentz gauge, $A^{\alpha}{}_{,\alpha} = 0$, we have the sourceless wave equation for A^{α}: $\Box A^{\alpha} = 0$. Correspondingly, a simple solution of the wave equation (2.10.11) for $h_{\alpha\beta}$ is a plane wave. By choosing the z-axis as the propagation axis of the plane wave, we then have

$$\left(\frac{\partial^2}{\partial z^2} - \frac{\partial^2}{\partial t^2} \right) h_{\alpha\beta} = 0 \qquad (2.10.12)$$

where $h_{\alpha\beta} = h_{\alpha\beta}(z \pm t)$, that is, $h_{\alpha\beta}$ is a function of $(z \pm t)$, where $c \equiv 1$. From expression (2.10.6), it follows that the Lorentz condition (2.10.8) and the simple form (2.10.11) of the vacuum field equation for $h_{\alpha\beta}$ are invariant for any infinitesimal transformation $x'^{\alpha} = x^{\alpha} + \xi^{\alpha}$, if ξ^{α} is a solution of $\Box \xi^{\alpha} = 0$. Here gravity is similar to electromagnetism where, with the gauge transformation $A'^{\alpha} = A^{\alpha} + \phi^{,\alpha}$, if $\Box\phi = 0$, the Lorentz condition is preserved, $A^{\alpha}{}_{,\alpha} = A'^{\alpha}{}_{,\alpha} = 0$, and we still have $\Box A'^{\alpha} = 0$. Therefore, by performing an infinitesimal coordinate transformation, with the four components of ξ^{α} solutions of $\Box \xi^{\alpha} = 0$, for a plane gravitational wave, $h_{\alpha\beta} = h_{\alpha\beta}(z \pm t)$, it is possible to satisfy the four conditions: $h_{i0} = 0$ and $h \equiv h^{\sigma}{}_{\sigma} = 0$; that is, the trace of $h_{\alpha\beta}$ equal to zero. Since in this gauge we have $h^{\alpha}{}_{\beta} - \frac{1}{2}\delta^{\alpha}{}_{\beta} h = h^{\alpha}{}_{\beta}$, the Lorentz gauge condition becomes simply $h^{\sigma}{}_{\alpha,\sigma} = 0$. Therefore, for the weak field plane gravitational wave $h_{\alpha\beta}(z \pm t)$, and more generally for any weak field gravitational wave, linear superposition of plane waves, in this gauge, from $h^{\sigma}{}_{\alpha,\sigma} = 0$, we can set $h_{00} = 0$.

Summarizing in this **gauge**, called **transverse-traceless** (transverse because the wave is orthogonal to its direction of propagation), we have

$$h^{\mathrm{TT}}_{\alpha 0} = 0, \text{ i.e., } h^{\mathrm{TT}}_{\alpha\beta} \text{ has spatial components only,} \qquad (2.10.13)$$

and

$$h^{\mathrm{TT}} \equiv h^{\mathrm{TT}\alpha}{}_{\alpha} = 0, \text{ i.e., } h^{\mathrm{TT}}_{\alpha\beta} \text{ is traceless,} \qquad (2.10.14)$$

and

$$h^{\mathrm{TT}k}{}_{i,k} = 0, \text{ i.e., } h^{\mathrm{TT}}_{ij} \text{ is transverse.} \qquad (2.10.15)$$

Finally, from expressions (2.10.13), (2.10.14), and (2.10.15), for the plane wave $h^{\mathrm{TT}}_{\alpha\beta}(z \pm t)$ described by equation (2.10.12), apart from integration constants, in the transverse-traceless gauge we get $h^{\mathrm{TT}}_{zz} = h^{\mathrm{TT}}_{zx} = h^{\mathrm{TT}}_{zy} = 0$, and a solution to $\Box h^{\mathrm{TT}}_{\alpha\beta} = 0$ is

$$\begin{aligned} h^{\mathrm{TT}}_{xx} &= -h^{\mathrm{TT}}_{yy} = A_+ e^{-i\omega(t-z)} \\ h^{\mathrm{TT}}_{xy} &= h^{\mathrm{TT}}_{yx} = A_\times e^{-i\omega(t-z)} \end{aligned} \tag{2.10.16}$$

where as usual we take the real part of these expressions, with all the other components of $h^{\mathrm{TT}}_{\alpha\beta}$ equal to zero to first order. This expression (2.10.16) describes a plane gravitational wave as a perturbation of the spacetime geometry, traveling with speed c. This perturbation of the spacetime geometry corresponds to the curvature perturbation $R_{i0j0} = -R_{i0jz} = R_{izjz} = -\frac{1}{2} h^{\mathrm{TT}}_{ij,00}$ traveling with speed c on the flat background, where i and j are 1 or 2.

In this simple case of a weak field, plane gravitational wave, in the transverse-traceless coordinate system (2.10.13)–(2.10.15), one can easily verify that test particles originally at rest in the flat background $\eta_{\alpha\beta}$ before the passage of the gravitational wave will remain at rest *with respect to the coordinate system* during the propagation of the gravitational wave. In fact, from the geodesic equation (2.4.13), to first order in $h^{\mathrm{TT}}_{\alpha\beta}$, we have

$$\frac{Du^\alpha}{ds} \cong \frac{du^\alpha}{ds} = 0. \tag{2.10.17}$$

However, the *proper distance* between the two test particles at rest in x^i and $x^i + dx^i$ is given by $dl^2 = g_{ik} dx^i dx^k$. Therefore, since $g_{ik} = \eta_{ik} + h_{ik}$ changes with time, the proper distance between the test particles will *change* with time during the passage of the gravitational wave. For a plane wave propagating along the z-axis in the transverse-traceless gauge, the proper distance between particles in the xy-plane is given by

$$dl = \left[\left(1 + h^{\mathrm{TT}}_{xx}\right) dx^2 + \left(1 - h^{\mathrm{TT}}_{xx}\right) dy^2 + 2h^{\mathrm{TT}}_{xy} dx dy \right]^{\frac{1}{2}}. \tag{2.10.18}$$

For the particles A, B, and C of figure 2.11, on a circumference with center at $x^\alpha = 0$, with coordinates

$$x^i_A \equiv (l, 0, 0); \quad x^i_B \equiv \left(\frac{l}{\sqrt{2}}, \frac{l}{\sqrt{2}}, 0 \right); \quad \text{and} \quad x^i_C \equiv (0, l, 0) \tag{2.10.19}$$

from the expression (2.10.18) for dl and from the expression (2.10.16) for h^{TT}_{xx} and h^{TT}_{xy}, we immediately find the behavior of the proper distance between test particles on a circumference due to the passage of a plane gravitational wave perpendicularly to the circumference, behavior that is shown in figure 2.11. Case I, $A_+ \neq 0$ and $A_\times = 0$, and case II, $A_+ = 0$ and $A_\times \neq 0$, describe two waves with polarizations at 45° one from the other. Of course one can get the same result by using the geodesic deviation equation (see § 3.6.1).

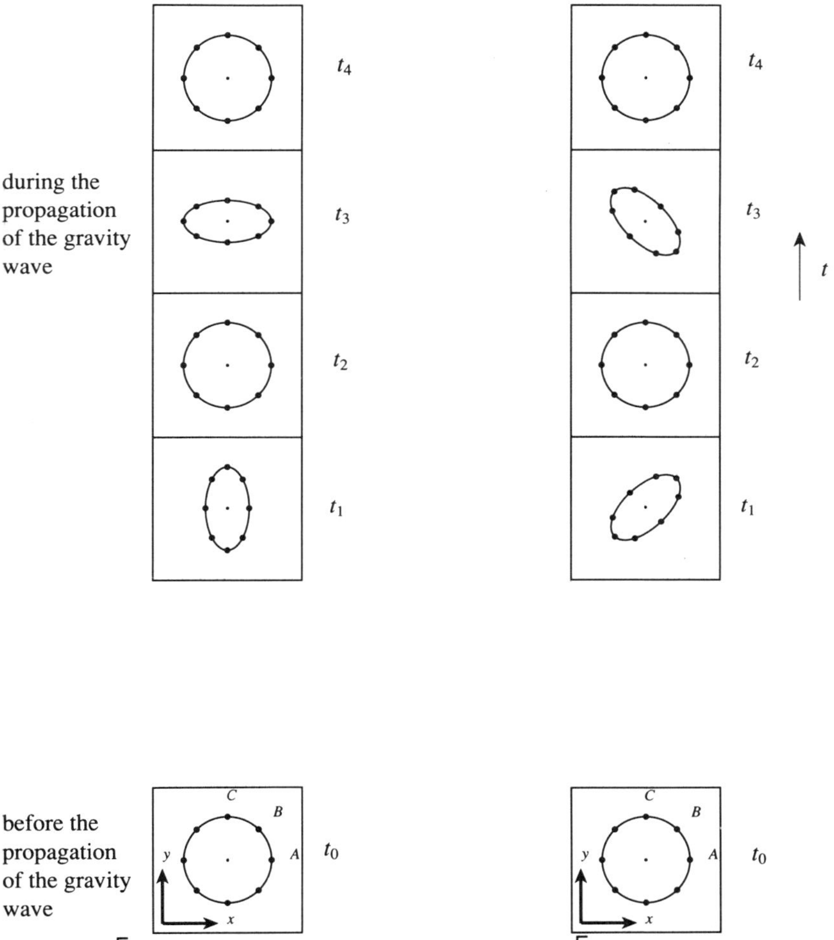

$$\begin{cases} A_+ \neq 0 \\ A_\times = 0 \end{cases}$$

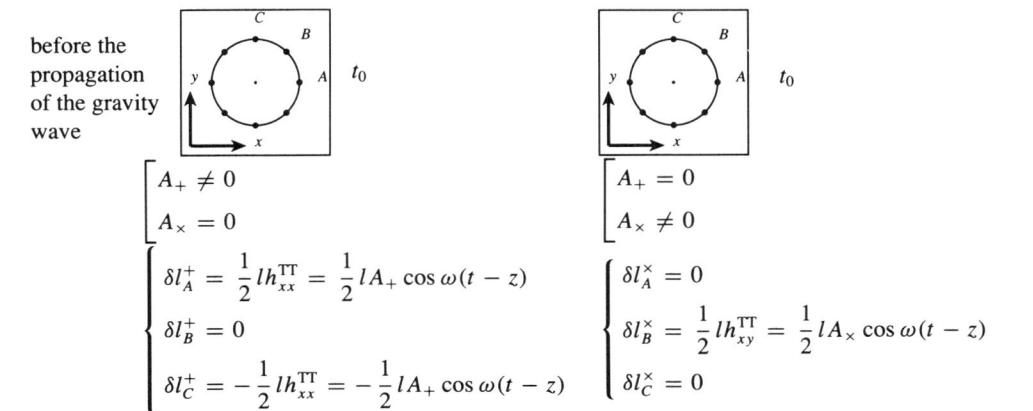

$$\begin{cases} \delta l_A^+ = \dfrac{1}{2} l h_{xx}^{TT} = \dfrac{1}{2} l A_+ \cos \omega (t - z) \\ \delta l_B^+ = 0 \\ \delta l_C^+ = -\dfrac{1}{2} l h_{xx}^{TT} = -\dfrac{1}{2} l A_+ \cos \omega (t - z) \end{cases}$$

$$\begin{cases} A_+ = 0 \\ A_\times \neq 0 \end{cases}$$

$$\begin{cases} \delta l_A^\times = 0 \\ \delta l_B^\times = \dfrac{1}{2} l h_{xy}^{TT} = \dfrac{1}{2} l A_\times \cos \omega (t - z) \\ \delta l_C^\times = 0 \end{cases}$$

FIGURE 2.11. Effect of weak plane gravitational wave, propagating along the z-axis, on the proper distance between a ring of test particles in the xy-plane.

We observe that in general relativity there are gravitational wave pulses that after their passage leave test particles slightly displaced from their original position for a very long time compared to the duration of the pulse (the pulse is characterized by a nonzero curvature tensor); thus, after the propagation of such gravitational-wave pulse, the position of the test particles may represent a record of the passage of the gravitational wave. This phenomenon is sometimes called **position-coded memory** and may be a linear effect[122–125] or an effect due to nonlinear terms[126] in the Einstein field equation. Gravitational-wave pulses with a **velocity-coded memory** have been also inferred in general relativity.[127]

By applying to a plane gravitational wave the definition (2.7.18) for the pseudotensor of the gravitational field,[120] in the TT gauge (2.10.13)–(2.10.15), after some calculations one gets:[11]

$$t^{\mathrm{GW}}_{\alpha\beta} = \frac{1}{32\pi}\left\langle h^{\mathrm{TT}}_{ij,\alpha} h^{\mathrm{TT}ij}{}_{,\beta}\right\rangle \tag{2.10.20}$$

where $\langle \quad \rangle$ means average over a region of several wavelengths. In particular, applying this expression for $t^{\mathrm{GW}}_{\alpha\beta}$ to the case of the plane gravitational wave (2.10.16), traveling along the z-axis with $h_{xx} = -h_{yy} = A_+ \cos\omega(t - z)$ and $h_{xy} = h_{yx} = A_\times \cos\omega(t - z)$, we get

$$t^{\mathrm{GW}}_{zz} = t^{\mathrm{GW}}_{tt} = -t^{GW}_{tz} = \frac{1}{32\pi}\omega^2\left(A^2_+ + A^2_\times\right), \tag{2.10.21}$$

that is, the *energy-momentum pseudotensor for a plane gravitational wave* propagating along the z-axis, averaged over several wavelengths. From section 2.7 we find that the expression (2.10.21) represents the flux of energy carried by a plane gravitational wave propagating along the z-axis.

Finally, we give the so-called **quadrupole formula** for the outgoing flux of gravitational wave energy emitted by a system characterized by a weak gravitational field and slow motion, that is, such that its size, R, is small with respect to the reduced wavelength $\frac{\lambda}{2\pi} \equiv \lambdabar$ of the gravitational waves emitted: $R \ll \frac{\lambda}{2\pi}$. The transverse and traceless linearized metric perturbation for gravitational waves in the wave zone, $r \gg \lambdabar$, and where the background curvature can be ignored,[118,137] has been calculated to be:[5,11,51,118,136,137]

$$h^{\mathrm{TT}}_{ij} = \frac{2}{r}\frac{\partial^2}{\partial t^2}\left[I\!\!\!-^{\mathrm{TT}}_{ij}(t - r)\right] + O\left(\frac{1}{r^2}\frac{\partial}{\partial t}I\!\!\!-^{\mathrm{TT}}_{ij}\right) \tag{2.10.22}$$

where $t - r$ is the retarded time, r the distance to the source center, t the proper time of a clock at rest with respect to the source, and $I\!\!\!-^{\mathrm{TT}}_{ij}$ the transverse (with respect to the radial direction of propagation of the gravitational waves) and traceless part of the mass quadrupole moment of the source. For a source characterized by a weak gravitational field and small stresses, the symmetric **reduced quadrupole moment** (traceless), of the source mass density ρ, is given by[11]:

$$\not{I}_{ij} = \int \rho(x_i x_j - \frac{1}{3}\delta_{ij} r^2) d^3x. \tag{2.10.23}$$

We can expand in powers of $\frac{1}{r}$ the Newtonian gravitational potential U, generated by this source, as a function of the reduced quadrupole moment $\not{I}_{ij}$. By a suitable choice of origin at the source, we have

$$U = \frac{M}{r} + \frac{3}{2}\frac{\not{I}_{ij} n^i n^j}{r^3} + O\left(\frac{1}{r^4}\right) \tag{2.10.24}$$

where $n^i \equiv \frac{x^i}{r}$. By inserting the transverse and traceless metric perturbation (2.10.22) in the expression (2.10.20) for the flux of energy carried by a gravitational wave and by integrating over a sphere of radius r, we then get the rate of gravitational-wave energy from the source crossing, in the wave zone, a sphere of radius r at time t:

$$\begin{aligned} \frac{dE}{dt} &= \int t^{0r} r^2 d\Omega = -\int t^{00} r^2 d\Omega \\ &= \frac{1}{5}\left\langle \sum_{ij}\left[\frac{\partial^3}{\partial t^3}\not{I}_{ij}(t-r)\right]^2\right\rangle \equiv \frac{1}{5}\left\langle \dddot{\not{I}}_{ij}\dddot{\not{I}}^{ij}\right\rangle \end{aligned} \tag{2.10.25}$$

where $d\Omega = \sin\theta d\theta d\phi$ and $\langle \quad \rangle$ means an average over several wavelengths.

From this formula for the emission of gravitational-wave energy due to the time variations of the quadrupole moment, one can calculate the time decrease of the orbital period of some binary star systems. This general relativistic theoretical calculation agrees with the observed time decrease of the orbital period of the **binary pulsar PSR 1913+1916** (see § 3.5.1).

Geons

In the 1950s one of us[71] found an interesting way to treat the concept of body in general relativity. An object can, in principle, be constructed out of gravitational radiation or electromagnetic radiation, or a mixture of the two, and may hold itself together by its own gravitational attraction. The gravitational acceleration needed to hold the radiation in a circular orbit of radius r is of the order of c^2/r. The acceleration available from the gravitational pull of a concentration of radiant energy of mass M is of the order GM/r^2. The two accelerations agree in order of magnitude when the radius r is of the order

$$r \sim GM/c^2 = (0.742 \times 10^{-28}\ \text{cm/g})M. \tag{2.10.26}$$

A collection of radiation held together in this way is called a **geon** (gravitational electromagnetic entity) and is a purely classical object. It has nothing whatsoever directly to do with the world of elementary particles. Its structure can be treated entirely within the framework of classical geometrodynamics, provided that a size is adopted for it sufficiently great that quantum effects do not come into play. Studied from a distance, such an object presents the same kind of gravitational attraction as any other mass. Moreover, it moves through space as a unit, and undergoes deflection by slowly varying fields of force just as does any other mass. Yet nowhere inside the geon is there a place where one can put a finger and say "here is mass" in the conventional sense of mass. In particular, for a geon made of pure gravitational radiation—**gravitational geon**—there is no local measure of energy, yet there is global energy. The gravitational geon owes its existence to a localized—but everywhere regular—curvature of spacetime, and to nothing more.

In brief, a geon is a collection of electromagnetic or gravitational-wave energy, or a mixture of the two, held together by its own gravitational attraction, that describes *mass without mass*.

REFERENCES CHAPTER 2

1. A. Einstein, Zur allgemeinen Relativitätstheorie, *Preuss. Akad. Wiss. Berlin, Sitzber.* 778–86 (1915).
2. A. Einstein, Zur allgemeinen Relativitätstheorie (Nachtrag), *Preuss. Akad. Wiss. Berlin, Sitzber.* 799–801 (1915).
3. A. Einstein, Erklärung der Perihelbewegung des Merkur aus der allgemeinen Relativitätstheorie, *Preuss. Akad. Wiss. Berlin, Sitzber.* 47:831–39 (1915).
4. A. Einstein, Die Feldgleichungen der Gravitation, *Preuss. Akad. Wiss. Berlin, Sitzber.* 844–47 (1915).
5. A. Einstein, Näherungsweise Integration der Feldgleichungen der Gravitation, *Preuss. Akad. Wiss. Berlin, Sitzber.* 688–96 (1916).
6. A. Einstein, Hamiltonsches Prinzip und allgemeine Relativitätstheorie, *Preuss. Akad. Wiss. Berlin, Sitzber.* 1111–16 (1916).
7. A. Einstein, The Foundation of the General Theory of Relativity, *Ann. Physik* 49:769–822 (1916).
8. A. Einstein, Kosmologische Betrachtungen zur allgemeinen Relativitätstheorie, *Preuss. Akad. Wiss. Berlin, Sitzber.* 142–52 (1917).
9. A. Einstein et al., *"The Principle of Relativity": A Collection of Original Papers on the Special and General Theory of Relativity by A. Einstein, H.A. Lorentz, H. Weyl and H. Minkowski*, with notes by A. Sommerfeld, trans. W. Perret and G.B. Jeffery (Dover, New York, 1952).
10. A. Einstein, *The Meaning of Relativity*, 5th ed. (Princeton University Press, Princeton, 1955).

11. C. W. Misner, K. S. Thorne, and J. A. Wheeler, *Gravitation* (Freeman, San Francisco, 1973); see also refs. 12–21.

12. C. Møller, *The Theory of Relativity* (Oxford University Press, London, 1972).

13. J. L. Synge, *Relativity: The General Theory* (North-Holland, Amsterdam, 1960).

14. J. A. Wheeler, *Geometrodynamics* (Academic Press, New York, 1962).

15. A. Z. Petrov, *Einstein Spaces*, trans. R. F. Kelleher (Pergamon Press, Oxford, 1969).

16. W. Rindler, *Essential Relativity: Special, General, and Cosmological* (Van Nostrand, New York, 1969).

17. L. D. Landau and E. M. Lifshitz, *The Classical Theory of Fields*, 3d rev. ed. (Addison-Wesley, Reading, MA, and Pergamon, London, 1971).

18. S. Weinberg, *Gravitation and Cosmology* (Wiley, New York, 1972).

19. S.W. Hawking and G.F.R. Ellis, *The Large Scale Structure of Space-time* (Cambridge University Press, Cambridge, 1973).

20. D. Kramer, H. Stephani, M. MacCallum, and E. Herlt, *Exact Solutions of Einstein's Field Equations* (VEB Deutscher Verlag der Wissenschaften, Berlin, 1980).

21. C.M. Will, *Theory and Experiment in Gravitational Physics*, rev. ed. (Cambridge University Press, Cambridge, 1993).

22. G. Galilei, *Discorsi e dimostrazioni matematiche intorno a due nuove scienze* (Elzevir, Leiden, 1638); trans. H. Crew and A. de Salvio as *Dialogues Concerning Two New Sciences* (Macmillan, New York, 1914); reprint (Dover, New York, 1954); see also ref. 23.

23. G. Galilei, *Dialogo dei due massimi sistemi del mondo* (Landini, Florence, 1632); trans. S. Drake as *Galileo Galilei: Dialogue Concerning the Two Chief World Systems—Ptolemaic and Copernican* (University of California Press, Berkeley and Los Angeles, 1953).

24. I. Newton, *Philosophiae naturalis principia mathematica* (Streater, London, 1687); trans. A. Motte (1729) and revised by F. Cajori as *Sir Isaac Newton's Mathematical Principles of Natural Philosophy and His System of the World* (University of California Press, Berkeley and Los Angeles, 1934; paperback, 1962).

25. B. Bertotti and L.P. Grishchuk, The strong equivalence principle, *Class. Quantum Grav.* 7:1733–45 (1990).

26. K. Nordtvedt Jr., Equivalence principle for massive bodies, I: Phenomenology, *Phys. Rev. 169*: 1014–16 (1968).

27. K. Nordtvedt Jr., Equivalence principle for massive bodies, II: Theory, *Phys. Rev. 169*: 1017–25 (1968).

28. P.L. Bender, D.G. Currie, R.H. Dicke, D.H. Eckhardt, J.E. Faller, W.M. Kaula, J.D. Mulholland, H.H. Plotkin, S.K. Poultney, E.C. Silverberg, D.T. Wilkinson, J.G. Williams, and C.O. Alley, The lunar laser ranging experiment, *Science* 182:229–38 (1973).

29. V. Fock, *The Theory of Space, Time and Gravitation* (Moscow, 1961); 2d rev. ed., trans. N. Kemmer (Pergamon Press, Oxford, 1966).

30. J.L. Anderson, *Principles of Relativity Physics* (Academic Press, New York, 1967).

31. K.F. Gauss, Disquisitiones generales circa superficies curvas, in *Karl Friedrich Gauss Werke*, vol. 4, 217–58; trans. J.C. Morehead and A.M. Hiltebeitel as *General Investigations of Curved Surfaces of 1827 and 1825*, reprint (Raven Press, New York, 1965).

32. N.I. Lobačevskij, *Novye Načala Geometrij s Polnoj Teoriej Parallel'nyh*, Učënye Zapiski Kazanskogo Universiteta (1835–1838).

33. G.F.B. Riemann, Über die Hypothesen, welche der Geometrie zu Grunde liegen, in *Gesammelte Mathematische Werke* (1866); reprint of 2d ed., ed. H. Weber (Dover, New York, 1953); see also the translation by W.K. Clifford, *Nature* 8:14–17, 36–37 (1873).

34. G. Ricci Curbastro, Resumé de quelques travaux sur les systèmes variables de fonction, *Bull. Sc. Mathematiques* 16:167–89 (1892).

35. G. Ricci Curbastro and T. Levi-Civita, Méthods du calcul différentiel absolu et leurs applications, *Math. Ann.* 54:125–201 (1901; also published by Blanchard, Paris, 1923); see also ref. 34.

36. E.B. Christoffel, Über die Transformation der homogenen Differentialausdrücke zweiten Grades, *J. Reine Angew. Math.* (Crelle) 70:46–70 (1869).

37. A. Schild, Lectures on general relativity theory, in *Relativity Theory and Astrophysics: I. Relativity and Cosmology; II. Galactic Structure; III. Stellar Structure*, ed. J. Ehlers (American Mathematical Society, Providence, RI, 1967), 1–105.

38. T. Levi-Civita, *Lezioni di calcolo differenziale assoluto*, compiled by E. Persico (Rome, Stock, 1925); trans. M. Long as *The Absolute Differential Calculus* (Dover, New York, 1977).

39. M. Spivak, *A Comprehensive Introduction to Differential Geometry*, 2d ed. (Publish or Perish, Berkeley, 1979).

40. S. Kobayashi and K. Nomizu, *Foundations of Differential Geometry* (Wiley-Interscience, New York, 1963).

41. B. Schutz, *Geometrical Methods of Mathematical Physics* (Cambridge University Press, Cambridge, 1980).

42. Y. Choquet-Bruhat, and C. DeWitt-Morette, with M. Dillard-Bleick, *Analysis, Manifold and Physics* (North-Holland, Amsterdam, 1982).

43. See also C.J.S. Clarke and F. de Felice, *Relativity on Curved Manifolds* (Cambridge University Press, Cambridge, 1990).

44. J.D. Jackson, *Classical Electrodynamics* (Wiley, New York, 1962).

45. D. Hilbert, Die Grundlagen der Physik, *Königl. Gesell. Wiss. Göttingen, Nachr., Math.-Phys. Kl.* 395–407 (1915); see also *Math. Ann.* 92:1–32 (1924).

46. J.W. York, Jr., Boundary terms in the action principles of general relativity, *Found. Phys.* 16:249–57 (1986).

47. A. Palatini, Deduzione invariantiva delle equazioni gravitazionali dal principio di Hamilton, *Rend. Circ. Mat. Palermo* 43:203–12 (1919).

48. L. Bianchi, Sui simboli a quattro indici e sulla curvatura di Riemann, *Rend. della R. Acc. dei Lincei* 11:3–7 (1902).

49. According to T. Levi-Civita (ref. 38, p. 182 of Dover ed.), G. Ricci first stated these identities without proof. Later, in 1902, L. Bianchi rediscovered and published them with a proof.

50. For a derivation of the geodesic equation of motion of a compact body, see, e.g., the approximation method by A. Einstein, L. Infeld, and B. Hoffmann, The gravitational equations and the problem of motion, *Ann. Math.* 39:65–100 (1938); see also ref. 51.

51. T. Damour, The problem of motion in Newtonian and Einsteinian gravity, in *300 Years of Gravitation*, ed. S.W. Hawking and W. Israel (Cambridge University Press, Cambridge, 1987), 128–98.

52. T. Levi-Civita, Sur l'écart géodésique, *Math. Ann.* 97:291–320 (1926).

53. I. Ciufolini, Generalized geodesic deviation equation, *Phys. Rev. D* 34:1014–17 (1986).

54. I. Ciufolini and M. Demianski, How to measure the curvature of space-time, *Phys. Rev. D* 34:1018–20 (1986).

55. K. Yano, *The Theory of Lie Derivatives and Its Applications* (North-Holland, Amsterdam, 1955).

56. S. Helgason, *Differential Geometry and Symmetric Spaces* (Academic Press, New York, 1962).

57. W. Killing, Über die Grundlagen der Geometrie, *J. Reine Angew. Math.* 109:121–86 (1892).

58. H. Takeno, *The Theory of Spherically Symmetric Space-times*, Scientific Reports of the Research Institute for Theoretical Physics, Hiroshima Univ., No. 3 (1963).

59. K. Schwarzschild, Über das Gravitationsfeld eines Massenpunktes nach der Einsteinschen Theorie, *Sitzber. Deut. Akad. Wiss. Berlin, Königl. Math.-Phys. Tech.* 189–96 (1916).

60. G.D. Birkhoff, *Relativity and Modern Physics* (Harvard University Press, Cambridge, 1923).

61. R.P. Kerr, Gravitational field of a spinning mass as an example of algebraically special metrics, *Phys. Rev. Lett.* 11:237–38 (1963).

62. E.T. Newman, E. Couch, K. Chinnapared, A. Exton, A. Prakash, and R. Torrence, Metric of a rotating, charged mass, *J. Math. Phys.* 6:918-19 (1965).

63. R.H. Boyer and R.W. Lindquist, Maximal analytic extension of the Kerr metric, *J. Math. Phys.* 8:265–81 (1967).

64. H. Reissner, Über die Eigengravitation des elektrischen Feldes nach der Einsteinschen Theorie, *Ann. Physik* 50:106–20 (1916).

65. G. Nordstrøm, On the energy of the gravitational field in Einstein's theory, *Proc. Kon. Ned. Akad. Wet.* 20:1238–45 (1918).

66. J. Lense and H. Thirring, Über den Einfluss der Eigenrotation der Zentralkörper auf die Bewegung der Planeten und Monde nach der Einsteinschen Gravitationstheorie, *Phys. Z.* 19:156–63 (1918).

67. H. Thirring and J. Lense, trans. B. Mashhoon, F.W. Hehl, and D.S. Theiss, as On the gravitational effects of rotating masses: The Thirring-Lense papers, *Gen. Rel. Grav.* 16:711–50 (1984); see also refs. 63 and 64.

68. H. Thirring, Über die Wirkung rotierender ferner Massen in der Einsteinschen Gravitationstheorie, *Phys. Z.* 19:33–39 (1918).

69. H. Thirring, Berichtigung zu meiner Arbeit: Über die Wirkung rotierender ferner Massen in der Einsteinschen Gravitationstheorie, *Phys. Z.* 22:29–30 (1921).

70. D. Ivanenko and A. Sokolov, *Klassische Feldtheorie*, (Akademie Verlag, Berlin, 1953); section on 2 graviton $\rightarrow$ electron pair.

71. J.A. Wheeler, Geons, *Phys. Rev.* 97:511–36 (1955)

72. R. Arnowitt, S. Deser, and C.W. Misner, The dynamics of general relativity, in *Gravitation: An Introduction to Current Research*, ed. L. Witten (Wiley, New York, 1962), 227–65.

73. J.W. York, Jr., Energy and momentum of the gravitational field, in *Essays in General Relativity*, ed. F.J. Tipler (Academic Press, New York, 1980), 39–58.

74. R. Penrose, Quasi-local mass and angular momentum in general relativity, *Proc. Roy. Soc. Lond. A* 381:53–63 (1982).

75. R.M. Schoen and S.-T. Yau, On the proof of the positive mass conjecture in general relativity, *Comm. Math. Phys.* (*Germany*) 65:45–76 (1979).

76. R.M. Schoen and S.-T. Yau, Proof of the positive-action conjecture in quantum relativity, *Phys. Rev. Lett.* 42:547–48 (1979).

77. R.M. Schoen and S.-T. Yau, Positivity of the total mass of a general space-time, *Phys. Rev. Lett.* 43:1457–59 (1979).

78. R.M. Schoen and S.-T. Yau, The energy and the linear momentum of space-times in general relativity, *Comm. Math. Phys.* 79:49–51 (1981).

79. R.M. Schoen and S.-T. Yau, Proof of the positive mass theorem. II, *Comm. Math. Phys.* 79:231–60 (1981).

80. S. Deser and C. Teitelboim, Supergravity has positive energy, *Phys. Rev. Lett.* 39:249–52 (1977).

81. E. Witten, A new proof of the positive energy theorem, *Comm. Math. Phys.* 80:381–402 (1981).

82. E. Witten, Positive energy and Kaluza-Klein theory, in *10th Int. Conf. on General Relativity and Gravitation*, ed. B. Bertotti (Reidel, Dordrecht, 1984), 185–97.

83. Y. Choquet-Bruhat, Positive-energy theorems, in *Relativité, Groupes et Topologie II*, ed. B.S. DeWitt and R. Stora (Elsevier, Amsterdam, 1984), 739–85.

84. J.A. Wheeler, *Physics and Austerity, Law Without Law* (Anhui Science and Technology Publications, Anhui, China, 1982).

85. J.A. Wheeler, Gravitation as geometry. II, in *Gravitation and Relativity*, ed. H.-Y. Chiu and W.F. Hoffmann (Benjamin, New York, 1964), 65–89.

86. E.H. Spanier, *Algebraic Topology* (McGraw-Hill, New York, 1966).

87. M. Göckeler and T. Schücker, *Differential Geometry, Gauge Theories, and Gravity* (Cambridge University Press, Cambridge, 1987).

88. W.A. Miller, The geometrodynamic content of the Regge equations as illuminated by the boundary of a boundary principle, in *Between QUANTUM and COSMOS: Studies and Essays in Honor of John Archibald Wheeler*, ed. W.H. Zurek, A. van

der Merwe, and W. A. Miller (Princeton University Press, Princeton, 1988), 201–27; reprinted from *Foundations of Physics* (1986); see also ref. 89.

89. A. Kheyfets, The boundary of a boundary principle: A unified approach, in *Between QUANTUM and COSMOS: Studies and Essays in Honor of John Archibald Wheeler*, ed. W. H. Zurek, A. van der Merwe, and W. A. Miller (Princeton University Press, Princeton, 1988), 284–98.
90. F. W. Hehl and J. D. McCrea, Bianchi identities and the automatic conservation of energy-momentum and angular momentum in general-relativistic field theories, in *Between QUANTUM and COSMOS: Studies and Essays in Honor of John Archibald Wheeler*, ed. W. H. Zurek, A. van der Merwe, and W. A. Miller (Princeton University Press, Princeton, 1988), 256–82.
91. É. Cartan, *Leçon sur la géométrie des espaces de Riemann*, 2d ed. (Gauthier-Villars, Paris, 1963).
92. J. A. Wheeler, Our universe: The known and the unknown, address before the American Association for the Advancement of Science, New York, 29 December 1967, in *Am. Scholar* 37:248–74 (1968) and *Am. Sci.* 56:1–20 (1968).
93. J. R. Oppenheimer and H. Snyder, On continued gravitational collapse, *Phys. Rev.* 56:455–59 (1939).
94. J. A. Wheeler, The superdense star and the critical nucleon number, in *Gravitation and Relativity*, ed. H.-Y. Chiu and W. F. Hoffman (Benjamin, New York, 1964), 195–230.
95. J. A. Wheeler, Geometrodynamics and the issue of the final state, in *Relativity Groups and Topology*, ed. C. DeWitt and B. DeWitt (Gordon and Breach, New York, 1964), 315–520.
96. B. K. Harrison, K. S. Thorne, M. Wakano, and J. A. Wheeler, *Gravitation Theory and Gravitational Collapse* (University of Chicago Press, Chicago, 1965).
97. R. Ruffini and J. A. Wheeler, Introducing the black hole, *Phys. Today* 24:30–36 (1971).
98. R. Penrose, Black holes, *Sci. Am.* 226:38–46 (1972).
99. C. DeWitt and B. DeWitt, eds., *Black Holes*, Les Houches Summer School, Grenoble 1972 (Gordon and Breach, New York, 1973).
100. J. A. Wheeler, The black hole, in *Astrophysics and Gravitation: Proc. 16th Solvay Conf. on Physics*, Brussels, September 1973, ed. R. Debever (Éditions de l'Université de Bruxelles, Brussels, 1974), 279–316.
101. K. S. Thorne, The search for black holes, *Sci. Am.* 231:32–43 (1974).
102. A. S. Eddington, A comparison of Whitehead's and Einstein's formulas, *Nature* 113:192 (1924).
103. D. Finkelstein, Past-future asymmetry of the gravitational field of a point particle, *Phys. Rev.* 110:965–67 (1958).
104. M. D. Kruskal, Maximal extension of Schwarzschild metric, *Phys. Rev.* 119:1743–45 (1960).
105. G. Szekeres, On the singularities of a Riemannian manifold, *Publ. Mat. Debrecen* 7:285–301 (1960).

106. A. Einstein and N. Rosen, The particle problem in the general theory of relativity, *Phys. Rev.* 48:73–77 (1935).
107. R.W. Fuller and J.A. Wheeler, Causality and multiply-connected space-time, *Phys. Rev.* 128:919–29 (1962).
108. M.S. Morris, K.S. Thorne, and U. Yurtsever, Wormholes, time machines, and the weak energy condition, *Phys. Rev. Lett.* 61:1446–49 (1988).
109. S.W. Hawking, The chronology protection conjecture, in *Proc. 6th Marcel Grossmann Meeting on General Relativity*, Kyoto, June 1991, ed. H. Sato and T. Nakamura (World Scientific, Singapore), 3–13.
110. S.W. Hawking, Particle creation by black holes, *Comm. Math. Phys.* 43:199–220 (1975).
111. R. Penrose, Gravitational collapse: The role of general relativity, *Riv. Nuovo Cimento* 1:252–76 (1969).
112. R. Penrose, Gravitational collapse and space-time singularities, *Phys. Rev. Lett.* 14:57–59 (1965).
113. S.W. Hawking, The occurrence of singularities in cosmology. III. Causality and singularities, *Proc. Roy. Soc. Lond. A* 300:187–201 (1967).
114. R.P. Geroch, Singularities, in *Relativity*, ed. S. Fickler, M. Carmeli, and L. Witten (Plenum Press, New York, 1970), 259–91.
115. S.W. Hawking and R. Penrose, The singularities of gravitational collapse and cosmology, *Proc. Roy. Soc. Lond. A* 314:529–48 (1970).
116. B.S. DeWitt, Quantum gravity: The new synthesis, in *General Relativity: An Einstein Centenary Survey*, ed. S.W. Hawking and W. Israel (Cambridge University Press, Cambridge, 1979), 680–893.
117. J. Weber, *General Relativity and Gravitational Waves* (Wiley-Interscience, New York, 1961).
118. K.S. Thorne, *Gravitational Radiation* (Cambridge University Press, Cambridge), to be published.
119. R.A. Isaacson, Gravitational radiation in the limit of high frequency. I. The linear approximation and geometrical optics, *Phys. Rev.* 166:1263–71 (1968).
120. R.A. Isaacson, Gravitational radiation in the limit of high frequency. II. Nonlinear terms and the effective stress tensor, *Phys. Rev.* 166:1272–80 (1968).
121. L.P. Grishchuk and A.G. Polnarev, Gravitational waves and their interaction with matter and fields, in *General Relativity and Gravitation, One Hundred Years After the Birth of Albert Einstein*, ed. A. Held (Plenum Press, New York and London, 1980), 393–434.
122. Ya.B. Zel'dovich and A.G. Polnarev, Radiation of gravitational waves by a cluster of superdense stars, *Sov. Astron. AJ* (*USA*) 18:17–23 (1974); *Astron. Zh.* (*USSR*) 51:30–40 (1974).
123. S.J. Kovács, Jr., and K.S. Thorne, The generation of gravitational waves. IV. Bremsstrahlung, *Astrophys. J.* 224:62–85 (1978).
124. V.B. Braginsky and L.P. Grishchuk, Kinematic resonance and memory effect in free-mass gravitational antennas, *Sov. Phys.—JETP* 62:427–30 (1985); *Zh. Eksp. Teor. Fiz.* 89:744–50 (1985).

125. V.B. Braginsky and K.S. Thorne, Gravitational-wave bursts with memory and experimental prospects, *Nature* 327:123–25 (1987).

126. D. Christodoulou, Nonlinear nature of gravitation and gravitational-wave experiments, *Phys. Rev. Lett.* 67:1486–89 (1991).

127. L.P. Grishchuk and A.G. Polnarev, Gravitational wave pulses with velocity-coded memory, *Sov. Phys.—JETP* 69:653–57 (1990); *Zh. Eksp. Theor. Fiz.* 96:1153–60 (1989).

128. R.P. Feynman, The principle of least action in quantum mechanics, Ph.D. dissertation, Princeton University (1942).

129. R.P. Feynman and A.R. Hibbs, *Quantum Mechanics and Path Integrals* (McGraw-Hill, New York, 1965).

130. M. Oda, P. Gorenstein, H. Gursky, E. Kellogg, E. Schreier, H. Tananbaum, and R. Giacconi, X-ray pulsations from Cygnus X-1 observed from UHURU, *Astrophys. J. Lett.* 166:L1–L7 (1971).

131. J. Ehlers, Contributions to the relativistic mechanics of continuous media, *Akad. Wiss. Lit. Mainz Abh. Math. Nat. Kl.* 793–837 (1961).

132. J. Ehlers, General relativity and kinetic theory, in *General Relativity and Cosmology, Proc. Course 47 of the Int. School of Physics "Enrico Fermi"*, ed. R.K. Sachs (Academic Press, New York, 1971), 1–70.

133. H. Bondi, Plane gravitational waves in general relativity, *Nature* 179:1072–73 (1957).

134. H. Bondi and W.H. McCrea, Energy transfer by gravitation in Newtonian theory, *Proc. Cambridge Phil. Soc.* 56:410–13 (1960).

135. H. Bondi, Some special solutions of the Einstein equations, in *Lectures on General Relativity*, Brandeis 1964 Summer Institute on Theoretical Physics, vol. 1, ed. A. Trautmann, F.A.E. Pirani, and H. Bondi (Prentice-Hall, Englewood Cliffs, NJ, 1965), 375–459.

136. A. Einstein, Über Gravitationswellen, *Preuss. Akad. Wiss. Berlin, Sitzber.* 8, 154–167 (1918).

137. K.S. Thorne, Multipole expansions of gravitational radiation, *Rev. Mod. Phys.* 52:299–339 (1980).

138. S.-W. Kim, Quantum effects of Lorentzian wormhole, *Proc. 6th Marcel Grossmann Meeting on General Relativity*, Kyoto, June 1991, ed. H. Sato and T. Nakamura (World Scientific, Singapore, 1992), 501–3.

139. U.P. Frolov and I.D. Novikov, Physical effects in wormholes and time machines, *Phys. Rev. D* 42:1057–65 (1990).

140. J. Friedman, M.S. Morris, I.D. Novikov, F. Echeverria, G. Klinkhammer, K.S. Thorne, and U. Yurtsever, Cauchy problem in spacetimes with closed timelike curves, *Phys. Rev. D* 42:1915–30 (1990).

141. I.D. Novikov, Time machine and self-consistent evolution in problems with self-interaction, *Phys. Rev. D* 45:1989–94 (1992).

142. J. Casares, P.A. Charles, and T. Naylor, A 6.5-day periodicity in the recurrent nova V404 Cygni implying the presence of a black hole, *Nature* 355:614–17 (1992).

143. R. A. Remillard, J. E. McClintock, and C. D. Bailyn, Evidence for a black hole in the X-ray binary Nova Muscae 1991, *Astrophys. J. Lett.* 399:L145–149 (1992).

144. Ya. B. Zel'dovich and I. D. Novikov, *Relativistic Astrophysics*, vol. 1, *Stars and Relativity* (University of Chicago Press, Chicago, 1971).

145. A. Ashtekar, *Lectures on Non-Perturbative Canonical Gravity* (World Scientific, Singapore, 1991); see also ref. 146.

146. C. Rovelli, Ashtekar formulation of general relativity and loop-space non-perturbative quantum gravity: a report, *Class. Quantum Grav.* 8:1613–75 (1991).

147. H. C. Ford et al., Narrowband *HST* images of M87: Evidence for a disk of ionized gas around a massive black hole, *Astrophys. J. Lett.* 435:L27–30 (1994). See also: R. J. Harms et al., ibid:L35–38.

3

Tests of Einstein Geometrodynamics

3.1 INTRODUCTION

In this chapter we review the main experimental tests of Einstein geometrodynamics, including some experiments that have been proposed. For each of the fundamental assumptions and consequences of general relativity, we describe the experimental tests, or null experiments, that support it. This approach may be useful to show which parts of the theory are, at present, more or less strongly confirmed by the experiments.

We first discuss the experimental tests of the **equivalence principle**, in its various forms, from the **uniqueness of free fall** for test particles, or *Galilei equivalence principle*, to the very strong form. In particular, one may regard some important features of Einstein theory as consequences or parts of the equivalence principle, among them, the **gravitational time dilation of clocks** or **gravitational redshift** (§ 3.2.2); the constancy in time and space of the gravitational "constant *G*," and of other physical "constants" (§ 3.2.3); and the local Lorentz invariance of the physical laws in the freely falling frames, that is, the invariance of physical laws for Lorentz transformations in the local freely falling frames (§ 3.2.4). In section 3.2.5 we describe the **Lunar Laser Ranging** test of the **very strong equivalence principle**, testing, among other effects, the equal contribution of the gravitational binding energy to the inertial mass and to the gravitational mass of celestial bodies in agreement with the very strong equivalence principle (§ 2.1). In section 3.2.1 we also discuss some hypothetical deviations from the classical, weak field, *inverse square law of gravity* due to hypothetical Yukawa-like terms, and we describe the main experiments that have placed strong limits on the existence of these deviations which might violate the **weak equivalence principle**.

We then discuss the experimental tests of another basic assumption of geometrodynamics: the **geometrical structure**. After a description of the experimental support for the existence of **space curvature** based on trajectories of photons, particles, and bodies, and in particular based on half of the total effect of **bending of electromagnetic waves** (§ 3.4.1) and half of the total effect of **time delay of electromagnetic waves** (§ 3.4.2), we discuss some possible experimental evidence for the **Riemannian geometrical structure** of spacetime, and we describe a new test of Einstein geometrodynamics, the **de Sitter or geodetic precession**, and its theoretical implications (3.4.3). We observe that the **local Lorentzian** geometrical structure of **spacetime** has some experimental support in some tests of the equivalence principle.

We then discuss what distinguishes general relativity from other metric theories of gravity (besides the very strong equivalence principle): the **field equation.**

We recall here, that a **metric theory** may be defined as a theory such that: (1) the spacetime is a *Lorentzian manifold*, (2) test particles move on spacetime *geodesics*, and (3) the *equivalence principle in the medium strong form*, or *Einstein equivalence principle*, is satisfied. Therefore, any metric theory of gravity has the same spacetime structure, Lorentzian manifold, and formally the same equation of motion for test particles as general relativity: the geodesic equation (spacetime geometry "tells" mass-energy how to "move" in the same way as it does in Einstein theory). However, the field equations, and the resulting spacetime metric, of other metric theories are different from the Einstein equation (mass-energy "tells" spacetime geometry how to curve in a different way from general relativity). In section 3.5.1 we discuss the **pericenter advance** of a test body; we also briefly mention the analyses of the dynamics of planets and spacecraft in the solar system, of the Earth satellite **LAGEOS**, and of the **binary pulsar PSR 1913+16**. All these observations may be interpreted as experimental verifications of the geodesic equation of motion and of the solution of the Einstein field equation. Finally in section 3.6 we describe the main experimental efforts to detect and measure a fundamental prediction of general relativity: **gravitational waves.**

General relativity demands for the mathematical determination of the spacetime geometry, not only the field equation, but also fixation of proper **initial conditions** and boundary conditions. In particular, according to the ideas of Einstein, to solve for the spacetime geometry (see chap. 5), as a part of the initial conditions one may require the spacetime manifold to be spatially compact (i.e., spatially closed and bounded; Einstein interpretation of the Mach principle in general relativity, essentially this: "Inertia here arises from mass there"). In this way one avoids having to give boundary conditions of asymptotic flatness (see chap. 5). Therefore, in chapters 4 and 6 we discuss some conceivable experiments to test a general relativistic interpretation of Mach's principle, among them: the proposed measurement of the **gravitomagnetic field** with an orbiting gyroscope and with laser ranged satellites and some experimental limits on any rotation of mass-energy in the universe relative to the local inertial frames based on the isotropy of the cosmic background radiation and on tests of the de Sitter precession.

In conclusion, we stress that, from all the various experiments and measurements of three-quarters of century, *Einstein geometrodynamics*[1] *triumphs as the classical (nonquantized) theory of gravitation*[2–6] (see table 3.1).

Table 3.1. Main *solar system* tests of Einstein general relativity and gravitation

PRINCIPLE OR PHENOMENON TESTED	QUANTITY OR PARAMETER DETERMINED	1994 EXPERIMENTAL LIMIT	METHOD (see text)
Medium Strong Equivalence Principle: • Weak Equivalence Principle	$\frac{\delta a}{a}$ (relative "acceleration" of two masses) (= 0 in G.R.)	$\left\lvert\frac{\delta a}{a}\right\rvert \lesssim 10^{-11}$ (field of Sun) $\lesssim 10^{-12}$ (field of Sun) $\lesssim 5 \times 10^{-10}$ (field of Earth) $\lesssim 10^{-11}$ (field of Earth)	Torsion Balance (1964)[7] Torsion Balance (1972),[8] Lunar Laser Ranging (1990)[258] Free-Fall (1987)[9] Rotating Torsion Balance (1989–1990)[10,11]
• Gravitational Time Dilation of Clocks Due to a Mass (Gravitational Redshift)	α: $\frac{\Delta\nu}{\nu} \equiv (1+\alpha)\Delta U$ ($\alpha = 0$ in G.R.)	$\lvert\alpha\rvert \lesssim 2 \times 10^{-4}$ (field of Earth) $\lesssim 10^{-2}$ (field of Earth) $\cong -0.01 \pm 0.02$ (field of Sun) $\lesssim 10^{-2}$ (field of Sun) $\lesssim 10^{-2}$ (field of Saturn)	Hydrogen-Maser Clock on Rocket (1979, 1980)[12,13] Redshift of γ-rays (1960), (1965)[75,76] Redshift of Infrared Radiation from Sun (1991)[253] Redshift of Radio Signals from Galileo (1993)[263] Redshift of Radio Signals from Voyager 1 (1990)[17]
Space Curvature Generated by a Mass	γ parameter (= 1 in G.R.)	$\gamma \cong 1 \pm 0.002$ Radio Time Delay $\cong 1.0002 \pm 0.002$ Deflection of Light $\cong 1 \pm 0.02$ de Sitter effect	Mars Vikings (1979)[18] VLBI-Quasars (1991)[19] Lunar Laser Ranging (1987),[166] (1988),[167,23] (1991)[257]
Nonlinearity of the Gravitational Interaction (see §3.5)	β parameter *in the standard post-Newtonian gauge* (= 1 in G.R.)	Assuming $\alpha = 0$ and $J_{2\odot} \cong 2 \times 10^{-7}$ $\beta \cong 1 \pm 0.003$ Mercury Perihelion $\cong 0.99 \pm 0.01$ Advance $\cong 0.9999 \pm 0.0006$ (assuming the other post-Newtonian parameters, apart from γ, equal to zero)	Mercury Radar Measurements (1976),[21] (1990)[106] plus Mariner 10 and Venus Ranging (1991)[22] Lunar Laser Ranging (1988),[23] (1991),[257] (1994)[279]
Very Strong Equivalence Principle	η parameter: Nordtvedt Effect (= 0 in G.R.) .	$\eta \cong -0.0005 \pm 0.0011$ $\eta \cong -0.0001 \pm 0.0015$	Lunar Laser Ranging (1989),[23] (1990),[106] (1991),[257] (1994)[279]
	$\frac{\dot G}{G}$ (= 0 in G.R.) .	$\frac{\dot G}{G} \cong (2 \pm 4) \times 10^{-12}\ \mathrm{yr}^{-1}$, and $\cong (-2 \pm 10) \times 10^{-12}\ \mathrm{yr}^{-1}$ $\cong (0.0 \pm 2.0) \times 10^{-12}\ \mathrm{yr}^{-1}$ $\cong (0.1 \pm 10) \times 10^{-12}\ \mathrm{yr}^{-1}$	Mars Viking Landers (1983),[105] (1990)[106] Mercury and Venus Ranging (1991)[22] Lunar Laser Ranging (1991)[257]
	$\frac{\nabla G}{G}$ (spatial gradient of G) (= 0 in G.R.).	$\frac{\nabla G}{G} \lesssim 3 \times 10^{-10}\ \mathrm{AU}^{-1}$	Planetary Data Analyses (1988)[24]
	$\frac{\lvert\delta G\rvert}{G}$ (spatial anisotropy of G) (= 0 in G.R.).	$\frac{\lvert\delta G\rvert}{G} \lesssim 2 \times 10^{-12}$	Satellite Laser Ranging (1993)[122]

3.2 TESTS OF THE EQUIVALENCE PRINCIPLE

In this section we describe the main experiments that test a fundamental assumption of general relativity and of other metric theories of gravity with a spacetime curvature: the equivalence principle, in its various aspects and formulations.

In section 3.2.1 we first review the experimental tests of the **weak** form of the **equivalence principle**, or *uniqueness of free fall* of test particles (see § 2.1), at the base of most viable gravitational theories. From the inclined tables and pendulum experiments of Galilei, to proposed missions with artificial Earth satellites, this is one of the most extensively tested and testable principles of physics. In section 3.2.1 we also briefly discuss a reanalysis of the classical Eötvös experiment and some related experiments.

In section 3.2.2 we then present what may be thought of as one of the most important observable consequences of the **medium strong equivalence principle** or *Einstein equivalence principle*: the **gravitational time dilation of clocks** or **gravitational redshift**, essentially, the slower rate of standard clocks close to a mass relative to standard clocks far from the mass. This is a classical test of general relativity, common, however, to any theory of gravity satisfying the medium strong equivalence principle and the conservation of energy. From the gravitational redshift one may infer some basic properties characteristic of gravitation. The gravitational dilatation of clock times gives, in fact, some evidence for the existence of a "time curvature" (Schild 1960,1962,1967)[25,26] and gives some evidence that local freely falling frames are local Lorentzian frames (medium strong equivalence principle). Finally, it may be interpreted as a test that the equivalence principle is independent of position in space (local position invariance).

Implicit in our definition (§ 2.1) of the equivalence principle are (1) independence of the results of local experiments from the *velocity* of the local laboratory: *local Lorentz invariance*, "in *every* freely falling frame," and (2) independence of the results of the local experiments from the *location* of the local laboratory in spacetime, or independence of the equivalence principle from position in space and time: *local position invariance*, "for *every* event in spacetime." Therefore, in section 3.2.3 we describe the experimental support of local position invariance, that is, the constancy in time and space of physical laws and physical "constants." In section 3.2.4 we then discuss the experimental tests supporting the local Lorentz invariance: the Hughes-Drever and related experiments, that is, the so-called tests of the isotropy of inertia.

Finally we present in section 3.2.5 the experimental test of the **very strong equivalence principle**, violation of which would produce the so-called *Nordtvedt effect*: the *Lunar Laser Ranging* experiment.

3.2.1. The Weak Equivalence Principle

The **weak equivalence principle**, or *uniqueness of free fall* for test particles, or *Galilei equivalence principle* is at the base of most *viable theories of gravitation*.

The methods used or proposed to test the weak equivalence principle are: *free fall* and free motion of different bodies; *pendulum* experiments measuring the hypothetical dependence of the period of the oscillations of the pendulum on the various substances and masses of the pendulum bob; *torsion balance* experiments by equilibrium position or by frequency methods (see below); floating balls (see below); comparison of weights of selected materials at different locations (see ref. 44); gravitational-wave detectors (see ref. 69); and free fall of gravity gradiometers made with two masses of different materials (see below).

The first systematic **tests of the uniqueness of free fall** date back to Galileo Galilei.[27] He tested the weak equivalence principle in two different ways: inclined tables to "dilute gravity" and pendula to minimize friction. He synthesized his results with the famous gedanken experiment, dropping different objects from the tower of Pisa. In his *Discorsi e dimostrazioni matematiche intorno a due nuove scienze* Galilei describes how, after having tried with freely falling objects and with inclined tables, he concluded that the best way (at his time) to test the equivalence principle was with pendula. He then reports how he performed the experiment: "and finally I have taken two balls, one of lead and one of cork, the former more than one hundred times heavier than the latter, and I have tied each of them to two very thin equal strings, four or five "braccia" long,* tied up above; then I have removed both balls from the perpendicular state, I have left them go at the same instant of time and, going down on the circumferences of the circles described by the equal strings, their semidiameters, they went through the perpendicular line, then, along the same way they came back, and reiterating one hundred times their going there and back, they have sensibly shown, that the heavier goes so slightly faster than the lighter, that not in one hundred vibrations, nor in one thousand, does it advance of a minimum interval of time, but they go at the very same rate. One can also see the effect of the medium, that by causing some obstacle to the motion, it reduces much more the vibrations of the cork than those of the lead, but does not make them more or less frequent."

The accuracy of the Galilei pendulum experiments (about 1610) has been estimated by some authors to be $\sim 2 \times 10^{-3}$. Later, Christian Huygens[28] and Isaac Newton[29] improved the pendulum experiment. Using different substances Newton reached (about 1680) an accuracy of $\sim 10^{-3}$. Bessel (1832)[30] repeated the pendulum experiment with an accuracy $\sim 2 \times 10^{-5}$, and later Southerns (1910)[31] with an accuracy of $\sim 5 \times 10^{-6}$ and Potter (1923),[32] with an accuracy of $\sim 2 \times 10^{-5}$ to 3×10^{-6}.

*One "braccio" was slightly more than 50 cm.

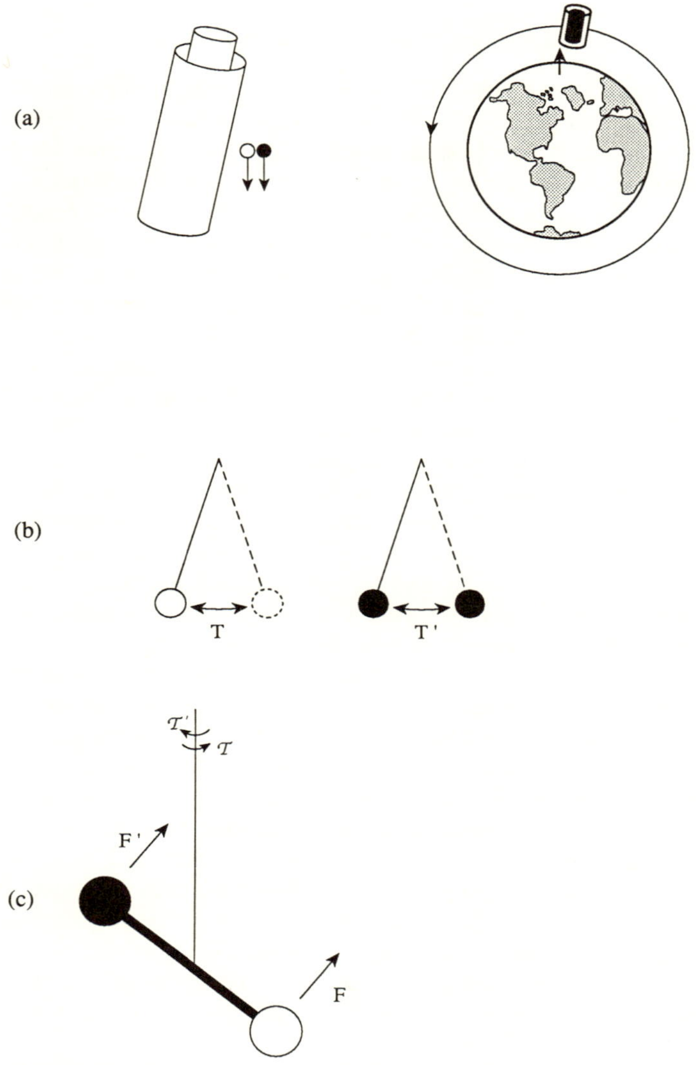

FIGURE 3.1. Tests of the weak equivalence principle. (a) Free-fall experiments; (b) pendulum experiments; and (c) torsion balance experiments (see § 3.2.1).

Highly accurate **experiments testing the weak equivalence principle** are the torsion balance experiments (see fig. 3.1). Masses of different materials are placed at the ends of a torsion balance. If the weak equivalence principle is violated, the masses will be subjected to different accelerations, and consequently there will be a measurable, net torque acting on the system. The classical Eötvös experiments (1889 and 1922),[33,34] using a torsion balance, confirmed the weak equivalence principle with an accuracy of $\sim 5 \times 10^{-9}$ (see below for a brief discussion of a 1986 reanalysis of the experiment). A torsion balance was also used in the experiments of Zeeman (1918)[35] with accuracy $\sim 3 \times 10^{-8}$, Renner (1935)[36] with accuracy $\sim 2 \times 10^{-9}$, Roll, Krotkov and Dicke (1964)[7] in the field of the Sun, with accuracy $\sim 10^{-11}$, and Braginsky and Panov (1972)[8] in the field of the Sun, with accuracy $\sim 10^{-12}$. These last two are among the most accurate laboratory tests of the weak equivalence principle as of 1992 (in the field of the Sun). The Lunar Laser Ranging experiment described in section 3.2.5 is the other highly accurate test of the weak (and very strong) equivalence principle; it tests the equality of acceleration of Earth and Moon in the field of the Sun; Lunar Laser Ranging has been calculated[258] to test the weak equivalence principle with an accuracy of the order of 10^{-12}. Other experiments performed are: Koester (1976)[37] testing free fall, with accuracy $\sim 3 \times 10^{-4}$, Worden (1982)[38] with magnetic suspension, with accuracy $\sim 10^{-4}$, and Keiser and Faller (1979)[39] using copper and tungsten floating on water, with accuracy $\sim 4 \times 10^{-11}$. We also mention the proposed NASA-ESA (European Space Agency) experiment, called STEP (Satellite Test of the Equivalence Principle),[40,41] to test free fall of different objects orbiting Earth inside a drag-free spacecraft. The accuracy of this experiment should be of the order of $\sim 10^{-17}$. Another experiment proposed to test the weak equivalence principle with an accuracy of about 4×10^{-14} would use a freely falling differential accelerometer, that is, a gradiometer measuring the difference in the accelerations of two nearby masses of different materials (which form the instrument itself). The gradiometer would be freely falling inside an evacuated cryostat in free fall from a ballon at about 40 km of altitude (see below and refs. 42 and 70). In 1987 Niebauer, McHugh, and Faller[9] performed a "Galilei tower" experiment. They dropped pieces of uranium and copper in evacuated towers. The objects were laser ranged by using mirrors. This experiment confirmed the weak equivalence principle in the field of Earth with an accuracy of $\sim 5 \times 10^{-10}$. In 1989–1990, using a rotating torsion balance, Heckel et al.,[10,11] confirmed the weak equivalence principle in the field of Earth, for copper and beryllium, and aluminum and beryllium, with an accuracy of about $\sim 10^{-11}$ (Eöt-Wash experiment).

At this point, it may be interesting to discuss briefly some hypothetical violations of the weak equivalence principle, as suggested by a reanalysis of the Eötvös experiment by Fischbach et al. (1986),[43] after the Australian mines measurements of G (see below).[44,45] Following some unified field theories (for a review see Fujii, 1991),[46] they have suggested the existence of an *intermediate-range new force* coupled to baryon number or hypercharge. Even though this hypothetical force may not necessarily be intrinsic to gravity, it is usually written as a deviation from classical gravity:

$$U = \frac{G_{\infty} m_1 m_2}{r}\left(1 + \alpha e^{\frac{-r}{\lambda}}\right) \tag{3.2.1}$$

where G_{∞} is the value of the gravitational constant to infinity and for $\lambda \gg r$, differentiating this expression, we get $G_{lab} \cong G_{\infty}(1+\alpha)$ (see § 3.2.3). Geophysical measurements of the gravitational constant G, performed in mines in Australia[44,45] (1981–1986), seemed to give a value of $G \sim 1\%$ larger than the usual laboratory measurements. These mine experiments seemed to suggest a value of $\alpha \cong -8 \times 10^{-3}$ and $\lambda \cong 200$ m. For these reasons, the possible existence of a new force coupled to baryon number or hypercharge was suggested.[43] Because of this coupling to baryon number, objects of different composition would fall toward Earth with different accelerations. Their accelerations would differ according to their baryon number-to-mass ratio $\frac{B}{\mu}$, or

$$\frac{\Delta a}{g} \cong f\left(\alpha, \frac{\lambda}{R_{\oplus}}, \frac{B_{\oplus}}{\mu_{\oplus}}\right) \times \left[\frac{B_1}{\mu_1} - \frac{B_2}{\mu_2}\right] \tag{3.2.2}$$

where Δa is the difference in the accelerations, g the gravity acceleration, B_i the total baryon number of the object i, μ_i the mass m_i in units of atomic hydrogen, and f a function of α, λ and other fundamental constants and Earth parameters ($R_{\oplus}$, $\mu_{\oplus}$, and $B_{\oplus}$ are radius, mass, and total baryon number of Earth). According to these authors,[43] their reanalysis of the Eötvös data, regarding $\frac{\Delta a}{g}$, seemed to show a linear correlation between the experimental values of $\frac{\Delta a}{g}$ and $\Delta\left(\frac{B}{\mu}\right) \equiv \left(\frac{B_1}{\mu_1} - \frac{B_2}{\mu_2}\right)$, in agreement with equation (3.2.2) and with the mine results. Furthermore, a few subsequent experiments were interpreted as suggesting some deviations from the weak equivalence principle.[47] These experiments were performed with a hollow copper sphere floating and almost totally submerged in distilled water inside a copper-lined tank at the top of a cliff,[48] and with a torsional pendulum made with a beryllium-aluminum ring near a large mass.[49] There were, in addition, claims of deviations from the inverse square law. These were obtained with a La Coste-Romberg gravimeter placed at the base and at different levels of a 600 m high television transmission tower[50] and with several measurements of G, with an accuracy of approximately 10^{-3}, performed with a gravimeter inside a 2 km ice borehole in Greenland,[51] from 200 m to 1.6 km. However, in 1988 Bizzeti et al.[52] repeated the floating-sphere experiment (with different materials) on a side of the Vallombrosa mountain, near Florence, with a null result: the residual acceleration of the solid nylon sphere used had various random orientations, with absolute values of less than $2.4 \times 10^{-9}\text{cm/s}^2$ (see also the subsequent reinterpretations of the results of the floating copper sphere by its author[53] and the subsequent Greenland determinations of G). The torsional pendulum experiment[49] has also been repeated using a copper-polyethylene ring with a null result.[54] Furthermore, Thomas et al. (1989)[55–57] have repeated the tower experiment, with a null result. They have measured gravity at twelve heights on a 465 m high tower in Nevada, using two La Coste-Romberg gravimeters. In addition they have measured gravity at 281 locations on the ground. Their observations have been consistent with the weak field, classical gravity predictions. Furthermore, in 1990, the analysis of the previous gravimeter measurements on the tall television tower was revised by its authors.[58] They included detailed topographic information and concluded that in fact there is no evidence for non-Newtonian, weak field, gravity, contrary to their first analysis of their measurements. Their final result is a limit for $|\alpha|$ constrained to be less than 10^{-3} for $\lambda > 100$ m. Finally, we report the results of a 1989 measurement by Speake et al.[59] who repeated the tower experiment in Colorado. They have measured the gravity acceleration using La Coste-Romberg gravimeters at eight different heights on a 300 m meteorological tower. After having taken into account tides, drift, gravimeter screw errors, and tower motions, their measurements confirmed the classical inverse square law for gravity with a rms of the residuals of only $\cong 10^{-7}\text{m s}^{-2}$, within the accuracy of the experiment. Therefore, this experiment places further *limits on* the existence of *any deviation from the inverse square law of gravity*. Furthermore, in all these experiments the distribution of the masses, all around the site of the measurements, plays a key role and has to be accurately known. Indeed, an anomalous mass distribution can explain the Eötvös experiment reanalysis, the mines measurements, and the tower and Greenland results. The apparent deviations from the inverse square law of the ice borehole Greenland experiment are well accounted for by unexpected geological features in the rock below the ice. See the discussion in Thomas and Vogel (1990).[259]

Most of the experiments, and especially most of the experiments that have been repeated, did not show any deviation from classical gravity and have placed limits on the existence of such deviations; we mention here some of these measurements only. The great accuracy (10^{-11} to 10^{-12}) of the 1964 Princeton and of the 1972 Moscow experiments cannot tell anything about a short-range fifth force coupled to baryon number, because they tested the weak equivalence principle in the field of the faraway Sun. However, as we mentioned, the "Galilei tower" experiment of Niebauer, McHugh, and Faller[9] had an accuracy of 5×10^{-10} and confirmed the validity of the weak equivalence principle in the field of Earth, for different substances (uranium and copper). The Eöt-Wash experiment of Adelberger et al.,[10,11] using a rotating torsion balance, also confirmed the weak equivalence principle in the field of Earth and has given a null result for a composition-dependent force for copper and beryllium and aluminum and beryllium with an accuracy of about $\sim 10^{-11}$. We also mention the Stubbs, et al. (1987)[60] experiment. They performed a torsion balance experiment on a hillside, with different materials, two berillium and two copper test bodies. The existence of a short-range fifth force coupled to baryon number should have given a measurable torsion. The null result placed limits on a single Yukawa-type term: $|\alpha| \lesssim 2 \times 10^{-4}$, for ranges $250\,\text{m} \lesssim \lambda \lesssim 1400\,\text{m}$, and $|\alpha| \lesssim 1 \times 10^{-3}$, for ranges $30\,\text{m} \lesssim \lambda \lesssim 250\,\text{m}$. In an Eötvös-type experiment, Fitch et al. (1988)[61] performed a torsion balance measurement on the slopes of a mountain in Montana, using different arrangements of copper and polyethylene. Their limits on a single Yukawa-type term were $\alpha_0\lambda = (-0.04 \pm 0.07)$ m, for $25\text{ m} < \lambda < 400$ m; $\alpha_0\lambda = (-0.05 \pm 0.09)$ m, at $\lambda = 1600$ m; and $|\alpha_0| < 10^{-4}$, from 1600 to 5000 m, where, for baryon coupling, $\alpha_0 \equiv \alpha \frac{\mu_1}{B_1} \frac{\mu_2}{B_2}$. Other experiments with a null result, setting strong *limits on* the existence of *composition-dependent deviations from the inverse square law*, are: Stubbs et al. (1989)[62] null result, using a beryllium-aluminum torsion balance close to a 10^3 kg lead source; Kuroda and Mio (1989)[63] null result, using objects of different composition freely falling in a vacuum chamber; Bennet (1989)[64] null result, using a copper-lead torsion balance near a perturbing mass of about 2×10^8 kg of flowing water of the Snake River in Washington, turned on or off; Cowsik et al. (1990)[65] null result, using a torsion balance driven in resonance by a set of masses; Zumberge et al. (1991),[66] using the ocean as an attracting mass to measure G over a distance scale of about 5 km and obtaining a value of G which agrees with the laboratory value; Carusotto et al. (1992)[262] null result, using a freely falling disk half of aluminum and half of copper; and various others. Moody and Paik (1993)[264] have performed a null test of the inverse square law of gravity testing Gauss law with a three-axis superconducting gravity gradiometer; they placed a 2σ limit of $\alpha = (0.9 \pm 4.6) \times 10^{-4}$ at $\lambda = 1.5$ m. Finally we mention the limits on deviations from the inverse square law using the variations in the water level of lakes,[67] and the laboratory constraints and the Earth-LAGEOS-Moon (LAGEOS[68] is a laser ranged satellite; see § 6.7) and planetary limits summarized in figures 3.2a and 3.2b.[69]

Following quantum theories of gravity, one may postulate the existence, together with gravitons, of spin-1 graviphotons and spin-0 graviscalars. Through coupling to fermions, they might give forces depending on the baryon number. Therefore some authors[45] have hypothesized that these fields might give two (or more) Yukawa-type terms of different signs, corresponding to repulsive graviphoton exchange and attractive graviscalar exchange, terms with conceivable ranges much longer than 200 m. The previous results might then be explained with an imperfect cancellation of two terms with different sign:

$$U = \frac{Gm_1m_2}{r}\left[1 \mp ae^{\frac{-r}{v}} \pm be^{\frac{-r}{s}} + \cdots\right]. \tag{3.2.3}$$

Some limits on two such terms have been placed by comparing data from Earth surface mines, Lunar Laser Ranging, and LAGEOS satellite (at an altitude of about 5900 km) laser ranging, thus restricting the range of these terms to less than a few hundred km. Furthermore, with the latest value of $GM_\oplus$ from

LAGEOS and with the data from Starlette, a laser ranged satellite orbiting Earth much lower than LAGEOS, at about 950 km of altitude, these limits should be improved. An experiment with a gradiometer[70] that should be able to measure gravity gradients with an accuracy of about 10^{-2} Eötvös $\equiv 10^{-11}\mathrm{s}^{-2}$ at room temperature is in preparation. Using these gradiometers in free fall near Earth one should be able to place some limits on the existence of a deviation from the standard inverse square law of gravity.

In summary, most of the experiments performed and repeated have confirmed the usual weak field inverse square law and have placed strong limits on any composition-dependent deviation from it (see figs. 3.2a and 3.2b). Moreover, the few experiments that showed some apparent deviations from the inverse square law have been well reinterpreted as due to some anomalous mass distributions around the site of the measurements.[259]

We conclude this section by observing that in the case of a hypothetical deviation from the weak field, classical inverse square law of gravity there are essentially two possibilities: a composition-dependent force, and a simple deviation from the weak field inverse square law of gravity independent of the composition. A composition-dependent force would not necessarily be intrinsic to the gravitational interaction; however, if intrinsic to gravity it would violate the weak equivalence principle. Nevertheless, practically all the experiments so far performed strongly confirm the weak equivalence principle (with accuracy of about $\sim 5 \times 10^{-10}$ to 10^{-11} in the field of Earth,[9–11] and with accuracy $\sim 10^{-11}$ to 10^{-12} in the field of the Sun (see also ref. 274)).[7,8] The other conceivability of a formal deviation from the inverse square law independent of the composition of the objects (α independent of the composition in expression 3.2.1) would not violate the weak equivalence principle, since it would affect all the test bodies in the same way, but might violate the very strong equivalence principle (§ 3.2.5). Such a deviation from the inverse square law would disagree with the very weak field limit predicted by general relativity: classical Newtonian gravitation.

3.2.2. Gravitational Redshift—Gravitational Time Dilation

We now discuss an important effect that may be considered as a consequence of the equivalence principle: the **time dilation of a clock in a gravitational field**, or **gravitational redshift**, and its experimental verifications.

The gravitational redshift may be derived in weak field from the *medium strong equivalence principle*, the *conservation of energy* and basic classical and quantum mechanics.[71] Let us assume, at the point A, a particle of mass m at rest in a local freely falling frame instantaneously at rest in the gravity field (see fig. 3.3). After falling from a height l, at the point B, the mass m has kinetic energy $\frac{1}{2}mv^2 = mgl$. From the equivalence principle the inertial mass is equal to the passive gravitational mass. At the point B, as viewed from a local freely

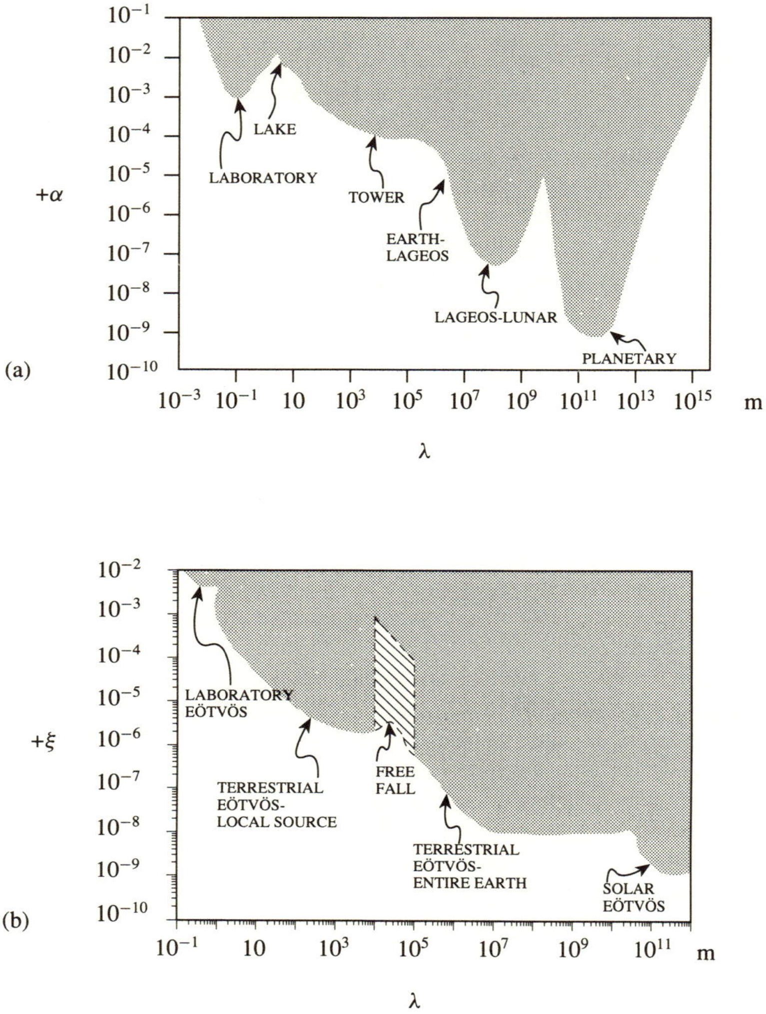

FIGURE 3.2. (a) 1991 constraints on the coupling constant α, as a function of the range λ, measuring composition-independent deviations from the weak field inverse square law. See also the limit $|\alpha| \lesssim 5.5 \times 10^{-4}$, at $\lambda = 1.5$ m, set by Moody and Paik, 1993 (§ 3.2.1). (b) Constraints on the coupling constant ξ, as a function of the range λ, measuring composition-dependent deviations from the weak field inverse square law, with α in equation (3.2.1) replaced by $\alpha_{ij} = -\xi \left(\frac{B_i}{\mu_i}\right)\left(\frac{B_j}{\mu_j}\right)$. These hypothetical deviations depend on the composition of the interacting materials (indicated with i and j) and are of the type $\sim -\xi \left(\frac{B_i}{\mu_i}\right)\left(\frac{B_j}{\mu_j}\right) e^{-\frac{r}{\lambda}}$, where B is the total baryon number of an object and $\mu \equiv \frac{m}{m_H}$ its mass in units of atomic hydrogen; see section 3.2.1. The hatched region corresponds to values of λ where the limits on ξ depend on detailed models of Earth and are at present uncertain. (Adapted from Fischbach and Talmadge 1992[69] with the kind permission of Professor Fischbach.)

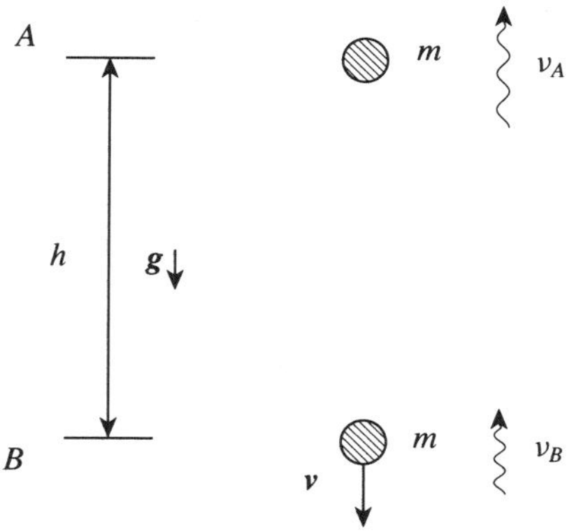

FIGURE 3.3. Redshift of electromagnetic waves in a gravitational field (gravitational time dilation of clocks); see section 3.2.2.

falling frame, instantaneously at rest in the gravity field, the total energy of the particle is thus the rest energy m, plus the kinetic energy $\frac{1}{2}mv^2$. Let us assume that, at this point B, the particle annihilates to form photons of total energy $h\nu_B$, which are absorbed by some heavy system which then reemits a single photon of the same total energy $h\nu_B$ propagating upward. From the conservation of energy, and from basic quantum mechanics, the frequency of the photon in a local freely falling frame instantaneously at rest is therefore given by the sum of the particle rest mass and kinetic energy: $\nu_B = (m + mgl)/h$. Let the photon propagate upward, returning to the initial point A, where it is eventually transformed back into particles with total energy $m' = h\nu_A$, where ν_A is the frequency of the photon at the point A, in a local freely falling frame momentarily at rest in the gravity field. From the conservation of energy, the final energy m' must be the same as the initial energy m, otherwise one would have a mechanism to create or destroy energy. Consequently: $m = h\nu_A$ and $m + mgl = h\nu_B$ or

$$\frac{\nu_A}{\nu_B} \cong 1 - gl, \tag{3.2.4}$$

that is, an electromagnetic wave has a frequency redshift, when propagating from a point nearby a mass to a more distant point, equal to $\frac{\Delta\nu}{\nu} = -gl$. One may therefore consider the experimental tests of the gravitational redshift as in part testing the medium strong equivalence principle and conservation of energy, both valid in any metric gravity theory.

Let us now derive the gravitational time dilation of clocks and the consequent electromagnetic redshift in a different way, using an approximate expression of the metric generated by a mass M, valid at the post-Newtonian order (the order of approximation beyond the Newtonian theory)[2,72,73] in metric theories of gravity (see § 3.7).

In every *stationary spacetime* (§ 2.6) we can find a coordinate system such that the metric is time independent, $g_{\alpha\beta,0} = 0$. The coordinate time required for electromagnetic waves to travel between any two points in the gravity field is obtained by using the condition $ds^2 = 0$, satisfied by photons in agreement with the equivalence principle (see below). In the coordinate system where $g_{\alpha\beta,0} = 0$, this travel coordinate time is independent from the time of emission. Therefore, if corresponding to one coordinate-fixed observer in A is emitted a wave crest at the coordinate time t_A and another wave crest at the coordinate time $t_A + \Delta t_A$, the amount of coordinate time for the two crests to reach a point B, following the same path, is the same coordinate time T for both, consequently the interval of coordinate time between detection of the two crests, at an arbitrary coordinate-fixed point B is $\Delta t_B = \Delta t_A$, that is, the interval of coordinate time corresponding to the two emissions in A and to the two detections in B of the two pulses is the same (see fig. 3.4).

Furthermore, if the *spacetime* is *static* (see § 2.6), we can find a coordinate system where the metric is time independent: $g_{\alpha\beta,0} = 0$ plus $g_{i0} = 0$, therefore the coordinate time T required for an electromagnetic signal to go from a coordinate point A to any other coordinate point B is the same as the coordinate time T for the signal to return from B to A (see below and fig. 3.5). Therefore, one can consistently define **simultaneity** on the manifold[74] between any two points using light signals between them (see formula 3.2.10 below). The observer in A sends a signal at coordinate time t_i, that is received by the observer in B at coordinate time $t_B = t_i + T$ and then reflected back and received in A at coordinate time $t_f = t_i + 2T$. The instant in A simultaneous to the instant $t_B = t_i + T$ may be defined as usual as $t_A \equiv \frac{t_f + t_i}{2} = t_i + T = t_B$, (see fig. 3.5). Therefore, if the spacetime is static, the coordinate time t can be used to define consistently simultaneous events, $t =$ constant, on the manifold, by the use of light signals between any two points.

Let us now see how the proper time measured by standard clocks[140] at arbitrary points in the gravity field is related to coordinate time. From the medium strong equivalence principle, locally, in the freely falling frames, we have

$$ds^2 = g_{\alpha\beta} dx^\alpha dx^\beta = d\tau^2 \tag{3.2.5}$$

where τ is the proper time measured by a clock at rest in a freely falling frame. Using the line element (2.6.35), that is, the metric generated by a nonrotating spherical mass, in the freely falling frames momentarily at rest in the gravity field, at the lowest order, we then have from equation (3.2.5)

$$\Delta\tau = (-g_{00})^{1/2} \Delta t \cong (1 - U)\Delta t \tag{3.2.6}$$

where U is the classical gravitational potential. From this expression we see that the coordinate time of metric (2.6.35) is the same as the proper time measured

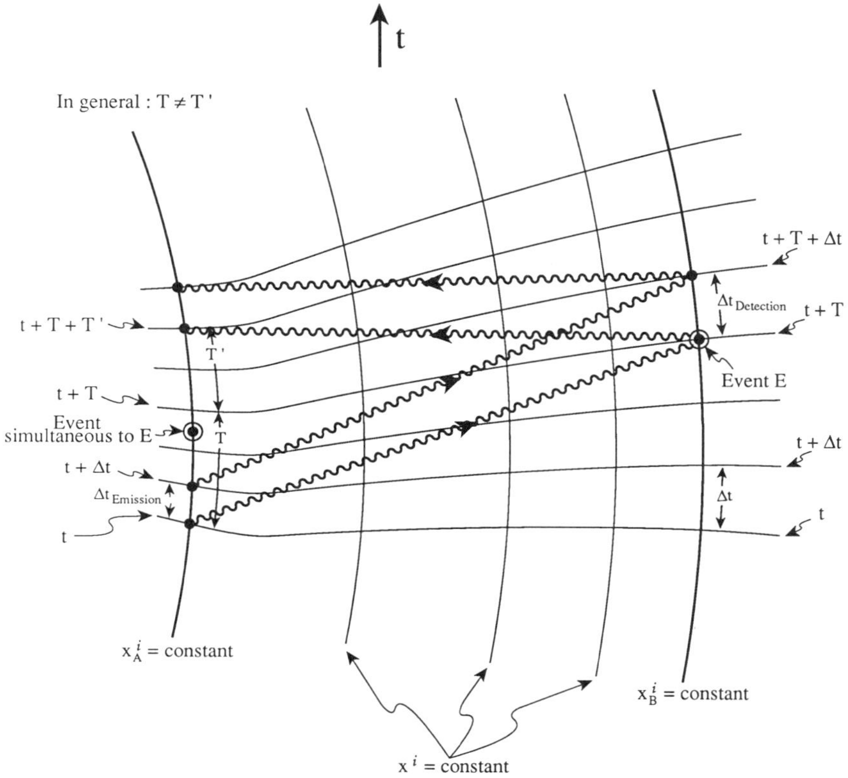

FIGURE 3.4. Coordinate time in a stationary spacetime (in a coordinate system where $g_{\alpha\beta,0} = 0$ and $g_{0i} \neq 0$).

by standard clocks of observers at infinity:

$$\Delta\tau_\infty = [-g_{00}(r \to \infty)]^{1/2}\Delta t = \Delta t. \qquad (3.2.7)$$

Since an interval of coordinate time corresponds to the same interval of coordinate time between simultaneous events at any standard point of the stationary manifold, a phenomenon that takes a proper time $\Delta\tau_A \cong (-g_{00}(A))^{1/2}\Delta t$, in A, as seen by one observer in B takes the proper time $\Delta\tau_B \cong (-g_{00}(B))^{1/2}\Delta t$, or

$$\Delta\tau_B = \left(\frac{g_{00}(B)}{g_{00}(A)}\right)^{1/2}\Delta\tau_A \cong \Delta\tau_A(1 + U_A - U_B) \equiv \Delta\tau_A(1 - \Delta U). \quad (3.2.8)$$

Therefore if $r_A > r_B$, $\Delta U \equiv U_B - U_A > 0$, and the duration of a phenomenon measured by the proper time of standard clocks in B, $\Delta\tau_B$, appears longer when measured by the proper time of standard clocks in A, $\Delta\tau_A$, that is, standard

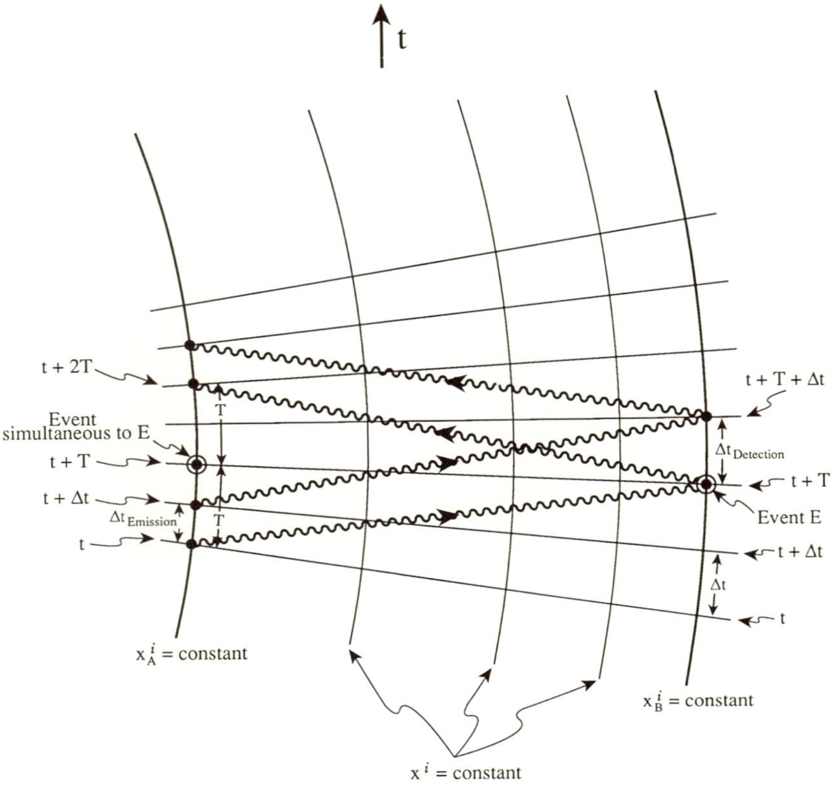

FIGURE 3.5. Coordinate time in a static spacetime (in a coordinate system where $g_{\alpha\beta,0} = 0$ and $g_{0i} = 0$)

clocks near a mass appear to go slower as observed by standard clocks far from the mass. This is the time dilation of clocks in a gravity field.

We can now easily rederive the redshift formula (3.2.4). If the observer in B emits n wave crests in the proper time $\Delta\tau_B$, corresponding to the coordinate time Δt, $\Delta\tau_B \cong (1 - U_B)\Delta t$, the observer in A will receive the n crests, in the same interval of coordinate time Δt, corresponding to the proper time, $\Delta\tau_A \cong (1 - U_A)\Delta t$. Therefore, with the usual definition of frequency of an electromagnetic wave $\nu = \frac{n}{\Delta\tau}$, we have

$$\frac{\nu_A}{\nu_B} \equiv 1 - \frac{\Delta\nu}{\nu_B} = 1 - \Delta U \tag{3.2.9}$$

where $\Delta\nu \equiv \nu_B - \nu_A$. This is the gravitational redshift of electromagnetic waves, consequence of the time dilation of clocks in a gravity field.

The gravitational time dilation of clocks has been tested by measuring the frequency shift of electromagnetic waves propagating in a gravity field and by comparing the rate at which standard clocks "tick" at different altitudes.

Among the **redshift experiments**, high accuracy has been reached by the Pound-Rebka-Snider experiment (1960–1965),[75,76] using the Mössbauer effect and measuring the frequency shift of γ-rays ascending a 22.6 m tower at Harvard. With this experiment the effect (3.2.9) has been tested with an accuracy of $\sim 10^{-2}$. Other measurements have used the gravitational redshift of electromagnetic waves from white dwarfs. However, in this case one needs a binary system to determine the mass of the white dwarf from the orbital motion. Redshift of H lines from 40 Eridani B has been measured by Popper (1954),[77] $\cong 1.2 \pm 0.3$ (1 corresponds to the theoretical general relativistic value 3.2.9); and redshift from Sirius B, $\cong 1.07 \pm 0.2$, measured by Greenstein et al. (1971).[78] The redshift of Na lines from the Sun has been measured as $\cong 1.05 \pm 0.05$ by Brault (1962);[14] see also Blamont and Roddier (1961);[79] and the redshift of K lines from the Sun, $\cong 1.01 \pm 0.06$, by Snider (1972).[15,16] The difference in wavelength between infrared oxygen triplet lines in the Sun and in the laboratory has been measured by LoPresto et al. (1991),[253] with a redshift result of $\cong 0.99 \pm 0.02$. A millisecond pulsar may in the future provide a $\sim 10^{-3}$ test of the gravitational redshift of the pulsar pulses due to the mass of the Sun and thus changing with the Earth-Sun distance.[106]

The most accurate **gravitational time dilation measurement** is, so far, the NASA, Vessot and Levine (1979–1980),[12,13] clock experiment, called Gravity Probe A (GPA). Two hydrogen-maser clocks, one on a rocket at about 10,000 km altitude and one on the ground, have been compared. The accuracy reached in measuring the shift, in agreement with the theoretical prediction (3.2.8), is 2×10^{-4}. We also recall the measurements by Jenkins (1969),[83] using crystal clocks on the GEOS-1 Earth satellite, with 9×10^{-2} accuracy; by Hafele-Keating (1972),[80,81] using cesium-beam clocks on jets, which measured $\cong 0.9 \pm 0.2$; and by Alley (1979),[82] using rubidium clocks on jets, with 2×10^{-2} accuracy. We observe that the high stability clocks on the Global Positioning System (GPS) satellites, at an altitude of about 20,200 km, regularly measure the gravitational time dilation relatively to the clocks on the ground. Furthermore a 1% accuracy test of the gravitational redshift of radio waves, due to the gravitational field of Saturn, has been obtained[17] with analyses of radio frequency measurements generated by the Deep Space Network (DSN). Radio waves were received by the Voyager 1 spacecraft during its encounter with Saturn. A 1% accuracy test of the solar gravitational redshift of radio signals transmitted from the Galileo spacecraft (1989) to DSN stations was then obtained was then obtained by Krisher, Morabito, and Anderson (1993).[263] Finally we mention a proposed interplanetary experiment.[84–86] The idea is to compare high-precision clocks on the ground with similar clocks orbiting in the solar system and passing near the Sun, at a few solar radii; the accuracy reachable with this space experiment should be of the order of 10^{-9} of ΔU and should test second order, $\sim \Delta U^2$, gravitational time dilation effects.

Let us further refine the definition of **simultaneity** in general relativity. We now consider a general spacetime, in an arbitrary coordinate system, and an observer with arbitrary four-velocity u^α. If a pointlike event has coordinates x^α and another pointlike event has coordinates $x^\alpha + dx^\alpha$, then an observer in x^α, with four-velocity u^α, measures between the two events (see § 4.5) a proper spatial distance $dl = \left[(g_{\alpha\beta} + u_\alpha u_\beta)dx^\alpha dx^\beta\right]^{1/2}$ and an interval of proper time $d\tau = \left(u_\alpha dx^\alpha u_\beta dx^\beta\right)^{1/2}$, as one may simply derive by using the equivalence principle and basic special relativity (see § 4.5). Therefore, the observer measures two events to be locally simultaneous if $d\tau = 0$, that is, if $u_\alpha dx^\alpha = u^\alpha dx_\alpha = 0$. Furthermore, if the observer is at rest in the coordinate system, the only component of its four-velocity different from zero is u^0, and the condition of simultaneity between two events becomes $dx_0 = 0$, that is, $g_{00}dx^0 + g_{0i}dx^i = 0$, or

$$dx^0 = -\frac{g_{0i}dx^i}{g_{00}}. \tag{3.2.10}$$

Therefore, the event at x^i with coordinate time x^0 *is not* simultaneous with the event at $x^i + dx^i$ with coordinate time x^0, but *is* simultaneous to the event at $x^i + dx^i$ with coordinate time $x^0 + dx^0$, where $dx^0 = -\frac{g_{0i}dx^i}{g_{00}}$, that is, to the event at $x^i + dx^i$ with coordinate time $x^0 - \frac{g_{0i}dx^i}{g_{00}}$ (see fig. 3.6). Given two events, one at x^i with coordinate time x^0, and the other at $x^i + dx^i$ with coordinate time $x^0 + \delta x^0$, the infinitesimal interval of coordinate time, at x^i, between x^0, and between the event, at x^i, simultaneous to the event $x^0 + \delta x^0$ at $x^i + dx^i$, that is, the event $\left[\left(x^0 + \delta x^0 + \frac{g_{0i}dx^i}{g_{00}}\right), x^i\right]$ (see fig. 3.6), is $\left(x^0 + \delta x^0 + \frac{g_{0i}dx^i}{g_{00}}\right) - x^0 = \delta x^0 + \frac{g_{0i}dx^i}{g_{00}}$. The corresponding interval of proper time measured at x^i is then $\delta\tau = \sqrt{-g_{00}}\left(\delta x^0 + \frac{g_{0i}dx^i}{g_{00}}\right)$.

The condition (3.2.10) of simultaneity between two nearby events corresponds to sending a light pulse from x^i to $x^i + dx^i$, then reflecting it back to x^i, and to defining at x^i the event simultaneous to the event of reflection at $x^i + dx^i$, as that event with coordinate time, at x^i, halfway between the pulse departure time and its arrival time back at x^i.

Using the condition (3.2.10), one can synchronize clocks at different locations, however, if g_{0i} is different from zero (for $i \neq 0$) the condition (3.2.10) of simultaneity along a path involves dx^i, that is, the synchronization between two clocks at different locations depends on the path followed to synchronize the clocks. Therefore, if g_{0i} is different from zero, by returning to the initial point along a closed circuit one gets in general a nonzero result for simultaneity, that is, the integral $\oint \frac{g_{0i}dx^i}{g_{00}}$ is in general different from zero, and consequently, one cannot consistently define simultaneity along a closed path.

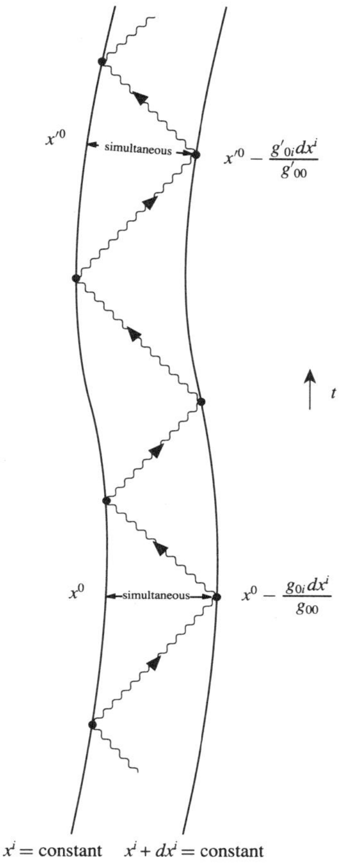

FIGURE 3.6. Simultaneity between nearby events.

This effect, admits a simple physical interpretation: to define the simultaneity between two events one uses light pulses, a different result for simultaneity, depending from the spatial path chosen, means an interval of time for a pulse to go to a point different from the interval of time for the pulse to return back to the initial point along the same spatial path. This effect depends on the path chosen and is due to the nonzero g_{0i} components (see figs. 3.4 and 3.7). In general, for a metric tensor with nondiagonal components $(0i)$ different from zero, for example in the case of a gravitomagnetic field (see chap. 6) described by a metric of the type (2.6.36), the interval of time for light to travel in a circuit is different if the light propagates in the circuit in one sense or the other.

For example, around a central body with angular momentum J, the spacetime is described, in the weak field limit, by the metric (2.6.39), and the difference in

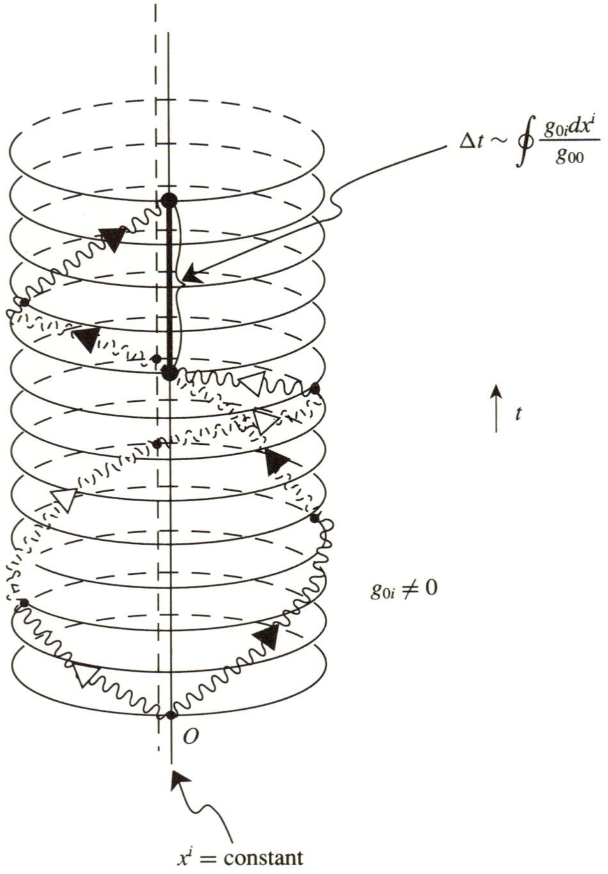

FIGURE 3.7. Two pulses of radiation counterpropagating in a circuit where $g_{0i} \neq 0$. For a gravitomagnetic field $g_{0i} \sim \frac{J}{r^2}$ one has $\oint \frac{g_{0i}}{g_{00}} dx^i \sim \frac{J}{r}$ and for a circuit rotating with angular velocity $\dot{\Omega}$: $\oint \frac{g_{0i}}{g_{00}} dx^i \sim \dot{\Omega} r^2$.

time between clockwise and counterclockwise travel of the light is proportional to $\oint \frac{g_{0i}}{g_{00}} dx^i \sim J/r$ (see § 6.9). Furthermore, even in the flat spacetime of special relativity, when one examines the propagation of light in a rotating circuit, from the point of view of noninertial observers comoving with the circuit (see fig. 6.14), one finds a difference in the total propagation time of counterrotating light pulses proportional to $\oint \frac{g_{0i}}{g_{00}} dx^i$, where g_{0i} is different from zero because of the transformation from the inertial frame to the rotating frame. This is the so-called Sagnac effect (§ 6.12). By performing a transformation from an inertial frame (corresponding quantities with an (i) on their symbol) to a frame rotating with angular frequency $\dot{\Omega}$ in the xy-plane, $\overset{(i)}{\phi} = \phi + \dot{\Omega}t$ (where $\dot{\Omega}R < 1$), we have $\overset{(i)}{x} = x \cos \dot{\Omega}t - y \sin \dot{\Omega}t$, $\overset{(i)}{y} = x \sin \dot{\Omega}t + y \cos \dot{\Omega}t$ and $\overset{(i)}{z} = z$. Therefore, the metric $\eta_{\alpha\beta}$ transforms to

$$\begin{aligned} ds^2 = &-(1 - \dot{\Omega}^2 r^2)dt^2 + \delta_{ik}dx^i\, dx^k \\ &- 2\dot{\Omega}y\, dx\, dt + 2\dot{\Omega}x\, dy\, dt, \end{aligned} \tag{3.2.11}$$

that is, with cylindrical coordinates, $x = r \cos \phi$ and $y = r \sin \phi$,

$$\begin{aligned} ds^2 = &-(1 - \dot{\Omega}^2 r^2)dt^2 + dr^2 + r^2 d\phi^2 \\ &+ 2\dot{\Omega}r^2 d\phi\, dt + dz^2 \end{aligned} \tag{3.2.12}$$

where dt represents an interval of proper time measured by inertial, nonrotating observers and $d\tau = (1 - \dot{\Omega}^2 r^2)^{1/2} dt = 0$ represents an interval of proper time measured by observers rotating with the circuit with $dx^j = 0$. If a pulse of light is propagating in the same sense of rotation of a circular circuit of radius R, thus with speed $R \frac{d\phi}{dt} \cong 1$, from $ds^2 = 0$, $r \equiv R =$ constant, $z =$ constant, and if the angular velocity is low, $\dot{\Omega}R \ll 1$, we have to first order,

$$d\tau^2 \cong dt^2 \cong R^2 d\phi^2 + 2\dot{\Omega}R^2 d\phi\, dt, \tag{3.2.13}$$

and,

$$\oint dt \cong \oint (1 + \dot{\Omega}R) R\, d\phi, \tag{3.2.14}$$

that is,

$$T_R \cong 2\pi R(1 + \dot{\Omega}R). \tag{3.2.15}$$

Therefore, the corotating light pulse will return to the emission point after a time

$$T_R = T(1 + \dot{\Omega}R) \tag{3.2.16}$$

where $T \equiv 2\pi R$. However, the counterrotating light pulse will return to the emission point after a shorter time (see fig. 6.14):

$$T_{CR} = T(1 - \dot{\Omega}R). \tag{3.2.17}$$

We shall further discuss this effect in chapter 6, where we describe optical gyroscopes (§ 6.12), the "generalized Sagnac effect," and the proposed measurement of the gravitomagnetic field using optical gyroscopes (§ 6.9). We observe here only that the Sagnac effect can be easily understood and derived, for low velocities, $\dot{\Omega}R \ll 1$, from the special relativistic principle of constancy of the speed of light, equal to $c \equiv 1$ in every inertial frame. The inertial, nonrotating, observers external to the rotating circuit can apply the special relativistic principle of constancy of the speed of light. Therefore, from the point of view of these external inertial observers both light pulses, rotating and counterrotating with the circuit, will return to the initial emission point, at the same time T. However, at that time the rotating observer in correspondence of the emission point has traveled by an amount $\dot{\Omega}RT$. Therefore he has already detected the counterrotating light pulse at the time $T_{CR} = T - \dot{\Omega}RT$, and he will detect the rotating light pulse at the time $T_R = T + \dot{\Omega}RT$.

Finally we give a useful generalized **formula for the gravitational time dilation of clocks** (i.e., the gravitational redshift), valid in a general spacetime, in an arbitrary frame and for any two observers (emitter and detector) moving with arbitrary four-velocities.

Let us assume that an observer with four-velocity u^α emits n waves corresponding to the events x^α (beginning of emission) and $x^\alpha + dx^\alpha$ (end of emission). Another arbitrary observer with four-velocity $u^{*\alpha}$ will then detect the n waves, corresponding to the events of coordinates $x^{*\alpha}$ (beginning of detection) and $x^{*\alpha} + dx^{*\alpha}$ (end of detection). Therefore, the proper time, corresponding to the interval of emission, measured by the first observer is $d\tau = |u^\alpha dx_\alpha|$, and the proper time, corresponding to the interval of detection, measured by the other observer is $d\tau^* = |u^{*\alpha} dx^*_\alpha|$. The ratio of these intervals of proper time measured by the two observers is then

$$\frac{d\tau^*}{d\tau} = \frac{u^{*\alpha}\, dx^*_\alpha}{u^\alpha\, dx_\alpha}, \tag{3.2.18}$$

and therefore the ratio of the measured frequencies is

$$\frac{\nu^*}{\nu} = \frac{(n/d\tau^*)}{(n/d\tau)} = \frac{g_{\alpha\beta}u^\alpha\, dx^\beta}{g^*_{\alpha\beta}u^{*\alpha}\, dx^{*\beta}}. \tag{3.2.19}$$

By the definition of wave vector, $k^\alpha \equiv \frac{dx^\alpha}{dp}$, if the parameter p along the null geodesics is such that the same interval dp corresponds to the interval of emission $d\tau$ and to the interval of detection $d\tau^*$ that is, $dp_{\text{emission}} = dp_{\text{detection}}$, as for example when $p \equiv t$ is the coordinate time in a time-independent metric and both observers are at rest in the coordinate system (figs. 3.4 and 3.5), we have

$$\frac{\nu^*}{\nu} = \frac{u_\alpha k^\alpha}{u^*_\alpha k^{*\alpha}}. \tag{3.2.20}$$

In the case of a diagonal metric and of both observers at rest relative to the coordinate system, we get the standard expression (3.2.8)

$$\frac{\nu^*}{\nu} = \left(\frac{g_{00}}{g^*_{00}}\right)^{\frac{1}{2}} \frac{dx^0}{dx^{*0}}. \tag{3.2.21}$$

3.2.3. The Equivalence Principle and Spacetime Location Invariance

In chapter 2, section 1, when we have formulated the equivalence principle in the medium strong and in the very strong form, we have stated that in a sufficiently small neighborhood of *every spacetime event*, in every local freely falling frame, the laws of physics are the standard laws of special relativity. Therefore, we have assumed that, locally, in every freely falling frame, the laws of physics are independent of the spacetime location of an event. It is a consequence of this assumption that all the fundamental physical "constants" $e, h, c, m_e, m_p, m_\pi \ldots$ must be time independent. In particular, if we postulate the validity of the very strong equivalence principle, this must be true for all the physical laws, and all the physical constants, including gravity and the gravitational "constant" G. However, one may consider a violation of the very strong equivalence principle, as in the Jordan-Brans-Dicke theory[87–93] where only the medium form is valid. In this case, G may change in time and/or space. On the contrary, according to the very strong formulation of the equivalence principle, satisfied in general relativity, G must be constant in time and in space. (For the limits on the time and space variations of the nongravitational "constants," e, h, c, m_e, m_p, m_π, etc., see reference 2.) Here we discuss the conceivability of *time and space variations of the gravitational "constant"* G.

The best-known examples of theories with G changing in time are the *Dirac cosmology*[94,95] and the *Jordan-Brans-Dicke theory*.[87–93] Motivated by Mach's idea[96] that inertial force arises from the acceleration of the object in question relative to the other masses in the universe,[97,98] one might think that the inertial mass M_i and the gravitational mass M_g of an object may depend from the distribution of the other masses in the universe. However, we cannot directly measure the value of M_g, active, but only the product GM_g of the active gravitational mass with the gravitational constant. Therefore one might think that G might change according to the change in distribution and motion of mass-energy in the universe, for example, due to its expansion. In other words, G might be imagined to depend on some universal function determined by the energy distribution in the universe. In the Brans-Dicke theory,[90–93] such a function is a scalar field ϕ, generated by the energy-momentum tensor through the wave equation

$$\Box \phi \equiv \phi^{;\alpha}{}_{;\alpha} = \frac{8\pi}{3+2\omega} T \equiv \frac{8\pi}{3+2\omega} T^{\alpha}{}_{\alpha}, \tag{3.2.22}$$

and the average cosmological value $\bar{\phi}$ of this scalar field turns out to be proportional to the inverse of the gravitational constant G:

$$\bar{\phi} \sim \frac{1}{G}. \tag{3.2.23}$$

In the Brans-Dicke theory the scalar field ϕ and its derivatives, together with the energy-momentum tensor $T^{\alpha\beta}$, determine the metric $g_{\alpha\beta}$ through the field equations. In the Brans-Dicke theory, as ϕ changes in time due to the expansion of the universe, so also the gravitational "constant" G changes with time. Whereas Dirac cosmology,[94,95] for values of the Hubble constant of the order of $H_0 \cong (10^{10}\text{yr})^{-1}$, predicts a rate of change $\frac{\dot{G}}{G} \cong -3 \times \frac{10^{-10}}{\text{year}}$, in disagreement with the observations (see below), in the Brans-Dicke theory, depending on the values of the constant ω and of the Hubble time, G decreases at a much lower rate.

Experimental upper limits on the rate of change of G with time, $\frac{\dot{G}}{G}$, can be determined in various ways. One method is to measure a conceivable change of the orbital period of planets in the solar system. If G decreases with time, the strength of the gravitational interaction decreases too, and consequently, if the orbital angular momentum is constant, the orbital period of the planets increases proportionally to $\frac{1}{G^2}$. These gravitational clocks may therefore be compared with nongravitational atomic clocks on Earth, unaffected by changes in G. By this line of reasoning, the limit $|\frac{\dot{G}}{G}| \lesssim 4 \times \frac{10^{-10}}{\text{year}}$ has been placed via use of passive radar data (to determine the distance) from Mercury and Venus by Shapiro et al. 1971;[99] later improved to $\left|\frac{\dot{G}}{G}\right| \lesssim 1.5 \times \frac{10^{-10}}{\text{year}}$ with the active Mariner 9, Mars orbiter, data by Reasenberg and Shapiro (1976,1978)[100,101] and by Anderson et al. (1978).[102] Moreover, the orbit of the Moon around Earth can also be used to place an upper bound on $\frac{\dot{G}}{G}$. Indeed, using Lunar Laser Ranging, Williams et al. (1978)[103] have obtained the limit $|\frac{\dot{G}}{G}| \lesssim 3 \times \frac{10^{-11}}{\text{year}}$, and Müller et al. (1991)[257] have obtained the limit $\frac{\dot{G}}{G} \cong (0.1 \pm 10) \times \frac{10^{-12}}{\text{year}}$. Using radar data from transponders on Vikings, Mars orbiters and landers, we then have $|\frac{\dot{G}}{G}| \lesssim 3 \times \frac{10^{-11}}{\text{year}}$ by Reasenberg (1983)[104], $\frac{\dot{G}}{G} \cong (2 \pm 4) \times \frac{10^{-12}}{\text{year}}$ by Hellings et al. (1983),[105] and $\frac{\dot{G}}{G} \cong (-2 \pm 10) \times \frac{10^{-12}}{\text{year}}$ by Shapiro and CfA–Center for Astrophysics (1990).[106] Finally, with combined data from Mariner 10, radar ranging to Mercury and Venus ranging, $\frac{\dot{G}}{G} \cong (0.0 \pm 2.0) \times \frac{10^{-12}}{\text{year}}$ was obtained by Anderson et al. (1991).[22] Limits on $\frac{\dot{G}}{G}$ have also been obtained with binary pulsar data analyses (see, for example, ref. 2). One can also determine the time history of the rotational orientation of Earth relative to the Moon and to the Sun. If G were to change with time, the orbital period of Earth around the Sun and the orbital period of Moon around Earth would also change in time in proportion to $\frac{1}{G^2}$. On the other hand, the rotation of Earth should be substantially less affected by small changes in G. This reasoning has been applied to analyses of ancient Sun eclipses and has given values of $\frac{\dot{G}}{G}$ from $\sim -1 \times \frac{10^{-11}}{\text{year}}$ to $\sim -1 \times \frac{10^{-10}}{\text{year}}$.[107] However, the uncertainties associated with the analysis of ancient eclipses are large and difficult to evaluate.[108] We also mention the limit placed on $\frac{\dot{G}}{G}$, by analyses of the thermonuclear time evolution of the Sun[109,110] which give[2] the limit $|\frac{\dot{G}}{G}| \lesssim 1 \times \frac{10^{-10}}{\text{year}}$. Finally, it is interesting to consider the proposal to measure via radar the position of a transponder orbiting Mercury.[111,112] This method should have a ranging accuracy[112] of a few centimeters and consequently should reach accuracies of the order of $3 \times \frac{10^{-13}}{\text{year}}$ in measuring $\frac{\dot{G}}{G}$; see Anderson et al. (1978),[102] Bender et al. (1989),[111] and Vincent and Bender (1990).[112]

Also conceivable is a variation of G in space, for example, a spatial anisotropy of G or a variation due to some unknown term, of the type given by expression (3.2.1). For $\lambda \gg r$, differentiating expression (3.2.1), we have

$$G_{\text{lab}}(r) = G_{\infty}\left[1 + \alpha\left(1 + \frac{r}{\lambda}\right)e^{-\frac{r}{\lambda}}\right] \approx G_{\infty}(1 + \alpha) \qquad (3.2.24)$$

where G_{lab} is the value of G measured in a laboratory using nearby masses, and G_{∞} is the value of G measured far from a body generating the gravity field. Regarding spatial changes of G and deviations from the usual inverse square law we refer to the discussion of section 3.2.1. It is interesting to note that the Brans-Dicke theory[90–93] predicts a change of G_{lab} of the type

$$G_{\text{lab}}(r) \cong G_{\infty}\left[1 - \frac{U}{\omega + 2}\right] \qquad (3.2.25)$$

where ω is the constant appearing in equation (3.2.22); however, it is important to stress that $\omega \gtrsim 620$ from radar time delay experiments,[18] light deflection measurements,[19,20] and especially from Lunar Laser Ranging analyses[113–116,106,257] (§ 3.2.5).

The solar system limit (1988) on the spatial gradient of G, from analyses of planetary data, has been calculated to be:[24] $\frac{\nabla G}{G} \lesssim 3 \times 10^{-10} AU^{-1}$. A limit on a conceivable spatial anisotropy of the gravitational interaction in the solar system has been calculated[117,118] to be about 10^{-13} by capitalizing on the alignment (difference of about 5 arcdegrees) of the Sun spin axis with the solar system angular momentum (after about 4.5 billion years). A new limit to a conceivable spatial anisotropy of the gravitational interaction of about $\frac{|\delta G|}{G} \lesssim 2 \times 10^{-12}$ has been set by analyzing satellite and LAGEOS laser ranging data.[122]

3.2.4. The Equivalence Principle and Lorentz Invariance

In our formulation (§ 2.1) of the medium strong equivalence principle, we have assumed that the special relativistic laws are recovered in a sufficiently small spacetime neighborhood, in *every local freely falling frame*. In other words, at each spacetime event the equivalence principle must be satisfied independently of the *velocity* of the freely falling frame. However, according to some interpretations[97,98] of Mach's principle,[96] even in a local freely falling frame, the inertial properties of a body, in particular its inertial mass, might change in different directions depending on the distribution of the masses around and depending on the direction of the acceleration. For example, the inertia of a body might be imagined to be different when it moves in the direction of the center of our galaxy, where there is a large distribution of mass, compared to its inertia when it moves perpendicularly to this direction. It is also conceivable that the inertial mass of a body could change when it moves relatively to an hypothetical preferred frame of the universe determined by the mean cosmic distribution of mass-energy.

A test, interpreted[97,98] by some authors as measuring hypothetical anisotropies in the inertial mass (however, see Dicke 1964),[91,119] is the Hughes et al. (1960)[120] and Drever (1961)[121] experiment, using Li^7 nuclei. The ground state of Li^7 has spin $\frac{3}{2}$ and therefore in the presence of a magnetic field splits into four energy levels which, if there is rotational symmetry, are equally spaced. Consequently, the transitions between these four levels have the same energy. However, if there are preferred directions in space, that is, if rotational symmetry is broken, the four energy levels would not be equally spaced. Since Earth rotates, the orientation of the experimental apparatus changes direction relatively to an external mass distribution with a period of $\cong$ 24 hours, therefore one should measure lines with energy separation changing in time with one day period. However, such a phenomenon has not been observed, and Hughes et al.[120] have set the limit $\cong 1.7 \times 10^{-16}$ eV on the maximum possible change in the energy levels. According to this limit, this null experiment has been interpreted (however, see different interpretations and discussions in the previously mentioned refs. 91 and 119) to place strong constraints[2,108] on a hypothetical anisotropy of the inertial mass of the nuclei.

Later experiments, with increased accuracy[122–127] further confirmed the isotropy of physical laws, in local freely falling frames, independently of their velocity. In particular there have been set **limits**[123,124,126] **on a Lorentz-invariance–violating contribution** to the energy levels of a bound system, such as a nucleus, of about 10^{-28} of the binding energy per nucleon, and limits[125,127] on anisotropies in the speed of light of respectively $\frac{\Delta c}{c} \lesssim 3 \times 10^{-9}$ and $\frac{\Delta c}{c} \lesssim 3.5 \times 10^{-7}$.

3.2.5. The Very Strong Equivalence Principle

In chapter 2, section 1, we have distinguished between different formulations of the **equivalence principle**, among them, the medium strong and the **very strong** forms. In the very strong form the equivalence principle is extended, locally, to all the laws of physics, including gravitation itself, whereas in the medium form the equivalence principle is locally valid for all the laws of physics, except for gravitation. In section 3.2.3 we have discussed a conceivable violation of the very strong equivalence principle: the variation of G in space and in time, as in the Brans-Dicke theory[90–93] and in the Dirac cosmology.[94,95]

A violation of the very strong equivalence principle would be a different contribution of the gravitational binding energy of a body to its inertial mass M_{inertial} and to its passive gravitational mass $M_{\text{gravitational}}$. Such a phenomenon, called the **Nordtvedt effect**,[128–130] may be written in the weak field limit:

$$M_{\text{gravitational}} = M_{\text{inertial}} + \eta\Omega \tag{3.2.26}$$

where η is a parameter measuring the size of this effect and Ω is the classical gravitational binding energy. In the PPN formalism[2] (see § 3.7) the η parameter is a linear combination of seven of the ten standard PPN parameters; in particular, neglecting the parameters corresponding to preferred-frame, preferred-location, and nonconservative theories, η is a combination of β and γ: $\eta = 4\beta - \gamma - 3$. Since the gravitational binding energy of a standard object is extremely small: $\Omega \cong -\frac{1}{2}\int \rho U d^3x \sim -\frac{M^2}{R}$, that is, $\frac{\Omega}{M} \sim -\frac{M}{R}$, one would not expect to be able to test the very strong equivalence principle through the Nordtvedt effect. However, in 1968, K. Nordtvedt[130] proposed measuring hypothetical violations of the very strong equivalence principle and the parameter η, which quantifies this effect, by using Lunar Laser Ranging. The gravitational binding energy of Earth is of course different from that of the Moon (indeed for a standard spherical body $\Omega \sim -\frac{M^2}{R}$). If the very strong equivalence principle suffers such a breakdown, the classical gravitational acceleration of Moon toward Sun—that is,

$$\boldsymbol{a}_{☽} \cong -\frac{M_\odot}{r^2}\frac{m_{☽\,\text{gravitational}}}{m_{☽\,\text{inertial}}}\hat{\boldsymbol{r}}_{☽} \cong -\frac{M_\odot}{r^2}\left(1+\eta\frac{\Omega_{☽}}{m_{☽}}\right)\hat{\boldsymbol{r}}_{☽} \tag{3.2.27}$$

where $\hat{\boldsymbol{r}}_{☽}$ is the unit vector from Sun to Moon—would be different from the gravitational acceleration of Earth toward Sun:

$$\boldsymbol{a}_\oplus \cong -\frac{M_\odot}{r^2}\left(1+\eta\frac{\Omega_\oplus}{m_\oplus}\right)\hat{\boldsymbol{r}}_\oplus \tag{3.2.28}$$

where $\hat{\boldsymbol{r}}_\oplus$ is the unit vector from Sun to Earth. Therefore, there would be an anomalous relative acceleration, between Earth and Moon,

$$\delta\boldsymbol{a}_{☽\oplus} \cong \eta\left(\frac{\Omega_\oplus}{m_\oplus} - \frac{\Omega_{☽}}{m_{☽}}\right)\frac{M_\odot}{r^2}\hat{\boldsymbol{r}}_\oplus \tag{3.2.29}$$

plus the classical, solar tidal perturbations on the Moon orbit. The anomalous effect would be about $|\delta\boldsymbol{a}_{☽\oplus}| \cong 2.6\eta \times 10^{-10}\text{cm/s}^2$. Using this anomalous acceleration $\delta\boldsymbol{a}$, one can derive the expressions for the change of the orbital angular momentum of the Moon around Earth and for the change of the Earth-Moon relative distance. The change in the relative distance between Earth and the Moon is proportional to $|\delta\boldsymbol{a}|$; it would cause an anomalous polarization of the Moon orbit around Earth, directed toward the Sun (see fig. 3.8). Using the

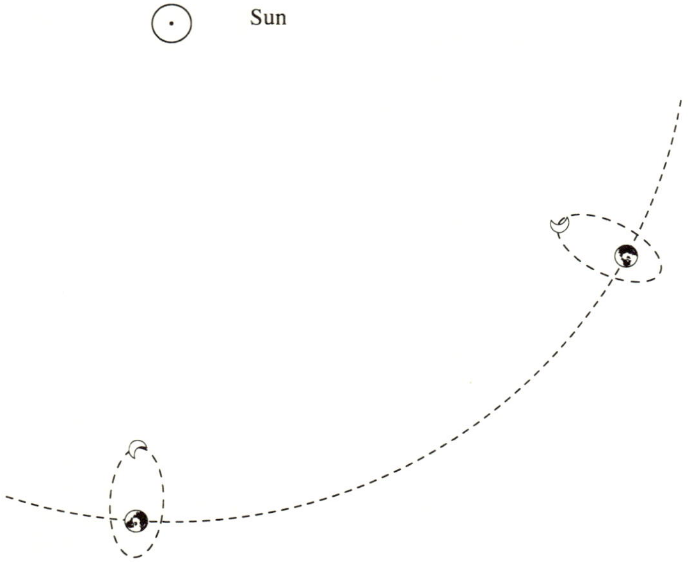

FIGURE 3.8. A hypothetical violation of the very strong equivalence principle, or Nordtvedt effect, and the consequent distortion of the Moon orbit, around Earth, in the direction of the Sun (Sun-Earth-Moon system not in scale).

numerical values $\frac{\Omega_\oplus}{m_\oplus} \cong -4.6 \times 10^{-10}$ and $\frac{\Omega_☽}{m_☽} \cong -0.2 \times 10^{-10}$, one calculates that this anomalous polarization would have an amplitude of about 9 m times η, and a frequency equal to the orbital frequency of Moon around Earth minus the orbital frequency of Earth around Sun (the synodic month).

Of course this effect is very small. However in 1969, via the Apollo 11 mission, a system of high-precision *optical corner reflectors* was placed *on the Moon's Sea of Tranquility*. Since then, other corner reflectors have been placed on the Moon (Apollo 14, Apollo 15, Luna 17, Luna 21). By the use of short laser pulses[113,114] it is possible to measure with high accuracy the ranges from the emitting lasers on Earth to the reflectors on Moon. The typical travel time is approximately 2.5 s, round trip Earth-Moon, and the accuracy is a fraction of a nanosecond, or less than ± 15 cm in one-way range. This technique is applied to the accurate study of the motion of the Moon and Earth. Areas such as: geodynamics, geodesy, astronomy, lunar science, gravitational physics, and relativity have been studied with this technique.[116] Some of the applications in relativity have put strong limits[115] on the validity of the Jordan-Brans-Dicke scalar-tensor theory ($\omega \gtrsim 620$) through M_g/M_i measurement, and limits on the time decrease of the gravitational constant[257] ($|\dot{G}/G| \lesssim 10^{-11}\mathrm{yr}^{-1}$). Since the classical tidal perturbations of the Sun on the Moon orbit should be known with sufficient accuracy, it has been possible to put a **limit** of about 0.1% **on the**

Nordtvedt effect. We have, in fact, the results: $\eta = 0 \pm 0.03$ by Williams et al. (1976);[131] $\eta = 0.001 \pm 0.015$ by Shapiro et al. (1976);[21] $\eta = 0.003 \pm 0.004$ by Dickey et al. (1988);[23] $\eta = 0.000 \pm 0.005$ by Shapiro et al. (1990),[106] and

$$\eta = -0.0001 \pm 0.0015 \tag{3.2.30}$$

by Müller et al. (1991),[257] and $\eta = -0.0005 \pm 0.0011$ by Dickey et al. (1994).[279] These limits correspond[258] to a test of the weak equivalence principle with accuracy of the order of 10^{-12}. Since in Brans-Dicke theory $\gamma = \frac{1+\omega}{2+\omega}$ and $\beta = 1$,[2] the limit (3.2.30) corresponds to $\omega \gtrsim 620$. Other tests of the Nordtvedt effect might be made by planetary measurements and a much less accurate test with LAGEOS data analyses.

3.3 ACTIVE AND PASSIVE GRAVITATIONAL MASS

According to the weak equivalence principle the inertial mass M_i of a particle is equal to the **passive gravitational mass** M_{gp} (see § 3.2.1). Furthermore, in general relativity, and in other theories with momentum conservation, the passive gravitational mass is equal to the **active gravitational mass** M_{ga}, that is, the mass that generates the field $g_{\alpha\beta}$. In the expression of the energy-momentum tensor of a perfect fluid, $T^{\alpha\beta} = (\varepsilon + p)u^\alpha u^\beta + p g^{\alpha\beta}$, in the Einstein field equation, $G^{\alpha\beta} = \chi T^{\alpha\beta}$, the density of mass-energy ε, in addition to other sources, acts as active gravitational mass density to generate the field $g^{\alpha\beta}$ (see § 4.5); however, in the equation of motion, $T^{\alpha\beta}{}_{;\beta} = 0$, a consequence of the field equation, ε may also be thought to have the function of passive gravitational mass and inertial mass. This may be seen simply from the equation of motion (geodesic) of a particle of a pressureless, $p = 0$, perfect fluid (see § 2.4). Then, in general relativity, $M_i = M_{gp} = M_{ga}$. However, if M_{gp} is different from M_{ga}, one can immediately see, from the weak field inverse square gravity law, that the "action," that is, the gravity force exerted by a body A on a body B, is not equal to the "reaction," that is, to the gravity force of B on A, and consequently there is no conservation of momentum and energy.

A test of the equivalence between active and passive gravitational mass is the Kreuzer (1968)[132] experiment. It measured the gravitational field generated by substances of different composition. The experiment was performed with a Cavendish balance, using two substances, one mainly composed of fluorine and one mainly of bromine. The two substances had the same passive gravitational mass, as measured in the Earth gravity field, that is, the same weight. However, since they are of different compositions, they could potentially have different active gravitational masses. The result of this **null experiment**, in agreement with general relativity, was[132]

$$\left| 1 - \frac{(M_{ga}/M_{gp})_{\text{fluorine}}}{(M_{ga}/M_{gp})_{\text{bromine}}} \right| \lesssim 5 \times 10^{-5}. \tag{3.3.1}$$

Furthermore, since there is a different distribution of iron and aluminum inside the Moon, the corresponding centers of mass have a different location; therefore different ratios of active and passive gravitational mass for these elements would cause a self-acceleration of the Moon. By analyzing its motion using Lunar Laser Ranging (LLR), Bartlett and van Buren (1986)[133] have set the limit

$$\left| 1 - \frac{(M_{ga}/M_{gp})_{\text{aluminum}}}{(M_{ga}/M_{gp})_{\text{iron}}} \right| \lesssim 4 \times 10^{-12}. \tag{3.3.2}$$

3.4 TESTS OF THE GEOMETRICAL STRUCTURE AND OF THE GEODESIC EQUATION OF MOTION

In the next sections we describe some important measurements that may be interpreted as tests of the **geometrical structure** of Einstein general relativity. As we have discussed in section 2.2, besides the equivalence principle, the other basic assumption of Einstein geometrodynamics and other metric theories is the existence of a dynamical spacetime curvature (in general), as opposed to the flat spacetime of special relativity. As we have previously observed, some authors[25,26] have pointed out that the time dilation of clocks in a gravitational field, or gravitational redshift of electromagnetic waves, might be thought to imply the existence of a "time curvature." However, in the next sections we describe the *direct* measurements of three effects in part due to the space curvature in the solar system: the bending of electromagnetic waves, the time delay of electromagnetic waves, and the de Sitter effect[134,135] (geodetic precession of gyroscopes). Furthermore, the binary pulsar PSR 1913+16 (§ 3.5.1) gives an astrophysical test of time delay of radio waves, and extragalactic gravitational lensing (§ 3.4.1) gives an astrophysical test of bending of electromagnetic waves. Of course, other measured effects, such as the Mercury perihelion advance and the binary pulsar periastron advance, are also in part due to space curvature in addition to other relativistic contributions (see §§ 3.5. and 3.5.1).

Another basic hypothesis regarding the geometrical structure of Einstein general relativity is the **pseudo-Riemannian character of the spacetime manifold**[136–138] (see § 2.2), that is, the hypothesis that the spacetime metric $g_{\alpha\beta}$ is symmetric, $g_{\alpha\beta} = g_{\beta\alpha}$ (plus, $\det(g_{\alpha\beta}) \neq 0$, or $\boldsymbol{g}$ nondegenerate, and $g_{\alpha\beta}dx^{\alpha}dx^{\beta}$ positive, negative, or null); we discuss this at the end of section 3.4.3. We finally observe that the **Lorentzian character**[136,137] **of the Riemannian manifold**, $g_{\alpha\beta} \rightarrow \eta_{\alpha\beta}$, that is, the hypothesis that the metric is at an event, in suitable coordinates, the Minkowski metric $\eta_{\alpha\beta}$ (with signature +2), is extensively supported by all the experiments which test the medium strong equivalence principle and the local validity of the laws and of the structure of special relativity.

3.4.1. Gravitational Deflection of Electromagnetic Waves

There are two known tests of the existence of space curvature generated by a central mass based on trajectories of photons: the **deflection of electromagnetic waves** and the **time delay of radio waves**. Both measurements may be interpreted as tests of the existence and of the amount of space curvature generated by a mass M and in part (see below) of the equivalence principle.

Let us consider the expression of a spatially spherically symmetric metric (2.6.21), derived in section 2.6 by assuming spherical symmetry only and without any use of the Einstein field equation. In addition we assume time independence of this metric:

$$ds^2 = -e^{m(r)}dt^2 + e^{n(r)}dr^2 + r^2(d\theta^2 + \sin^2\theta\, d\phi^2) \tag{3.4.1}$$

The equation of motion of a photon is given by the geodesic equation (see § 2.4)

$$\frac{d^2x^\alpha}{d\sigma^2} + \Gamma^\alpha_{\mu\nu}\frac{dx^\mu}{d\sigma}\frac{dx^\nu}{d\sigma} = 0. \tag{3.4.2}$$

Integrating this equation we get the constants of motion (we have chosen the coordinate system so that the initial conditions for the photon satisfy $\theta|_{\sigma=0} = \frac{\pi}{2}$ and $\frac{d\theta}{d\sigma}|_{\sigma=0} = 0$; then by symmetry $\theta = \frac{\pi}{2}$ and $\frac{d\theta}{d\sigma} = 0$ for every σ, and the photon propagates in the equatorial plane):

$$\begin{aligned} \theta &= \frac{\pi}{2} \\ r^2\frac{d\phi}{d\sigma} &= L \\ \text{and} & \\ \frac{dt}{d\sigma} &= \frac{K}{e^{m(r)}}. \end{aligned} \tag{3.4.3}$$

These constants of motion are obtained from the $\alpha = 2$, $\alpha = 3$, and $\alpha = 0$ components of the geodesic equation.

We observe that whenever the metric $\mathbf{g}$ is independent from a coordinate x^β, along a geodesic $x^\alpha(\sigma)$ the corresponding covariant component of the four-velocity $u_\beta = g_{\beta\mu}\frac{dx^\mu}{d\sigma}$ is constant. This follows from the geodesic equation: $g^{\alpha\beta}\frac{Du_\beta}{d\sigma} = g^{\alpha\beta}\left(\frac{du_\beta}{d\sigma} - \frac{1}{2}g_{\mu\nu,\beta}u^\mu u^\nu\right) = 0$. The constants (3.4.3) correspond to $u_\alpha =$ constant, for $\alpha = 3$ and $\alpha = 0$.

Instead of the $\alpha = 1$ component of the geodesic equation we use the condition that the world line of a photon has null arc-length, $ds^2 = 0$ (in agreement with the equivalence principle):

$$g_{\alpha\beta}\frac{dx^\alpha}{d\sigma}\frac{dx^\beta}{d\sigma} = 0 \tag{3.4.4}$$

or

$$-e^{m(r)}\left(\frac{dt}{d\sigma}\right)^2 + e^{n(r)}\left(\frac{dr}{d\sigma}\right)^2 + r^2\left(\frac{d\theta}{d\sigma}\right)^2 + r^2\sin^2\theta\left(\frac{d\phi}{d\sigma}\right)^2 = 0. \tag{3.4.5}$$

We can use the integrals of motion to obtain a differential equation for $r(\sigma)$:

$$\left(\frac{dr}{d\sigma}\right)^2 + \frac{L^2}{r^2 e^n} = \frac{K^2}{e^m e^n}, \tag{3.4.6}$$

and using the constants of motion (3.4.3) again, we can obtain a differential equation for $r(\phi)$:

$$\left(\frac{dr}{r^2 d\phi}\right)^2 + \frac{1}{r^2 e^n} = \frac{1}{b^2 e^m e^n} \tag{3.4.7}$$

where $b \equiv \frac{L}{K}$. Using the Schwarzschild[139] metric (2.6.35) for $e^{m(r)}$ and $e^{n(r)}$, one can solve equation (3.4.7), obtaining, at the order beyond Newtonian theory, the known result of a total deflection angle of $\Delta\phi = \frac{4M_\odot}{r_\odot}$, for light emitted by a far source, passing close to the surface of the Sun and finally detected by a far observer (see below and fig. 3.9). We recall that the Schwarzschild metric was derived in section 2.6 assuming spatial spherical symmetry plus the Einstein field equation.

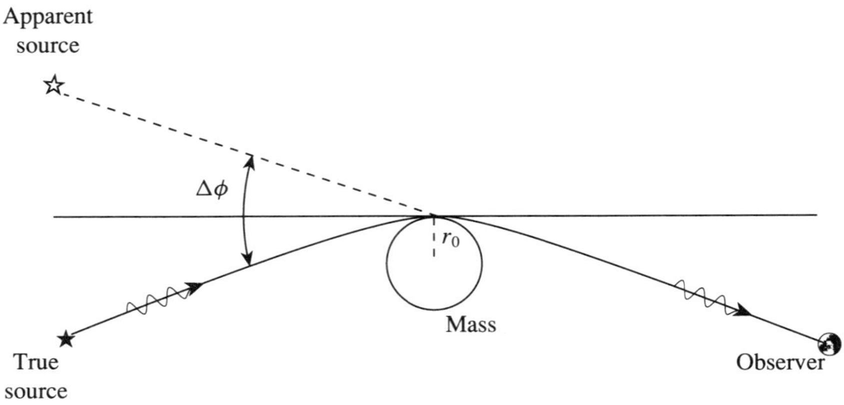

FIGURE 3.9. Deflection of electromagnetic waves by a central mass. $\Delta\phi$ is the total angle of deflection.

However, here we do not yet assume the validity of the Einstein equation, and we derive the deflection angle for electromagnetic waves using a more general expression for a spherically symmetric metric generated by a central mass M, that, even though approximate, is valid for different field equations, provided that the spacetime is pseudo-Riemannian (i.e., is valid for different metric theories of gravity). This expression of the spherically symmetric metric, written at the order beyond the classical gravitational theory and called post-Newtonian, contains the two parameters β and γ, which describe different geometrical solutions generated by a central mass M in different gravity theories (for simplicity, we do not include here other, less interesting, post-Newtonian parameters). This metric way be simply obtained by expanding in terms of $\frac{M}{r}$ the Schwarzschild metric, up to the order beyond the Newtonian approximation, then by multiplying the terms of this expansion by dimensionless parameters in order to distinguish between the post-Newtonian limits of different metric theories of gravity (see appendix 3.7 for the complete expression of the post-Newtonian–parametrized PPN metric).[2,72,73] We thus have in nonisotropic coordinates

$$ds^2 = -\left(1 - 2\frac{M}{r} + 2(\beta - \gamma)\frac{M^2}{r^2}\right)dt^2 + \left(1 + 2\gamma\frac{M}{r}\right)dr^2 + r^2 d\theta^2 + r^2 \sin^2\theta \, d\phi^2 \tag{3.4.8}$$

where in general relativity $\beta = \gamma = 1$. Using this expression for $e^{m(r)}$ and $e^{n(r)}$, we can rewrite equation (3.4.7) retaining only the post-Newtonian order corrections:

$$M^2\left(\frac{dr}{r^2 d\phi}\right)^2 + \frac{M^2}{r^2}\left(1 - 2\gamma\frac{M}{r}\right) = \frac{M^2}{b^2}\left(1 + 2\frac{M}{r} - 2\gamma\frac{M}{r}\right). \tag{3.4.9}$$

By introducing the dimensionless variable $y \equiv \frac{M}{r}$, we have

$$\left(\frac{dy}{d\phi}\right)^2 + y^2(1 - 2\gamma y) = \frac{M^2}{b^2}(1 + 2y - 2\gamma y). \tag{3.4.10}$$

On differentiating this equation with respect to ϕ and assuming $\frac{dy}{d\phi} \neq 0$, we find

$$\frac{d^2 y}{d\phi^2} + y = 3\gamma y^2 + (1 - \gamma)\frac{M^2}{b^2}. \tag{3.4.11}$$

Note that $y \equiv \frac{M}{r} \ll 1$ and $\frac{M}{b} \ll 1$, since b is at the lowest order the classical radius of closest approach. Then, to second order in $\frac{M}{b}$, equation (3.4.11) yields the solution

$$y \equiv \frac{M}{r} = \frac{M}{b}\sin\phi + \frac{M^2}{b^2}(1 + \gamma\cos^2\phi) \tag{3.4.12}$$

where the radius of closest approach of the photons to the central body occurs for $\phi = \frac{\pi}{2}$:

$$r_0 \cong b\left[1 - \frac{M}{b}\right]. \tag{3.4.13}$$

From expression (3.4.12) of r as a function of ϕ, for $r \to \infty$, we have the two solutions $\phi \cong 0 - (1+\gamma)\frac{M}{b}$ and $\phi \cong \pi + (1+\gamma)\frac{M}{b}$, and the total change in ϕ is then $\Delta\phi_{\mathrm{T}} = \pi + 2(1+\gamma)\frac{M}{b}$ (see fig. 3.9). For a straight line, one would have $\Delta\phi_0 = \pi$. Therefore, the difference is the total relativistic **deflection angle of electromagnetic waves** passing close to the surface of the Sun:

$$\begin{aligned} \Delta\phi &= 2(1+\gamma)\frac{M_\odot}{r_0} \\ &\cong \frac{1}{2}(1+\gamma)1''.75 \qquad \text{for} \quad r_0 = R_\odot \end{aligned} \tag{3.4.14}$$

where we have used expression (3.4.13). This expression for the deflection of photons may be thought of as the sum of two effects (see expression 3.4.8 for the geometry generated by a mass M). One contribution, $2\gamma\frac{M_\odot}{r_0}$, is the bending due to the space curvature, due to the g_{rr} component of the metric (3.4.8); space curvature generated by a central mass may be measured at the post-Newtonian order, in the standard PPN approximation, by the parameter γ. Indeed, the *spatial* components of the Riemann curvature tensor are proportional to the parameter γ, to post-Newtonian order $R_{ijkl} \propto \gamma\frac{M}{r^3}$. The other contribution to the deflection of electromagnetic waves, $2\frac{M_\odot}{r_0}$, may be understood and derived as a consequence of the equivalence principle, special relativity, and classical (Newtonian) gravitational theory. From special relativity and from the equivalence principle we conclude that a photon must feel the gravity field of a central body. According to classical mechanics, one can then calculate the total deflection angle of a photon due to the gravitational attraction of the central body, $\Delta\phi^{EP} = 2\frac{M_\odot}{r_0}$. This is half of the total general relativistic deflection.[141]

The deflection of electromagnetic waves, and consequently the existence of space curvature measured, in the standard PPN approximation, by the parameter γ, have as of today been determined with a relative accuracy of the order of 10^{-3}. Among the **optical measurements** of the deflection of photons, we recall the expeditions of Eddington and Dyson[142] to the islands of Sobral, South America, and Principe, Africa, during the solar eclipse of May 1919, confirming the general relativistic predictions with about 30% accuracy, and the optical measurement of the Texas Mauritanian Eclipse Team in Mauritania,[143] during the solar eclipse of June 1973:

$$\frac{1}{2}(1+\gamma) = 0.95 \pm 0.11 \qquad \text{optical measurement} \tag{3.4.15}$$

($\gamma = 1$ in Einstein geometrodynamics).

Long baseline interferometry and Very Long Baseline Interferometry (VLBI) have dramatically improved the accuracy of photon **deflection measurements**. The **Very Long Baseline Array (VLBA)**, a network of new radio antennas distributed at various sites on Earth, is now operating with increased accuracy.[277] With VLBI one measures the arrival time of radio signals from distant radio sources, quasars, using radio astronomy antennas at different locations on Earth.[144,145] The signals, recorded on tapes at different stations, are then cross-correlated to determine the arrival times of the same signal at the various locations. Knowing the direction of the radio sources, using a sufficient number of quasars and at least three antennas far from each other, one can then determine, with high accuracy, the components of the vectors connecting the stations and then the orientation of Earth relative to the distant stars (that is, relative to an asymptotic quasi-inertial frame, if the distant quasars are not rotating with respect to the asymptotic quasi-inertial frames, in agreement with a strong general relativistic interpretation of the Mach principle; see § 3.4.3 and chap. 5). Furthermore, from the difference in the arrival time of the radio signals at different locations, that is, from the difference in the phase of the radio waves at different stations, one can also determine the angular position of a distant star. According to Einstein general relativity, the apparent angular separation of quasars changes when their radio signals pass close to the Sun. VLBI can measure this separation to an accuracy of better than 10^{-1} milliarcsec.[145]

We stress that VLBI has been, is, and will be extremely important in a number of high precision[145] general relativistic measurements, from the deflection of radio waves,[19,20] with an accuracy of the order of 10^{-3} (see below), to another test of Einstein geometrodynamics, the measurement of the de Sitter or geodetic effect, obtained with two methods:[166,167,23,257] by comparing Lunar Laser Ranging (LLR)[113–116] with VLBI data, and LLR with planetary measurements (see § 3.4.3). We also stress the importance of VLBI to measure with high-accuracy, the Earth orientational parameters relative to the distant quasars. These parameters, the X and Y coordinates of the pole and UT1 (the rotational orientation of Earth), determined with VLBI with an accuracy of about 1 milliarcsec per year or less,[145] are necessary in order to measure the gravitomagnetic field using laser ranged satellites (SLR; see § 6.7 on LAGEOS III).

Among the various **measurements of the deflection of radio waves and microwaves** by the Sun, confirming the general relativistic predictions, we mention here the results of Muhleman et al. (1970),[146] and Seielstad et al. (1970),[147] with an accuracy of about 10^{-1}, and the measurements of Fomalont and Sramek (1975–1977),[148–150] with an accuracy of about 10^{-2}. We also report the VLBI result of Robertson, Carter, and Dillinger (1991)[19] (see also Robertson and Carter 1984):[20]

$$\frac{1}{2}(1+\gamma) = 1.0001 \pm 0.001 \qquad \text{VLBI radio measurement.} \tag{3.4.16}$$

Using VLBI, Truehaft and Lowe (1991)[255] have measured radio wave deflection by Jupiter with an accuracy of about 5×10^{-1}.

Finally we mention space optical measurements of the deflection of light with the satellite Hipparcos (see ref. 275), launched in 1989 by the European Space Agency (ESA), to measure with accuracy of nearly a milliarcsec the position of stars, a future "Cornerstone" project to be launched by ESA consisting of an interferometric astrometry mission in space with 10^{-2} milliarcsec accuracy or better, and the proposed space experiment by NASA, called Precision Optical INTerferometry in Space (POINTS),[151] to orbit a space interferometer with $\sim 10^{-3}$ milliarcsec accuracy in angular separation measurements. POINTS is an Earth-orbiting experiment using two optical starlight interferometers to measure the angular position of stars with high precision. The angular separation $\Delta\Theta$ between two stars, with an angle of about $\Delta\Theta \cong 90°$, should be measured with a precision of about 5 μarcsec. This precision should allow the detection of higher order relativistic light deflection by the Sun. We have just seen that the first-order electromagnetic wave deflection of about $1''\!.75$ has already been measured with about 10^{-3} accuracy using VLBI; POINTS should improve this measurement by about 2 or 3 orders of magnitude. The higher order light deflection has been calculated to be 11 μarcsec; one can write 9 μarcsec of this bending of light as a function of the standard PPN parameters γ and β, the remaining 2 μarcsec arise from a never-measured term in $g_{\alpha\beta}$, that may be quantified using a new parameter λ (not one of the ten standard PPN parameters). It is interesting to observe that the angular momentum of the Sun (the Sun gravitomagnetic field) also produces a very tiny deflection of light of about 1 μarcsec.

A dramatic example of the bending of light was discovered in 1979[152] with the double-image quasar Q0957+561, interpreted as two images of the same quasar due to the bending of electromagnetic waves by a large unseen mass. This phenomenon has been called **gravitational lensing**. After this double-image quasar, several examples of gravitational lensing have been discovered, some of which by using the Hubble space telescope (see pictures 3.1 and 3.2). Possible gravitational microlensing effects due to dark bodies have also been reported (see § 4.2).

3.4.2. Delay of Electromagnetic Waves

The other high-precision test of space curvature is the **delay in the propagation of radio waves** traveling near the Sun, or **Shapiro time delay**, proposed in 1964.[153,154]

One easily derives this effect using the metric (3.4.8) written in isotropic coordinates; with the transformation: $r \cong r_{\text{iso}}\left(1 + \gamma \frac{M}{r_{\text{iso}}}\right)$, from expression (3.4.8) we immediately get at the post-Newtonian order (see § 3.7), dropping

PICTURE 3.1. Gravitational lens G2237+0305, or Einstein Cross. The light from a quasar at a distance of approximately 8 billion light years is bent (see § 3.4.1) by the mass of a galaxy at a distance of about 400 million light years. The bending produces four images of the same quasar as observed from Earth. The diffuse central object is the bright core of the galaxy. The photograph was taken by the ESA Faint Object Camera on the NASA Hubble Space Telescope.

the subscript "iso,"

$$g_{00} \cong -\left(1 - \frac{2M}{r} + 2\beta \frac{M^2}{r^2}\right) \quad \text{and}$$

$$g_{ik} \cong \delta_{ik} \left(1 + 2\gamma \frac{M}{r}\right), \text{ or} \tag{3.4.17}$$

$$dl^2 \cong (1 + 2\gamma \frac{M}{r})(dx^2 + dy^2 + dz^2).$$

PICTURE 3.2. A pair of L-shaped images with mirror symmetry obtained with the NASA Hubble Space Telescope. Due to their high degree of symmetry and identical colors, they are believed to be two images of the same very distant galaxy. This effect should arise from gravitational lensing due to the mass of the cluster of galaxies AC114. The two objects in the center of the image are thought to be unrelated galaxies in the cluster. The observations were made with the Hubble's Wide Field Camera (courtesy of Prof. Richard Ellis of Durham University, 1992).

We then choose these quasi-Cartesian coordinates such that the emitting and the reflecting body have the same $y \equiv r_0$ and $z = 0$ coordinates (see fig. 3.10), but a different x coordinate. The deflection of light, with respect to the coordinate "line" $y =$ constant and $z =$ constant (see fig. 3.10), makes a contribution to the travel time delay that is negligible because it is of higher-order $\sim \left(\frac{M}{r}\right)^2$. Thus, from the condition of null arc-length, $ds^2 = 0$, along the world line of photons, we have:

$$g_{00}dt^2 \cong -g_{11}dx^2 \tag{3.4.18}$$

and

$$\frac{dx}{dt} \cong \pm\sqrt{\frac{-g_{00}}{g_{11}}}. \tag{3.4.18$'$}$$

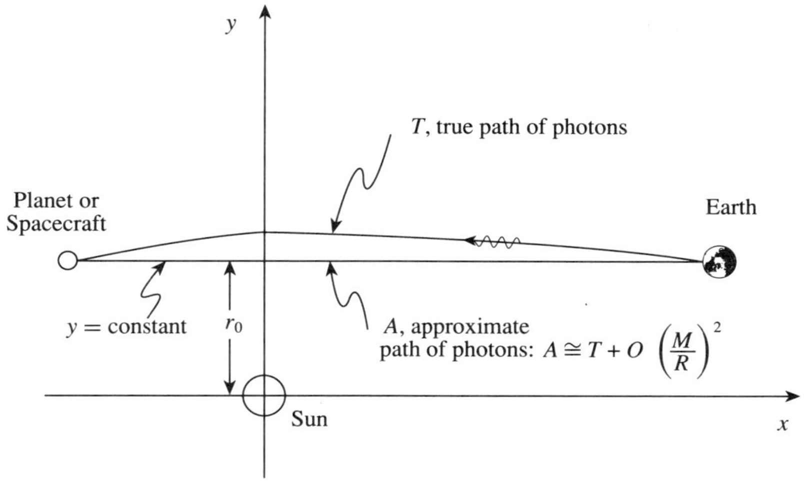

FIGURE 3.10. Geometry of the time delay of electromagnetic waves in the field of the Sun, in the round trip between Earth and a planet or a spacecraft with a transponder. Isotropic coordinates are used, and deflection of the electromagnetic waves is also shown (not to scale).

Integrating this expression from x to $x = 0$, corresponding to the radius r_0 of closest approach (see fig. 3.10), we thus get

$$\Delta t(r, r_0) = \int_0^x \left(1 + (1 + \gamma)\frac{M}{\sqrt{x^2 + r_0^2}}\right) dx$$
$$\cong x + (1 + \gamma)M \ln\left(\frac{x + \sqrt{x^2 + r_0^2}}{r_0}\right) \tag{3.4.19}$$
$$= x + (1 + \gamma)M \sinh^{-1} \frac{x}{r_0}.$$

This is the interval of coordinate time that it takes for a radio pulse to travel from r, for example, corresponding to Earth, to the point of closest approach to the Sun, r_0 (neglecting the travel time delay due to Earth mass). The first term of expression (3.4.19) is the time that it takes for an electromagnetic pulse to travel from r to r_0 in the absence of the central mass, $M = 0$. The corresponding interval of proper time measured by a clock on Earth is then given by $\Delta\tau(r, r_0) = (-g_{00\oplus})^{\frac{1}{2}} \times \Delta t(r, r_0)$. Therefore the interval of proper time that it takes for a radio pulse to travel from Earth to a planet, and back to Earth (fig. 3.10; neglecting the travel time delay due to the planet and to the

Earth masses), when the pulse passes close to the edge of the Sun, $r_0 \cong R_\odot$, is

$$\Delta\tau_\oplus \cong 2\left(1-\frac{M_\odot}{r_\oplus}\right)\left[x_\oplus + (1+\gamma)M_\odot \ln\left(\frac{x_\oplus+\sqrt{x_\oplus^2+R_\odot^2}}{R_\odot}\right)\right.$$
$$\left. + |x_P| + (1+\gamma)M_\odot \ln\left(\frac{|x_P|+\sqrt{x_P^2+R_\odot^2}}{R_\odot}\right)\right]$$
$$= 2(1-\frac{M_\odot}{r_\oplus})\left[x_\oplus + |x_P|\right.$$
$$\left. + (1+\gamma)M_\odot\left(\sinh^{-1}\frac{x_\oplus}{R_\odot} + \sinh^{-1}\frac{|x_P|}{R_\odot}\right)\right]. \qquad (3.4.20)$$

In expression (3.4.18) we see the contribution of the space curvature which changes the proper length to be traveled by the photons: $dl = \sqrt{g_{11}}\, dx$. When the Sun is far from the "line" of propagation of the pulses one has $g_{ik} \cong \eta_{ik}$ and $dl \cong dx$. The coordinate speed of light is in general different from 1, $\frac{dx}{dt} = \sqrt{-\frac{g_{00}}{g_{11}}}$, but is 1 if $g_{\alpha\beta} = \eta_{\alpha\beta}$. Of course, the speed of light measured by local inertial observers, with standard clocks and standard rods, is always $\frac{dl}{d\tau} = 1$, in agreement with the equivalence principle.

In order to measure the relativistic time delay for the photons to go and come back, when the planet is at superior conjunction (see fig. 3.10) and the pulses propagate near the edge of the Sun, one would need to know with great accuracy the time that it would take for the photons to go and come back in the absence of the Sun. However, what can be practically measured is the rate of change with proper time of the travel time of the photons, corresponding to the rate of change of the distance of the Earth-planet "line" from the Sun. By taking the derivative with respect to τ of expression (3.4.20) of $\Delta\tau$, subtracting the known classical part and neglecting smaller terms in $\frac{r_0}{r_\oplus} \ll 1$ and $\frac{r_0}{r_P} \ll 1$, and in $(\frac{dr_\oplus}{d\tau} / \frac{dr_0}{d\tau}) \ll 1$, and $(\frac{dr_P}{d\tau} / \frac{dr_0}{d\tau}) \ll 1$ (planetary orbits with small eccentricity), we then find the calculated **relativistic rate of change of travel proper time**:

$$\left.\frac{d(\Delta\tau)}{d\tau}\right|_{\substack{\text{Radio}\\\text{Time}\\\text{Delay}}} \cong -4(1+\gamma)\frac{M_\odot}{r_0}\frac{dr_0}{d\tau} \sim -(1+\gamma)\ \mu\text{sec/hr} \qquad (3.4.21)$$

where we have used $r_0 \sim 7.10^{10}$ cm and the typical rate $\frac{dr_0}{d\tau} \sim 10^6$cm/s. One can then compare this theoretical value for the rate of change of the travel proper time of radio pulses with the measured value.

One can also derive formula (3.4.21) by using the metric element (3.4.8) written in standard nonisotropic coordinates. The condition of null arc-length along the world line of photons, together with (3.4.5), gives

$$\left(\frac{dr}{dt}\right)^2 + \frac{e^{2m}}{e^n}\frac{b^2}{r^2} - \frac{e^m}{e^n} = 0. \tag{3.4.22}$$

From expression (3.4.13), at the post-Newtonian order, the radius of closest approach of the photons to the central body is $r_0 \cong b\left[1 - \frac{M}{b}\right]$ and therefore $b \cong r_0\left[1 + \frac{M}{r_0}\right]$. Integrating equation (3.4.22) from r to r_0, one then gets the interval of coordinate time required for the photons to travel from r to r_0

$$\Delta t(r, r_0) = \int_r^{r_0} \sqrt{\frac{e^n}{e^m - (e^{2m}b^2/r^2)}}\, dr. \tag{3.4.23}$$

Finally, by writing this expression from $r_\oplus$ to r_P and back to $r_\oplus$, and by taking its derivative with respect to τ, one again finds formula (3.4.21).

There are essentially two different methods to measure the radar echo delay: passive reflection, using planets, and active reflection, using transponders, either freely orbiting in the solar system, orbiting around a planet or on its surface. With the first method, using Mercury or Venus, the limiting accuracy is due to the uncertainties in the knowledge of the topography of the planets. A way to overcome this uncertainty is to get a "delay-Doppler-mapping" of the planets at the inferior conjunction, measuring time of arrival and frequency of the signals. However, using planets as passive reflectors, the limiting accuracy has been calculated to be about $\sim 5\mu$sec of time (corresponding to 1.5 km). The other possibility is to reflect electromagnetic signals using transponders. In the case of a spacecraft orbiting in the solar system, the errors are associated with the uncertainties in the spacecraft accelerations due to solar wind, solar radiation pressure, and spacecraft control devices. The uncertainties in the accelerations can give an error up to $\sim 0.1\mu$sec. Such *measurements* have been performed with Mariner 6 and 7. Measurements have been also performed with Mariner 9 orbiting Mars and with the Viking spacecraft orbiting Mars or on its surface. The limit to this accuracy arises, however, from dispersion of the electromagnetic waves by the solar corona, even though dual frequency ranging can reduce this error.

The main determinations of $\frac{1}{2}(1+\gamma)$ via radar echo delay have been the following: $\cong 0.9 \pm 0.2$, by the use of radio time delay measurements to Mercury and Venus, by Shapiro (1968);[155] $\cong 1.015 \pm 0.05$, with Mercury and Venus, by Shapiro et al. (1971);[99] $\cong 1.003 \pm 0.03$ and $\cong 1.000 \pm 0.03$, by the use of the Mariner 6 and 7 spacecraft, by Anderson et al. (1975);[156] $\cong 1.00 \pm 0.02$ with Mariner 9, by Reasenberg and Shapiro (1977)[157] and Anderson et al. (1978);[102] and

$$\frac{1}{2}(1+\gamma) = 1 \pm 0.001 \qquad \text{Viking spacecraft} \tag{3.4.24}$$

by Reasenberg et al. (1979)[18] (see also Shapiro et al. (1977)[158] and Cain et al. (1978)).[159] Krisher et al. (1991)[256] have obtained $\frac{1}{2}(1+\gamma) \cong 1 \pm 0.015$ by analyzing the roundtrip radio range measurements to the Voyager 2 spacecraft, generated by the Deep Space Network (DSN) during solar conjunction, with a distance Earth–Voyager 2 of about 3×10^9 km.

Finally the binary pulsar **PSR 1913+16**[160] and especially[260] the **millisecond pulsar 1937+21** give a further test of the time delay of the pulsar radio signals, and of other important general relativistic effects (see § § 3.5.1 and 2.10, and ref. 2). The measured relativistic parameters are: time delay in propagation of radio pulses (range and shape),[265] time dilation or gravitational redshift, mean rate of periastron advance, and rate of decrease of orbital period. Using PSR 1937+21, the time delay of radio signals has been tested[260] with accuracy of about 3×10^{-2}.

In conclusion, we recall that from the experimental tests of the gravitational time dilation of clocks (§ 3.2.2) we had a $\sim 2 \times 10^{-4}$ test of the lowest order (beyond Minkowskian metric) contribution of a mass to the g_{00} component of the metric $\boldsymbol{g}$. Therefore, from expressions (3.4.14) and (3.4.21), the experimental tests of the deflection of electromagnetic waves and of the radio time delay may be considered as strong tests, with $\sim 2 \times 10^{-3}$ accuracy, of the existence of space curvature generated by a mass, of the correct amount predicted by the Einstein field equation (measured in weak field by the parameter γ).

3.4.3. [de Sitter or Geodetic Effect and Lense-Thirring Effect]

In this section we describe a further test of Einstein geometrodynamics: the test of the **de Sitter effect or geodetic (or geodesic) precession**[134,135] and some theoretical aspects of its measurement. Let us first derive the expression for the precession of a gyroscope moving in a general, though weak, gravitational field.

Let us consider a spacelike four-vector S^α which represents the angular momentum vector or spin vector (§ 6.10) of a test gyroscope which travels along a timelike curve $x^\alpha(s)$ with tangent vector u^α. We then have $S^\alpha u_\alpha = 0$. In agreement with special relativistic kinematics and with the medium strong equivalence principle, the spin vector S^α obeys the **Fermi-Walker transport**[161] along the curve

$$S^\alpha{}_{;\beta} u^\beta = u^\alpha (a^\beta S_\beta) \equiv u^\alpha (u^\beta{}_{;\gamma} u^\gamma S_\beta) \tag{3.4.25}$$

where $a^\beta \equiv u^\beta{}_{;\gamma} u^\gamma$ is the four-acceleration of the test gyroscope. If the timelike curve is a geodesic we have $u^\beta{}_{;\gamma} u^\gamma = 0$ (however, see eq.6.10.4), and therefore

$$S^\alpha{}_{;\beta} u^\beta = 0. \tag{3.4.25$'$}$$

In this case the Fermi-Walker transport is just parallel transport along the geodesic.

Let us now consider the approximate, post-Newtonian, metric element generated by a mass distribution, with energy density ε and energy current density $j^i = \varepsilon v^i$. We first transform to isotropic coordinates the approximate, post-Newtonian and parametrized, central mass solution (3.4.8), with the transformation $r \cong r_{\text{iso}}(1 + \gamma \frac{M}{r_{\text{iso}}})$, yielding expression (3.4.17). We then generalize this expression to an arbitrary, weak field, mass distribution by replacing $\frac{M}{r}$ with U, where U is the standard Newtonian potential, solution of $\Delta U = -4\pi\varepsilon$. Finally, we introduce the nondiagonal components of the metric tensor, $g_{0i} \ll 1$, generated by mass currents (see chap. 6), of order $g_{0i} \sim Uv$. We thus have at post-Newtonian order (for a rigorous derivation of the *PPN metric*,[72,73] see § 3.7 and ref. 2):

$$ds^2 = -(1-2U+2\beta U^2)\, dt^2+(1+2\gamma U)\delta_{ik}dx^i dx^k+2\mu g_{0i}dt\, dx^i \quad (3.4.26)$$

where β and γ are the two PPN parameters (equal to 1 in general relativity) introduced in (3.4.8) and in section 3.7.

For simplicity, in expression (3.4.26) we have set equal to zero the other PPN parameters characteristic of preferred-frame, preferred-location, and nonconservative theories, but we have introduced the **new parameter** μ which measures the strength of the g_{0i} term sometimes called the *gravitomagnetic* vector potential, or magnetogravitational, or gyromagnetic vector potential. In general relativity $\mu = 1$. This new gravitomagnetic parameter μ is not one of the standard ten PPN parameters, nor is it a function of these ten parameters. It measures the g_{0i} field generated by mass-energy currents relative to other masses—the *intrinsic gravitomagnetic field*—but not merely the g_{0i} field generated by motion on a static background; there does not yet exist any direct experimental determination of μ, apart from some indirect astrophysical evidence (see below). In other words, μ quantifies the contribution to the spacetime curvature of the mass-energy currents relative to other masses, that is, μ tests intrinsic gravitomagnetism, in general relativity and conceivable alternative metric theories of gravity. At this point it is necessary to observe that the g_{0i} term generated by mass-energy currents relative to other masses, for example, by the angular momentum of a body, and the g_{0i} term generated in a frame moving on a static background are two basically *different* phenomena, as is rigorously shown in section 6.11. In the former case, we have the "new" phenomenon of general relativity, called gravitomagnetism, that mass-energy currents relative to other masses generate spacetime curvature, as we show in section 6.11 by using some curvature invariants. This phenomenon corresponds to the never directly detected and measured Lense-Thirring, frame dragging effect, the third term of equation (3.4.38) below. In the latter case the g_{0i} term is due to local Lorentz boosts on a static background; for a gyroscope this phenomenon corresponds to the de Sitter effect, the second term of equation (3.4.38), already measured since 1987–1988. Therefore, these two phenomena are fundamentally different. Furthermore, the complete expression of the g_{0i} term in the PPN formalism (with a linear dependence of g_{0i} on the mass-energy currents ρv^i) is dependent on the already well-determined γ and α_2 parameters. However, in section 3.7 it is rigorously shown that the PPN formalism does not describe the post-Newtonian limit of every conceivable metric theory of gravity. In principle, one can have an, a priori, infinite number of post-Newtonian terms and of new parameters generated in hypothetical alternative metric theories of gravity; for example in g_{0i} one can have post-Newtonian terms nonlinear in the mass-energy currents. Therefore, *in general*, the post-Newtonian term g_{0i} *cannot* a priori be written as a function of the parameters γ and α_2 only, and we have introduced the new parameter μ to quantify gravitomagnetism. In other words, one might write at the post-Newtonian order $g_{0i} = g_{0i}^{(L)} + \mu g_{0i}^{(G)}$, where $g_{0i}^{(L)}$ is the part of g_{0i} due to local Lorentz invariance and has already been tested through the measurement of the parameters γ and α_2, whereas $g_{0i}^{(G)}$ is the intrinsic gravitomagnetic "potential" never directly measured.

In general relativity, for a stationary distribution of matter, we can write g_{0i} at the post-Newtonian order[1] (§ 6.1):

$$g_{0i} \cong -4V_i \qquad \text{where} \quad \Delta V_i = -4\pi\varepsilon v_i. \tag{3.4.27}$$

Asymptotically, for $r \to \infty$, the metric (3.4.26) becomes the Minkowski metric $\eta_{\alpha\beta}$.

Let us construct, at each point along the timelike world-line of a particle, a **local orthonormal frame**[162] $\boldsymbol{\lambda}_{(\mu)}$: $g_{\alpha\beta}\lambda^{\alpha}_{(\nu)}\lambda^{\beta}_{(\mu)} = \eta_{\nu\mu}$. The index between parentheses, $\mu\epsilon(0, 1, 2, 3)$, is the label of each vector of the **local tetrad**, or **local vierbein**. One raises and lowers these indices between parentheses with $\eta^{\alpha\beta}$ and $\eta_{\alpha\beta}$ and, as usual, the standard indices with $\mathbf{g}$. The timelike vector of this tetrad is by definition the four-velocity: $\lambda^{\alpha}_{(0)} \equiv u^{\alpha}$. By construction, the spatial axes are nonrotating with respect to the asymptotic frame where $g_{\alpha\beta} \to \eta_{\alpha\beta}$. In other words, the spatial axes are nonrotating with respect to a local orthonormal tetrad at rest in the asymptotically flat coordinate system of the metric (3.4.26). Therefore, the spatial axes are not Fermi-Walker transported along the timelike world line, but are obtained *at each point* with a pure Lorentz boost—with no spatial rotations—between a local frame at rest in the coordinate system of the asymptotically flat, post-Newtonian, metric (3.4.26), and an observer moving along the curve $x^{\alpha}(s)$ with four-velocity u^{α}.

To obtain the expression of this local orthonormal and comoving tetrad, we first define a local orthonormal tetrad, $\boldsymbol{\theta}^{[\alpha]}$, at rest in the coordinate system of expression (3.4.26)

$$ds^2 = \eta_{\alpha\beta}\boldsymbol{\theta}^{[\alpha]}\boldsymbol{\theta}^{[\beta]} \tag{3.4.28}$$

where

$$\boldsymbol{\theta}^{[0]} = \theta^{[0]}_{\sigma}\boldsymbol{dx}^{\sigma} \equiv (1 - U + (\beta - \frac{1}{2})U^2)\boldsymbol{dx}^0 - \mu g_{0l}\boldsymbol{dx}^l$$
$$\boldsymbol{\theta}^{[i]} = \theta^{[i]}_{\sigma}\boldsymbol{dx}^{\sigma} \equiv (1 + \gamma U)\boldsymbol{dx}^i . \tag{3.4.29}$$

The components of a tensor with respect to an orthonormal tetrad are labeled with indices within parentheses, lowered and raised[162] with the Minkowski tensor $\boldsymbol{\eta}$, and $\theta^{\alpha}_{[\beta]} = g^{\alpha\sigma}\eta_{\beta\rho}\theta^{[\rho]}_{\sigma}$.

Then, we locally transform, with a pure Lorentz boost $\Lambda^{[\alpha]}_{(\beta)}$, from this local orthonormal tetrad to a local orthonormal tetrad comoving with the particle with four-velocity u^{α}. The result of these two transformations is

$$\boldsymbol{\lambda}_{(\beta)} = \lambda^{\sigma}_{(\beta)}\frac{\boldsymbol{\partial}}{\boldsymbol{\partial x}^{\sigma}} \equiv \theta^{\sigma}_{[\rho]}\Lambda^{[\rho]}_{(\beta)}\frac{\boldsymbol{\partial}}{\boldsymbol{\partial x}^{\sigma}} \tag{3.4.30}$$

where $\frac{\boldsymbol{\partial}}{\boldsymbol{\partial x}^{\sigma}}$ is the coordinate basis and $\boldsymbol{\lambda}_{(\beta)}$ is the orthonormal comoving basis. We write with $v^{[i]}$ the velocity of the test particle as measured using the tetrad $\boldsymbol{\theta}^{[\alpha]}$, and with v^i its velocity using the coordinate basis $\boldsymbol{dx}^{\alpha}$, since $u^{[\alpha]} = \theta^{[\alpha]}_{\beta}u^{\beta}$, $u^i = v^i u^0$, $u^{[i]} = v^{[i]}u^{[0]}$, and $u^0 \cong 1 + \frac{1}{2}v^2 + U$, $u^{[0]} \cong 1 + \frac{1}{2}v^2$, we have $v^{[i]} \cong v^i(1 + (1+\gamma)U)$.

A low-velocity Lorentz transformation, from a local orthonormal tetrad $\boldsymbol{\theta}^{[\alpha]}$ at rest in the coordinate system (indices within square parentheses label vectors of this tetrad) to a local orthonormal tetrad $\boldsymbol{\lambda}^{(\alpha)}$ moving with velocity $v^{[i]}$ with respect to $\boldsymbol{\theta}^{[\alpha]}$ (indices within round parentheses label vectors of this other tetrad), is described by the **post-Galilean transformation** (Chandrasekhar and Contopoulos):[163]

$$x^{(i)} = x^{[i]} - \left(1 + \frac{1}{2}v^2\right)v^{[i]}x^{[0]} + \frac{1}{2}x^{[k]}v_{[k]}v^{[i]} + O(v^4)\cdot x$$
$$x^{(0)} = \left(1 + \frac{1}{2}v^2 + \frac{3}{8}v^4\right)x^{[0]} - \left(1 + \frac{1}{2}v^2\right)x^{[k]}v_{[k]} + O(v^5)\cdot t. \tag{3.4.31}$$

Therefore, at the post-Newtonian order,

$$\Lambda^{[0]}_{(0)} \cong 1 + \frac{1}{2}v^2 + \frac{3}{8}v^4$$
$$\Lambda^{[0]}_{(i)} = \Lambda^{[i]}_{(0)} \cong v^{[i]}\left(1 + \frac{1}{2}v^2\right) \cong v^i\left(1 + \frac{1}{2}v^2 + (1+\gamma)U\right) \tag{3.4.32}$$
$$\Lambda^{[i]}_{(k)} = \Lambda^{[k]}_{(i)} \cong \delta_{ik} + \frac{1}{2}v^{[i]}v^{[k]} \cong \delta_{ik} + \frac{1}{2}v^i v^k,$$

and $\lambda^{\alpha}_{(\beta)} = \theta^{\alpha}_{[\rho]}\Lambda^{[\rho]}_{(\beta)}$ is given by

$$\begin{aligned}
\lambda^{0}_{(0)} &= 1 + \frac{1}{2}v^2 + U + O(U^2) = u^0 \\
\lambda^{0}_{(i)} &= v^i\left[1 + \frac{1}{2}v^2 + (2+\gamma)U\right] + \mu g_{0i} \\
\lambda^{i}_{(0)} &= v^i\left[1 + \frac{1}{2}v^2 + U\right] = u^i \\
\lambda^{i}_{(j)} &= (1 - \gamma U)\delta_{ij} + \frac{1}{2}v^i v^j
\end{aligned} \tag{3.4.33}$$

where v^i is the velocity of the observer moving along the world-line $x^{\alpha}(s)$, as measured in the post-Newtonian coordinate system of expression (3.4.26).

The spatial axes $\boldsymbol{\lambda}_{(i)}$ of this local frame may be thought of as physically realized by three orthonormal telescopes, always pointing toward the same distant stars fixed with respect to the asymptotic inertial frame where $g_{\alpha\beta} \longrightarrow \eta_{\alpha\beta}$ (however, see the end of this section and chap. 4); the timelike axis of this local tetrad is the four-velocity of the observer moving along $x^{\alpha}(s)$: $\lambda^{\alpha}_{(0)} = u^{\alpha}$.

The vector S^{α} may be thought of as physically realized by the spacelike *angular momentum vector of a spinning particle or "test gyroscope."* Since S^{α} is a spacelike vector, $S^{\alpha}u_{\alpha} = 0$, in the local frame $\boldsymbol{\lambda}_{(\alpha)}$ we get $S^{(0)} = 0$, and assuming negligible nongravitational torques on the test gyroscope, from the conservation of angular momentum, we have $S_{\alpha}S^{\alpha} = S_{(i)}S^{(i)} = \text{constant}$. We then have

$$\frac{DS_{(i)}}{ds} \equiv \frac{D(S_{\alpha}\lambda^{\alpha}_{(i)})}{ds} = \frac{DS_{\alpha}}{ds}\lambda^{\alpha}_{(i)} + S_{\alpha}\frac{D\lambda^{\alpha}_{(i)}}{ds} = S_{\alpha}\frac{D\lambda^{\alpha}_{(i)}}{ds} \tag{3.4.34}$$

where from Fermi-Walker transport, $\frac{DS_{\alpha}}{ds} = u_{\alpha}\left(\frac{Du^{\sigma}}{ds}S_{\sigma}\right)$, and therefore, from $g_{\alpha\beta}\lambda^{\alpha}_{(0)}\lambda^{\beta}_{(i)} = 0$, $\frac{DS_{\alpha}}{ds}\lambda^{\alpha}_{(i)} = 0$. From the transformations (3.4.33), the components of S^{α} as functions of $S^{(\alpha)}$ are:

$$S^0 = v^i S_{(i)}, \quad S^i = (1 - \gamma U)S_{(i)} + \frac{1}{2}v^i v^j S_{(j)}. \tag{3.4.35}$$

From $\frac{dv_i}{ds} = a_i + U_{,i}$, from the metric (3.4.26), and from (3.4.34), we then have *at the post-Newtonian order*

$$\frac{d\left(S_\alpha \lambda^\alpha_{(i)}\right)}{ds} = \frac{D\left(S_\alpha \lambda^\alpha_{(i)}\right)}{ds} \cong S_0\left(\frac{d\lambda^0_{(i)}}{ds} + \Gamma^0_{k0}\lambda^k_{(i)}\right)$$
$$+ S_j\left(\frac{d\lambda^j_{(i)}}{ds} + \Gamma^j_{k0}\lambda^k_{(i)} + \Gamma^j_{00}\lambda^0_{(i)} + \Gamma^j_{lk}\lambda^l_{(i)}v^k\right) \cong \tag{3.4.36}$$
$$S_{(l)}\left[\frac{1}{2}\left(v_i a_l - v_l a_i\right) - \left(\frac{1}{2} + \gamma\right)\left(v_i U_{,l} - v_l U_{,i}\right)\right.$$
$$\left. + \frac{1}{2}\mu\left(g_{0l,i} - g_{0i,l}\right)\right]$$

or

$$\frac{dS_{(i)}}{ds} \equiv \epsilon_{(i)(j)(k)}\dot{\Omega}_{(j)}S_{(k)}$$
$$\dot{\Omega}_{(j)} = \epsilon_{(j)(l)(m)}\left(-\frac{1}{2}v_l a_m + \left(\frac{1}{2} + \gamma\right)v_l U_{,m} - \frac{1}{2}\mu g_{0m,l}\right) \tag{3.4.37}$$

where $\epsilon_{(i)(j)(k)}$ is the Levi-Civita pseudotensor, and in standard vector notation:

$$\frac{d\boldsymbol{S}}{ds} \equiv \dot{\boldsymbol{\Omega}} \times \boldsymbol{S}$$
$$\dot{\boldsymbol{\Omega}} = -\frac{1}{2}\boldsymbol{v} \times \boldsymbol{a} + \left(\frac{1}{2} + \gamma\right)\boldsymbol{v} \times \boldsymbol{\nabla}U - \frac{1}{2}\mu\boldsymbol{\nabla} \times \boldsymbol{h} \tag{3.4.38}$$

that one may rewrite

$$\dot{\boldsymbol{\Omega}} = -\frac{1}{2}\boldsymbol{v} \times \frac{d\boldsymbol{v}}{dt} + (1 + \gamma)\boldsymbol{v} \times \boldsymbol{\nabla}U - \frac{1}{2}\mu\boldsymbol{\nabla} \times \boldsymbol{h} \tag{3.4.38$'$}$$

where:

$$\frac{d\boldsymbol{v}}{dt} = \boldsymbol{a} + \boldsymbol{\nabla}U$$

and where $\boldsymbol{h} \equiv \left(g_{01}, g_{02}, g_{03}\right)$ and $\boldsymbol{S} \equiv \left(S_\alpha \lambda^\alpha_{(i)}\right)$.

The first term of equation (3.4.38) is the **Thomas precession**, due to the noncommutativity of the nonaligned Lorentz transformations and to the non-gravitational acceleration $\boldsymbol{a}$. The second term is the **de Sitter or geodetic precession**,[134,135] which may be interpreted (eq. 3.4.38′) as due to a part (contributing with the factor $-\frac{1}{2}$) analogous to the Thomas precession, arising from the noncommutativity of the nonaligned Lorentz transformations of special relativity and from the gravitational acceleration $\boldsymbol{\nabla}U$, that is, due to Fermi-Walker transport and to the *gravitational acceleration* (that might be derived even in the flat spacetime of special relativity), plus a part (contributing with the factor $(1 + \gamma)$) due to Fermi-Walker transport and to the *spacetime curvature* of general relativity, that is, to the medium strong equivalence principle and to the space curvature (valid even for zero total acceleration $\frac{dv}{dt}$). This effect was discovered in 1916 by de Sitter[134] (see also Fokker 1920).[135] The third term is the **Lense-Thirring[164,165] precession** of an orbiting gyroscope, due to the gravitomagnetic potential $\boldsymbol{h}$ generated, for example, by the angular momentum of a central body (see chap. 6).

The **de Sitter effect** for the Earth-Moon system was **experimentally tested** by Bertotti, Ciufolini, and Bender in 1987,[166] with an accuracy of the order of 10%; by Shapiro, Reasenberg, Chandler, and Babcock in 1988,[167] with an accuracy of about 2%; by Dickey et al. in 1989,[23] with an accuracy of about 2%; and by Müller et al. in 1991,[257] and Dickey et al. in 1994,[279] with an accuracy of about 1%. The methods used in the paper by Bertotti, Ciufolini, and Bender (1987)[166] are two: analysis of the Lunar Laser Ranging data with respect to the very well defined planetary frame (through spacecraft measurements), and comparison of the Earth rotational orientation as measured with two methods, locally, with *Lunar Laser Ranging* (LLR),[113–116] relative to the Moon perigee, and, nonlocally, with *Very Long Baseline Interferometry* (VLBI;[144,145] see § 3.4.1) relative to the distant quasars; the accuracy of this second approach is of the order of 10% of the de Sitter effect. The method used by the other groups is a detailed analysis of the LLR data in the planetary frame.

In this section we discuss the method of comparing the rotational orientation of Earth as measured with LLR and with VLBI. This comparison has some theoretical implications, regarding the "nonrotation" of the universe.

Let us first define two frames comoving with Earth (see fig. 3.11). The first is a *local Fermi frame*, with spatial axes Fermi-Walker transported[161] along the Earth world line and realizable by three gyroscopes comoving with Earth. The second is a "*nonlocal*" *frame* (or "*astronomical*" *frame*), with spatial axes fixed relative to an asymptotic quasi-inertial frame, where $g_{\alpha\beta} \rightarrow\approx \eta_{\alpha\beta}$. It might be realized physically by three telescopes comoving with Earth and always pointing toward the same distant stars (however, see below). This is the local orthonormal frame $\boldsymbol{\lambda}_{(\mu)}$, which we have just defined with formula (3.4.33). According to the de Sitter effect, the gyroscopes precess relative to the asymptotic inertial frames and therefore relative to the "nonlocal" frame, and to the telescopes, with angular velocity $\dot{\boldsymbol{\Omega}}^{\text{de Sitter}}$. Two measurements of an angular velocity vector $\dot{\boldsymbol{V}}$ in the two frames, local and nonlocal, are then related by

$$\dot{\boldsymbol{V}}_{\text{nonlocal}} - \dot{\boldsymbol{V}}_{\text{local}} = \dot{\boldsymbol{\Omega}}^{\text{de Sitter}}. \qquad (3.4.39)$$

A physical realization of this angular velocity vector is, for example, the angular velocity vector $\dot{\boldsymbol{\Omega}}_{\oplus}$ of Earth (fig. 3.11). In general relativity, measurements of $\dot{\boldsymbol{\Omega}}_{\oplus}$ in the two frames, (a) local and (b) nonlocal, are related by equation

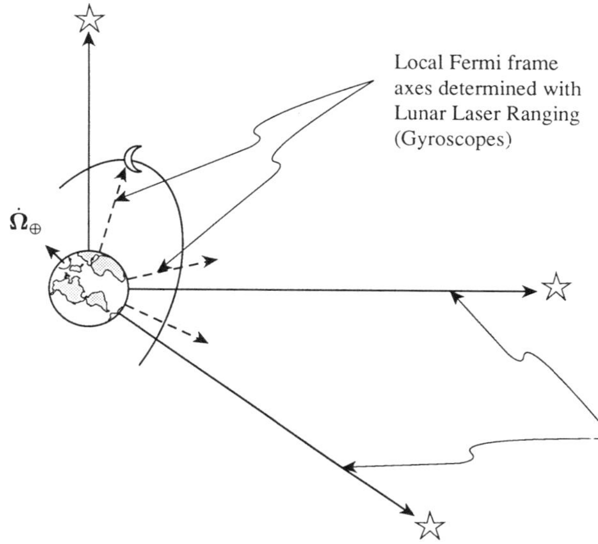

FIGURE 3.11. The angular velocity vector, $\dot{\Omega}_\oplus$, of Earth evaluated in (a) the local Fermi frame determined by observing Moon with Lunar Laser Ranging, and in (b) the asymptotic frame determined by observing quasars with Very Long Baseline Interferometry.

(3.4.39). Measurement (b) of $\dot{\mathbf{\Omega}}_\oplus$ can be realized practically by measuring the Earth rotational parameters with VLBI, measuring the arrival time at different locations on Earth of radio signals from quasars, that is, by determining the Earth orientation relative to the distant stars (nonlocal measurements of $\dot{\mathbf{\Omega}}_\oplus$), and measurement (a) with LLR, relative to the "nearby" Moon and, in particular, relative to the well determined Moon perigee (local measurements of $\dot{\mathbf{\Omega}}_\oplus$). Thus the line from Earth to Moon perigee, after removal by theory of known perigee motions and perturbations, may be thought of as one of the axes of a Fermi frame comoving with Earth. From formula (3.4.39), we therefore have for the difference of the values of $\dot{\mathbf{\Omega}}_\oplus$ measured in the two frames

$$\dot{\mathbf{\Omega}}_\oplus^{\text{VLBI}} - \dot{\mathbf{\Omega}}_\oplus^{\text{LLR}} = \dot{\mathbf{\Omega}}^{\text{de Sitter}}. \qquad (3.4.40)$$

At this point it is however necessary to observe that, in practice, with LLR, the Earth rotational orientation is determined relative to the Moon perigee after theoretical removal of known Moon perigee perturbations, with respect to an asymptotic inertial frame, including general relativistic perturbations and in particular including the *theoretical* general relativistic *value* of $\dot{\mathbf{\Omega}}^{\text{de Sitter}}$. Therefore the value of the Earth angular velocity practically given with LLR is $\dot{\mathbf{\Omega}}_\oplus^{\text{LLR}*} \equiv \dot{\mathbf{\Omega}}_\oplus^{\text{LLR}} + \dot{\mathbf{\Omega}}^{\text{de Sitter}}$, where the first contribution $\dot{\mathbf{\Omega}}_\oplus^{\text{LLR}}$ is measured with

LLR and the second contribution $\dot{\Omega}^{\text{de Sitter}}$, is, a priori, theoretically assumed to be valid (plus other known theoretical corrections). On the contrary, with VLBI, the Earth angular velocity is simply its *measured value* relative to the distant stars. Therefore, from formula (3.4.40), according to general relativity, we should have

$$\dot{\Omega}_{\oplus}^{\text{VLBI}} - \dot{\Omega}_{\oplus}^{\text{LLR}*} = \mathbf{0}. \tag{3.4.41}$$

The uncertainty in the value of UT1 (the rotational orientation of Earth) obtained with LLR is due to measurement uncertainties and to errors in the calculated, classical value of the Moon perigee rate, and is of the order of 1 milliarcsec/yr. The uncertainty in UT1 obtained with VLBI is less than 1 milliarcsec/yr. A comparison of the UT1 rates obtained with the two techniques, reveals agreement to better than 1.5 milliarcsec/yr: $\text{UT1}^{\text{LLR}*} - \text{UT1}^{\text{VLBI}} \cong 0 \pm 1.5$ milliarcsec/yr (see eq. 3.4.41). Combining these three figures, we conclude that a precession of the *local, "Earth-Moon Fermi frame"* relative to the distant stars takes place,[166] about the normal to the ecliptic, at $\cong$ 19.2 milliarcsec/yr with a $\sim$ 10% uncertainty, in agreement with the de Sitter effect predicted by general relativity.

What does it imply to have this test of the de Sitter effect? One may think of the de Sitter–geodetic effect as originating from two sources. A part of the effect is similar to the Thomas precession of an electron around a nucleus, due to the noncommutativity of the nonaligned Lorentz transformations of special relativity. Around a loop, the result is a rotation of the spatial axes; in the case of a freely falling gyroscope, with gravitational acceleration ∇U and orbital velocity $\boldsymbol{v}$, it amounts to $-\frac{1}{2}(\boldsymbol{v} \times \nabla U)$. The other part of the effect is due to the *spacetime curvature* of general relativity, which contributes with the factor $(1+\gamma)$, even for zero total acceleration $\frac{d\boldsymbol{v}}{dt}$ (see expression 3.4.38′). The sum is therefore $\dot{\Omega}^{\text{de Sitter}} = \left(\frac{1}{2} + \gamma\right)\left(\boldsymbol{v} \times \nabla U\right)$. The measurement of the de Sitter effect is thus a further test of the existence of space curvature generated by a mass in agreement with the Einstein field equation.

Finally, it is interesting to observe that the test of the de Sitter effect obtained by analyzing LLR observations relative to the distant stars frame, defined by the quasars, may be interpreted as a weak test of the absence of rotation of the distant stars frame in the universe with respect to the local gyroscopes. Indeed, the general relativistic prediction (3.4.38) refers to an asymptotic inertial frame, but, experimentally, it has been measured with VLBI relative to a frame defined by the quasars. However, in agreement with a general relativistic interpretation of the Mach principle (see chap. 4), there is no measured difference between the two frames.

It is now interesting to discuss some experimental evidence in favor of symmetric "structures" on the spacetime manifold versus nonsymmetric "structures".[168–176] An example is a nonsymmetric connection. A non-Riemannian, nonsymmetric connection $\Gamma^{\alpha}_{\mu\nu}$ differs by a tensor $K_{\mu\nu}{}^{\alpha}$ from the symmetric

Christoffel symbols $\{ {}^{\alpha}_{\mu\nu} \}$ generated in the usual way by the symmetric metric $g_{\alpha\beta} = g_{(\alpha\beta)}$:

$$\Gamma^{\alpha}_{\mu\nu} = \{ {}^{\alpha}_{\mu\nu} \} - K_{\mu\nu}^{\cdot\ \cdot\ \alpha}. \tag{3.4.42}$$

The non-Riemannian part of the connection $K_{\mu\nu}^{\cdot\ \cdot\ \alpha}$, is sometimes called the contorsion tensor and depends on the symmetric metric $g_{\alpha\beta}$ and on the anti-symmetric tensor $\Gamma^{\alpha}_{[\mu\nu]} = \frac{1}{2}\left(\Gamma^{\alpha}_{\mu\nu} - \Gamma^{\alpha}_{\nu\mu}\right)$, called the Cartan *torsion* tensor (see § 2.8).

In most theories endowed with a non-Riemannian connection such as the *Einstein-Cartan theory*,[168–172] the torsion does not *directly* affect the propagation of light and test particles and does not propagate in vacuum. Therefore, the classical relativity tests, such as the bending of light, the radio time delay and the perihelion advance, cannot directly test the existence of a non-Riemannian connection. The torsion might, however, affect spinning particles and indirectly act on light and test particles through the field equations which determine the spacetime metric. Some authors[177] have assumed that the propagation of the spin S^{α} of a particle might be regulated by Fermi-Walker transport, $S^{\alpha}{}_{;\beta}u^{\beta} = u^{\alpha}(u^{\beta}{}_{;\gamma}u^{\gamma}S_{\beta})$, relative to the asymmetric connection $\Gamma^{\alpha}_{\mu\nu}$. In that case torsion might change the orientation of S^{α}; indeed the term $\Gamma^{\alpha}_{[\mu\nu]}S^{\mu}u^{\nu}$ is, in general, different from zero and might contribute to such a change of S^{α}. In particular the torsion tensor has been decomposed in three parts: a vector, a purely antisymmetric part, and a part due to a spinning fluid of "Weyssenhoff-type." It has been shown[177] that these last two only might contribute to the rate of change of the vector $\boldsymbol{S}$, according to the formula $\frac{d\boldsymbol{S}}{ds} = \boldsymbol{K} \times \boldsymbol{S}$, where $\boldsymbol{K}$ is a tensor depending on the torsion. Therefore, in some hypothetical theories with torsion propagating in vacuum,[178] the torsion of spacetime might in principle be measured by observing the rate of change of $\boldsymbol{S}$ or the precession of spinning particles. However, as we have mentioned, torsion can also *indirectly* act on light and spinless test particles through the field equations that determine the metric $g_{(\alpha\beta)}$.

An example of a *nonsymmetric* theory of gravity is the *Moffat theory*.[179–181] In this theory spacetime is described by a nonsymmetric Hermitian tensor, $g_{\mu\nu} = g_{(\mu\nu)} + g_{[\mu\nu]}$, and parallel displacement of a vector and the curvature tensor $R^{\alpha}{}_{\beta\mu\nu}$ (derived in the usual way by parallel transport of a vector around a loop) are defined by a nonsymmetric Hermitian affine connection $\Gamma^{\mu}_{\alpha\beta} = \Gamma^{\mu}_{(\alpha\beta)} + \Gamma^{\mu}_{[\alpha\beta]}$. The field equations contain this nonsymmetric connection. The solution to the field equations of the 1979 version of the Moffat theory, describing the metric of a spherically symmetric spacetime generated by a mass M (corresponding to the Schwarzschild solution of Einstein theory), is[179]

$$ds^2 = -(1 - \frac{l^4}{r^4})(1 - \frac{2M}{r})dt^2 + (1 - \frac{2M}{r})^{-1}dr^2 + r^2 d\Omega^2$$

and

$$g_{[10]} = -i\,\frac{l^2}{r^2} \qquad (3.4.43)$$

where $i \equiv \sqrt{-1}$ and l^2 is a parameter proportional to some measure of the mass of the central body or of its baryon number times some coupling constant; l has dimensions of length. The motion of test particles takes place in the geometry (3.4.43) and is therefore influenced by the additional parameter l^2; some consequent, strong experimental *limits on* the 1979 version of the *Moffat theory* are described in section 3.5.1.

We conclude this section on the geometrical structure of general relativity by briefly recalling the experimental support of the other basic geometrical assumption of Einstein general relativity: the spacetime pseudo-Riemannian manifold is a Lorentzian manifold. In other words, spacetime is a four-dimensional pseudo-Riemannian manifold and the signature of the metric $\boldsymbol{g}$ is $+2$ (or -2, depending on the definition); that is, the metric $\boldsymbol{g}$ is at an event, in suitable coordinates, Minkowskian, with signature $(-1, +1, +1, +1)$. The experimental tests[2] of the equivalence principle (in local freely falling frames the physics of special relativity is valid), reviewed in section 3.2, and the local experimental tests of special relativity test the local Minkowskian character of the spacetime manifold.

3.4.4. Other Tests of Space Curvature

It is necessary to stress that measurements of the *deflection and time delay of electromagnetic waves* and the test of the *de Sitter effect* are not the only confirmations of the existence of *space curvature* generated by a mass according to the Einstein field equation. Other tests are the *perihelion advance of Mercury* (§ 3.5.1), *Lunar Laser Ranging* (§ 3.2.5), the *pericenter advance* of other orbiting bodies, such as *LAGEOS* (§ 3.5.1), and the observations of some *astrophysical binary systems*, such as the binary pulsar *PSR 1913+16* with its relativistic parameters: time dilation or gravitational redshift, periastron advance, time delay in propagation of pulses, and rate of change of orbital period (see § 3.5.1). Some of these effects are only in part due to the space curvature, measured at the post-Newtonian order by the parameter γ: for example, $\frac{2}{3}$ of the relativistic pericenter advance is due to the space curvature (see § 3.5.1). Pericenter advance, Moon motion, etc., arise from a combination of phenomena: from the gravitational time dilation of clocks, from space curvature, and from the "nonlinearity" of the gravitational interaction in the PPN gauge (see next section). Therefore it may be difficult to interpret the observable effects as a verification of a single aspect of general relativity, such as space curvature (however, see § 3.5). (It is interesting to observe[182] that the perihelion advance of Mercury might be explained, from a formal point of view, by a deviation from the weak field limit, inverse square law, that is, $\frac{1}{r^{2.00000016}}$ instead of $\frac{1}{r^2}$. However, this deviation is not in agreement with the other relativistic measurements.) Furthermore, present-day determination of most of these other relativistic effects does not reach the level of accuracy of the gravitational time dilation of clocks and of the deflection and time delay of electromagnetic waves.

3.5 OTHER TESTS OF EINSTEIN GENERAL RELATIVITY, OF THE EQUATIONS OF MOTION, AND OF THE FIELD EQUATIONS

We have seen that in Einstein general relativity spacetime is a Lorentzian manifold, whose curvature is determined via the field equation by the distribution and motion of mass-energy in the universe. For this reason, we call Einstein general relativity geometrodynamics.

So far, we have discussed the experimental support of two basic assumptions of general relativity: the *equivalence principle* and the curved pseudo-Riemannian, *Lorentzian, structure of the spacetime* manifold. These two assumptions (the equivalence principle in the medium strong form and the curved Lorentzian spacetime), together with the equations of motion, $T^{\alpha\beta}{}_{;\beta} = 0$, are satisfied, however, by definition, by any *metric theory of gravity*, such as the theories of Jordan-Brans-Dicke,[87–93] Ni,[183] Rosen,[184,185] etc. The structure of the spacetime geometry and the way "spacetime geometry tells mass-energy how to move" are the same in every metric theory. However, what distinguishes general relativity from other metric theories is the way "mass-energy tells spacetime geometry how to curve," that is, the **field equations**.

Apart from some attempts to modify the spacetime geometrical structure, admitting, for example, the existence of asymmetric connections and torsion (see § 3.4.3), most attempts toward alternative classical theories of gravity have kept the same geometrical structure and the same equations of motion (for the energy-momentum tensor) as general relativity, and have instead modified the field equation. In other metric theories of gravity, the field equations also contain additional scalar, vector, and tensor fields.

However, the triumph of general relativity has been not only a triumph of the idea of the dynamical Riemannian structure of spacetime, but also a triumph of the Einstein field equation[186,187] (see § 2.3), derived in 1915 by Hilbert and Einstein in two different ways (Hilbert[188] used a variational principle and Einstein[186,187] the requirement that the conservation laws for momentum and energy for both gravitational field and mass-energy be satisfied as a direct consequence of the field equations). *Einstein geometrodynamics*, differently from most other gravity theories, has the important and beautiful property that *the equations of motion are a direct mathematical consequence of the Bianchi identities:*[189–190] $G^{\alpha\beta}{}_{;\beta} = 0$ *and of the field equation*, $G^{\alpha\beta} = \chi T^{\alpha\beta}$, *only* (see § 2.4).

Regarding the experimental verifications of the Einstein field equation, we observe that the gravitational time dilation of clocks, the deflection of electromagnetic waves, the time delay of radio waves, and the de Sitter effect test not only the equivalence principle and the existence of space curvature, but also the *amount* of spacetime curvature generated by a mass M, in agreement with the Einstein field equation.

One of the important features of the general relativistic *field equation*, different from the classical Poisson equation, is its *nonlinearity. In the standard post-Newtonian gauge*, the nonlinear contribution of M to the metric tensor $g_{\alpha\beta}$ may be seen, at the post-Newtonian order, in the approximate metric element (3.4.8) where the g_{00} component of the metric tensor is a function of $\left(\frac{M}{r}\right)^2$. The amount of nonlinearity in the superposition law for gravity is measured, at the post-Newtonian order and *in the standard post-Newtonian gauge*, by the parameter β, however this interpretation of the meaning of β is *gauge dependent.*[2] For example, in the coordinate system of expression (3.4.8) the nonlinear term is absent in general relativity. So far, we have not discussed any experimental verification of this phenomenon. However, in the next section we review a famous classical test of general relativity: the precession of the perihelion of Mercury and, in general, the precession of the pericenter of an orbiting test particle and its relativistic dynamics. At the end of the next section we also briefly describe the 1974 binary pulsar PSR 1913+16. We observe that the post-Newtonian corrections to the motion of a test particle are determined by a combination of various effects. For example, the weak field pericenter advance (eq. 3.5.17 below) might be interpreted as a combination of gravitational time dilation, space curvature and nonlinear contributions of a mass to the solution $g_{\alpha\beta}$ of the field equation (in the standard PPN gauge). Now gravitational time dilation of clocks has been measured with an accuracy of about 2×10^{-4} via the Vessot-Levine experiment,[12,13] and the space curvature (γ) with an accuracy of about 2×10^{-3} via radio time delay[18] and via the deflection of electromagnetic waves.[19] Therefore, if we assume the other PPN parameters equal to zero, we may regard the determination of the perihelion advance of Mercury as a *measurement of* β. In other words, since the other two effects are much better determined with other measurements, the perihelion advance may also be regarded as a further test of the field equations.

In particular, *neglecting any parameter of preferred-frame, preferred-location, and nonconservative theories*, and considering the existing limits on γ, from tests of the very strong equivalence principle with Lunar Laser Ranging[257,279] and the corresponding limit (3.2.30), one finds a value of β equal to 1 with an accuracy of less than 10^{-3}.

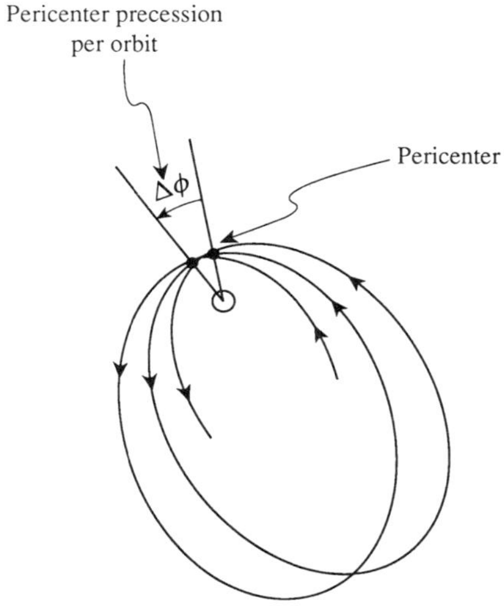

FIGURE 3.12. The pericenter advance of an orbiting test particle.

3.5.1. Pericenter Advance

Let us first derive the expression of the *pericenter advance of a test particle* orbiting a central mass M (see fig. 3.12).

From the spherically symmetric, time-independent, metric (3.4.8) and from the geodesic equation, (3.4.2), $\frac{Du^\alpha}{ds} = 0$, parametrized with the arc-length s, we get the constants of motion (3.4.3):

$$\theta = \frac{\pi}{2}; \; r^2 \frac{d\phi}{ds} \equiv L \quad \text{and} \quad \frac{dt}{ds} \equiv K e^{-m(r)}, \tag{3.5.1}$$

and from the condition that the geodesic be parametrized with the arc-length s, or proper time τ, that is, that the tangent vector to the geodesic, $u^\alpha \equiv \frac{dx^\alpha}{ds}$, have norm equal to -1, $u^\alpha(s)u_\alpha(s) = -1$, we have

$$g_{\alpha\beta} \frac{dx^\alpha}{ds} \frac{dx^\beta}{ds} = -1. \tag{3.5.2}$$

We then have

$$-e^{m(r)}\left(\frac{dt}{ds}\right)^2 + e^{n(r)}\left(\frac{dr}{ds}\right)^2 + r^2\left(\frac{d\theta}{ds}\right)^2 + r^2 \sin^2\theta\left(\frac{d\phi}{ds}\right)^2 = -1, \tag{3.5.3}$$

and from equations (3.5.1), we get

$$e^{-m(r)}K^2 - e^{n(r)}\frac{L^2}{r^4}\left(\frac{dr}{d\phi}\right)^2 - \frac{L^2}{r^2} = 1. \tag{3.5.4}$$

In terms of the spherically symmetric, approximate expression (3.4.8) for the general metric generated by a central mass M,

$$e^m \cong \left(1 - \frac{2M}{r} + \frac{2(\beta - \gamma)M^2}{r^2}\right) \quad \text{and} \quad e^n \cong \left(1 + \frac{2\gamma M}{r}\right). \tag{3.5.5}$$

Neglecting terms of order higher than $(\frac{M}{r})^2$, defining $y \equiv \frac{1}{r}$, observing that at the lowest order $K^2 \cong 1 + O\left(\frac{M}{r}\right)$ and $L^2 \cong r^2 v_\phi^2 \left(1 + O\left(\frac{M}{r}\right)\right) \sim Mr\left(1 + O\left(\frac{M}{r}\right)\right)$, and dividing expression (3.5.4) by L^2, we get

$$\left(\frac{dy}{d\phi}\right)^2 + y^2 = \frac{(K^2 - 1)}{L^2} + \frac{2MK^2}{L^2} y + \frac{2\gamma M(1 - K^2)}{L^2} y - 2(\beta + \gamma - 2)\frac{M^2K^2}{L^2} y^2 + 2\gamma M y^3. \tag{3.5.6}$$

On differentiating this equation with respect to ϕ and by considering $\frac{dy}{d\phi} \neq 0$, we obtain

$$\frac{d^2y}{d\phi^2} + y = \frac{MK^2}{L^2} + \gamma\frac{M}{L^2}(1 - K^2) - 2(\beta + \gamma - 2)\frac{M^2K^2}{L^2} y + 3\gamma M y^2. \tag{3.5.7}$$

In classical mechanics the differential equation for $\frac{d}{d\phi}\left(\frac{1}{r}\right)$ is

$$\frac{d^2y}{d\phi^2} + y = \frac{M}{L^2} \tag{3.5.8}$$

which has the classical general solution

$$y_{(0)} = \frac{M}{L^2}(1 + e\cos(\phi - \overline{\phi})) \tag{3.5.9}$$

where e is the eccentricity ($e > 1$ corresponds to a hyperbola, $e = 1$ to a parabola, $e < 1$ to an ellipse, and $e = 0$ to a circle) and we set the initial value $\overline{\phi} = 0$. The post-Newtonian relativistic equation (3.5.7) for $\frac{d^2y}{d\phi^2}$ agrees with the classical equation (3.5.8) plus smaller post-Newtonian terms. Therefore, one may write the solution to the relativistic equation (3.5.7)

$$y = y_{(0)} + y_{(p)} \tag{3.5.10}$$

where $y_{(p)}$ is a small relativistic correction to the classical solution (3.5.9). By inserting this expression for y into equation (3.5.7), and neglecting higher order terms, we get

$$\frac{d^2 y_{(p)}}{d\phi^2} + y_{(p)} = (\gamma - 1)(1 - K^2)\frac{M}{L^2} + (4 + \gamma - 2\beta)\frac{M^3}{L^4} + 2(2 + 2\gamma - \beta)\frac{M^3}{L^4} e \cos\phi + 3\gamma \frac{M^3}{L^4} e^2 \cos^2\phi. \tag{3.5.11}$$

As one can easily verify, a particular solution of this equation is

$$y_{(p)} = \frac{M}{L^2}\left[(1 - K^2)(\gamma - 1) + (4 + \gamma - 2\beta)\frac{M^2}{L^2} + (2 + 2\gamma - \beta)\frac{M^2}{L^2} e\phi \sin\phi + 3\gamma \frac{M^2}{L^2} e^2 \left(\frac{1}{2} - \frac{1}{6}\cos 2\phi\right)\right]. \tag{3.5.12}$$

In this expression we have a constant part, corresponding to a small constant shift of the radial coordinate r, a small periodical term $\sim \cos 2\phi$, and a term increasing as $\phi \sin\phi$. This last named term is the only one observable in ordinary weak field systems. Therefore, neglecting the other terms, we may rewrite the solution of equation (3.5.7) in the form

$$y \cong \frac{M}{L^2}\left(1 + e\cos\phi + (2 + 2\gamma - \beta)\frac{M^2}{L^2} e\phi \sin\phi\right). \tag{3.5.13}$$

Since $\frac{M^2}{L^2} \sim \frac{M}{r}$, we have

$$y \equiv \frac{1}{r} \cong \frac{M}{L^2}\left\{1 + e\cos\left[\left(1 - (2 + 2\gamma - \beta)\frac{M^2}{L^2}\right)\phi\right]\right\}. \tag{3.5.14}$$

The point of minimum r, that is, the pericenter, corresponds to $\cos\left[\left(1 - (2 + 2\gamma - \beta)\frac{M^2}{L^2}\right)\phi\right] = 1$, and therefore to:

$$\phi = \frac{2n\pi}{\left(1 - (2 + 2\gamma - \beta)\frac{M^2}{L^2}\right)} \cong 2n\pi + 2(2 + 2\gamma - \beta)n\pi\frac{M^2}{L^2}. \tag{3.5.15}$$

Consequently, each orbit undergoes a pericenter advance,

$$\Delta\phi = 2(2 + 2\gamma - \beta)\pi\frac{M^2}{L^2}. \tag{3.5.16}$$

Moreover, from classical mechanics, at the lowest order, we have $\frac{M}{L^2} = \frac{1}{a(1-e^2)}$, where a is the semimajor axis. Therefore, we finally get:

$$\Delta\phi = \frac{6\pi M}{a(1 - e^2)} \times \frac{(2 + 2\gamma - \beta)}{3}. \tag{3.5.17}$$

This is the expression of the *pericenter advance* of a test particle orbiting a central mass, valid at the post-Newtonian order. If we substitute $\gamma \equiv \beta \equiv 1$, corresponding to general relativity, we get

$$\Delta\phi = \frac{6\pi M}{a(1-e^2)}. \tag{3.5.18}$$

Expression (3.5.17) is useful to test different metric theories of gravity with different field equations. It shows the three weak field relativistic contributions to the pericenter advance: gravitational time dilation anomaly contributing with the factor 2, space curvature anomaly contributing with the factor 2γ, and $-\beta$ that *in the standard PPN gauge* might be interpreted as measuring nonlinearity of gravitational interaction (see § 3.5).

Using the general relativistic values $\beta = \gamma = 1$, the mass of the Sun $M_\odot = 1.989 \times 10^{33}$g $= (1.47664 \pm 0.00002) \times 10^5$cm, the semimajor axis $a = 0.3871$AU(1AU $= 1.495985 \times 10^8$km), and the eccentricity of the Mercury orbit[191] $e_☿ = 0.205615$, we find the **perihelion advance of Mercury**:

$$\dot\omega_☿ = 0''\!.10352/_{\text{revolution}}, \tag{3.5.19}$$

or, since the Mercury orbital period is $P_☿ = 0.24085$ year:

$$\dot\omega_☿ = 42''\!.98/_{\text{century}} \tag{3.5.20}$$

this advance of the Mercury perihelion has been *experimentally confirmed* with analyses of optical data of the last three centuries and, since 1966, with radar observations. After removal of the various classical planetary perturbations the results are, per century, $\dot\omega_☿ = 43''\!.20 \pm 0''\!.86$ by Shapiro et al. (1972);[192] $\dot\omega_☿ = 43''\!.11 \pm 0''\!.22$ by Shapiro et al. (1976)[21] and by Anderson et al. (1978);[102] $\dot\omega_☿ = (1.0034 \pm 0.0033) \times 42''\!.98$ by Anderson et al. (1991);[22] and $\dot\omega_☿ = (1.000 \pm 0.002) \times 42''\!.98$ by Shapiro et al. (1990).[106]

Unfortunately, the interpretation of the perihelion advance of Mercury as a test of general relativity is complicated by the uncertainty in the quadrupole moment of the Sun which also contributes to the perihelion advance. Several other factors may also affect the amount of perihelion advance. The *relativistic pericenter advance* (3.5.17), plus the **quadrupole, J_2, contribution of the central body**, plus the *gravitomagnetic contribution*[193] due to the angular momentum J of the central body (see chap. 6), plus, as an example of an alternative theory of gravity, the *Moffat theory contribution*[180] (see § 3.4.3), assuming the PPN parameters other than β and γ equal to zero, in the slow motion, weak field approximation, is

$$\Delta\omega = \frac{6\pi M}{a(1-e^2)}\left[\frac{1}{3}(2+2\gamma-\beta) + J_2\frac{R^2}{2Ma(1-e^2)} - \frac{l^4(1+\frac{1}{4}e^2)}{M^2a^2(1-e^2)^2} - \frac{4}{3}\frac{J/M}{(Ma(1-e^2))^{1/2}}\right] \tag{3.5.21}$$

where M, J_2, R, and J are mass, quadrupole coefficient, radius, and angular momentum of the central body, and l is the Moffat theory parameter proportional to some measure of the square root of the mass of the central body or of the square root of its baryon number. The orbit lies in the equatorial plane $\theta = \frac{\pi}{2}$.

From the last term of expression (3.5.21), we expect that the gravitomagnetic effect for Mercury is less than 10^{-3} times smaller than the standard general relativistic perihelion advance (de Sitter 1916).[193] Furthermore, from the first two terms of expression (3.5.21) and from the 1976 experimental uncertainties in $\dot{\omega}$, one can easily calculate that, in the case of Mercury orbiting Sun, any value of $J_{2\odot}$ larger than about 3×10^{-6} would disagree with the general relativistic prediction of $42''.98$ per century. If one assumes that Sun rotates with uniform angular velocity that produces the solar oblateness, by measuring Sun's angular velocity one can infer the small value $J_{2\odot} \sim 10^{-7}$, leaving Mercury's precession in complete agreement with Einstein geometrodynamics. However, in 1966, Dicke and Goldenberg[194] evaluated $J_{2\odot}$ from *measurements* of the apparent optical oblateness of the Sun. Using their measured values for the polar and the equatorial radii, they inferred a value of $J_{2\odot} \cong 2.5 \pm 0.2 \times 10^{-5}$. This value of the **Sun quadrupole coefficient** makes the observations of $\dot{\omega}$ for Mercury disagree with the predictions of general relativity. In 1974, Hill et al.,[195] by measuring Sun's optical oblateness, obtained $J_{2\odot} \cong 0.1 \pm 0.4 \times 10^{-5}$. Other analyses[196] of apparent oscillations of the Sun were interpreted as giving information on its internal rotation, and led to the following values: $J_{2\odot} \cong (5.5. \pm 1.3) \times 10^{-6}$ by Hill et al. (1982);[197] $J_{2\odot} \cong (3.6$ to $1.2) \times 10^{-6}$ by Gough (1982);[198] and $J_{2\odot} \cong (5.0$ to $1.6) \times 10^{-6}$ by Campbell et al. (1983).[199] Dicke et al. (1987),[200] with another measurement of the optical oblateness, got the value $J_{2\odot} \cong 4.7 \times 10^{-6}$. However, it is important to stress that according to more recent determinations of $J_{2\odot}$, with Doppler analyses of the oscillations of the Sun to get information about its differential rotation, the following values have been obtained: $J_{2\odot} \cong (1.7. \pm 0.4) \times 10^{-7}$, by Duvall et al. (1984);[201] $J_{2\odot} \cong 1.7. \times 10^{-7}$ by Christensen et al. (1985);[202] and finally $J_{2\odot} \cong 1.7. \times 10^{-7}$ by Brown et al. (1989).[203] According to these values of $J_{2\odot}$ of about 1.7×10^{-7} the observed perihelion advance of Mercury is in agreement with the general relativistic predictions. Nevertheless, a definitive value of $J_{2\odot}$ has still to be determined. For such a purpose a mission, called Solar Probe (or STARPROBE, or VULCAN)[204–206] was proposed to measure $J_{2\odot}$ with a spacecraft very close to the Sun. The accuracy in measuring $J_{2\odot}$ with this method should be of the order of 10^{-8}. A mission with a lander on Icarus[207] (one of the Earth-crossing asteroids with a perihelion distance of 0.187 AU), and a mission with a small Mercury orbiter[111,112] have been also proposed for relativity tests and $J_{2\odot}$ measurements, with estimated accuracies in measuring $J_{2\odot}$ respectively of the order of 10^{-7} and of the order of 10^{-8} to 10^{-9}. Laboratory observations have also been proposed[208] to measure the frequency of modes of solar oscillations with $l = 1$, to get information on the rotation of the very core of the Sun from the $l = 1$ rotational frequency splitting.

The disagreement between the value of $J_{2\odot}$ obtained from Sun's optical oblateness and the observations of Mercury's perihelion advance might conceivably be explained in some alternative theories of gravity, such as the Jordan-Brans-Dicke theory or the 1983 version of the nonsymmetric Moffat theory. However, in order to explain this apparent disagreement by using the Jordan-Brans-Dicke theory, one should assume a value of the Brans-Dicke parameter ω in strong disagreement with the determinations of ω from Lunar Laser Ranging, deflection of electromagnetic waves, and radio time delay, that give $\omega \gtrsim 620$ (§ 3.2.5).[2,5] Furthermore, the pericenter advance predictions of the 1983 Moffat theory disagree with the LAGEOS and with the Moon perigee observations. Preliminary analyses of the **LAGEOS laser ranging** data show in fact a good agreement with the general relativistic theoretical predictions of the pericenter advance of a test particle.[209] However, using the Moffat theory, in order to explain the apparent disagreement between the perihelion advance $\dot{\omega}$ and the optical determinations of $J_{2\odot}$, one has to assume a value of the Moffat parameter $l_\odot$ in disagreement with the indirect *limits* on $l_\odot$ from $l_\oplus$ for Earth obtained from Lunar Laser Ranging determinations of the Moon perigee rate, and especially from laser ranging determinations of the LAGEOS perigee rate.[209] Furthermore, the nonsymmetric Moffat theory violates the weak equivalence principle;[254] limits on the Moffat parameter for Earth $l_\oplus$ can thus be set[2] by using the null tests of the weak equivalence principle in the field of Earth.

Analyses of solar system orbits are regularly performed by JPL-Caltech and by the Harvard Smithsonian Center for Astrophysics by including general relativistic parameters. The post-Newtonian parameters γ and β are then obtained by fitting the measured orbits of planets and spacecraft, including in the analyses radio time delay and bending of electromagnetic waves. This method gives values of the post-Newtonian parameters β and γ in good agreement with general relativity.[104,210]

Finally, we mention that **binary pulsars**, in particular the binary pulsar **PSR 1913+16** discovered in 1974 by Hulse and Taylor,[160] give other remarkable tests of the general relativistic dynamics and indirect evidence for the existence of gravitational waves.[211] Other binary pulsars have been later discovered; in particular PSR 2127+11C (1990) and PSR 1534+12 (1990) should give improved general relativistic tests (for a detailed description of the binary pulsars as tests of general relativity see, for example, ref. 2 and 265). PSR 1913+16 consists of a pulsar orbiting around another compact object, a neutron star (or a white dwarf, or a black hole).[2,265] The period of the pulsar is 59 millisec and its orbital period 2.79×10^4 s. The rate of change of the orbital period is $\dot{P} \cong -2.4 \times 10^{-12}$. The parameters of the binary system are obtained from analyses of the arrival times of the signals from the pulsar. The relativistic parameters observable are the periastron advance $\dot{\omega} \cong 4.227$ degrees/year, the time dilation or gravitational redshift, the time delay of electromagnetic waves (range and shape),[265] and the rate of decrease of the orbital period, explained by the loss of energy by gravitational radiation. The binary pulsar has given the first *indirect evidence for* the existence of *gravitational waves*.[211,2,4] Indeed, the observed decrease, $\dot{P} \cong -2.4 \times 10^{-12}$, agrees with the value of $\dot{P}$ due to loss of energy via gravitational waves as predicted by the general relativistic formula for quadrupole radiation (see § 2.10). The agreement between observed and calculated general relativistic values of $\dot{P}$ is better than 1%. Other conceivable causes of orbital period decrease: tidal energy dissipation, other forms of energy loss, changes of orbital period by acceleration due to a third body, etc., have also been studied.[2] Since the general relativistic periastron advance of the binary pulsar is generated by the masses of both bodies rotating around each other, Nordtvedt has suggested that the periastron advance of PSR 1913+16 should be also considered as indirect astrophysical evidence for gravitomagnetism. The measured decrease of the pulsar orbital period puts strong constraints on several gravitational theories. For example the 1983 nonsymmetric Moffat theory[180] and the Rosen bimetric theory[185] predict dipole gravitational radiation, whereas in general relativity there is no dipole gravitational radiation. This implies a flux of energy different from the prediction of the general relativistic quadrupole formula and consequently a faster, or slower, decrease of the orbital period, in disagreement with the observations.[212]

3.6 [PROPOSED TESTS TO MEASURE GRAVITATIONAL WAVES AND GRAVITOMAGNETIC FIELD]

In this section we shall briefly describe the main experiments proposed to measure a fundamental phenomenon of geometrodynamics: **gravitational waves** (see § 2.10). For a detailed treatment of this broad topic we suggest references 213, 214, and 215.

Here we summarize five methods that have been proposed to directly detect and measure gravitational waves, we have already mentioned (§§ 3.5.1 and 2.10) the astronomical observations of loss of energy from the binary pulsar attributable to gravitational radiation: **resonant bar detectors**[216–219,261] **on Earth** (§ 3.6.1); **laser interferometers**[214,215,220–223,232,252,266–272] **on Earth** (LIGO, i.e., Laser-Interferometer Gravitational-wave Observatory) (§ 3.6.2); **space microwave interferometers**[224,225] (MIGO, Michelson millimeter wave Interferometer Gravitational wave Observatory) (§ 3.6.3); **laser interferometers in space**[227–232] (LISA, or LAGOS, Laser Gravitational-wave Observatory in Space (§ 3.6.4) and LINE–SAGITTARIUS[226,273] (§ 3.6.5)); and **Doppler tracking in space**[234–238] (§ 3.6.6).

In geometrodynamics, to prove the existence and the uniqueness of a spacetime solution of the field equation, we have to specify the proper initial conditions on a Cauchy surface (see chap. 5). Therefore, general relativity is defined not only by (1) equivalence principle, (2) Lorentzian spacetime, and (3) Einstein field equation, but also by (4) suitable initial conditions. We refer the reader to chapters 4, 5 and 6 for discussions of the relations of the general relativistic formulation of the initial-value problem with the interpretations of the origin of inertia in Einstein geometrodynamics, with the *dragging of inertial frames* and with the *gravitomagnetic field*, and in regard to the experiments proposed to test the dragging of inertial frames and the existence of the gravitomagnetic field.

3.6.1. [Resonant Detectors]

In section 2.5 we have derived the fundamental equation describing the relative acceleration of two test particles as a function of the Riemann tensor: the *geodesic deviation equation*. Furthermore, in section 2.10 we have seen that when a plane gravitational wave, treated there as a small perturbation of a Minkowskian background, is propagating through two nearby test particles, the proper distance between the two particles changes, in general, with time due to the gravitational wave. Of course, this result can also be derived from the equation of geodesic deviation.

Let us first generalize the geodesic deviation equation to the case of two particles connected by an elastic support, idealized, for example, as a *spring* (see fig. 3.13). In a local inertial frame $\overset{(i)}{x}{}^{\alpha}$, the equation of motion of a particle of mass m, subjected to a nongravitational force represented by a four-vector $\overset{(i)}{F}{}^{\alpha}$, is just $\frac{d\overset{(i)}{u}{}^{\alpha}}{ds} = \frac{\overset{(i)}{F}{}^{\alpha}}{m}$. In a general frame we then have the equation $\frac{Du^{\alpha}}{ds} = \frac{F^{\alpha}}{m}$, that is, the geodesic equation with a force term. In section 2.5 we have obtained the geodesic deviation equation by calculating the difference between the geodesic equations of two nearby test particles. In the case of external forces acting on the two particles we then get (see fig. 3.13)

$$\frac{D^2\delta x^{\alpha}}{ds^2} + R^{\alpha}{}_{\beta\mu\nu}u^{\beta}\delta x^{\mu}u^{\nu} = \frac{f^{\alpha}}{m} \tag{3.6.1}$$

where f^{α} is the covariantly defined[216] difference between the nongravitational forces acting on the two particles of mass m. In the case of the harmonic oscillator of figure 3.13, one can write f^{α} as the sum of a restoring force $-K^{\alpha}{}_{\beta}\xi^{\beta}$ and of a damping force $-D^{\alpha}{}_{\beta}\frac{D\xi^{\beta}}{ds}$, where ξ^{β} is the small displacement from the equilibrium position $\xi^{\beta}(t) = \delta x^{\beta}(t) - l^{\beta}_{(0)}$, and where the tensors $K^{\alpha}{}_{\beta}$ and $D^{\alpha}{}_{\beta}$ are typical of the spring.[213]

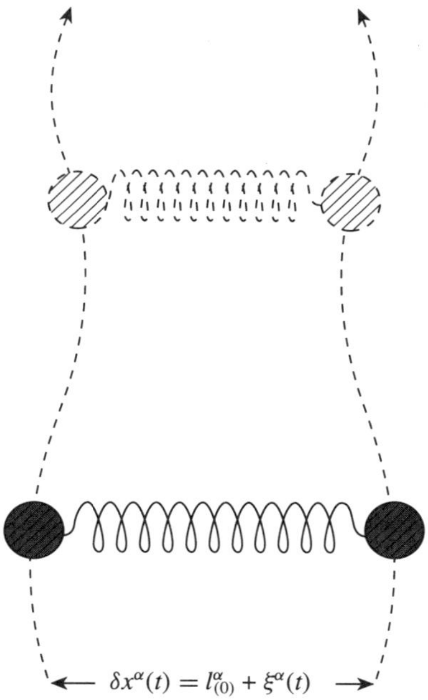

FIGURE 3.13. Two particles of mass m connected by a spring and their world lines during the passage of a gravitational wave (see also § 2.10 and fig. 2.10). δx^α is the total separation between the two masses, $l^\alpha_{(0)}$ their equilibrium separation, and ξ^α their small displacement from the equilibrium position (see eq. 3.6.5).

We now analyze a system that is small (in comparison with the typical radius of curvature of the gravitational field, see § 2.5), that is freely falling, and that consists of two masses connected by a spring. We may consider this system as a test particle in the gravity field, and we may define a *local inertial Fermi frame* (see § 2.1) along the world line of the system's center of mass. The spatial axes are Fermi-Walker transported and may be physically represented by three gyroscopes (see § 3.4.3). In this local inertial system the metric is $\eta_{\alpha\beta}$. However, if the system is at rest on the surface of Earth, we may perform a local transformation from the inertial Fermi frame $\overset{(i)}{x}{}^\alpha$, momentarily at rest, to the *local frame* x^α on Earth. Its axes are also Fermi-Walker transported. At the lowest order, locally, we can write the transformation[1] (see eq. 3.4.31)

$$\begin{aligned} \overset{(i)}{x}{}^0 &\cong x^0 + a_k x^k x^0 + O(a^2x^3) + O(Rx^3) \\ \overset{(i)}{x}{}^k &\cong x^k + \frac{1}{2}a^k\left(x^0\right)^2 + O(a^2x^3) + O(Rx^3) \end{aligned} \qquad (3.6.2)$$

where a^i are the spatial components of the nongravitational acceleration of the system center of mass, and R and x indicate the typical magnitude of some components of $R_{\alpha\beta\gamma\delta}$ and x^α. Therefore, in this frame, called *local proper reference frame*, the metric is at the lowest order

$$ds^2 \cong -\left(1 + 2a_i x^i\right)dt^2 + \delta_{ik} dx^i dx^k \\ + [O(a^2x^2) + O(Rx^2)]dx^\alpha dx^\beta. \tag{3.6.3}$$

In section 2.10 we derived the Riemann tensor of a plane gravitational wave, the perturbation of a flat background, propagating along one spatial axis, $R_{i0j0} = -\frac{1}{2}\ddot{h}_{ij}^{\mathrm{TT}}$ (see also ref. 276). Let us consider a gravitational wave with a reduced wavelength $\lambda\!\!\!^{-} \equiv \frac{\lambda}{2\pi}$ much longer then the characteristic dimension $l^\alpha_{(0)}$ of our system, $\lambda\!\!\!^{-} \gg l_0$. For a proper choice of the local inertial frame, this expression of the Riemann tensor is still valid in this frame. Furthermore, by applying the above transformation (3.6.2) from the local inertial frame to the local proper frame, at the lowest order we still have

$$R_{i0j0} = \overset{(i)}{R}_{i0j0}(1 + O(\mathbf{a}\cdot\mathbf{x}) + O(at) + O(h)) = -\frac{1}{2}\ddot{h}_{ij}^{\mathrm{TT}} \tag{3.6.4}$$

where for a local frame on Earth, in geometrized units, $|a_i x^i| \ll 1$, and a and h indicate the typical magnitude of a^i and h_{ij}.

Therefore, we may rewrite equation (3.6.1) in the local proper reference frame. By neglecting here, for simplicity, all the driving forces except those due to the spring and gravitational waves (neglecting therefore terms with $R^i_{\oplus 0j0}$ and a^i), we then get the time-dependent part of $\delta x^i(t) = l^i_{(0)} + \xi^i(t)$:

$$\frac{d^2\xi^i}{dt^2} = \frac{1}{2}\ddot{h}_{ij}^{\mathrm{TT}} l^j_{(0)} - \frac{D^i{}_j}{m}\frac{d\xi^j}{dt} - \frac{K^i{}_j}{m}\xi^j. \tag{3.6.5}$$

According to expression (3.6.3), at the lowest order, $\xi^i(t)$ represents the change of the proper distance between the two particles, as a function of the proper time measured at the origin, $x^i = 0$, of the local proper frame, due to the gravitational-wave perturbation h_{ij}. If the two particles are freely falling, and no spring connects them, we have

$$\frac{d^2\xi^i}{dt^2} = -R^i{}_{0j0} l^j_{(0)} = \frac{1}{2}\ddot{h}_{ij}^{\mathrm{TT}} l^j_{(0)}, \tag{3.6.6}$$

and we then get again the results of section 2.10 for the change of the proper distance between two test particles due to a gravitational wave. We now assume that the two masses be connected by an elastic support, idealized for example as a massless spring (fig. 3.13) with angular frequency of the fundamental mode ω_0. For a resonant oscillator oriented along the x-axis, we write $K^x{}_x \equiv k = \omega_0^2 m$ and $D^x{}_x \equiv D = \frac{\omega_0 m}{Q}$, where Q is the so-called *quality factor*, its inverse measures internal energy dissipation (for a typical aluminum bar cooled at low temperatures,[219] $Q > 10^6$). We thus have for two masses along the x-axis originally at $-\frac{L}{2}$ and $+\frac{L}{2}$, with $l^i_{(0)} = (L, 0, 0)$:

$$\ddot{\xi} + \frac{\omega_0}{Q}\dot{\xi} + \omega_0^2 \xi = \frac{1}{2} L \ddot{h} \tag{3.6.7}$$

where, for simplicity, we have considered only the polarization state $h_+(t)$, written here $h(t)$ (the other polarization state $h_\times(t)$ is rotated by $\frac{\pi}{4}$ with respect to $h_+(t)$; see § 2.10). Here the small displacement is $\xi^i = (\xi, 0, 0)$, and $\frac{2Q}{\omega_0} = \tau_0$ is the typical damping time of the system.

By inserting in this equation $\xi(t) = \frac{1}{2\pi}\int_{-\infty}^{+\infty}\tilde{\xi}(\omega)e^{i\omega t}d\omega$, where $\tilde{\xi}(\omega)$ is the Fourier transform of $\xi(t)$: $\tilde{\xi}(\omega) = \int_{-\infty}^{+\infty}\xi(t)e^{-i\omega t}dt$, and by using the Fourier transform $\tilde{h}(\omega)$ of $h(t)$, we get: $\omega^2\tilde{\xi}(\omega) - i\frac{\omega_0}{Q}\omega\,\tilde{\xi}(\omega) - \omega_0^2\tilde{\xi}(\omega) = \frac{L}{2}\omega^2\tilde{h}(\omega)$. Then, we can write the solution of equation (3.6.7):

$$\xi(t) = \frac{1}{2\pi}\int_{-\infty}^{+\infty}\frac{L}{2}\,\frac{\omega^2\tilde{h}(\omega)e^{i\omega t}}{(\omega^2 - \omega_0^2) - i\frac{\omega_0}{Q}\omega}\,d\omega. \tag{3.6.8}$$

By integration, we then have

$$\begin{aligned}\xi(t) &= i\frac{L}{2}\left(\frac{\omega_1^2\tilde{h}(\omega_1)e^{i\omega_1 t}}{\omega_1 - \omega_2} + \frac{\omega_2^2\tilde{h}(\omega_2)e^{i\omega_2 t}}{\omega_2 - \omega_1}\right)\\ &\cong \frac{iL}{4\omega_0}\left(\omega_1^2\tilde{h}(\omega_1)e^{i\omega_1 t} - \omega_2^2\tilde{h}(\omega_2)e^{i\omega_2 t}\right)\end{aligned} \tag{3.6.9}$$

where $\omega_{1,2} \cong i\frac{\omega_0}{2Q} \pm \omega_0$ are the poles of the integrand, neglecting higher order terms in $\frac{1}{Q}$. When[219] $\tilde{h}(\omega_1) = \tilde{h}(\omega_2)$, we then get

$$\xi(t) = -\frac{L}{2}e^{-\left(\frac{\omega_0}{2Q}t\right)}\tilde{h}(\omega_1)\omega_0 \sin\omega_0 t + O\left(\frac{1}{Q}\right). \tag{3.6.10}$$

So far we have considered two particles connected by a spring. A **gravitational-wave resonant detector** is usually a cylindrical bar of length L. The small change $\xi(t)$ in the length of the whole bar at the fundamental resonance angular frequency ω_0 can be described by the solution (3.6.10) of the equation of a harmonic oscillator, with resonance angular frequency ω_0, with a supplementary $\frac{4}{\pi^2}$ factor obtained by solving the problem of a continuous bar:[219]

$$\xi(t) = \frac{-2L}{\pi^2}e^{-\left(\frac{\omega_0}{2Q}t\right)}\tilde{h}(\omega_1)\omega_0 \sin\omega_0 t + O\left(\frac{1}{Q}\right). \tag{3.6.11}$$

Therefore, with a resonant antenna we measure the Fourier component of the metric perturbation near the antenna resonance frequency ω_0. The typical damping time of the resonant detector is $\frac{2Q}{\omega_0}$. For an oscillation of amplitude ξ_0 and angular frequency ω_0, the energy of the bar has been calculated to be equal to the energy of a harmonic oscillator with mass $M/2$:

$$E_A = \frac{M}{4}\omega_0^2\xi_0^2 \tag{3.6.12}$$

where M is the mass of the antenna. For a gravitational-wave perturbation of duration $\tau \ll \frac{2Q}{\omega_0}$, from expression (3.6.11), we then get

$$E_A = \frac{1}{4}M\omega_0^2\Big(\frac{2L}{\pi^2}\omega_0|\tilde{h}(\omega_0)|\Big)^2 = \frac{M\omega_0^2|\tilde{h}(\omega_0)|^2 v^2}{\pi^2} \tag{3.6.13}$$

where v is the velocity of sound in the antenna, $v = \frac{L\omega_0}{\pi}$, ω_0 is the angular frequency of the fundamental mode of the antenna, and L is its length.

We recall that in section 2.10 we have derived the effective energy-momentum pseudotensor $t^{\alpha\beta}$ for a gravitational-wave perturbation of $\eta_{\alpha\beta}$ in the transverse-traceless gauge $t^{\rm GW}_{\mu\nu} = \frac{1}{32\pi}\langle h^{\rm TT}_{ij,\mu}h^{\rm TT\,ij}{}_{,\nu}\rangle$, where $\langle\ \ \rangle$ means average over several wavelengths. For a weak, monochromatic, plane gravitational wave traveling along the z-axis, the flux of energy per unit area is

$$t^{\rm GW}_{00} = \frac{1}{16\pi}\Big\langle|\dot{h}_+(t)|^2 + |\dot{h}_\times(t)|^2\Big\rangle. \tag{3.6.14}$$

(In ordinary units this expression of $t^{\rm GW}_{00}$ has to be multiplied by $\frac{c^3}{G}$). When, for simplicity, one polarization state only, $h(t)$ equal to $h_+(t)$ or $h_\times(t)$, comes into play, with its Fourier transform $\tilde{h}(\omega) = \int_{-\infty}^{+\infty} h(t)e^{-i\omega t}dt$, then the total incident energy per unit area is

$$I = \int_{-\infty}^{+\infty} t^{\rm GW}_{00}(t)dt = \frac{1}{16\pi}\int_{-\infty}^{+\infty}|\dot{h}(t)|^2 dt$$
$$= \frac{1}{8\pi}\int_0^{+\infty}\omega^2|\tilde{h}(\omega)|^2 d\nu$$

where we have used Parseval's relation: $\int_{-\infty}^{+\infty}|h(t)|^2dt = \int_{-\infty}^{+\infty}|\tilde{h}(\omega)|^2d\nu$, where $\nu = \omega/2\pi$, and where $f(\omega) \equiv \frac{1}{8\pi}\omega^2|\tilde{h}(\omega)|^2$ is called *spectral energy density*, energy per unit frequency interval per unit area carried by the gravitational waves (frequencies from 0 to $+\infty$). One may then define[219] the classical *resonance integral of the absorption cross section*, $\sum_0$, as the energy gained by the bar (expression 3.6.13; for optimal direction and polarization of the incident gravitational wave) divided by the spectral energy density at the resonance frequency ω_0:

$$\sum\nolimits_0 \equiv \frac{E_A}{f(\omega_0)} = \frac{8}{\pi}v^2M = 4.2\times 10^{-25}\Big(\frac{v}{5400\ \text{m/s}}\Big)^2\Big(\frac{M}{2300\ \text{kg}}\Big)\ \text{m}^2\text{Hz}.$$

(In ordinary units this expression of $\sum_0$ has to be multiplied by $\frac{G}{c^3}$). Therefore, $\sum_0$ for a resonant detector is larger with a greater mass M and a higher speed of sound v.

It is necessary to consider[219] all the possible sources of disturbance that can compete with the gravitational-wave signals, thermal, electrical, acoustic, cosmic rays, seismic and electromagnetic noise. The thermal noise, or Brownian noise, is essentially the antenna vibration induced by the Brownian motion of the atoms at a temperature T. The electronic noise is associated with the amplifier and related electronics, and the seismic noise is induced by the environment. To reduce the Brownian noise the antenna is cooled to very low temperature; to reduce the seismic noise the antenna is usually suspended by wires, and mechanical filters are used.

The minimum uptake of gravitational-wave energy ("energy innovation") in principle detectable for a burst is governed by the energy associated with the noise

$$E_{\min} = kT_{\text{eff}} \tag{3.6.15}$$

where T_{eff} is the so-called *effective temperature* of the system and is a function of the thermodynamical temperature of the detector T and of other parameters characteristic of the mechanical and of the electrical parts of the system, among which are the inverse $\frac{1}{Q}$ of the mechanical quality factor and the so-called amplifier noise temperature. Of course, to be detectable, the energy transmitted by a gravitational wave to the antenna must be larger than the energy associated with the noise $E_A > kT_{\text{eff}}$. Therefore, by expression (3.6.13), the *minimum energy variation* in principle detectable is

$$E_{\min} = \frac{1}{\pi^2} v^2 M \omega_0^2 |\tilde{h}(\omega_0)|^2_{\min} = kT_{\text{eff}} \tag{3.6.16}$$

and the lowest detectable gravitational wave amplitude is

$$|\tilde{h}(\omega_0)|_{\min} = \frac{L}{v^2}\sqrt{\frac{kT_{\text{eff}}}{M}}. \tag{3.6.17}$$

For a *short sinusoidal burst* of gravitational radiation, for example a short sinusoidal wave of amplitude h_0, angular frequency ω_0 and duration τ : $h(t) = h_0 \cos \omega_0 t$ for a short time τ, one has $|\tilde{h}(\omega_0)| \approx h_0 \frac{\tau}{2}$, and the theoretical minimum amplitude h_0 detectable, corresponding to a short duration τ, is $h_{0_{\min}} \cong \frac{1}{\tau} \times \frac{2L}{v^2}\sqrt{\frac{kT_{\text{eff}}}{M}}$. For a typical antenna, $L \cong 3$ m, $M \cong 2300$ kg, and $v \cong 5400$ m/s for *Al* at about 4.2 K. Therefore, for a pulse of duration $\tau \cong 1$ millisec, an effective temperature $T_{\text{eff}} \cong 5$ mK corresponds to an h_0 or fractional change in dimensions of about 10^{-18}.

A gravitational-wave resonant detector is practically built as shown in figure 3.14. The mechanical oscillations of the bar induced by a gravitational wave are converted by an electromechanical transducer into electric signals which are amplified with a low-noise amplifier, such as a dc SQUID. Then, we have data analysis.

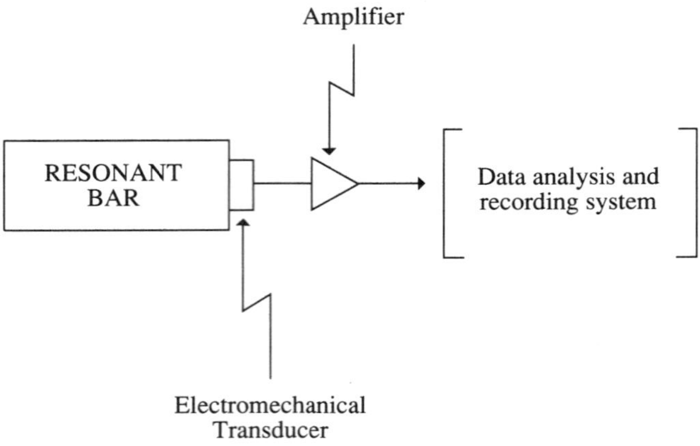

FIGURE 3.14. A simplified scheme of a resonant detector.

Resonant antennas were first built by J. Weber, around 1960, at the University of Maryland.[216] As of 1994, such gravitational-wave resonant detectors are operated by the following groups (alphabetical order): CNR of Frascati, University of Beijing, University of Guangzhou, University of Louisiana, University of Maryland, University of Moscow, University of Rome, Stanford University, University of Tokyo, University of Western Australia—Perth, plus a group at Princeton University studying Earth itself as a gravitational-wave resonant detector. In particular the University of Rome second-generation antenna (located at CERN, Geneva) reached, in 1985–1986, an effective temperature T_{eff} of about 15 mK corresponding, for a short burst of radiation, to an h_0 of about 10^{-18} and, in 1990,[218] an effective temperature of about 3 mK corresponding, for a short burst, to an h_0 between 10^{-18} and 10^{-19}. The ultimate sensitivity of these bar antennas[261] for a short burst of gravitational radiation has been estimated to be of the order of 10^{-20} or 10^{-21}. Bar detectors, usually 3 m long aluminum bars, work at a typical frequency of about 10^3 Hz.

3.6.2. [LIGO: Laser-Interferometer Gravitational-Wave Observatory]

Another type of Earth-based, gravitational-wave detector is the so-called **LIGO, Laser-Interferometer Gravitational-wave Observatory**,[214,215,220–223,232,252,266–272] operating in the range of frequencies between 10^4 Hz and a few 10 Hz. Various types of gravitational-wave laser-interferometer have been proposed, among which are the standard *Michelson and Fabry-Perot types*.

A *Michelson-type gravitational-wave laser interferometer* is essentially made of three masses, suspended by wires, at the ends of two orthogonal arms, as is shown in the simplified scheme of figure 3.15.

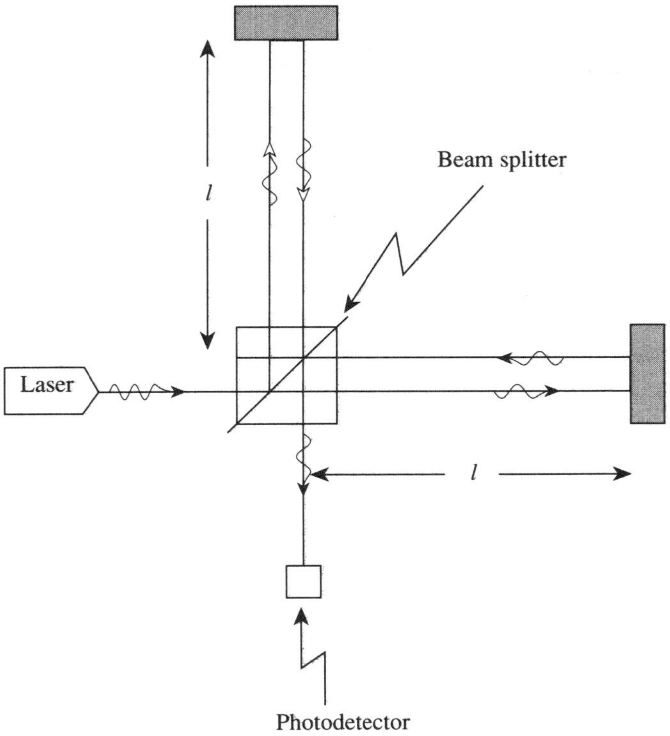

FIGURE 3.15. A simplified scheme of a Michelson laser interferometer.

Let us assume, for simplicity, that a gravitational wave is impinging perpendicularly to the system, along the z-axis. The proper length l of the arms of the laser interferometer will change according to the description of section 2.10. For gravitational waves with $\lambda\!\!\!^{-} = \frac{\lambda}{2\pi} \gg l$ and with polarization h_+, in a local proper reference frame with x- and y-axes coincident with the arms of the interferometer, we have

$$\begin{aligned} \delta l_x &= \frac{1}{2} l h_{xx}(t) \equiv \frac{1}{2} l h_+(t), \\ \delta l_y &= -\frac{1}{2} l h_+(t), \end{aligned} \tag{3.6.18}$$

and the difference between the proper lengths of the two arms is:

$$\delta l(t) = \delta l_x(t) - \delta l_y(t) = l h_+(t) \equiv l h(t). \tag{3.6.19}$$

Variations in h_{ij} due to a gravitational wave will, in turn, produce oscillations in $\delta l(t)$ and therefore oscillations in the relative phase of the laser light at the beamsplitter; they will finally produce oscillations in the intensity of the laser light measured by the photodetector. If the laser light travels back and forth between the test masses $2N$ times (N = number of round trips), then the variation of the difference between the proper lengths of the two arms is (assuming $Nl \ll \lambdabar$):

$$\Delta l = 2Nlh(t), \tag{3.6.20}$$

and therefore, the relative phase delay due to the variations in δl will be

$$\Delta\phi = \frac{\Delta l}{\lambdabar_{\mathrm{L}}} = \frac{2Nl}{\lambdabar_{\mathrm{L}}} h(t) \tag{3.6.21}$$

where $\lambdabar_{\mathrm{L}} = \frac{\lambda_{\mathrm{L}}}{2\pi}$ is the reduced wavelength of the laser light. Comparing $\Delta\phi$ with the precision of the photodetector one then obtains the *minimum* $h(t)$ *in principle detectable*, apart from the other important sources of noise, such as seismic noise, gravity field gradient noise, photon shot noise, thermal noise, etc. For a gravitational-wave burst of frequency ν, one then gets[214,215]

$$h_{\min} = 7.2 \times 10^{-21} \times \frac{50}{N} \times \frac{1\ \mathrm{km}}{l} \left[\frac{\lambdabar_{\mathrm{L}}}{0.082\mu\mathrm{m}} \times \frac{10\ \mathrm{watt}}{P \times \eta} \times \frac{\nu}{1000\mathrm{Hz}} \right]^{\frac{1}{2}} \tag{3.6.22}$$

where P is the laser output power, and η is the efficiency of the photodetector.

For most of the fundamental limiting factors of these Earth-based detectors, such as seismic noise, photon shot noise, etc., the displacement noise is essentially independent of the arm length l, according to formula (3.6.22). Therefore, by increasing l one increases the sensitivity of the LIGO detectors.

Prototype laser interferometers of a few meters arm length have been built at MIT (1.5 m) and Caltech (40 m). The Caltech laser interferometer reaches a sensitivity[214,215] to bursts of gravitational radiation of the order of $h \sim 10^{-17}$ at about 1000 Hz. In Europe, at the Max-Planck-Institut at Garching, Munich, a laser interferometer with 30 m arm length, and one at Glasgow with 10 m arm length, have sensitivities to bursts of gravitational radiation of the order of 10^{-17} at about 1000 Hz. A 10 m laser interferometer has been built in Japan (TENKO-10). The near future MIT and Caltech interferometers (US Ligo),[272] with 4 km arm lengths, should reach sensitivities[214,215] to bursts of gravitational radiation of the order of $h \sim 10^{-20}$ to 10^{-21} between 1000 and 100 Hz. A 20 m Fabry-Perot laser interferometer, one delay line of 100 m (TENKO-100) and one underground of 3 km arm length are planned in Japan. The University of Glasgow and the Max-Planck-Institutes for Quantum Optics and for Astrophysics of Garching are planning to built underground a 600 m laser interferometer (GEO-600).[252,271] The University of Western Australia has proposed a large laser interferometer (AIGO), likewise the University of Guangzhou. Finally, INFN of Pisa together with the University of Paris-Sud at Orsay are building a 3 km interferometer (VIRGO)[223,270] that should reach frequencies of operation as low as a few 10 Hz, using special filters to eliminate the seismic noise at these lower frequencies. The ultimate burst sensitivity for all of the above large interferometers is currently estimated[215] to be of the order of 10^{-22} or 10^{-23} at frequencies near 100 Hz.

3.6.3. [MIGO: Michelson Millimeter Wave Interferometer Gravitational- Wave Observatory]

Space is the site of another proposed experiment to detect gravitational waves, the **Michelson millimeter wave Interferometer Gravitational-wave Observatory (MIGO).**[224,225]

This enterprise aims to use three spacecraft in geostationary orbits around Earth to form a Michelson interferometer in space.[224] The system would be operated from a ground station that would provide a high-stability frequency standard and the control uplink and downlink data signals (see fig. 3.16). MIGO should operate in the range of gravitational-wave frequencies between about 10^{-1} Hz and about 10^{-3} Hz. MIGO, according to a study,[225] should approach a sensitivity to bursts of gravitational radiation of about 10^{-18}, by using a microwave interferometer with a frequency of about 10^{11} Hz ($\lambda \cong 3 \times 10^{-3}$ m) and arm lengths of about 60,000 km. However, for a periodic gravitational wave the sensitivity should increase as the square root of the number of periods of the observed signal.

The main limitation to the sensitivity of MIGO is the stability of any Earth-based frequency standard. However, if the arm lengths were exactly equal any frequency instability would cancel out. Thus, the contribution of frequency instabilities to limit the sensitivity should be reduced by a factor $\frac{\delta L}{L}$, where δL is

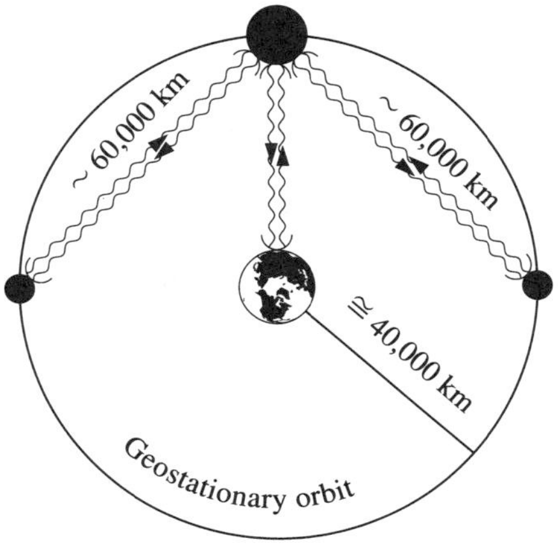

FIGURE 3.16. MIGO: Michelson millimeter wave Interferometer Gravitational wave space Observatory.

the difference between the arm lengths and L is the total arm length. For example, with a frequency stability of 2×10^{-15} and a relative arm length control, $\frac{\delta L}{L}$, of 5×10^{-4} one should in principle reach a *sensitivity* sufficient to detect a pulse of gravitational radiation of $\sim 10^{-18}$. Perturbations in the arm lengths would be caused by such influences as tidal pulls, as well as nongravitational forces, such as radiation pressure, solar winds, etc. The errors resulting from nongravitational accelerations should be eliminated with drag-free systems (§ 6.12) or by accurately measuring these perturbations with high-quality accelerometers as proposed in the MIGO experiment.[225] Other error sources would be receiver thermal noise; thermal and mechanical instabilities of the spacecraft antenna, of the feed components and of the spacecraft itself; transponder loop response, etc.

Tests of the Nordtvedt effect, of gravitational time dilation of clocks, of time delay of electromagnetic waves and of isotropy of the speed of light with MIGO have also been proposed.[225]

3.6.4. [LISA or LAGOS: Laser Gravitational-Wave Observatory in Space]

Below about 10 Hz the sensitivity of Earth-based gravitational-wave detectors is limited by gravity gradient variations. Even for perfect isolation of a detector from seismic and ground noise, an Earth-based detector would still be affected by the time changes in the gravity field due to density variations in Earth and its atmosphere. Due to this source of noise the sensitivity has been calculated to worsen as roughly the inverse fourth power of the frequency.[228]

Therefore, to avoid this type of noise and to reduce noise from other sources, one should use an interferometer far from Earth and with very long arms. Indeed, to detect gravitational waves in the range of frequencies between 10^{-5} Hz and 1 Hz, it has been proposed[227–231] (Faller and Bender 1984) to orbit in the solar system a **space interferometer made of three spacecraft** at a typical distance between about 1 and 10 million km (see fig. 3.17).

The central spacecraft should have a circular orbit around the Sun with one year period. By properly choosing the position of the other two spacecraft, one could keep a right angle between the two arms of the interferometer and keep the distances, between the two end spacecraft and the central one, for example at about 10^7 km. Indeed, the two end spacecraft should orbit around the central one with one year period in a plane tilted by $60°$ with respect to the ecliptic plane. It has been shown that with this configuration the variations in the arm lengths are less than $\pm$ 0.2%, over a ten year period, corresponding to relative velocities of less than 10 m/s. Each of the three test masses of **LAGOS**, in each of the three spacecraft, will be freely floating within a cavity. The laser light is received by 30 cm diameter telescopes, and to make its intensity detectable when reflected back over such distances (corresponding to a 10^7 km arm, only $\sim$

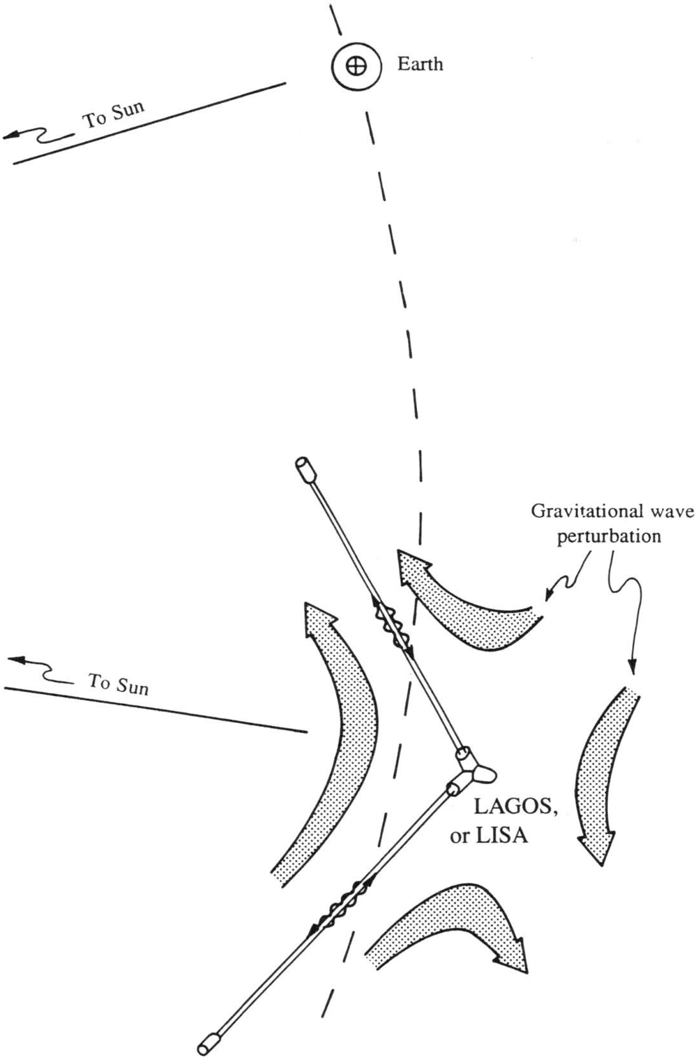

FIGURE 3.17. The ultimate observatory for gravitational waves. LAGOS: Laser Gravitational-wave Observatory in Space[227–231], or LISA.

10^{-9} of the transmitted power will be received by the telescopes), a "coherent transponder approach" should be used. This system will allow laser light to be sent back in a phase-coherent way from a laser in each of the two end spacecraft (corresponding to an arm length of about 10^7 km, the lasers should have a power of about 1 watt). Since there should not be unknown solar system gravitational perturbations of the arm lengths at frequencies between 10^{-5} and 1 Hz and since the test masses are freely falling because of drag-free systems, any unmodeled variation of the difference of the LAGOS arm lengths δl at these frequencies should be a measure of the oscillations of the arms proper length due to gravitational waves with frequencies between 10^{-5} and 1 Hz. Although the phase measurement system and the thermal stability are essential requirements, it is the main technological challenge of this experiment[227–231] to keep very small the spurious accelerations of the test masses. The drag-free system should be able to reduce the spurious accelerations below about 10^{-13} $(\text{cm/s}^2)/\sqrt{\text{Hz}}$ from 10^{-5} to 10^{-3} Hz. Furthermore, one should be able to measure the changes of the difference of the LAGOS arm lengths to 5×10^{-11} $\text{cm}/\sqrt{\text{Hz}}$ from 10^{-3} to 1 Hz.

Considering all the error sources, it has been calculated[232] that, for periodic gravitational waves, with an integration time of about one year, LAGOS should reach a *sensitivity* able to detect amplitudes of about $h \sim 10^{-24}$, in the range of frequencies between 10^{-3} and 10^{-2} Hz, amplitudes from about $h \sim 10^{-20}$ to about 10^{-24} between 10^{-4} and 10^{-3} Hz, and amplitudes from about $h \sim 10^{-22}$ to about 10^{-24} between 1 and 10^{-2} Hz. For stochastic gravitational waves and for gravitational-wave bursts, LAGOS should reach a sensitivity of about $h \sim 10^{-21}$ at about 10^{-3} Hz, and from about $h \sim 10^{-18}$ to about $\sim 10^{-21}$ between 10^{-4} and 10^{-3} Hz and between 1 and 10^{-3} Hz.

Therefore, comparing the LAGOS sensitivity to the predicted theoretical amplitudes[232,233] of gravitational radiation at these frequencies, LAGOS should be able to detect gravitational waves from galactic binaries, including ordinary main-sequence binaries, contact binaries, cataclysmic variables, close white dwarf binaries, and neutron star binaries. The LAGOS sensitivity should also allow detection of possible gravitational pulses from the in-spiral of compact objects into supermassive black holes in Active Galactic Nuclei (AGNs) and from collapse of very massive objects to form black holes.

LAGOS and the LINE–SAGITTARIUS proposal described in the next section have been combined in a single project of ESA (European Space Agency), called **LISA (Laser Interferometer Space Antenna)**. LISA–LAGOS will be launched by ESA as a "Cornerstone" mission.

3.6.5. [SAGITTARIUS-LINE: Laser Interferometer Near Earth]

Another Earth-orbiting experiment proposed (1991)[226,273] to detect gravitational waves is **SAGITTARIUS (Spaceborne Astronomical Gravitational-**

wave Interferometer Testing Aspects of Relativity and Investigating Unknown Sources), also called **LINE (Laser Interferometer Near Earth)**.

This enterprise would use four, or six, Earth orbiting spacecraft to form a laser interferometer in space. The spacecraft would be in circular geocentric orbits (retrograde and long-term stable) at about 600,000 km altitude; they would be at the three vertices of an equilateral triangle. The arm lengths of this space interferometer would be about 1 million km. Two close spacecraft correspond to the central mass of the laser interferometer of figure 3.15, the other two spacecraft correspond to its end masses, and each of them would be tracked by one of the central spacecraft. A configuration has also been proposed[273] with two spacecraft at each vertex of the equilateral triangle (see fig. 3.18). Each satellite would be equipped with a very accurate accelerometer and with ion thrusters capable of very low thrust (for example, the so-called FEEPs, Field Emission Electric Propulsion units). The accelerometers, with a sensitivity of

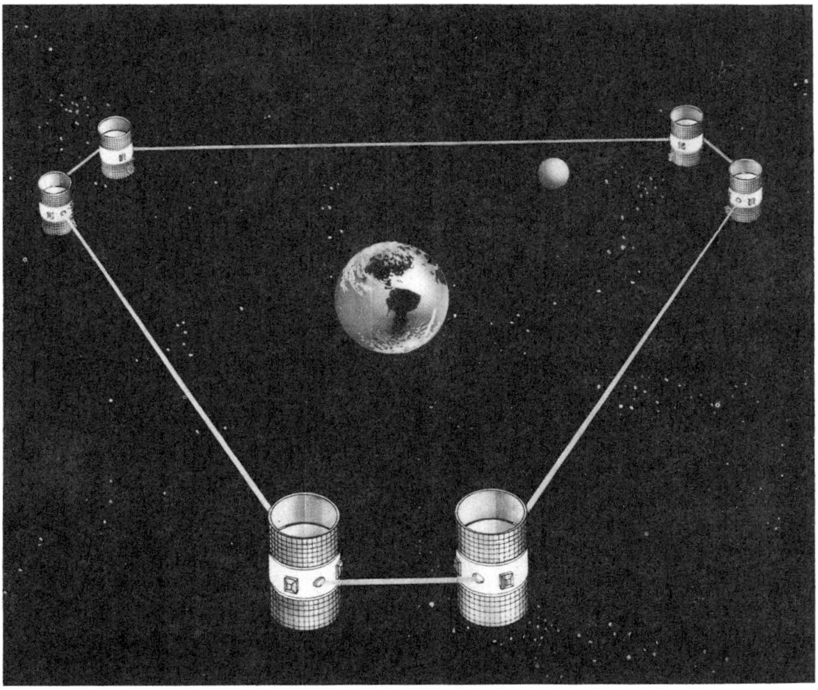

FIGURE 3.18. An interesting proposal for detection of gravitational waves in space. SAGITTARIUS-LINE,[226,273] Laser Interferometer near Earth.

about 3×10^{-13} cm s^{-2} Hz$^{-\frac{1}{2}}$, would govern the thrusters. Laser signals propagating in one arm of the interferometer, generated from one of the central spacecraft and actively transponded back from the corresponding end spacecraft, would be compared with the original signals at the central spacecraft. When a gravitational wave passes by at the same time, for example perpendicularly to the arm, a Doppler shift $\delta\nu$ of the frequency ν_0 of the laser pulse would be produced, $\frac{\delta\nu}{\nu_0} \sim h$, where h is the gravitational-wave metric perturbation of section 2.10. From the time integral of the difference between the two Doppler signals from the two arms one would then get the interferometer relative phase: $\frac{d(\delta\phi)}{dt} = 2\pi(\delta\nu_1 - \delta\nu_2)$.

For periodic gravitational waves and with an integration time of about four months, SAGITTARIUS has been estimated to reach a *sensitivity* corresponding to amplitudes of the order of $h \sim 10^{-23}$ in the range of frequencies between about 10^{-1} and 10^{-3} Hz.

The main noise sources limiting the sensitivity of SAGITTARIUS have been estimated to be laser shot noise, position readout noise, and nongravitational perturbations of the spacecraft.

According to the general relativistic formulae, SAGITTARIUS will be able to detect gravitational waves from some *known* galactic binary stars with frequencies of about 10^{-3} Hz and amplitudes of about $h \sim 10^{-21}$, and gravitational waves from some probable sources, such as neutron star binaries, black hole binaries, and white dwarf binaries, and also from some exotic astrophysical sources such as supermassive black holes in galactic nuclei.

3.6.6. [Interplanetary Doppler Tracking]

Let us assume that an electromagnetic signal is sent from a point A, for example on Earth, to another point B, for example on a spacecraft orbiting in the solar system, and then reflected back to the original point A. If at the same time a gravitational wave with a wavelength of the order of the Earth-spacecraft distance were to propagate between point A and the reflection point B, then oscillations in the **Doppler shift** of the received **electromagnetic radiation** would be produced[234–238] **by the gravitational wave**. The oscillations in the Doppler shift have a relative magnitude of $\frac{\delta\nu}{\nu} \sim h^{TT}$, where ν is the frequency of the electromagnetic wave and h is the metric perturbation (2.10.13)–(2.10.15) in the transverse-traceless gauge.

Doppler tracking of spacecraft in the solar system[239,240] in practice uses a high-stability frequency standard on Earth (hydrogen-maser clock) to control the frequency of the emitted electromagnetic wave; after the wave is reflected back from the spacecraft with a transponder in a phase coherent way, the frequency of the wave received on Earth is again compared with the frequency of the clock. Therefore, the stability of the clock is a fundamental limitation to the accuracy of the measurement. Interplanetary Doppler tracking of spacecraft works in the range of gravitational-wave frequencies between about 10^{-1} and 10^{-4} Hz. Preliminary measurements have already been done with the Viking, Voyager, Pioneer 10 and 11 spacecraft, and with the ULYSSES and GALILEO spacecraft, and will be performed with the CASSINI mission.[239,240,278]

The main error source in this type of experiment is due to the density fluctuations of the interplanetary plasma, producing fluctuations in the dispersion of the radio waves and therefore oscillations in the Doppler shift. However, since this effect is inversely proportional to the square of the frequency of the radio wave it could be reduced by using higher frequency tracking, that is, X-band tracking (10 GHz), both uplink, to the spacecraft, and downlink, from the spacecraft, with GALILEO, and K-band tracking (30 GHz) with CASSINI. This source of error could also be substantially reduced with dual frequency tracking.

With the GALILEO and the CASSINI missions the *sensitivity* to bursts of gravitational radiation should be, respectively, better than $\sim 10^{-14}$ and better than $\sim 10^{-15}$, in the range of frequencies between 10^{-2} and 10^{-4} Hz. Using Doppler tracking the future sensitivity to bursts of gravitational radiation[232] should be of the order of $h \sim 10^{-16}$ or 10^{-17}.

3.7 [APPENDIX: METRIC THEORIES AND PPN FORMALISM]

According to some physicists (see review in ref. 2) to test general relativity versus other gravitational theories, one might consider only **metric theories of gravitation**. (This assumption is based on the so-called **Schiff conjecture**[2] that in any viable theory of gravity the weak equivalence principle implies the medium strong equivalence principle.) The metric theories of gravitation, as explained in section 3.1, have the same spacetime structure (Lorentzian manifold) and the same equations of motion for test particles (geodesics) as general relativity. However, they have different field equations.

Eddington (1922),[241] Robertson (1962)[242], and Schiff (1967)[243,244] introduced the method to expand the Schwarzschild metric at the order beyond Newtonian theory (post-Newtonian) and then to multiply each post-Newtonian term by a dimensionless parameter, to be experimentally determined. Nordtvedt (1968),[129] Will (1971),[245] and Will and Nordtvedt (1972)[72,73] have then developed the PPN formalism, a powerful and useful tool for testing general relativity and alternative metric theories. The **PPN formalism**,[2,72,73,129,130,241–248] is a **post-Newtonian parametrized expansion** of the metric tensor $\boldsymbol{g}$ and of the energy-momentum tensor $\boldsymbol{T}$ in terms of well-defined classical potentials.

In order to find a general form of the post-Newtonian metric $\boldsymbol{g}$ generated by the matter fields, to describe metric theories beyond the Newtonian approximation, without knowing any particular field equations, one may establish the following criteria (here the matter is idealized as a fluid,[2] however in Nordtvedt (1968)[129] the matter is described by a collection of mass elements):

(1) Weak gravitational fields and slow motions, that is: $U = \int \frac{\rho(\boldsymbol{x}')d^3x'}{|\boldsymbol{x}-\boldsymbol{x}'|}$ = Newtonian potential, v^2 = (matter velocity)^2, $\frac{p}{\rho} = \left(\frac{\text{pressure}}{\text{rest mass density}}\right)$, and $\Pi \equiv \frac{\varepsilon-\rho}{\rho}$ = (specific internal energy density), where ε = total mass-energy density, are—all four—quantities small compared to one (in geometrized units) of order "ϵ^2"

$$U \sim v^2 \sim \frac{p}{\rho} \sim \Pi \sim \epsilon^2 \ll 1 \tag{3.7.1}$$

and from the slow motion assumption[2]

$$\frac{|\partial/\partial t|}{|\partial/\partial x|} \sim \epsilon^1. \tag{3.7.2}$$

(2) One can write the metric tensor $\boldsymbol{g}$ and the energy-momentum tensor $\boldsymbol{T}$ at the post-Newtonian order as a linear combination of small potentials, such as the Newtonian potential U (and its square), determined through classical relations by the distribution and by the motion of matter. In the PPN formalism, the metric, $\boldsymbol{g}$, and the tensor $\boldsymbol{T}$ are expanded up to the post-Newtonian order without including higher order terms, that is, without post-post-Newtonian terms. Therefore, as one can see from the geodesic equation of motion, one needs g_{00} up to order ϵ^4, g_{0i} up to order ϵ^3, and g_{ik} up to order ϵ^2, and consistently with $\boldsymbol{\nabla} \cdot \boldsymbol{T} = \boldsymbol{0}$, for a perfect fluid with $T^{\alpha\beta} = (\varepsilon + p)u^\alpha u^\beta + pg^{\alpha\beta}$, one needs T^{00} up to order $\rho\epsilon^2$, T^{0i} up to order $\rho\epsilon^3$, and T^{ik} up to order $\rho\epsilon^4$.

(3) In terms of coordinates such that the metric $\boldsymbol{g}$ is dimensionless, any term appearing in the expansion of $\boldsymbol{g}$ must be dimensionless.

(4) For simplicity, the metric $\boldsymbol{g}$, and therefore any quantity it contains, is assumed to be a function of ρ, ε, p, and $\boldsymbol{v}$ only, but not of their spatial gradients (for a justification of this assumption see ref. 1).

(5) Since one wants to find the metric $\boldsymbol{g}$ generated by an isolated system, one assumes the existence of a set of frames, the quasi-Lorentzian frames (and the quasi-Cartesian coordinate systems), where the metric asymptotically becomes the Minkowski metric $\boldsymbol{\eta}$. Consequently, in these frames any quantity appearing in $\boldsymbol{g} - \boldsymbol{\eta}$ must go to zero as the spatial distance between an event and the isolated system goes to infinity: $\boldsymbol{g} - \boldsymbol{\eta} \to \boldsymbol{0}$ as $r \to \infty$

(6) The general expression of the PPN metric should be valid in any quasi-Lorentzian frame. Then, the functional dependence of the post-Newtonian metric $\boldsymbol{g}^{\text{PPN}}$ on the classical potentials $U\ldots$, and on the frame velocity $\boldsymbol{w}$ (for hypothesis $\sim \epsilon^1$) relative to a hypothetical preferred frame of the universe (postulated in some metric theories), should

be the same in any quasi-Lorentzian frame, that is, the metric $\boldsymbol{g}^{\mathrm{PPN}}$ should be *formally* invariant (not necessarily Lorentz invariant; see § 2.6) as a function of the potentials $U\ldots$, and of the frame velocity $\boldsymbol{w}$, for every *low-velocity* $v \sim \epsilon^1$ inhomogeneous Lorentz transformation. In particular, the general expression of the PPN metric should be formally invariant (1) for spatial translations of the origin of the quasi-Cartesian coordinate system, (2) for spatial rotations of the coordinates, and (3) for post-Galilean transformations, that is, low-velocity, $v \sim \epsilon^1$, boosts (see § 3.4.3). A way to satisfy (1) and (2) is to require that every term in the expansion of g_{00} be a scalar (at the post-Newtonian order) for spatial rotations, spatial translations, and low-velocity boosts, and every term in the expansions of g_{0i} and of g_{ik} be, respectively, a vector a tensor under spatial rotations, spatial translations, and low-velocity boosts. (For the PPN metric $\boldsymbol{g}^{\mathrm{PPN}}$ in anisotropic model universes, see ref. 246).

In agreement with these conditions, one can write g_{ik} as a linear combination of $U\delta_{ik} \sim \epsilon^2$ and U_{ik}, where

$$U_{ik} \equiv \int \frac{\rho'(x_i - x_i')(x_k - x_k')}{|\boldsymbol{x} - \boldsymbol{x}'|^3} d^3x' \sim \epsilon^2. \tag{3.7.3}$$

Indeed, $U\delta_{ik}$ and U_{ik} are of order ϵ^2, are dimensionless, go to zero as r goes to infinity, and behave as three-tensors for spatial translations, for spatial rotations (in a quasi-Cartesian coordinate system), and for low-velocity boosts of a quasi-Lorentzian frame. Similarly, one can write g_{0i} in terms of V_i and W_i, or other more complex potentials:

$$\begin{aligned} V_i &\equiv \int \frac{\rho v_i'}{|\boldsymbol{x} - \boldsymbol{x}'|} d^3x' \sim \epsilon^3, \\ W_i &\equiv \int \frac{\rho \boldsymbol{v}' \cdot (\boldsymbol{x} - \boldsymbol{x}')(x_i - x_i')}{|\boldsymbol{x} - \boldsymbol{x}'|^3} d^3x' \sim \epsilon^3. \end{aligned} \tag{3.7.4}$$

Indeed, V_i and W_i are of order ϵ^3 and satisfy conditions (1)–(6). Finally, one can write g_{00} in terms of U^2, ϕ_1, ϕ_2, ϕ_3, ϕ_4, A, B, ϕ_W, or other more complex potentials:

$$\begin{aligned} \phi_1 &\equiv \int \frac{\rho' v'^2}{|\boldsymbol{x} - \boldsymbol{x}'|} d^3x', \qquad \phi_2 \equiv \int \frac{\rho' U'}{|\boldsymbol{x} - \boldsymbol{x}'|} d^3x', \\ \phi_3 &\equiv \int \frac{\rho' \Pi'}{|\boldsymbol{x} - \boldsymbol{x}'|} d^3x', \qquad \phi_4 \equiv \int \frac{p'}{|\boldsymbol{x} - \boldsymbol{x}'|} d^3x', \\ A &\equiv \int \frac{\rho' [\boldsymbol{v}' \cdot (\boldsymbol{x} - \boldsymbol{x}')]^2}{|\boldsymbol{x} - \boldsymbol{x}'|^3} d^3x', \\ B &\equiv \int \frac{\rho'}{|\boldsymbol{x} - \boldsymbol{x}'|} (\boldsymbol{x} - \boldsymbol{x}') \cdot \frac{d\boldsymbol{v}'}{dt} d^3x', \end{aligned} \tag{3.7.5}$$

$$\phi_W \equiv \int \rho'\rho'' \frac{\boldsymbol{x}-\boldsymbol{x}'}{|\boldsymbol{x}-\boldsymbol{x}'|^3} \cdot \left(\frac{\boldsymbol{x}'-\boldsymbol{x}''}{|\boldsymbol{x}-\boldsymbol{x}''|} - \frac{\boldsymbol{x}-\boldsymbol{x}''}{|\boldsymbol{x}'-\boldsymbol{x}''|} \right) d^3x'd^3x''.$$

Indeed, these potentials are of order ϵ^4 and satisfy conditions (1)–(6). These *classical potentials* can be calculated explicitly, by the use of their definitions (3.7.3), (3.7.4), and (3.7.5), once the distribution and the motion of mass-energy are known. They can also be expressed in the equivalent, useful differential form:

$$\begin{aligned}
&\Delta U = -4\pi\rho, \quad U_{ik} = \chi_{,ik} + \delta_{ik}U, \\
&\Delta\chi = -2U, \quad \Delta V_i = -4\pi\rho v_i, \quad \Delta\phi_1 = -4\pi\rho v^2, \\
&\Delta\phi_2 = -4\pi\rho U, \quad \Delta\phi_3 = -4\pi\rho\Pi, \quad \Delta\phi_4 = -4\pi p, \qquad (3.7.6) \\
&\Delta\phi_W = -2\Delta(U^2) + 3\Delta\phi_2 + 2\chi_{,ij}U_{,ij}\,, \\
&A + B = \chi_{,00} + \phi_1, \quad V_{i,i} = -U_{,0}\,, \quad \chi_{,0i} = V_i - W_i,
\end{aligned}$$

where $\chi \equiv -\int \rho'|\boldsymbol{x}-\boldsymbol{x}'|d^3x'$.

It is now necessary to observe that the post-Newtonian metric $\boldsymbol{g}^{\mathrm{PPN}}$ might contain not only the terms introduced above, but also any other term satisfying conditions (1)–(6), for example any "combination" of the potentials (3.7.3), (3.7.4), and (3.7.5) satisfying conditions (1)–(6) (see below). Therefore, these terms are a priori only a small subset of all the conceivable terms that might appear in $\boldsymbol{g}^{\mathrm{PPN}}$.[249,250]

Finally, from the *formal* invariance (at the post-Newtonian order) of the **metric** $\boldsymbol{g}^{\mathrm{PPN}}$ as a function of the potentials (3.7.3), (3.7.4), and (3.7.5) for post-Galilean transformations, that is, for low-velocity, $v^2 \sim \epsilon^2$, boosts, one *may* write in every quasi-Lorentzian frame,[2] moving with a velocity w^i relative to a hypothetical preferred frame of the universe[2,183] postulated in some metric theories, in the *standard post-Newtonian parametrized gauge*[2]

$$\begin{aligned}
g_{00} = {} & -1 + 2U - 2\beta U^2 - 2\xi\phi_W + (2\gamma + 2 + \alpha_3 + \zeta_1 - 2\xi)\phi_1 \\
& + 2(3\gamma - 2\beta + 1 + \zeta_2 + \xi)\phi_2 \\
& + 2(1 + \zeta_3)\phi_3 + 2(3\gamma + 3\zeta_4 - 2\xi)\phi_4 \\
& - (\zeta_1 - 2\xi)A - (\alpha_1 - \alpha_2 - \alpha_3)\boldsymbol{w}^2 U - \alpha_2 w^i w^k U_{ik} \\
& + (2\alpha_3 - \alpha_1)w^i V_i, \\
g_{0i} = {} & -\frac{1}{2}(4\gamma + 3 + \alpha_1 - \alpha_2 + \zeta_1 - 2\xi)V_i - \frac{1}{2}(1 + \alpha_2 - \zeta_1 + 2\xi)W_i \\
& - \frac{1}{2}(\alpha_1 - 2\alpha_2)w^i U - \alpha_2 w^k U_{ik}, \\
g_{ik} = {} & (1 + 2\gamma U)\delta_{ik}, \qquad (3.7.7)
\end{aligned}$$

and from the expression of the **energy-momentum tensor** $T^{\alpha\beta}$ of a perfect fluid, and from $\nabla \cdot T = 0$, one has at the post-Newtonian order

$$\begin{aligned} T^{00} &= \rho(1 + \Pi + v^2 + 2U), \\ T^{0i} &= \rho(1 + \Pi + v^2 + 2U + \frac{p}{\rho})v^i, \\ T^{ik} &= \rho v^i v^k (1 + \Pi + v^2 + 2U + \frac{p}{\rho}) + p\delta^{ik}(1 - 2\gamma U). \end{aligned} \tag{3.7.8}$$

In these expressions, the coefficients of the potentials (3.7.3), (3.7.4), and (3.7.5) are the ten standard **PPN parameters**, $\beta, \gamma, \alpha_1, \alpha_2, \alpha_3, \zeta_1, \zeta_2, \zeta_3, \zeta_4, \xi$, in the notation of Will-Nordtvedt[72,73] and Will.[2] γ measures space curvature produced by mass, and in this *standard PPN gauge* β is related to nonlinearity of mass contributions to the metric (see § 3.5). α_1, α_2, and α_3 measure preferred frame effects, $\zeta_1, \zeta_2, \zeta_3, \zeta_4$ violations of conservation of four-momentum, and ξ preferred location effects. *In general relativity $\beta = \gamma = 1$ and all the other parameters are zero.*

By means of expressions (3.7.7) and (3.7.8) for $\boldsymbol{g}^{\mathrm{PPN}}$ and $\boldsymbol{T}^{\mathrm{PPN}}$ it is possible to test different metric theories of gravity by comparing the experimental results with the theoretical predictions written as a function of the PPN parameters and then by determining experimental limits[2,4,6,106] on these parameters. Indeed, the post-Newtonian approximation of several metric theories of gravity corresponds to $\boldsymbol{g}^{\mathrm{PPN}}$, $\boldsymbol{T}^{\mathrm{PPN}}$, and $\nabla \cdot \boldsymbol{T}^{\mathrm{PPN}} = \mathbf{0}$, with a particular set of values of the PPN parameters; vice versa, a particular set of values of the PPN parameters may correspond to the post-Newtonian approximation of some metric theory.

However, even though the PPN formalism is a powerful and useful tool to describe metric theories of gravitation at the post-Newtonian order, in principle, one can imagine new alternative metric theories of gravitation that, at the post-Newtonian order, are not described by the PPN formalism, as is shown by the following counterexample of alternative metric theories[249,250].

Given, on the spacetime with physical metric tensor $g_{\alpha\beta}$, a distribution of mass-energy described by the physical fields ε, p, u^α ..., energy density, pressure, four-velocity ..., first solve for a tensor $\tilde{g}_{\alpha\beta}$ solution of field equations of the type of the Einstein field equation, with $\tilde{R}^{\alpha\beta}$, $\tilde{R}$, and $\tilde{T}^{\alpha\beta}$ built using $\tilde{\boldsymbol{g}}$, ε, p, u^α .... Then introduce auxiliary scalar, vector, and tensor fields $\tilde{U}$, $\tilde{V}^\alpha$, $\tilde{U}_{\alpha\beta}$ ... to be determined by solving equations of the type

$$\begin{aligned} \tilde{\Box}\tilde{U} &= f\left(\varepsilon, u^\alpha, \tilde{T}^{\alpha\beta}, \tilde{g}^{\alpha\beta}, \tilde{U} \ldots\right) \\ &\cdots \\ \tilde{\Box}\tilde{U}_{\alpha\beta} &= f_{\alpha\beta}\left(\varepsilon, u^\alpha, \tilde{T}^{\alpha\beta}, \tilde{g}^{\alpha\beta}, \tilde{U}_{\alpha\beta} \ldots\right) \end{aligned} \tag{3.7.9}$$

where the operator $\tilde{\Box} = \tilde{g}^{\alpha\beta}\tilde{\nabla}_\alpha\tilde{\nabla}_\beta$ is the d'Alambertian operator on a curved manifold, with covariant derivatives defined by using the tensor $\tilde{g}_{\alpha\beta}$, and where f and $f_{\alpha\beta}$ are, respectively, suitable scalar and tensor combinations of ε, $\tilde{u}^\alpha$, $\tilde{T}^{\alpha\beta}$ .... Finally, the physical metric tensor $g_{\alpha\beta}$ of the spacetime is given by

$$g_{\alpha\beta} = \tilde{g}_{\alpha\beta} + h_{\alpha\beta}(\tilde{U}, \ldots, \tilde{U}_{\alpha\beta}) \tag{3.7.10}$$

where the tensor field $h_{\alpha\beta}$ is a suitable combination of the tensorial quantities $\tilde{U}$, $\tilde{V}^\alpha$, $\tilde{U}_{\alpha\beta}$ .... Then, by a proper choice of f, $f_{\alpha\beta\ldots}$, and taking the post-Newtonian limit for weak fields and low velocities, one finds: $\tilde{\Box}\tilde{U} \cong \Delta\tilde{U} \cong -4\pi\varepsilon$, and thus at the post-Newtonian order: $\tilde{U} \cong U =$ (Newtonian potential); $\tilde{U}^{\alpha\beta} \cong U^{\alpha\beta}$, etc.

Therefore, by a suitable choice of $h_{\alpha\beta}$, in addition to the standard post-Newtonian terms, one may get in the post-Newtonian expression of $g_{\alpha\beta}$ other terms containing *noninteger powers of the classical potentials* U, ϕ_2, χ ..., such as $g_{\alpha\beta} \cong g_{\alpha\beta}^{\text{general relativity}} + G_2\left(\frac{|\phi_4|}{U^2}\right)^{\frac{1}{2}} \chi_{,\alpha\beta\ldots}$, where G_2 is a constant.

This metric is *not described by the standard PPN expansion*[249,250] which contains integer powers of these classical potentials only. In this way one could in principle generate an infinite number of new terms in the post-Newtonian expression of the metric, and therefore one would in principle have an infinite number of new post-Newtonian parameters.

In conclusion, this counterexample of metric theories of gravity, which may also be reformulated with a Lagrangian,[251] shows that the **standard ten parameter PPN formalism**, though a very powerful and useful tool to test metric theories, **is not sufficient to describe every conceivable metric theory of gravity at the post-Newtonian order**, and one would *in principle* need an infinite set of *new terms* and *new parameters* to add to the standard ten parameter PPN formalism in order to describe the post-Newtonian approximation of any a priori conceivable metric theory of gravity.

REFERENCES CHAPTER 3

1. C.W. Misner, K.S. Thorne, and J.A. Wheeler, *Gravitation* (Freeman, San Francisco, 1973).
2. C.M. Will, *Theory and Experiment in Gravitational Physics*, rev. ed. (Cambridge University Press, Cambridge, 1993).
3. C.M. Will, The confrontation between general relativity and experiment: An update, *Phys. Rep.* 113:345–422 (1984).
4. C.M. Will, Experimental gravitation from Newton's Principia to Einstein's general relativity, in *300 Years of Gravitation*, ed. S.W. Hawking and W. Israel (Cambridge University Press, Cambridge, 1987), 80–127.

5. C.M. Will, General relativity at 75: How right was Einstein?, *Science* 250:770–76 (1990).
6. C.M. Will, The confrontation between general relativity and experiment: A 1992 update, *Int. J. Mod. Phys. D* 1:13–68 (1992).
7. P.G. Roll, R. Krotkov, and R.H. Dicke, The equivalence of inertial and passive gravitational mass, *Ann. Phys. (NY)* 26:442–517 (1964).
8. V.B. Braginsky and V.I. Panov, Verification of the equivalence of inertial and gravitational mass, *Zh. Eksp. Teor. Fiz.* 61:873–79 (1971); *Sov. Phys.—JETP* 34:463–66 (1972).
9. T.M. Niebauer, M.P. McHugh, and J.E. Faller, Galilean test for the fifth force, *Phys. Rev. Lett.* 59:609–12 (1987).
10. B.R. Heckel, E.G. Adelberger, C.W. Stubbs, Y. Su, H.E. Swanson, G. Smith, and W.F. Rogers, Experimental bounds on interactions mediated by ultralow-mass bosons, *Phys. Rev. Lett.* 63:2705–8 (1989).
11. E.G. Adelberger, C.W. Stubbs, B.R. Heckel, Y. Su, H.E. Swanson, G. Smith, J.H. Gundlach, and W.F. Rogers, Testing the equivalence principle in the field of the Earth: Particle physics at masses below 1 μeV?, *Phys. Rev. D* 42:3267–92 (1990).
12. R.F.C. Vessot and M.W. Levine, A test of the equivalence principle using a space-borne clock, *J. Gen. Rel. and Grav.* 10:181–204 (1979).
13. R.F.C. Vessot, L.W. Levine, E.M. Mattison, E.L. Blomberg, T.E. Hoffman, G.U. Nystrom, B.F. Farrel, R. Decher, P.B. Eby, C.R. Baugher, J.W. Watt, D.L. Teuber, and F.O. Wills, Test of relativistic gravitation with a space-borne hydrogen maser, *Phys. Rev. Lett.* 45:2081–84 (1980).
14. J.W. Brault, The gravitational redshift in the solar spectrum, Ph.D. dissertation, Princeton University Abstract; Gravitational redshift of solar lines, *Bull. Amer. Phys. Soc.* 8:28 (1962).
15. J.L. Snider, New measurement of the solar gravitational redshift, *Phys. Rev. Lett.* 28:853–56 (1972).
16. J.L. Snider, Comments on two recent measurements of the solar gravitational redshift, *Solar Phys.* 36:233–34 (1974).
17. T.P. Krisher, J.D. Anderson, and J.K. Campbell, Test of the gravitational redshift effect at Saturn, *Phys. Rev. Lett.* 64:1322–25 (1990).
18. R.D. Reasenberg, I.I. Shapiro, P.E. MacNeil, R.B. Goldstein, J.C. Breidenthal, J.P. Brenkle, D.L. Cain, T.M. Kaufman, T.A. Komarek, and A.I. Zygielbaum, Viking relativity experiment: Verification of signal retardation by solar gravity, *Astrophys. J. Lett.* 234:L219–L221 (1979).
19. D.S. Robertson, W.E. Carter, and W.H. Dillinger, New measurement of solar gravitational deflection of radio signals using VLBI, *Nature* 349:768–70 (1991).
20. D.S. Robertson and W.E. Carter, Relativistic deflection of radio signals in the solar gravitational field measured with VLBI, *Nature* 310:572–74 (1984).
21. I.I. Shapiro, C.C. Counselman III, and R.W. King, Verification of the principle of equivalence for massive bodies, *Phys. Rev. Lett.* 36:555–58 (1976).
22. J.D. Anderson, J.K. Campbell, R.F. Jurgens, E.L. Lau, X.X. Newhall, M.A. Slade III, and E.M. Standish, Jr., Recent developments in solar-system tests of general

relativity, in *Proc. 6th Marcel Grossmann Meeting on General Relativity*, Kyoto, June 1991, ed. H. Sato and T. Nakamura (World Scientific, Singapore, 1992), 353–55.

23. J.O. Dickey, X.X. Newhall, and J.G. Williams, Investigating relativity using Lunar Laser Ranging: Geodetic precession and the Nordtvedt effect, *Adv. Space Res.* 9:75–78 (1989); see also Relativistic gravitation, in Symp. 15 of the COSPAR 27th Plenary Meeting, Espoo, Finland, 18–29 July 1988.

24. C. Talmadge, J.–P. Berthias, R.W. Hellings, and E.M. Standish, Model-independent constraints on possible modifications of Newtonian gravity, *Phys. Rev. Lett.* 61:1159–62 (1988).

25. A. Schild, Gravitational theories of the Whitehead type and the principle of equivalence, in *Evidence for Gravitational Theories*, ed. C. Möller (Academic Press, New York, 1962).

26. A. Schild, Lectures on general relativity theory, in *Relativity Theory and Astrophysics: I. Relativity and Cosmology; II. Galactic Structure; III. Stellar Structure*, ed. J. Ehlers (American Mathematical Society, Providence, RI, 1967), 1–105.

27. G. Galilei, *Discorsi e dimostrazioni matematiche intorno a due nuove scienze* (Elzevir, Leiden, 1638); trans. H. Crew and A. de Salvio as *Dialogues Concerning Two New Sciences* (Macmillan, New York, 1914); reprint (Dover, New York, 1954).

28. C. Huygens, *Horologium oscillatorium sive de motu pendulorum ad horologia aptato demonstrationes geometricae*, (Parisiis, 1673) in Oeuvres Completes (Société Hollandaise des Sciences, La Haye, 1934) 18:69–368 and 17:1–237.

29. I. Newton, *Philosophiae naturalis principia mathematica* (Streater, London, 1687); trans. A. Motte (1729) and revised by F. Cajori as *Sir Isaac Newton's Mathematical Principles and His System of the World* (University of California Press, Berkeley and Los Angeles, 1934; paperback, 1962).

30. F.W. Bessel, Versuche über die Kraft, mit welcher die Erde Körper von verschiedener Beschaffenheit anzieht, *Ann. Physik und Chemie* (*Poggendorff*) 25:401–17 (1832).

31. L. Southerns, A determination of the ratio of mass to weight for a radioactive substance, *Proc. Roy. Soc. Lond.* 84:325–44 (1910).

32. H.H. Potter, Some experiments on the proportionality of mass and weight, *Proc. Roy. Soc. Lond.* 104:588–610 (1923).

33. R.V. Eötvös, Über die Anziehung der Erde auf verschiedene Substanzen, *Math. Naturw. Ber. aus. Ungarn* 8:65–68 (1889).

34. R.V. Eötvös, D. Pekár, and E. Fekete, Beiträge zum Gesetze der Proportionalität von Trägheit und Gravität, *Ann. Physik* 68:11–66 (1922).

35. P. Zeeman, Experiments on gravitation, *Proc. K. Akad. Amsterdam* 20 (4): 542–53 (1918).

36. J. Renner, Experimentelle Untersuchungen über die Proportionalität von Gravität und Trägheit Matematikai, *és Természettudományi Értesitö*, 53:569–70 (1935).

37. L. Koester, Verification of the equivalence of gravitational and inertial mass for the neutron, *Phys. Rev. D* 14:907–9 (1976).

38. P.W. Worden, Measurement of small forces with superconducting magnetic bearings, *Precision Eng.* 4:139–44 (1982).

39. G.M. Keiser and J.E. Faller, A new approach to the Eötvös experiment, *Bull. Am. Phys. Soc.* 24:579 (1979).

40. P.K. Chapman and A.J. Hanson, An Eötvös experiment in Earth orbit, in *Proc. Conf. on Experimental Tests of Gravitational Theories*, JPL Tech. Memo 33-499, ed. R.W. Davies (Jet Propulsion Laboratory, Pasadena, 1971), 228–35.

41. P.W. Worden, Jr., and C.F.W. Everitt, The gyroscope experiment. III. Tests of the equivalence of gravitational and inertial mass based on cryogenic techniques, in *Experimental Gravitation: Proc. Course 56 of the Int. School of Physics Enrico Fermi*, ed. B. Bertotti (Academic, New York, 1974), 381–402.

42. E.C. Lorenzini et al., Test of the weak equivalence principle in an Einstein elevator (1992), to be published.

43. E. Fischbach, D. Sudarsky, A. Szafer, C. Talmadge, and S.H. Aronson, Reanalysis of the Eötvös experiment, *Phys. Rev. Lett.* 56:3–6 (1986).

44. F. Stacey et al., Geophysics and the law of gravity, *Rev. Mod. Phys.* 59:157–74 (1987).

45. F. Stacey, G.J. Tuck, and G.I. Moore, Quantum gravity: Observational constraints on a pair of Yukawa terms, *Phys. Rev. D* 36:2374–80 (1987).

46. Y. Fujii, The theoretical background of the fifth force, *Int. J. Mod. Phys. A* 6:3505–57 (1991).

47. For various papers on the fifth force, see *Phys. Rev. Lett.* 56, no. 22 (1986).

48. P. Thieberger, Search for a substance-dependent force with a new differential accelerometer, *Phys. Rev. Lett.* 58:1066–69 (1987).

49. P.E. Boynton, D. Crosby, P. Ekstrom, and A. Szumilo, Search for an intermediate-range composition-dependent force, *Phys. Rev. Lett.* 59:1385–89 (1987).

50. D.H. Eckhardt, C. Jekeli, A.R. Lazarewicz, A.J. Romaides, and R.W. Sands, Tower gravity experiment: Evidence for non-Newtonian gravity, *Phys. Rev. Lett.* 60:2567–70 (1988).

51. M.E. Ander et al., Test of Newton's inverse-square law in the Greenland ice cap, *Phys. Rev. Lett.* 62:985–88 (1989).

52. P.G. Bizzeti, A.M. Bizzeti-Sona, T. Fazzini, A. Perego, and N. Taccetti, Search for a composition-dependent fifth force, *Phys. Rev. Lett.* 62:1901–4 (1989).

53. P. Thieberger, Thieberger replies, *Phys. Rev. Lett.* 62:2333 (1989).

54. P.E. Boynton and S.H. Aronson, New limits on the detection of a composition-dependent macroscopic force, in *New and Exotic Phenomena, Proc. 25th Rencontre de Moriond*, ed. O. Fackler and J. Trân Thanh Vân (Éditions Frontières, Gif-sur-Yvette, 1990), 207–24.

55. J. Thomas, P. Kasameyer, O. Fackler, D. Felske, R. Harris, J. Kammeraad, M. Millett, and M. Mugge, Testing the inverse-square law of gravity on a 465-m tower, *Phys. Rev. Lett.* 63:1902–5 (1989).

56. D.F. Bartlett and W.L. Tew, Possible effect of the local terrain on the North Carolina tower gravity experiment, *Phys. Rev. Lett.* 63:1531 (1989).

57. D.F. Bartlett and W.L. Tew, Terrain and geology near the WTVD tower in North Carolina: Implications for non-Newtonian gravity, *J. Geophys. Res. B* 95:17363–69 (1990).

58. C. Jekeli, D.H. Eckhardt, and A.J. Romaides, Tower gravity experiment: No evidence for non-Newtonian gravity, *Phys. Rev. Lett.* 64:1204–6 (1990).

59. C.C. Speake, T.M. Niebauer, M.P. McHugh, P.T. Keyser, J.E. Faller, J.Y. Cruz, J.C. Harrison, J. Mäkinen, and R.B. Beruff, Test of the inverse-square law of gravitation using the 300-m tower at Erie, Colorado, *Phys. Rev. Lett.* 65:1967–71 (1990).

60. C.W. Stubbs, E.G. Adelberger, F.J. Raab, J.H. Gundlach, B.R. Heckel, K.D. McMurry, H.E. Swanson, and R. Watanabe, Search for an intermediate-range interaction, *Phys. Rev. Lett.* 58:1070–73 (1987).

61. V.L. Fitch, M.V. Isaila, and M.A. Palmer, Limits on the existence of a material-dependent intermediate-range force, *Phys. Rev. Lett.* 60:1801–4 (1988).

62. C.W. Stubbs, E.G. Adelberger, B.R. Heckel, W.F. Rogers, H.E. Swanson, R. Watanabe, J.H. Gundlach, and F.J. Raab, Limits on composition-dependent interactions using a laboratory source: Is there a fifth force coupled to isospin?, *Phys. Rev. Lett.* 62:609–12 (1989).

63. K. Kuroda and N. Mio, Test of a composition-dependent force by a free-fall interferometer, *Phys. Rev. Lett.* 62:1941–44 (1989).

64. W.R. Bennet, Jr., Modulated-source Eötvös experiment at Little Goose Lock, *Phys. Rev. Lett.* 62:365–68 (1989).

65. R. Cowsik, N. Krishnan, S.N. Tandon, and S. Unnikrishnan, Strength of intermediate-range forces coupling to isospin, *Phys. Rev. Lett.* 64:336–39 (1990).

66. M.A. Zumberge, J.A. Hildebrand, J.M. Stevenson, R.L. Parker, A.D. Chave, M.E. Ander, and F.N. Spiess, Submarine measurement of the Newtonian gravitational constant, *Phys. Rev. Lett.* 67:3051–54 (1991).

67. G. Müller, W. Zürn, K. Lindner, and N. Rösch, Determination of the gravitational constant by an experiment at a pumped-storage reservoir, *Phys. Rev. Lett.* 63:2621–24 (1989).

68. M.H. Soffel, *Relativity in Astrometry, Celestial Mechanics and Geodesy* (Springer-Verlag, Berlin and Heidelberg, 1989).

69. E. Fischbach and C. Talmadge, Six years of the fifth force, *Nature* 356:207–15 (1992).

70. I. Ciufolini et al., Testing deviations from the inverse square law with a freely falling gravity gradiometer at balloon altitudes, *Nuovo Cimento C* 15:973–82 (1992).

71. A. Einstein, *The Meaning of Relativity*, 3d ed. (Princeton University Press, Princeton, 1950).

72. C.M. Will and K. Nordtvedt, Jr., Conservation laws and preferred frames in relativistic gravity. I. Preferred-frame theories and an extended PPN formalism, *Astrophys. J.* 177:757–74 (1972).

73. K. Nordtvedt, Jr., and C.M. Will, Conservation laws and preferred frames in relativistic gravity. II. Experimental evidence to rule out preferred-frame theories of gravity. *Astrophys. J.* 177:775–92 (1972).

74. L.D. Landau and E.M. Lifshitz, *The Classical Theory of Fields*, 3rd rev. ed. (Addison-Wesley, Reading, MA, and Pergamon, London, 1971).

75. R.V. Pound and G.A. Rebka, Jr., Apparent weight of photons, *Phys. Rev. Lett.* 4:337–41 (1960).

76. R.V. Pound and J.L. Snider, Effect of gravity on gamma radiation, *Phys. Rev. B* 140:788–803 (1965).

77. D.M. Popper, red shift in the spectrum of 40 Eridani B, *Astrophys. J.* 120:316–21 (1954).

78. J.L. Greenstein, J.B. Oke, and H.L. Shipman, Effective temperature, radius, and gravitational redshift of Sirius B, *Astrophys. J.* 169:563–66 (1971).

79. J.E. Blamont and F. Roddier, Precise observation of the profile of the Fraunhofer strontium resonance line: Evidence for the gravitational red shift on the Sun, *Phys. Rev. Lett.* 7:437–39 (1961).

80. J.C. Hafele and R.E. Keating, Around-the-world atomic clocks: Predicted relativistic time gains, *Science* 177:166–68 (1972).

81. J.C. Hafele and R.E. Keating, Around-the-world atomic clocks: Observed relativistic time gains, *Science* 177:168–70 (1972).

82. C.O. Alley, Relativity and clocks, in *Proc. 33d Annual Symposium on Frequency Control* (Electronic Industries Association, Washington, 1979).

83. R.E. Jenkins, A satellite observation of the relativistic Doppler shift, *Astron. J.* 74:960–63 (1969).

84. J. Jaffe and R.F.C. Vessot, Feasibility of a second-order gravitational redshift experiment, *Phys. Rev. D* 14:3294–3300 (1976).

85. K. Nordtvedt, Jr., A study of one- and two-way Doppler tracking of a clock on an arrow toward the Sun, in *Proc. Int. Meeting on Experimental Gravitation*, ed. B. Bertotti (Accademia Nazionale dei Lincei, Rome, 1977), 247–56.

86. R.F.C. Vessot, Past and future tests of relativistic gravitation with atomic clocks, in *Proc. 1st William Fairbank Int. Meeting on Relativistic Gravitational Experiments in Space,* Rome, 10–14 September 1990 (World Scientific, Singapore, 1994).

87. P. Jordan, Formation of the stars and development of the universe, *Nature* 164:637–40 (1949).

88. P. Jordan, *Schwerkraft und Weltfall* (Vieweg und Sohn, Brannschweig, 1955).

89. P. Jordan, Zum gegenwärtigen Stand der Diracschen kosmologischen Hypothesen, *Z. Physik* 157:112–21 (1959).

90. C. Brans and R.H. Dicke, Mach's principle and a relativistic theory of gravitation, *Phys. Rev.* 124:925–35 (1961).

91. R.H. Dicke, *The Theoretical Significance of Experimental Relativity* (Blackie and Son Ltd., London and Glasgow, 1964).

92. R.H. Dicke, Scalar-tensor gravitation and the cosmic fireball, *Astrophys. J.* 152:1–24 (1968).

93. R.H. Dicke, Mach's principle and invariance under transformation of units, *Phys. Rev.* 125:2163–67 (1962).

94. P.A.M. Dirac, The cosmological constants, *Nature* 139:323 (1937).

95. P.A.M. Dirac, New basis for cosmology, *Proc. Roy. Soc. Lond. A* 165:199–208 (1938).

96. E. Mach, *Die Mechanik in Ihrer Entwicklung Historisch-Kritisch Dargestellt* (Brockhaus, Leipzig, 1912); trans. T.J. McCormack with an introduction by Karl Menger, as *The Science of Mechanics* (Open Court, La Salle, IL, 1960).

97. G. Cocconi and E.E. Salpeter, A search for anisotropy of inertia, *Nuovo Cimento* 10:646–51 (1958).

98. G. Cocconi and E.E. Salpeter, Upper limit for the anisotropy of inertia from the Mössbauer effect, *Phys. Rev. Lett.* 4:176–77 (1960).

99. I.I. Shapiro, M.E. Ash, D.B. Campbell, R.B. Dyce, R.P. Ingalls, R.F. Jurgens, and G.H. Pettengill, Fourth test of general relativity: New radar result, *Phys. Rev. Lett.* 26:1132–35 (1971).

100. R.D. Reasenberg and I.I. Shapiro, Bound on the secular variation of the gravitational interaction, in *Atomic Masses and Fundamental Constants*, vol. 5, ed. J.H. Sanders and A.H. Wapstra (Plenum, New York, 1976), 643–49.

101. R.D. Reasenberg and I.I. Shapiro, A radar test of the constancy of the gravitational interaction, in *On the Measurement of Cosmological Variations of the Gravitational Constant*, ed. L. Halpern (University Presses of Florida, Gainesville, 1978), 71–86.

102. J.D. Anderson, M.S.W. Keesey, E.L. Lau, E.M. Standish, and X.X. Newhall, Tests of general relativity using astrometric and radiometric observations of the planets, *Acta Astronautica* 5:43–61 (1978).

103. J.G. Williams, W.S. Sinclair, and C.F. Yoder, Tidal acceleration of the Moon, *Geophys. Res. Lett.* 5:943–46 (1978).

104. R.D. Reasenberg, The constancy of G and other gravitational experiments, *Phil. Trans. Roy. Soc. Lond.* 310:227–38 (1983).

105. R.W. Hellings, P.J. Adams, J.D. Anderson, M.S. Keesey, E.L. Lau, E.M. Standish, V.M. Canuto, and I. Goldman, Experimental test of the variability of G using Viking lander ranging data, *Phys. Rev. Lett.* 51:1609–12 (1983).

106. I.I. Shapiro, Solar system tests of general relativity: Recent results and present plans, in *General Relativity and Gravitation, 1989, Proc. 12th Int. Conf. on General Relativity and Gravitation*, ed. N. Ashby, D.F. Bartlett, and W. Wyss (Cambridge University Press, Cambridge, 1990), 313–30.

107. D.R. Curott, Earth deceleration from ancient solar eclipses, *Astron. J.* 71:264–69 (1966).

108. S. Weinberg, *Gravitation and Cosmology* (Wiley, New York, 1972).

109. R.H. Dicke, Gravitational theory and observation, *Phys. Today* 20:55–70 (1967).

110. R.H. Dicke, Implications for cosmology of stellar and galactic evolution rates, *Rev. Mod. Phys.* 34:110–22 (1962).

111. P.L. Bender, N. Ashby, M.A. Vincent, and J.M. Wahr, Conceptual design for a Mercury relativity satellite, *Adv. Space Res.* 9:113–16 (1989).

112. M.A. Vincent and P.L. Bender, Orbit determination and gravitational field accuracy for a Mercury transponder satellite, *J. Geophys. Res.* 95 (21): 357–61 (1990).

113. P.L. Bender, R.H. Dicke, D.T. Wilkinson, C.O. Alley, D.G. Currie, J.E. Faller, J.D. Mulholland, E.C. Silverberg, H.E. Plotkin, W.M. Kaula, and G.J.F. MacDonald, The Lunar Laser Ranging experiment, in *Proc. Conf. on Experimental Tests of Gravitational Theories*, JPL Tech. Memo 33-499, ed. R.W. Davies (Jet Propulsion Laboratory, Pasadena, 1971), 178–81.
114. C.O. Alley, Laser ranging to retro-reflectors on the Moon as a test of theories of gravity, in *Quantum Optics, Experimental Gravity, and Measurement Theory*, ed. P. Meystre and M.O. Scully (Plenum Publishing, New York, 1983), 429–95.
115. P.L. Bender, D.G. Currie, R.H. Dicke, D.H. Eckhardt, J.E. Faller, W.M. Kaula, J.D. Mulholland, H.H. Plotkin, S.K. Poultney, E.C. Silverberg, D.T. Wilkinson, J.G. Williams, and C.O. Alley, The Lunar Laser Ranging experiment, *Science* 182:229–38 (1973).
116. J.O. Dickey, J.O. Williams, and X.X. Newhall, Fifteen years of Lunar Laser Ranging: Accomplishments and future challenges, in *Proc. 5th Int. Workshop on Laser Ranging Instrumentation*, Royal Greenwich Observatory, England, September 1984.
117. K. Nordtvedt, Jr., Probing gravity to the second post-Newtonian order and to one part in 10^7 using the spin axis of the Sun, *Astrophys. J.* 320:871–74 (1987).
118. K. Nordtvedt, Jr., Toward higher order tests of the gravitational interaction, in *Relativistic Gravitational Experiments in Space*, Annapolis, 28–30 June 1988, ed. R.W. Hellings (NASA Conf. Pub. 3046), 51–54.
119. R.H. Dicke, Experimental tests of Mach's principle, *Phys. Rev. Lett.* 7:359–60 (1961).
120. V.W. Hughes, H.G. Robinson, and V.B. Beltran-Lopez, Upper limit for the anisotropy of inertial mass from nuclear resonance experiments, *Phys. Rev. Lett.* 4:342–44 (1960).
121. R.W.P. Drever, A search for anisotropy of inertial mass using a free precession technique, *Phil. Mag.* 6:683–87 (1961).
122. I. Ciufolini and K. Nordtvedt, On the isotropy of Newtonian gravity (1993), to be published.
123. J.D. Prestage, J.J. Bollinger, W.M. Itano, and D.J. Wineland, Limit for spatial anisotropy by use of nuclear-spin-polarized ^{9}Be $^+$Ions, *Phys. Rev. Lett.* 54:2387–2390 (1985).
124. S.K. Lamoreaux, J.P. Jacobs, B.R. Heckel, F.J. Raab, and E.N. Fortson, New limit on spatial anisotropy from optically pumped ^{201}Hg and ^{199}Hg, *Phys. Rev. Lett.* 57:3125–28 (1986).
125. E. Riis, L.A. Andersen, N. Bjerre, O. Poulsen, S.A. Lee, and J.L. Hall, Test of the isotropy of the speed of light using fast–beam laser spectroscopy, *Phys. Rev. Lett.* 60:81–84 (1988).
126. T.E. Chupp, R.J. Hoare, R.A. Loveman, E.R. Oteiza, J.M. Richardson, M.E. Wagshul, and A.K. Thompson, Result of a new test of local Lorentz invariance: A search for mass anisotropy in ^{21}Ne, *Phys. Rev. Lett.* 63:1541–45 (1989).
127. T.P. Krisher, L. Maleki, G.F. Lutes, L.E. Primas, R.T. Logan, J.D. Anderson, and C.M. Will, Test of the isotropy of the one-way speed of light using hydrogen-maser frequency standard, *Phys. Rev. D* 42:731–34 (1990).

128. K. Nordtvedt, Jr., Equivalence principle for massive bodies. I. Phenomenology, *Phys. Rev.* 169:1014–16 (1968).

129. K. Nordtvedt, Jr., Equivalence principle for massive bodies. II. Theory. *Phys. Rev.* 169:1017–25 (1968).

130. K. Nordtvedt, Jr., Testing relativity with laser ranging to the Moon, *Phys. Rev.* 170:1186–87 (1968).

131. J.G. Williams et al., New test of the equivalence principle from Lunar Laser Ranging, *Phys. Rev. Lett.* 36:551–54 (1976).

132. L.B. Kreuzer, Experimental measurement of the equivalence of active and passive gravitational mass, *Phys. Rev.* 169:1007–12 (1968).

133. D.F. Bartlett and D. Van Buren, Equivalence of active and passive gravitational mass using the Moon, *Phys. Rev. Lett.* 57:21–24 (1986).

134. W. de Sitter, On Einstein's theory of gravitation and its astronomical consequences, *Mon. Not. Roy. Astron. Soc.* 77:155–84 and 481 (1916).

135. A. Fokker, The geodesic precession: A consequence of Einstein's theory of gravitation, *Proc. Kon. Ned. Akad. Wet.* 23 (5): 729–38 (1921).

136. S.W. Hawking and G.F.R. Ellis, *The Large Scale Structure of Space-time* (Cambridge University Press, Cambridge, 1973).

137. M. Spivak, *A Comprehensive Introduction to Differential Geometry* (Publish or Perish, Boston, 1970).

138. S. Kobayashi and K. Nomizu, *Foundations of Differential Geometry* (Wiley-Interscience, New York, 1963).

139. K. Schwarzschild, Über das Gravitationsfeld einer Kugel aus inkompressibler Flussigkeit nach der Einsteinschen Theorie, *Sitzber. Deut. Akad. Wiss. Berlin, Kl. Math.-Phys. Tech.* 424–34 (1916).

140. For a discussion of how, in principle, to build ideal clocks and rods using photons and freely falling test particles, see R.F. Marzke and J.A. Wheeler, Gravitation as geometry. I. the geometry of space-time and the geometrodynamical standard meter, in *Gravitation and Relativity*, ed. H.-Y. Chiu and W.F. Hoffman (W. A. Benjamin, New York, 1964), 40–64; see also J. Ehlers, F.A.E. Pirani, and A. Schild, The geometry of free-fall and light propagation, in *General Relativity, Papers in Honor of J.L. Synge*, ed. L. O'Raifeartaigh (Oxford University Press, London, 1972), 63–84.

141. J. Soldner (1801); see P. Lenard, Über die Ablenkung eines Lichtstrahls von seiner geradlinigen Bewegung durch die Attraktion eines Weltkörpers, an welchem er nahe vorbeigeht, von J. Soldner, 1801, *Ann. Physik* 65:593–604 (1921).

142. F.W. Dyson, A.S. Eddington, and C. Davidson, A determination of the deflection of light by the Sun's gravitational field, from observations made at the total eclipse of May 29, 1919, *Phil. Trans. Roy. Soc. Lond. A* 220:291–333 (1920).

143. Texas Mauritanian Eclipse Team, Gravitational deflection of light: solar eclipse of 30 June 1973. I. Description of procedures and final results, *Astron. J.* 81:452–54 (1976).

144. W. Cannon, The classical analysis of the response of a long baseline radio interferometer, *Geophys. J. Roy. Astron. Soc.* 53:503–30 (1978).

145. D. Robertson and W. Carter, Earth orientation determinations from VLBI observations, *Proc. Int. Conf. on Earth Rotation and the Terrestrial Reference Frame*, Columbus, OH, ed. I. Mueller (Dept. Geod. Sci. and Surv., Ohio State University, 1985), 296–306.

146. D.O. Muhleman, R.D. Ekers, and E.B. Fomalont, Radio interferometric test of the general relativistic light bending near the Sun, *Phys. Rev. Lett.* 24:1377–80 (1970).

147. G.A. Seielstad, R.A. Sramek, and K.W. Weiler, Measurement of the deflection of 9.602-GHz radiation from 3C279 in the solar gravitational field, *Phys. Rev. Lett.* 24:1373–76 (1970).

148. E.B. Fomalont and R.A. Sramek, A confirmation of Einstein's general theory of relativity by measuring the bending of microwave radiation in the gravitational field of the Sun, *Astrophys. J.* 199:749–55 (1975).

149. E.B. Fomalont and R.A. Sramek, Measurement of the solar gravitational deflection of radio waves in agreement with general relativity, *Phys. Rev. Lett.* 36:1475–78 (1976).

150. E.B. Fomalont and R.A. Sramek, The deflection of radio waves by the Sun, *Comm. Astrophys.* 7:19–33 (1977).

151. R.D. Reasenberg, Optical interferometers for tests of relativistic gravity in space, in *Relativistic Gravitational Experiments in Space*, Annapolis, 28–30 June 1988, ed. R.W. Hellings (NASA Conf. Pub. 3046), 155–62.

152. D. Walsh, R.F. Carswell, and R.J. Weymann, 0957+561 A, B: Twin quasistellar objects or gravitational lens?, *Nature* 279:381–84 (1979).

153. I.I. Shapiro, Fourth test of general relativity, *Phys. Rev. Lett.* 13:789–91 (1964).

154. D.O. Muhleman and P. Reichley, Effects of general relativity on planetary radar distance measurements, *JPL Space Programs Summary* 4 (37–39): 239–41 (1964).

155. I.I. Shapiro, Fourth test of general relativity: Preliminary results, *Phys. Rev. Lett.* 20:1265–69 (1968).

156. J.D. Anderson, P.B. Esposito, W. Martin, C.L. Thornton, and D.O. Muhleman, Experimental test of general relativity using time-delay data from Mariner 6 and Mariner 7, *Astrophys. J.* 200:221–33 (1975).

157. R.D. Reasenberg and I.I. Shapiro, Solar-system tests of general relativity, in *Proc. Int. Meeting on Experimental Gravitation*, ed. B. Bertotti (Accademia Nazionale dei Lincei, Rome, 1977), 143–60.

158. I.I. Shapiro, R.D. Reasenberg, P.E. MacNeil, R.B. Goldstein, J. Brenkle, D. Cain, T. Komarek, A. Zygielbaum, W.F. Cuddihy, and W.H. Michael, The Viking relativity experiment, *J. Geophys. Res.* 82:4329–34 (1977).

159. D.L. Cain, J.D. Anderson, M.S.W. Keesey, T. Komarek, P.A. Laing, and E.L. Lau, Test of general relativity with data from Viking orbiters and landers, *Bull. Am. Astron. Soc.* 10:396 (1978).

160. R.A. Hulse and J.H. Taylor, Discovery of a pulsar in a binary system, *Astrophys. J. Lett.* 195:L51–L53 (1975).

161. E. Fermi, Sopra i fenomeni che avvengono in vicinanza di una linea oraria, *Atti R. Accad. Lincei Rend. Cl. Sci. Fis. Mat. Nat.* 31:21–23, 51–52, 101–3 (1922).

162. J.L. Synge, *Relativity the General Theory*, 5th ed. (North-Holland, Amsterdam, 1976).

163. S. Chandrasekhar and G. Contopoulos, On a post-Galilean transformation appropriate to the post-Newtonian theory of Einstein, Infeld and Hoffmann, *Proc. Roy. Soc. Lond. A* 298:123–41 (1967).

164. J. Lense and H. Thirring, Über den Einfluss der Eigenrotation der Zentralkörper auf die Bewegung der Planeten und Monde nach der Einsteinschen Gravitationstheorie, *Phys. Z.* 19:156–63 (1918).

165. J. Lense and H. Thirring, On the gravitational effects of rotating masses: The Thirring-Lense papers, trans. B. Mashhoon, F.W. Hehl, and D.S. Theiss, *Gen. Rel. Grav.* 16:711–50 (1984).

166. B. Bertotti, I. Ciufolini, and P.L. Bender, New test of general relativity: Measurement of de Sitter geodetic precession rate for lunar perigee, *Phys. Rev. Lett.* 58:1062–65 (1987).

167. I.I. Shapiro, R.D. Reasenberg, J.F. Chandler, and R.W. Babcock, Measurement of the de Sitter precession of the Moon: A relativistic three-body effect, *Phys. Rev. Lett.* 61:2643–46 (1988).

168. É. Cartan, Sur une généralisation de la notion de courbure de Riemann et les espaces à torsion, *C.R. Acad. Sci. Paris* 174:593–95 (1922).

169. É. Cartan, Sur les variétés à connexion affine et la théorie de la relativité généralisée (premiére partie), *Ann. Éc. Norm. Sup.* 40:325–412 (1923).

170. É. Cartan, Sur les variétés à connexion affine et la théorie de la relativité généralisée (suite), *Ann. Éc. Norm. Sup.* 41:1–25 (1924).

171. É. Cartan, Sur les variétés à connexion affine et la théorie de la relativitée généralisée II, *Ann. Éc. Norm. Sup.* 42:17–88 (1925).

172. É. Cartan, *Sur les Variétés à Connexion Affine et la Théorie de la Relativitée Généralisée* (Gauthier Villars, 1955). See also *On Manifolds with an Affine Connection and the Theory of General Relativity*, trans. A. Magnon and A. Ashtekar (Bibliopolis, Napoli, 1986).

173. F.W. Hehl, P. von der Heyde, G.D. Kerlick, and J.M. Nester, General relativity with spin and torsion: Foundations and prospects, *Rev. Mod. Phys.* 48:393–416 (1976).

174. A. Trautman, On the Einstein-Cartan equations. I, *Bull. de l'Acad. Pol. des Sciences, Ser. Sci. Math. Astron. Phys.* 20:185–90 (1972).

175. A. Trautman, On the Einstein-Cartan equations. II, *Bull. de l'Acad. Pol. des Sciences, Ser. Sci. Math. Astron. Phys.* 20:503–6 (1972).

176. A. Trautman, On the Einstein-Cartan equations. III, *Bull. de l'Acad. Pol. des Sciences, Ser. Sci. Math. Astron. Phys.* 20:895–96 (1972).

177. W. Adamowicz and A. Trautman, The equivalence principle for spin, *Bull. de l'Acad. Pol. des Sciences, Ser. Sci. Math. Astron. Phys.* 23:339–42 (1975).

178. S. Hojman, M. Rosenbaum, M.P. Ryan, and L.C. Shepley, Gauge invariance, minimal coupling, and torsion, *Phys. Rev. D* 17:3141–46 (1978).

179. J.W. Moffat, New theory of gravitation, *Phys. Rev. D* 19:3554–58 (1979).

180. J. W. Moffat, Consequences of a new experimental determination of the quadrupole moment of the Sun for gravitation theory, *Phys. Rev. Lett.* 50:709–12 (1983).

181. J. W. Moffat and E. Woolgar, Motion of massive bodies: Testing the nonsymmetric gravitation theory, *Phys. Rev. D* 37:918–30 (1988).

182. W. Rindler, *Essential Relativity* (Springer-Verlag, New York, Heidelberg, and Berlin, 1977).

183. W.-T. Ni, A new theory of gravity, *Phys. Rev. D.* 7:2880–83 (1973).

184. N. Rosen, A bi-metric theory of gravitation, *J. Gen. Rel. and Grav.* 4:435–47 (1973).

185. N. Rosen, Bimetric gravitation theory on a cosmological basis, *J. Gen. Rel. and Grav.* 9:339–51 (1978).

186. A. Einstein, Zur allgemeinen Relativitätstheorie, *Preuss. Akad. Wiss. Berlin, Sitzber.*, 778–86 (1915).

187. A. Einstein, Zur allgemeinen Relativitätstheorie (Nachtrag), *Preuss. Akad. Wiss. Berlin, Sitzber.*, 799–801 (1915).

188. D. Hilbert, Die Grundlagen der Physik, *Königl. Gesell. Wiss. Göttingen, Nachr., Math.-Phys. Kl.* 395–407 (1915); see also *Math. Ann.* 92:1–32 (1924).

189. L. Bianchi, Sui simboli a quattro indici e sulla curvatura di Riemann, *Rend. della R. Acc. dei Lincei* 11:3–7 (1902); see also ref. 190.

190. T. Levi-Civita, *Lezioni di calcolo differenziale assoluto*, compiled by E. Persico (Stock, Rome, 1925); trans. M. Long, as *The Absolute Differential Calculus* (Dover, New York, 1977). T. Levi Civita credits G. Ricci Curbastro as having stated (around 1889) these identities for the first time, without a proof.

191. C. W. Allen, *Astrophysical Quantities*, 3d ed. (Athlone, London, 1976).

192. I. I. Shapiro, G. H. Pettengill, M. E. Ash, R. P. Ingalls, D. B. Campbell, and R. B. Dyce, Mercury's perihelion advance: Determination by radar, *Phys. Rev. Lett.* 28:1594–97 (1972).

193. W. de Sitter, Einstein's theory of gravitation and its astronomical consequences, *Mon. Not. Roy. Astron. Soc.* 76:699–728 (1916).

194. R. Dicke and H. Goldenberg, The oblateness of the Sun, *Astrophys. J. Suppl.* 27:131–82 (1974).

195. H. A. Hill, P. D. Clayton, D. L. Patz, A. W. Healy, R. T. Stebbins, J. R. Oleson, and C. A. Zanoni, Solar oblateness, excess brightness, and relativity, *Phys. Rev. Lett.* 33:1497–1500 (1974).

196. R. Leighton, R. W. Noyes, and G. W. Simon, Velocity fields in the solar atmosphere, *Astrophys. J.* 135:474–99 (1962).

197. H. Hill, R. J. Bos, and P. R. Goode, Preliminary determination of the Sun's gravitational quadrupole moment from rotational splitting of global oscillations and its relevance to tests of general relativity, *Phys. Rev. Lett.* 49:1794–97 (1982).

198. D. Gough, Internal rotation and gravitational quadrupole moment of the Sun, *Nature* 298:334–39 (1982).

199. L. Campbell, J. C. McDow, J. W. Moffat, and D. Vincent, The Sun's quadrupole moment and perihelion precession of Mercury, *Nature* 305:508–10 (1983).

200. R.H. Dicke, J.R. Kuhn, and K.G. Libbrecht, Is the solar oblateness variable? Measurements of 1985, *Astrophys. J.* 318:451–58 (1987).

201. T.L. Duvall, W.A. Dziembowski, P.R. Goode, D.O. Gough, J.W. Harvey, and J.W. Leibacher, Internal rotation of the Sun, *Nature* 310:22–25 (1984).

202. J. Christensen-Dalsgaard, D. Gough, and J. Toomre, Seismology of the Sun, *Science* 229:923–31 (1985).

203. T.M. Brown, J. Dalsgaard, W.A. Dziembowski, P. Goode, D.O. Gough, and C.A. Morrow, Inferring the Sun's internal angular velocity from observed p-mode frequency splittings, *Astrophys. J.* 343:526–46 (1989).

204. G. Colombo, GSFC *Symposium on the Sun and Interplanetary Medium, 1975* (NASA TM-X-71097, March 1976).

205. J.D. Anderson, G. Colombo, L.D. Friedman, and E.L. Lau, An arrow to the Sun, in *Proc. Int. Meeting on Experimental Gravitation*, ed. B. Bertotti (Accademia Nazionale dei Lincei, Rome, 1977), 393–422.

206. J.D. Anderson, Gravitational experiments on a solar probe mission: Scientific objectives and technology considerations, in *Relativistic Gravitational Experiments in Space*, Annapolis, 28–30 June 1988, ed. R.W. Hellings (NASA Conf. Pub. 3046), 148–54.

207. R.W. Hellings, Icarus lander, in *Relativistic Gravitational Experiments in Space*, Annapolis, 28–30 June 1988, ed. R.W. Hellings (NASA Conf. Pub. 3046), 141–43.

208. A. Cacciani et al., An experiment to measure the solar $l = 1$ rotational frequency splitting, in *Proc. Oji Int. Seminar on Progress of Seismology of the Sun and the Stars*, Hakone, Japan, 11–14 December 1989, *Lecture Notes in Physics*, ed. Y. Osaki and H. Shibahashi (1989).

209. I. Ciufolini and R.A. Matzner, Non-Riemannian theories of gravity and lunar and satellite laser ranging, *Int. J. Mod. Phys. A* 7:843–52 (1992).

210. R.W. Hellings, Testing relativity with solar system dynamics, in *Proc. 10th Conference on General Relativity and Gravitation*, ed. B. Bertotti, F. de Felice, and A. Pascolini, Padova, 4–9 July 1983 (Reidel, Dordrecht, 1984), 365–85.

211. R.V. Wagoner, Test for the existence of gravitational radiation, *Astrophys. J. Lett.* 196:L63–L65 (1975).

212. T.P. Krisher, 4U 1820–30 as a potential test of the nonsymmetric gravitational theory of Moffat, *Astrophys. J. Lett.* 320:L47–L50 (1987).

213. J. Weber, *General Relativity and Gravitational Waves* (Wiley-Interscience, New York, 1961).

214. K.S. Thorne, Gravitational Radiation, in *300 Years of Gravitation*, ed. S.W. Hawking and W. Israel (Cambridge University Press, Cambridge 1987), 330–458.

215. K.S. Thorne, Sources of gravitational waves and prospects for their detection, in *Recent Advances in General Relativity*, ed. A.I. Janis and J.R. Porter (Birkhäuser, Boston, Basel, and Berlin, 1992), 196–229; see also *Gravitational Radiation* (Cambridge University Press, Cambridge, 1993), to be published.

216. J. Weber, Detection and generation of gravitational waves, *Phys. Rev.* 117:306–13 (1960).

217. D.G. Blair, ed., *The Detection of Gravitational Waves* (Cambridge University Press, Cambridge, 1991).

218. E. Amaldi et al., Sensitivity of the Rome gravitational wave experiment with Explorer cryogenic resonant antenna operating at 2 K, *Europhys. Lett.* 12:5–11 (1990).

219. G. Pizzella, Resonant bar gravitational experiments, in *Proc. 12th Int. Conf. on General Relativity and Gravitation*, Boulder, July 1989, ed. N. Ashby, D.F. Bartlett, and W. Wyss (Cambridge University Press, Cambridge, 1990), 295–311.

220. M.E. Gertsenshtein and V.I. Pustovoit, On the detection of low frequency gravitational waves, *J. Exp. Theoret. Phys.* 43:605–7 (1962); *Sov. Phys.—JETP* 16:433–35 (1963).

221. F.A.E. Pirani, On the physical significance of the Riemann tensor, *Acta Phys. Pol.* 15:389–405 (1956).

222. G.E. Moss, L.R. Miller, and R.L. Forward, Photon-noise-limited laser transducer for gravitational antenna, *Appl. Opt.* 10:2495–98 (1971).

223. A. Giazotto, The Pisa experiments on seismic noise reduction and the Franco-Italian plans for a large interferometer, in *Int. Symposium on Experimental Gravitational Physics*, Guangzhou, China 3–8 August 1987, ed. P.F. Michelson (World Scientific, Singapore, 1988), 347–50.

224. A.J. Anderson, The space microwave interferometer and the search for cosmic background gravitational wave radiation, in *Relativistic Gravitational Experiments in Space*, Annapolis, 28–30 June 1988, ed. R.W. Hellings (NASA Conf. Pub. 3046), 75–79.

225. A.J. Anderson, J. Anderson, J. Fordyce, R. Hellings, T. Krisher, L. Maleki, C. Moreno, D. Sonnabend, and M. Vincent, A feasibility study for a millimeter-wave interferometer gravity-wave observatory (MIGO), *JPL-Caltech report* (August 1990).

226. R.W. Hellings, LF gravitational wave experiments in space, in *Proc. 6th Marcel Grossmann Meeting on General Relativity*, Kyoto, ed. H. Sato and T. Nakamura (World Scientific, Singapore, 1992), 203–212.

227. J.E. Faller and P.L. Bender, A possible laser gravitational wave experiment in space, in *Precision Measurement and Fundamental Constants II*, ed. B.N. Taylor and W.D. Phillips (NBS Spec. Pub. 617, 1984), 689.

228. P.L. Bender et al., Optical interferometer in space, in *Relativistic Gravitational Experiments in Space*, Annapolis, 28–30 June 1988, ed. R.W. Hellings (NASA Conf. Pub. 3046), 80–88.

229. J.E. Faller, P.L. Bender, J.L. Hall, D. Hils, R.T. Stebbins, and M.A. Vincent, An antenna for laser gravitational-wave observations in space, *Adv. Space Res.* 9:107–11 (1989).

230. R.T. Stebbins, P.L. Bender, J.E. Faller, J.L. Hall, D. Hils, and M.A. Vincent, A laser interferometer for gravitational wave astronomy in space, in *Proc. 5th Marcel Grossmann Conference*, Perth, August 1988, ed. D.G. Blair and M.J. Buckingham (World Scientific, Singapore, 1989).

231. P.L. Bender, J.E. Faller, D. Hils, and R.T. Stebbins, Disturbance reduction techniques for a laser gravitational wave observatory in space, in *Abstracts of Contributed Papers, 12th Int. Conf. on General Relativity and Gravitation*, Boulder, CO (2–8 July 1989).

232. Working Papers, Astronomy and Astrophysics Panel Reports (Nat. Acad. Press, Washington, 1991), V-17 and V-18.

233. D. Hils, P.L. Bender, and R.F. Webbink, Gravitational radiation from the galaxy, *Astrophys. J.* 360:75–94(1990).

234. V.B. Braginsky and M.E. Gertsenshtein, Concerning the effective generation and observation of gravitational waves, *Sov. Phys.–JETP Lett.* 5:287–89 (1967).

235. A.J. Anderson, Probability of long period (VLF) gravitational radiation, *Nature* 229:547–48 (1971).

236. R.W. Davies, Issues in gravitational wave detection with space missions, in *Trans. Int. Conf. on Gravitational Waves and Radiations*, Paris, 1973 (CNRS, Paris, 1974), 33–45.

237. F.B. Estabrook and H.D. Wahlquist, Response of Doppler spacecraft tracking to gravitational radiation, *Gen. Rel. and Grav.*, 6:439–47 (1975).

238. A.J. Anderson, Detection of gravitational waves by spacecraft Doppler data, in *Proc. Int. Meeting on Experimental Gravitation*, ed. B. Bertotti (Accademia Nazionale dei Lincei, Rome 1977), 235–46.

239. J.W. Armstrong, F.B. Estabrook, and H.D. Wahlquist, A search for sinusoidal gravitational radiation in the period range 30–2000 seconds: Advanced Doppler tracking experiments, *Astrophys. J.* 318:536–41 (1987).

240. J.W. Armstrong, Advanced Doppler tracking experiments, in *Relativistic Gravitational Experiments in Space*, Annapolis, 28–39 June 1988, ed. R.W. Hellings (NASA Conf. Pub. 3046), 70–74.

241. A.S. Eddington, *The Mathematical Theory of Relativity* (Cambridge University Press, Cambridge, 1922).

242. H.P. Robertson, Relativity and cosmology, in *Space Age Astronomy*, ed. A.J. Deutsch and W.B. Klemperer (Academic Press, New York, 1962), 228–35.

243. L.I. Schiff, General relativity: Theory and experiment, *J. Indust. Appl. Math.* 10:795–801 (1962).

244. L.I. Schiff, Comparison of theory and observation in general relativity, in *Relativity Theory and Astrophysics: I. Relativity and Cosmology*, ed. J. Ehlers (American Mathematical Society, Providence, RI, 1967).

245. C.M. Will, Theoretical frameworks for testing relativistic gravity. II. Parametrized post-Newtonian hydrodynamics and the Nordtvedt effect, *Astrophys. J.* 163:611–28 (1971).

246. K. Nordtvedt, Jr., Anisotropic Parametrized post-Newtonian gravitational metric field, *Phys. Rev. D* 14:1511–17 (1976).

247. G.W. Richter and R.A. Matzner, Second order contributions to gravitational deflection of light in the parametrized post-Newtonian formalism, *Phys. Rev. D* 26:1219–24 (1982).

248. G.W. Richter and R.A. Matzner, Second-order contributions to gravitational deflection of light in the parametrized post–Newtonian formalism. II. Photon orbits and deflection in three dimensions, *Phys. Rev. D* 26:2549–56 (1982).

249. I. Ciufolini, Theory and experiments in general relativity and other metric theories, Ph.D. dissertation, University of Texas at Austin (Ann Arbor, Michigan, 1984).

250. I. Ciufolini, New class of metric theories of gravity not described by the parametrized post-Newtonian (PPN) formalism, *Int. J. Mod. Phys. A* 6:5511–32 (1991).

251. M. Reisenberger, private communication (1991).

252. J. Hough, The Geo-project: A status report, in *Proc. 6th Marcel Grossmann Meeting on General Relativity*, Kyoto, June 1991, ed. H. Sato and T. Nakamura (World Scientific, Singapore, 1992), 192–94.

253. J.C. LoPresto, C. Schrader, and A.K. Pierce, Solar gravitational redshift from the infrared oxygen triplet, *Astrophys. J.* 376:757–60 (1991).

254. C.M. Will, Violation of the weak equivalence principle in theories of gravity with a nonsymmetric metric, *Phys. Rev. Lett.* 62:369–72 (1989).

255. R.N. Truehaft and S.T. Lowe, A measurement of planetary relativistic deflection, *Phys. Rev. Lett.* 62:369–72 (1989).

256. T.P. Krisher, J.D. Anderson, and A.H. Taylor, *Voyager 2* test of the radar time-delay effect, *Astrophys. J.* 373:665–70 (1991).

257. J. Müller, M. Schneider, M. Soffel, and H. Ruder, Testing Einstein's theory of gravity by analyzing Lunar Laser Ranging data, *Astrophys. J. Lett.* 382:L101–L103 (1991).

258. E.G. Adelberger, B.R. Heckel, G. Smith, Y. Su, and H.E. Swanson, Eötvös experiments, lunar ranging and the strong equivalence principle, *Nature* 347:261–63 (1990).

259. J. Thomas and P. Vogel, Testing the inverse-square law of gravity in boreholes at the Nevada test site, *Phys. Rev. Lett.* 65:1173–76 (1990).

260. J.H. Taylor, Astronomical and space experiments to test relativity, in *Proc. 11th Int. Conf. on General Relativity and Gravitation*, Stockholm, July 1986, ed. M.A.H. MacCallum (Cambridge University Press, Cambridge, 1987), 209–222.

261. W.O. Hamilton, Resonant bar gravity wave detectors-past, present and future, in *Proc. 6th Marcel Grossmann Meeting on General Relativity*, Kyoto, June 1991, ed. H. Sato and T. Nakamura (World Scientific, Singapore, 1992), 988–1009.

262. S. Carusotto et al., Test of *g* universality with a Galileo type experiment, *Phys. Rev. Lett.* 69:1722–25 (1992).

263. T.P. Krisher, D.D. Morabito, and J.D. Anderson, The Galileo solar redshift experiment, *Phys. Rev. Lett.* 70:2213–16 (1993).

264. M.V. Moody and H.J. Paik, Gauss's law test of gravity at short range, *Phys. Rev. Lett.* 70:1195–98 (1993).

265. J.H. Taylor, Testing relativistic gravity with binary and millisecond pulsars, in *Proc. 13th Int. Conf. on General Relativity and Gravitation*, Cordoba, Argentina, June-July 1992, ed. R.J. Gleiser, C.N. Kozameh, and O.M. Moreschi (Inst. of Phys. Pub., Bristol and Philadelphia, 1993), 287–94.

266. G.E. Moss, L.R. Miller, and R.L. Forward, Photon-noise-limited laser transducer for gravitational antenna, *Applied Optics* 10:2495–98 (1971).

267. R.L. Forward, Wideband laser-interfero meter gravitational-radiation experiment, *Phys. Rev. D* 17:379–90 (1978).

268. R.W.P. Drever, Gravitational wave astronomy, *Quarterly J. Roy. Astron. Soc.* 18:9–27 (1977).

269. R. Weiss, Gravitational radiation: The status of the experiments and prospects for the future, in *Proc. Battelle: Sources of Gravitational Radiation*, July-August 1978, ed. L. Smarr (Cambridge University Press, Cambridge, 1979), 7–35.

270. C.N. Man et al., The Virgo project: A progress report, in *Proc. 5th Marcel Grossman Meeting on General Relativity*, Perth, August 1988, ed. D.G. Blair and M.J. Buckingham (World Scientific, Singapore, 1989), 1787–1801.

271. K. Danzmann, Laser interferometric gravitational wave detectors, in *Proc. 13th Int. Conf. on General Relativity and Gravitation*, Cordoba, Argentina, June-July 1992, ed. R.J. Gleiser, C.N. Kozameh, and O.M. Moreschi (Inst. of Phys. Pub., Bristol and Philadelphia, 1993), 3–19.

272. R.E. Vogt, The U.S. LIGO project, in *Proc. 6th Marcel Grossmann Meeting on General Relativity*, Kyoto, June 1991, ed. H. Sato and T. Nakamura (World Scientific Singapore, 1992), 244–66.

273. P. Bender, A. Brillet, I. Ciufolini, K. Danzmann, R. Hellings, J. Hough, A. Lobo, M. Sandford, B. Schutz and P. Touboul, LISA, Laser Interferometer Space Antenna for gravitational wave measurements, ESA report, to be published (1994).

274. A. Franklin, *The Rise and Fall of the Fifth Force: Discovery, Pursuit and Justification in Modern Physics* (Amer. Inst. of Phys., New York, 1993).

275. A. Gould, Deflection of light by the Earth, *Astroph. J. Lett.* 414:L37–L40 (1993)

276. A.G. Wiseman and C.M. Will, Christodoulou's nonlinear gravitational-wave memory: Evaluation in the quadrupole approximation, *Phys. Rev. D.* 44:2945–49 (1991).

277. P.J. Napier et al., The Very Long Baseline Array, *Proc. IEEE* 82:658–72 (1994).

278. B. Bertotti et al., Search for gravitational wave trains with the spacecraft ULYSSES, to be published in *Astron. Astrophys.* (1994).

279. J.O. Dickey et al., Lunar Laser Ranging: a continuing legacy of the Apollo program, *Science* 265:482–90 (1994).

4

Cosmology, Standard Models, and Homogeneous Rotating Models

In this chapter we describe the standard **Friedmann cosmological models** that represent **spatially homogeneous and isotropic model universes**, that is, models which satisfy the so-called **cosmological principle** of spatial homogeneity and isotropy. We then discuss the relations between **spatial compactness** and **closure in time**, that is, the dynamics of a universe that expands and eventually recollapses. We show that a universe "closed in time" must be spatially compact, with spatial topology equivalent to connected sums of copies of (S^3/P_i) and $(S^2 \times S^1)$; we also discuss under which conditions the converse might be true. A Riemannian model universe spatially isotropic about every point is also spatially homogeneous, and, vice versa, a Riemannian spatially homogeneous model universe, spatially isotropic about one point is spatially isotropic about every other point. However, in the second part of this chapter we describe some model universes that are **spatially homogeneous** but not necessarily spatially isotropic. After a brief introduction on the classification of the spatially homogeneous cosmological models, the **Bianchi models**, we describe the famous **Gödel rotating models** (the first Gödel type is homogeneous in both space and time), the Ozsváth-Schücking solutions, and the **Bianchi IX cosmological models** studied by, among others, Shepley and Matzner, Shepley, and Warren. We finally study the relations of **local inertial frames** (and gyroscopes) to the global rotation of matter in some of these **rotating models of universe**.

4.1 THE UNIVERSE ON A LARGE SCALE

The universe, on a scale of less than a few tens of megaparsecs (1 Mpc $=$ 3.26×10^6 lt-yr), is observed to be inhomogeneous.[1] We see stars, galaxies, groups and clusters of galaxies, superclusters of clusters, etc.[2–11] We then observe a large structure made of voids,[4,12,13] holes, clouds, filaments,[5,14] walls[201] and knots, etc., by different authors called bubbliness,[15] filamentariness,[10] texture,[16] sponginess,[17] etc. Some of the most striking examples are superclusters with dimensions reaching tens of megaparsecs, the enormous "hole in space," a void of about 100–200 h^{-1} millions of lt-yr diameter (h is thought to be between

$\cong 0.3$ and 1 and represents the Hubble constant in units of 100 km s^{-1} Mpc^{-1}; see next section) beyond the constellation of Boötes the Shepherd,[12] the long filaments[5,14] of more than 100 h^{-1} millions of lt-yr, the conjectured "Great Attractor,"[18–20] the "Great Wall," a sheet of galaxies[201] of size $\sim 50 \times 100h^{-1}$ Mpc at the distance of about 70 h^{-1} Mpc, of the Coma cluster (see also the so-called Southern-Wall and the southern sky redshift survey, ref. 223), and the claimed periodicity in the redshift distribution of galaxies in the deep pencil-beam survey.[200] Deviations from the velocity versus distance law seem to be observed in the motion of elliptical and spiral galaxies.[19–20] Beside the isotropic cosmological expansion, these galaxies seem to have a peculiar velocity field that might be explained by the gravitational attraction induced by an unseen conjectured "Great Attractor" located beyond the Hydra-Centaurus supercluster. According to some theories,[18] the conjectured "Great Attractor" should have a mass of the order of $5.4 \times 10^{16} M_{\odot}$, comparable to the largest superclusters, and should be at a distance of about 44 h^{-1} Mpc (for different interpretations see ref. 209). The nature of the mass of the conjectured "Great Attractor" is, at this writing, still unknown. Theories range from a concentration of clusters of galaxies in the direction of the "Great Attractor," to black holes, to a loop of cosmic string. . . .

Nevertheless, on a larger scale of the order of a few hundred millions of lt-yr, the universe appears to be essentially uniform and homogeneous.[1] The matter in the universe might be described as "expanding smoke;" (see the map of galaxies: picture 4.1 and fig. 4.1). This *very large scale spatial homogeneity* of the universe is one of the observational bases for the mathematical development of the standard **Friedmann cosmological models.**[21,22]

The main other observational basis for these standard cosmological models is the *average isotropy* of the universe around us. We observe an average, very large scale, isotropy[1] in the distribution of matter and galaxies, about us, up to few 10^9 lt-yr in depth. We also observe a *very large scale* isotropy[1] in the value of the Hubble constant, that is, isotropy in the redshift, or recession velocity versus distance relation of the galaxies around us, apart from our local motion and local deviations of the "Great Attractor" type. We find isotropy in the distribution of radio-sources,[23–28] and no more than a few percent variation in the distribution and isotropy of the cosmic X-ray background.[29–34]

The **background cosmic microwave radiation**[35–39] has been observed isotropic on various angular scales, with experiments using balloons and spacecraft, and ground-based, with an upper limit to anisotropies of a few parts in 10^{-5} (apart from anisotropies due to Earth's motion). However, the satellite **COBE** (COsmic Background Explorer)[40-44] has observed (1992) a structure in the background cosmic microwave radiation at a level of $\approx 10^{-5}$ (see below and pictures 4.2–4.4, and ref. 224). Fluctuations in the cosmic microwave background, at a level *of the order* of 10^{-5}, have been also observed in a number of ground-based and balloon measurements at various angular scales (see e.g. ref. 225).

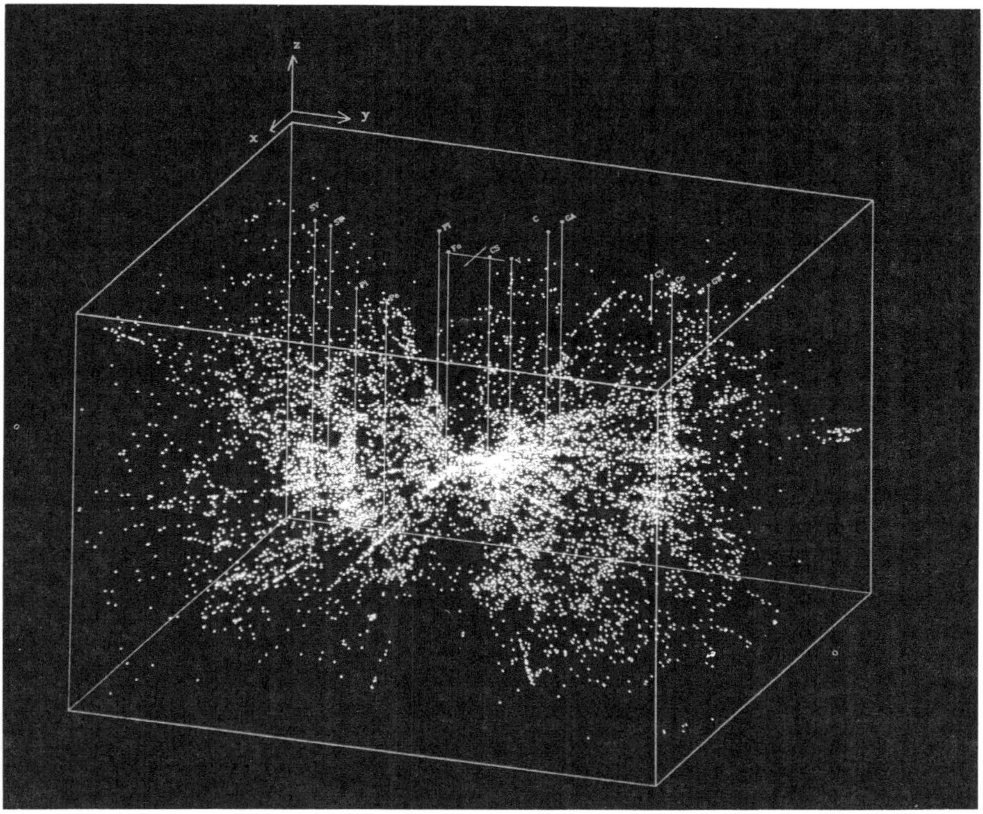

PICTURE 4.1. Our part of the universe. Distribution of 14,235 galaxies in the local neighborhood (in supergalactic coordinates). Our galaxy is located in the middle of the box (under the cross on the top plane). The box size is 200, 200, and 140 h^{-1} Mpc in the x, y, and z directions, respectively. Included are galaxies from the first 6° CfA slice, Southern-Sky Redshift Survey, and the "z-cat" catalog (galaxy apparent magnitudes brighter than 15.5). Symbols: CS = Coma-Sculptor cloud, V = Virgo cluster, Fo = Fornax cluster, PI = Pavo-Indus-Telescopium cloud, C = Centaurus cluster, GA = the "Great Attractor," CV = CfA Void, Co = Coma cluster, GW = the Great Wall, Pi = Pisces cluster, Pe = Perseus cluster, SV = a void in the southern sky, SW = a wall in the southern sky (courtesy of Changbom Park 1991).

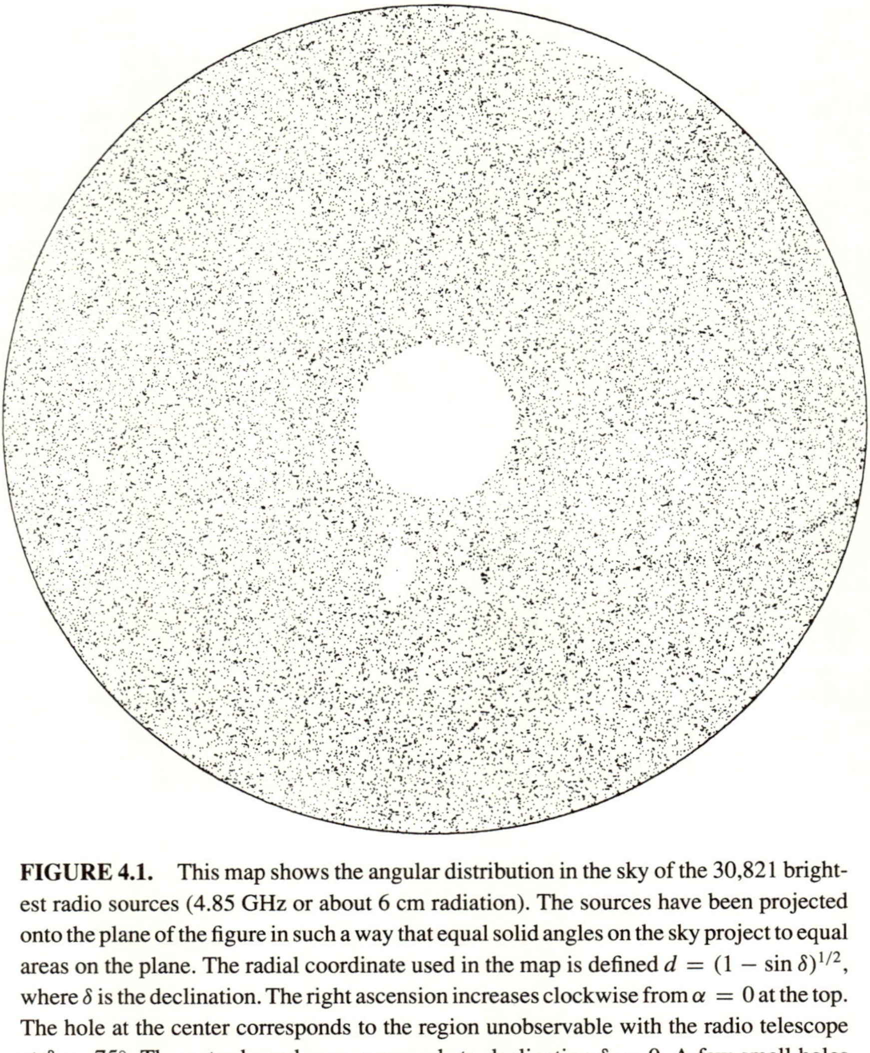

FIGURE 4.1. This map shows the angular distribution in the sky of the 30,821 brightest radio sources (4.85 GHz or about 6 cm radiation). The sources have been projected onto the plane of the figure in such a way that equal solid angles on the sky project to equal areas on the plane. The radial coordinate used in the map is defined $d = (1 - \sin\delta)^{1/2}$, where δ is the declination. The right ascension increases clockwise from $\alpha = 0$ at the top. The hole at the center corresponds to the region unobservable with the radio telescope at $\delta > 75°$. The outer boundary corresponds to declination $\delta = 0$. A few small holes appear where there are strong or extended sources in the galactic plane, located at about $\alpha = 19^h$ (from Gregory-Condon 1991,[193] reproduced here by permission, courtesy of J. J. Condon).

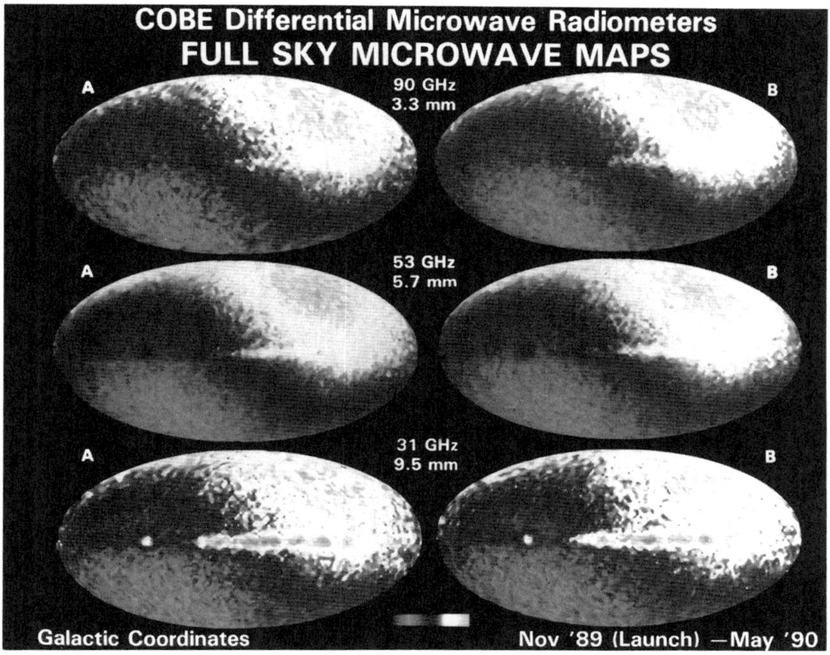

PICTURE 4.2. Images of Cosmic Microwave Background radiation constructed from preliminary data taken by the Differential Microwave Radiometers on NASA's Cosmic Background Explorer (COBE). Six images are shown corresponding to each of the two independent channels, marked "A" and "B," at each of the three frequencies: 31.5, 53, and 90 GHz. Galactic coordinates are used, with the plane of our Milky Way galaxy horizontal across the middle, and the galactic center at the center. The smooth variation on opposite sides of the sky, corresponding to a relative temperature variation $\frac{\Delta T}{T}$ of about 10^{-3}, known as the dipole anisotropy, is interpreted as due to the motion of the solar system relative to the cosmic microwave background; see § 4.1 (courtesy of NASA 1990).

According to standard models of the evolution of the universe[45,46] (see next sections), after an initial phase of thermodynamical equilibrium between matter and radiation, with very high temperatures and densities, the expansion of the universe reduced the temperature until eventually the plasma recombined, with a consequent reduction in the opacity, and the universe became transparent to radiation.[46] According to theoretical calculations this now freely propagating radiation had a blackbody spectrum.[47] The further expansion of the universe (see next sections) has redshifted this radiation, initially characterized by high temperatures, to a radiation with a blackbody spectrum corresponding to a temperature of a few degrees Kelvin. This cosmic blackbody radiation was

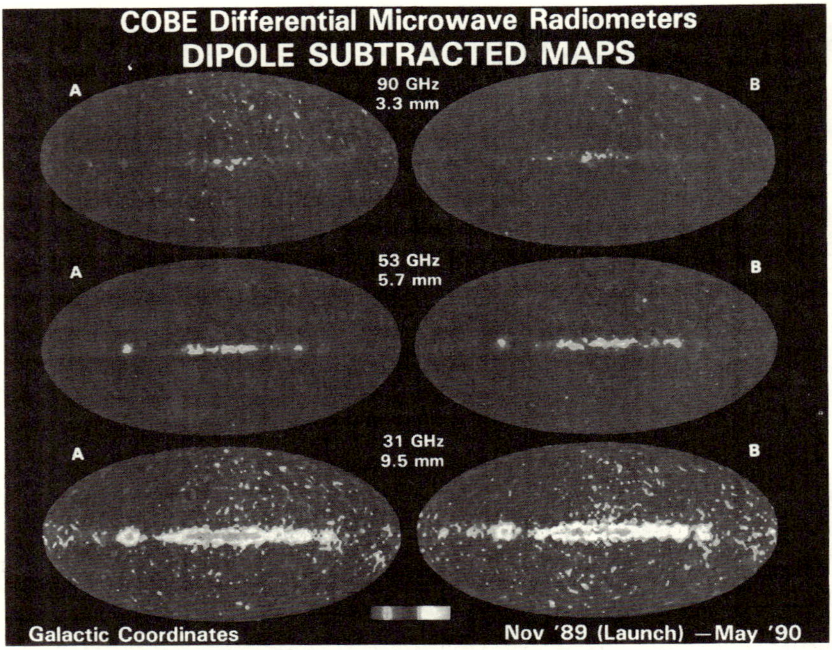

PICTURE 4.3. Same as picture 4.2, but after the dipole anisotropy due to the motion of the solar system has been subtracted from the maps. Radiation from our Milky Way galaxy can be seen in all the maps. The accuracy in detecting temperature variations $\frac{\Delta T}{T}$ corresponding to this 1990 image is $\sim 4 \times 10^{-5}$ (courtesy of NASA 1990).

predicted in 1946 by G. Gamow[47,48] (see also Alpher and Herman 1950[49] and Dicke et al. 1965[35]) and was detected in 1965 by Penzias and Wilson,[36] with a temperature of about 3 K. This cosmic blackbody microwave radiation has been observed[37–39] to be isotropic, in measurements at various wavelengths and over various integration angles, with an accuracy of about 10^{-4}–10^{-5}, apart from (1) the emission from the galactic plane and from (2) a dipole anisotropy of about $\frac{\Delta T}{T} \sim 10^{-3}$ due to the motion of our Local Group of galaxies. Nevertheless, COBE has observed anisotropies in the cosmic microwave background radiation at a characteristic level of about $\frac{\Delta T}{T} \approx 10^{-5}$; anisotropies of about the same *order of magnitude* have also been observed in ground-based and balloons experiments (see e.g. ref. 225).

The Cosmic Background Explorer satellite was launched in 1989 to observe conceivable irregularities in the blackbody cosmic background radiation, from microwave to far-infrared wavelengths. COBE carried three experiments; an absolute radiometer, to measure the infrared emission from the earliest galaxies, a spectrophotometer, to measure the spectrum of the cosmic background radia-

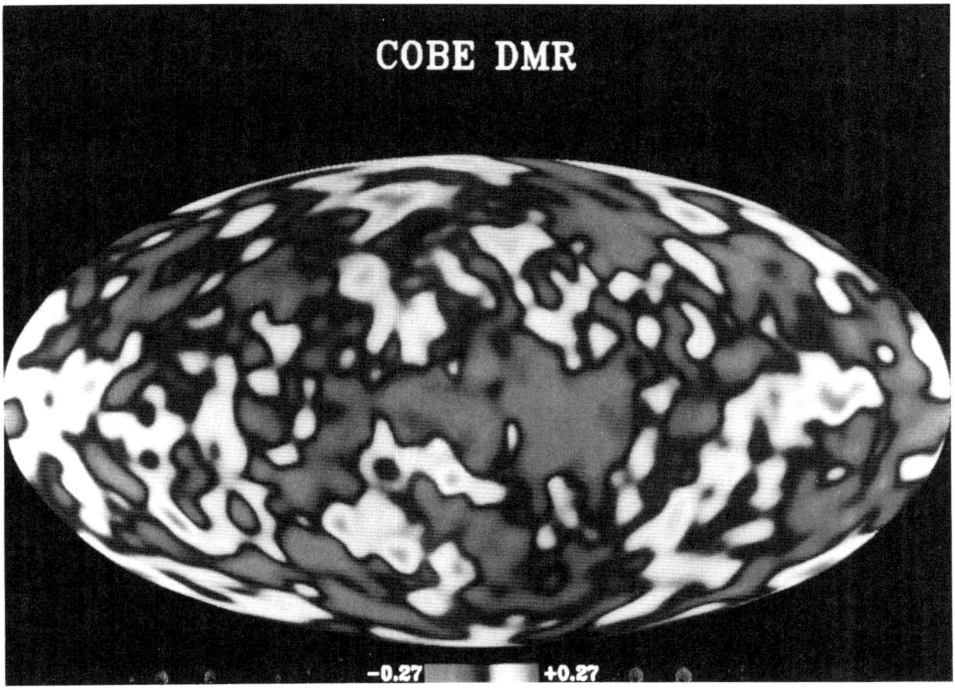

PICTURE 4.4. Microwave map (April 1992) of the whole sky made from one year of data taken by COBE Differential Microwave Radiometers (DMR). Dipole anisotropy due to the motion of the solar system and microwave emission from our Milky Way galaxy have been removed. The temperature variations $\frac{\Delta T}{T}$ shown correspond to about 10^{-5}. Most of the different intensity patches are the result of instrument noise, but computer analyses indicate that faint cosmic signals also are present with an estimated characteristic anisotropy of $\approx 6 \times 10^{-6}$. The Cosmic Microwave Background radiation observed by COBE corresponds, according to the standard models, to a time of emission of about 300,000 years after the big bang, that is, to a time of emission of the order of 15 billion years ago. The fluctuations observed may be consistent with a number of models of formation of galaxies; see § 4.1 (courtesy of NASA 1992).

tion, and differential microwave radiometers to observe any large angular scale anisotropies of the cosmic microwave background, that is, to measure small temperature differences between different directions in the sky at the three frequencies: 31.5, 53, and 90 GHz. The COBE observations have confirmed the Planck distribution of the cosmic background radiation at a temperature of $\cong 2.726 \pm 0.01$ K, from microwave to far-infrared wavelengths, with deviations from a blackbody spectrum of less than 3×10^{-4}.[215,216] Furthermore, only to a level of $\frac{\Delta T}{T} \cong 1.1 \times 10^{-5}$, and over integration angles of 7°, the COBE differential microwave radiometers have observed fluctuations in the isotropy of the cosmic microwave background radiation and a large angular scale ($\gtrsim 7°$) anisotropic structure in the cosmic radiation. After removal of the kinematical dipole and quadrupole anisotropy, due to local motion, and of the galactic emission, the COBE group has found a cosmic quadrupole anisotropy of about $Q_{RMS} = (13 \pm 4)\mu K$, or $(\frac{\Delta T}{T})_Q = (4.8 \pm 1.5) \times 10^{-6}$ from the first year of data, and $(\frac{\Delta T}{T})_Q = (2.2 \pm 1.1) \times 10^{-6}$ from the first two years of data[226] (however, see also other estimates, for example in refs. 213 and 214). Here, quadrupole anisotropy means a nonzero $l = 2$ term in the multipole expansion of the Cosmic Microwave Background temperature as a function of sky location: $T(\theta, \varphi) = \sum_l \sum_{m=-l}^{l} a_{lm} Y_{lm}(\theta, \varphi)$. The RMS quadrupole amplitude is calculated with the five $l = 2$ components. The large angular scale temperature anisotropies (see picture 4.4) observed by COBE, corresponding to radiation emitted about 300,000 years after the "big bang," suggest the existence of some form of dark matter (§ 4.2) and are consistent with a number of models of formation of galaxies (see also ref. 222).

In summary, on the average and to a substantial level (apart from small perturbations), the universe is observed to be isotropic about us, and on a very large scale of a few 10^8 lt-yr it appears to be essentially homogeneous.

In the next section we derive and study the general form of the Lorentzian metric which describes the overall dynamics of the universe under the mathematical assumptions of spatial homogeneity and spatial isotropy about one point, and therefore spatial isotropy about every other point, which is known as the **Friedmann-Robertson-Walker metric**.[21,22,50–52]

Regarding the study of the universe on scales of less than about 10^8 lt-yr we refer to other books[1,53] and papers[7–11] written on this topic. Two main competing theories deserve mention.

One is the "top-down" scenario or "pancake" theory of Zel'dovich et al.[7,54–56] According to this model, pancakes of gas of mass comparable to a supercluster were formed first, and then galaxies were formed by fragmentation of such pancakes. According to calculations based on this model, high-density regions were formed along lines, filaments and points, in agreement with the observed distribution of galaxies. The problem of the **missing mass**,[53] or **dark matter** (for the cosmological and the galactic missing mass problems, see § 4.2), and the problem of the agreement between the observed level of isotropy of the blackbody radiation (according to the "big-bang" theory emitted, or last scattered, some $\sim 10^5$ yr after the initial "big-bang") and the early formation of galaxies are explained in this *top-down model* by massive neutrinos, with a mass of a few electron-volts, which formed early condensations in the homogeneous plasma, in fact they should have decoupled from the primeval plasma before other particles. These hypothesized massive neutrinos, originally moving with relativistic speeds, are usually called *Hot Dark Matter*.

The other theory is the "*bottom-up*" *scenario* discussed by Lemaître[57] and supported by the work of Peebles et al.[10] In this scenario the galaxies were first formed by gravitational interaction and only afterwards was a hierarchy of clusters formed. Some versions of the bottom-up theory explain the missing mass problem and the formation of galaxies by assuming the existence of exotic particles,[58] such as "axions," "photinos," "selectrons," and "gravitinos," predicted by some unified field theories. This exotic form of dark matter, moving with velocities much slower than the massive neutrinos, is usually called *Cold Dark Matter*. *Mixed models*, hot plus cold dark matter, have also been proposed to try to fit contemporary observations.

4.2 HOMOGENEITY, ISOTROPY, AND THE FRIEDMANN COSMOLOGICAL MODELS

Derivation of a Spatially Homogeneous and Isotropic Metric and Its Properties

How can we find the metric of a spacetime which is spatially homogeneous and additionally has the feature that is spherically symmetric about one point; that is, it appears spatially isotropic from that point of observation? Alexander Friedmann found the metric of spatially homogeneous and spatially isotropic model universes in 1922,[21,22] but to arrive at the answer we start here by using a formalism of far-reaching power, the **Lie derivative**,[59–63] $\mathcal{L}_\xi \boldsymbol{T}$, of a tensor field $\boldsymbol{T}$ with respect to a vector field $\boldsymbol{\xi}$. This vector field $\boldsymbol{\xi}$ may be thought of as in imagination carrying points with coordinates x^α to nearby points with coordinates $x'^\alpha = x^\alpha + \varepsilon\xi^\alpha$, with $\varepsilon \ll 1$. Given a tensor field $\boldsymbol{T}(x^\alpha)$ one can then imagine at x'^α a new tensor field $\boldsymbol{T}'(x'^\alpha)$, defined by equation (4.2.2) below. By taking the difference between the original tensor $\boldsymbol{T}(x'^\alpha)$ at x'^α and the new tensor $\boldsymbol{T}'(x'^\alpha)$ also at x'^α, and dividing by ε, one then gets in the limit $\varepsilon \to 0$ the (so-called) Lie derivative

$$\left(\mathcal{L}_\xi \boldsymbol{T}\right)^{\alpha\cdots}{}_{\beta\cdots} = \xi^\sigma T^{\alpha\cdots}{}_{\beta\cdots,\sigma} - T^{\sigma\cdots}{}_{\beta\cdots}\xi^\alpha{}_{,\sigma} + T^{\alpha\cdots}{}_{\sigma\cdots}\xi^\sigma{}_{,\beta} + \cdots. \tag{4.2.1}$$

The Lie derivative has a simple interpretation. It measures how much a tensor field $\boldsymbol{T}(x^\alpha)$ deviates from being formally invariant under the infinitesimal transformation $x'^\alpha = x^\alpha + \varepsilon\xi^\alpha$, with $\varepsilon \ll 1$. We recall that formal invariance of $T^{\beta\cdot\cdot}{}_{\gamma\cdot\cdot}(x^\alpha)$ requires that the transformed tensor field $T'^{\beta\cdot\cdot}{}_{\gamma\cdot\cdot}(x'^\alpha)$ have the same functional dependence on the new coordinates as the original tensor field had on the old coordinates; that is, that $T^{\beta\cdot\cdot}{}_{\gamma\cdot\cdot}(y^\alpha \equiv x^\alpha) = T'^{\beta\cdot\cdot}{}_{\gamma\cdot\cdot}(y^\alpha \equiv x'^\alpha)$. Let us derive equation (4.2.1). For simplicity we just consider here the (1,1) tensor $T^\alpha{}_\beta$. The results generalize in an obvious way to higher rank tensors.

For $x'^{\alpha} = x^{\alpha} + \varepsilon\xi^{\alpha}$, we have

$$T'^{\alpha}{}_{\beta}(x') = \partial^{\alpha'}{}_{\sigma}\partial^{\rho}{}_{\beta'}T^{\sigma}{}_{\rho}(x)$$
$$= T^{\alpha}{}_{\beta}(x) + \varepsilon\xi^{\alpha}{}_{,\sigma}T^{\sigma}{}_{\beta}(x) - \varepsilon\xi^{\rho}{}_{,\beta}T^{\alpha}{}_{\rho}(x); \tag{4.2.2}$$

with a Taylor expansion, we also have

$$T^{\alpha}{}_{\beta}(x') = T^{\alpha}{}_{\beta}(x) + \varepsilon\xi^{\sigma}T^{\alpha}{}_{\beta,\sigma}(x). \tag{4.2.3}$$

By taking the difference between (4.2.3) and (4.2.2), and dividing by ε, in the limit $\varepsilon \rightarrow 0$, we get

$$\left(\mathcal{L}_{\xi}\boldsymbol{T}\right)^{\alpha}{}_{\beta} = \xi^{\sigma}T^{\alpha}{}_{\beta,\sigma} - \xi^{\alpha}{}_{,\sigma}T^{\sigma}{}_{\beta} + \xi^{\sigma}{}_{,\beta}T^{\alpha}{}_{\sigma}. \tag{4.2.4}$$

Therefore, when $\mathcal{L}_{\xi}\boldsymbol{T} = 0$, we have equality between (4.2.2) and (4.2.3), or

$$T'^{\alpha}{}_{\beta}(x') = T^{\alpha}{}_{\beta}(x') \tag{4.2.5}$$

or similarly

$$T'^{\alpha}{}_{\beta}(x) = T^{\alpha}{}_{\beta}(x),$$

that is, the tensor field is formally invariant under the infinitesimal transformation $x'^{\alpha} = x^{\alpha} + \varepsilon\xi^{\alpha}$, or the transformed tensor field $T'^{\alpha}{}_{\beta}$ has the same functional dependence on the x'^{α} that $T^{\alpha}{}_{\beta}$ had on the x^{α}.

The properties of the Lie derivative allow it to describe not only the symmetries of tensor fields on the manifold but also the symmetries of the manifold itself. Indeed, if $\mathcal{L}_{\xi}\boldsymbol{g} = \boldsymbol{0}$ the metric $\boldsymbol{g}$ is formally invariant under the transformation $x'^{\alpha} = x^{\alpha} + \varepsilon\xi^{\alpha}$ that may be thought of as "carrying" points with coordinates x^{α} to nearby points with coordinates x'^{α}.

In this case the transformation is called an **isometry** of $\boldsymbol{g}$, and the vector field ξ^{α} is called a **Killing vector**.[64] Killing vectors represent the isometries, that is, the symmetries of the manifold with metric $\boldsymbol{g}$. On taking the Lie derivative of $\boldsymbol{g}$, with respect to $\boldsymbol{\xi}$, we have

$$\mathcal{L}_{\xi}\boldsymbol{g} \equiv g_{\alpha\beta,\sigma}\xi^{\sigma} + g_{\sigma\beta}\xi^{\sigma}{}_{,\alpha} + g_{\alpha\sigma}\xi^{\sigma}{}_{,\beta}$$
$$= g_{\alpha\beta,\sigma}\xi^{\sigma} + \xi_{\beta,\alpha} - g_{\sigma\beta,\alpha}\xi^{\sigma} + \xi_{\alpha,\beta} - g_{\alpha\sigma,\beta}\xi^{\sigma} \tag{4.2.6}$$
$$= \xi_{\alpha,\beta} - \Gamma^{\sigma}_{\alpha\beta}\xi_{\sigma} + \xi_{\beta,\alpha} - \Gamma^{\sigma}_{\beta\alpha}\xi_{\sigma} = \xi_{\alpha;\beta} + \xi_{\beta;\alpha}.$$

This is the Lie derivative of the metric tensor. If the infinitesimal transformation $x'^{\alpha} = x^{\alpha} + \varepsilon\xi^{\alpha}$ is an isometry, then ξ^{α} is a Killing vector, and we have the Killing equation

$$\xi_{(\alpha;\beta)} = 0 \tag{4.2.7}$$

where the round brackets mean symmetrization (see the mathematical appendix).

The set of all the infinitesimal isometries, that is, the set of all the Killing vector fields on a Riemannian manifold, is a *Lie algebra*,[61–63] that is, a finite-dimensional vector space with a multiplicative operation $[\boldsymbol{\xi}_A, \boldsymbol{\xi}_B]$ such that

$$[\boldsymbol{\xi}_A, \boldsymbol{\xi}_B] = -[\boldsymbol{\xi}_B, \boldsymbol{\xi}_A] \tag{4.2.8}$$

$$[\boldsymbol{\xi}_A, [\boldsymbol{\xi}_B, \boldsymbol{\xi}_C]] + [\boldsymbol{\xi}_B, [\boldsymbol{\xi}_C, \boldsymbol{\xi}_A]] + [\boldsymbol{\xi}_C, [\boldsymbol{\xi}_A, \boldsymbol{\xi}_B]] = \mathbf{0}. \tag{4.2.9}$$

This last cyclic property is the so-called Jacobi identity. The multiplication operation of this Lie algebra is given by the Lie derivative $[\boldsymbol{\xi}_A, \boldsymbol{\xi}_B] \equiv \mathcal{L}_{\boldsymbol{\xi}_A}\boldsymbol{\xi}_B$, see § 4.4. The group of isometries on a Riemannian manifold (with a finite number of connected components) is a *Lie group*[60–63] (a group which itself is also a smooth manifold and in which xy^{-1} is a C^∞ mapping; see mathematical appendix). On a geodesically complete Riemannian manifold, the Lie algebra of the Lie group of isometries is given by the Lie algebra of all the infinitesimal isometries of the manifold, that is, all the Killing vector fields, generators of the Lie group.

The concepts of homogeneity and isotropy of an n-dimensional manifold are defined most naturally in terms of the isometry groups of that manifold.

A metric $\boldsymbol{g}$ is defined to be homogeneous if it admits an isometry group acting transitively on it, that is, for any two points on the manifold there is an element of this isometry group (that is, an isometry) that carries one of the two points into the other; in other words, the metric is formally invariant for "translations."

A metric $\boldsymbol{g}$ (of an n-dimensional manifold) is then defined to be isotropic about a point P if it admits an isometry group of all the isometries that leave the point fixed (isotropy group) and this isometry group is (isomorphic to) $\mathrm{SO}(n)$, the group of the n-dimensional rotations.

One can prove[46] that the maximum number of independent Killing vectors of a manifold of dimension N is $\frac{N(N+1)}{2}$. When this number is reached, the metric $\boldsymbol{g}$ is called *maximally symmetric*. A metric $\boldsymbol{g}$ is then maximally symmetric if and only if it is homogeneous and isotropic. In a three-dimensional maximally symmetric space there are six Killing vectors, three corresponding to translations and three to rotations.

In the previous section we have seen that the universe is observed to be spatially isotropic about us to a significant level and also, on a scale of a few hundred million lt-yr, essentially homogeneous. Following the concept of field equations of the type of the Einstein field equation (with metric as the field, and mass-energy as the source of field), we go on to assume that homogeneity and isotropy in the distribution and motion of the mass-energy in the universe imply the mathematical homogeneity and isotropy of the metric of the spacetime Lorentzian manifold. We then introduce the most general metric satisfying the mathematical requirements of spatial isotropy about one point, that is, spherical symmetry about one point, and spatial homogeneity. It will explicitly turn out that this metric is also isotropic about every other point.

The general expression for a spacetime metric that is spatially spherically symmetric about one point was derived in section 2.6 using the Killing vectors representing spatial spherical symmetry:

$$ds^2 = A(r,t)dt^2 + B(r,t)dr^2 + C(r,t)drdt + D(r,t)(d\theta^2 + \sin^2\theta d\phi^2). \tag{4.2.10}$$

We now require that the metric (4.2.10), **isotropic about the origin** of r, θ, ϕ, shall also be **spatially homogeneous**, that is, formally invariant under the transformations defined by the Killing vectors representing spatial translations. Let us first write the expression of the **Killing vectors representing spatial homogeneity**.

If a spacelike homogeneous hypersurface is flat, that is, if it has vanishing intrinsic curvature, its metric, in some coordinate system (x^1, x^2, x^3), will be formally invariant for translations along each of the three axes x^1, x^2, and x^3. In this case these isometries correspond to the three Killing vectors $\xi^i_{(n)} = \delta^i{}_n$, where $n \epsilon \{1, 2, 3\}$. With the standard transformation of section 2.6, $x^1 = \chi \sin\theta \cos\phi$, $x^2 = \chi \sin\theta \sin\phi$, and $x^3 = \chi \cos\theta$, we then have

$$\begin{aligned}
\boldsymbol{\xi}_I &= \left(\sin\theta\cos\phi, \frac{\cos\theta\cos\phi}{\chi}, -\frac{\sin\phi}{\chi\sin\theta}\right)\\
\boldsymbol{\xi}_{II} &= \left(\sin\theta\sin\phi, \frac{\cos\theta\sin\phi}{\chi}, \frac{\cos\phi}{\chi\sin\theta}\right)\\
\boldsymbol{\xi}_{III} &= \left(\cos\theta, -\frac{\sin\theta}{\chi}, 0\right).
\end{aligned} \tag{4.2.11}$$

If the metric is spatially nonflat, that is, if the spacelike homogeneous hypersurface has nonvanishing intrinsic curvature, by analogy with the non-flat homogeneous two-surfaces embedded in a flat three-space (that is, the two-sphere, with constant positive curvature, embedded in $\mathfrak{R}^3$, and the two-hyperboloid, with constant negative curvature, embedded in M^3, flat three-space with signature $(-1, +1, +1)$), we imagine the spacelike homogeneous hypersurface embedded in some four-dimensional flat manifold, with coordinates (x^1, x^2, x^3, x^4). The metric of the hypersurface should then be formally invariant for rotations about some point P of the flat four-manifold.

Indeed, in the case of a hypersurface embedded in $\mathfrak{R}^4$, the Killing vectors representing infinitesimal translations on the hypersurface, that is, representing the homogeneity of the hypersurface, are $\boldsymbol{\xi}_I = (x^4, -x^3, x^2, -x^1)$, $\boldsymbol{\xi}_{II} = (x^3, x^4, -x^1, -x^2)$, and $\boldsymbol{\xi}_{III} = (-x^2, x^1, x^4, -x^3)$; and each Killing vector is the composition of two infinitesimal four-dimensional rotations in $\mathfrak{R}^4$. However, for simplicity, let us just use the Killing vectors representing, on the homogeneous hypersurface, an infinitesimal translation in a neighborhood of the origin O, around which the metric is also spatially spherically symmetric. Such a translation may be obtained by a four-dimensional rotation that keeps fixed two axes not through $O \equiv (0, 0, 0, \text{constant})$. It translates all the points in a neighborhood of O. Such four-dimensional rotations take place in the planes (x^1, x^4), (x^2, x^4), and (x^3, x^4) and therefore correspond to the Killing vectors: $\xi^\alpha_I = (x^4, 0, 0, -x^1)$, $\xi^\alpha_{II} = (0, x^4, 0, -x^2)$, and $\xi^\alpha_{III} = (0, 0, x^4, -x^3)$. We can then perform the transformation $x^1 = a \sin\chi \sin\theta \cos\phi$, $x^2 = a \sin\chi \sin\theta \sin\phi$, $x^3 = a \sin\chi \cos\theta$, and $x^4 = a\cos\chi$, and we get the Killing vectors representing infinitesimal translations in a neighborhood of the point O on the spacelike hypersurface, expressed in the coordinates χ, θ, ϕ,

$$\boldsymbol{\xi}_I = \left(\sin\theta\cos\phi,\ \frac{\cos\chi}{\sin\chi}\cos\theta\cos\phi,\ -\frac{\cos\chi}{\sin\chi}\frac{\sin\phi}{\sin\theta}\right)$$

$$\boldsymbol{\xi}_{II} = \left(\sin\theta\sin\phi,\ \frac{\cos\chi}{\sin\chi}\cos\theta\sin\phi,\ \frac{\cos\chi}{\sin\chi}\frac{\cos\phi}{\sin\theta}\right)$$

$$\boldsymbol{\xi}_{III} = \left(\cos\theta,\ -\frac{\cos\chi}{\sin\chi}\sin\theta,\ 0\right). \tag{4.2.12}$$

In the case of a hypersurface embedded in M^4, Minkowski space, the four-dimensional rotations in the planes (x^1, x^4), (x^2, x^4) and (x^3, x^4) correspond to the Killing vectors: $\xi^\alpha_I = (x^4, 0, 0, x^1)$, $\xi^\alpha_{II} = (0, x^4, 0, x^2)$, and $\xi^\alpha_{III} = (0, 0, x^4, x^3)$. We can then perform the transformation $x^1 = a\sinh\chi\sin\theta\cos\phi$, $x^2 = a\sinh\chi\sin\theta\sin\phi$, $x^3 = a\sinh\chi\cos\theta$, and $x^4 = a\cosh\chi$, and we get the Killing vectors representing infinitesimal translations in a neighborhood of the point O on the spacelike hypersurface, expressed in the coordinates χ, θ, ϕ,

$$\boldsymbol{\xi}_I = \left(\sin\theta\cos\phi,\ \frac{\cosh\chi}{\sinh\chi}\cos\theta\cos\phi,\ -\frac{\cosh\chi}{\sinh\chi}\frac{\sin\phi}{\sin\theta}\right)$$

$$\boldsymbol{\xi}_{II} = \left(\sin\theta\sin\phi,\ \frac{\cosh\chi}{\sinh\chi}\cos\theta\sin\phi,\ \frac{\cosh\chi}{\sinh\chi}\frac{\cos\phi}{\sin\theta}\right) \tag{4.2.13}$$

$$\boldsymbol{\xi}_{III} = \left(\cos\theta,\ -\frac{\cosh\chi}{\sinh\chi}\sin\theta,\ 0\right).$$

To find the spatially spherically symmetric and homogeneous four-dimensional metric, we then use the general *spatially* spherically symmetric metric (2.6.13) (observing that the expression of the Killing vectors 2.6.8 is unchanged whether $r \equiv \chi$, or $r \equiv \sin\chi$, or $r \equiv \sinh\chi$) plus the Killing equation (4.2.7) and just one Killing vector $\boldsymbol{\xi}_{III}$ representing an infinitesimal translation in a neighborhood of O:

$$\boldsymbol{\xi}_{III} = (\cos\theta, -Y\sin\theta, 0) \tag{4.2.14}$$

where $Y \equiv \frac{1}{\chi}$, or $Y \equiv \frac{\cos\chi}{\sin\chi}$, or $Y \equiv \frac{\cosh\chi}{\sinh\chi}$, corresponding to the three cases (4.2.11), (4.2.12), and (4.2.13).

From the Killing equation (4.2.7) and equation (4.2.6), for $(\alpha, \beta) = (0, 0)$, we find that $g_{00} = f(t)$, that is, g_{00} is a function of t only; for $(\alpha, \beta) = (2, 2)$, we get respectively $\frac{g_{22,\chi}}{g_{22}} = \frac{2}{\chi}$, $\frac{g_{22,\chi}}{g_{22}} = 2\frac{\cos\chi}{\sin\chi}$, and $\frac{g_{22,\chi}}{g_{22}} = 2\frac{\cosh\chi}{\sinh\chi}$, and integrating $g_{22} = \chi^2 l^2(t)$, $g_{22} = \sin^2\chi l^2(t)$, and $g_{22} = \sinh^2\chi l^2(t)$, where $l^2(t)$ is a function of t only and $g_{22} = g_{33}/\sin^2\theta$; for $(\alpha, \beta) = (1, 2)$, we have $g_{11} = l^2(t)$; finally for $(\alpha, \beta) = (0, 2)$, we get $g_{01} = 0$. Therefore, the spatially spherically symmetric and homogeneous metric is $ds^2 = f(t)dt^2 + l^2(t)\big(d\chi^2 + X^2 d\Omega^2\big)$ where $X \equiv \chi$, or $X \equiv \sin\chi$, or $X \equiv \sinh\chi$, corresponding to the above three cases. With a transformation of the time coordinate t we may set $f(t) = -1$, and we have (also indicating with t the new time coordinate)

$$ds^2 = -dt^2 + R^2(t)\left(d\chi^2 + X^2 d\Omega^2\right). \tag{4.2.15}$$

This metric may be further rewritten with the transformation $r = \chi$, or $r = \sin\chi$, or $r = \sinh\chi$, corresponding to each of the above three cases:

$$ds^2 = -dt^2 + R^2(t)\left[\frac{1}{1 - kr^2}dr^2 + r^2 d\theta^2 + r^2\sin^2\theta d\phi^2\right] \tag{4.2.16}$$

where k is equal to 0, or +1, or −1. This is the **Friedmann-Robertson-Walker**[21,22,50–52] **metric** that we have obtained here under the hypotheses of (a) spatial homogeneity, that is, formal invariance of $\boldsymbol{g}$ under the transformations representing infinitesimal spatial "translations," and (b) spatial isotropy about one point, that is, formal invariance of $\boldsymbol{g}$ under the transformations (2.6.8) representing infinitesimal spatial rotations about one point.

By integrating the infinitesimal transformations defined by the Killing vectors representing spatial translations, we can carry the origin of our spatial coordinate system to any other point on the manifold. Of course, since we have derived the metric (4.2.16) under the hypothesis of formal invariance for spatial translations, we can rewrite the metric (4.2.16) in a form having the same functional dependence on the new coordinates, r', θ', ϕ', as it had on the old coordinates, r, θ, ϕ, that is, $\boldsymbol{g}'(\boldsymbol{x}') = \boldsymbol{g}(\boldsymbol{x}')$:

$$ds^2 = -dt^2 + R^2(t)\left[\frac{1}{1 - kr'^2}dr'^2 + r'^2 d\theta'^2 + r'^2\sin^2\theta' d\phi'^2\right]. \tag{4.2.17}$$

This transformability in the large has also been rigorously shown (Tolman 1934;[65] see also Raychaudhuri 1979)[66] via a finite coordinate transformation and via an embedding of the four-dimensional spacetime in a five-dimensional manifold (see also § 4.3).

An infinitesimal spatial rotation about the new origin is represented for the metric (4.2.17) by the Killing vectors (2.6.8) with r', θ', ϕ' in place of r, θ, ϕ. Therefore, since we have derived the metric (4.2.16) by using its formal invariance under infinitesimal spatial rotations about the origin of r, θ, ϕ represented by expression (2.6.8) (and under spatial translations), from expression (4.2.17) we have that this metric must also be invariant under infinitesimal spatial rotations about the new origin of r', θ', ϕ'. Since the new origin is arbitrary, the metric is isotropic about every point. Then, a three-dimensional Riemannian manifold that is *spatially isotropic about one point and spatially homogeneous* is also *spatially isotropic about every other point*. This explicitly shows a three-dimensional case of the general property that homogeneity plus isotropy about one point imply isotropy about every other point.[46]

Therefore, metric (4.2.16) is the metric of a manifold spatially homogeneous and spatially isotropic about every point. As we have already observed, a homogeneous and isotropic manifold has the maximum number of symmetries $\frac{N(N+1)}{2}$ and is thus called maximally symmetric.

On a very large scale the universe appears to be spatially isotropic (apart from small perturbations) and homogeneous (§ 4.1) but on smaller scales not isotropic and not homogeneous. How then can one most understandingly reconcile the observations with an idealized spatially isotropic and homogeneous model universe?[67–69] A possible approach[69] focuses on the so-called **fitting problem** (Ellis 1984):[68] one does not *a priori* and necessarily assume a spatially isotropic and homogeneous universe, but instead one searches for an isotropic and homogeneous model that would best fit the real universe.

First, one considers a "realistic lumpy" cosmological model, which consists of a model with physical quantities, such as the matter four-velocity, energy density, galaxy number-density, etc., describing the universe with its inhomogeneities. Then, one considers an idealized spatially isotropic and spatially homogeneous cosmological model. The problem is to determine a "best fit" between these two cosmological models: the "realistic lumpy" model and an idealized spatially isotropic and homogeneous model. One of the methods proposed is called the **null data approach to the fitting problem**.[69] With this method one considers a point of the "realistic lumpy" model, together with a properly defined four-velocity, and a corresponding point of an idealized Friedmann model, with its four-velocity. Then after fitting the four-velocities at the two points in the two models, and after a suitable average of the quantities of the "lumpy" model over some distance on the past null cone, one can find the parameters determining a "best-fit" spatially isotropic and homogeneous model universe, and then seek on the past null cone a best-fitting of this model to the "lumpy" model universe. Finally, one constructs the "best-fit" spacetime inside and outside the null cone from the initial data on it. Using this approach one also determines an averaging scale, such that over that length scale one gets an idealized isotropic and homogeneous model to a certain good approximation.

At every point P of a manifold we can define the **sectional curvature**[70] at P associated with a given orientation Σ_p, where the tangent plane Σ_p at P is determined by two independent vectors $\boldsymbol{v}_1$, $\boldsymbol{v}_2$ of the n-dimensional tangent space T_p at P:

$$K_{\Sigma_p} = \frac{R_{\alpha\beta\mu\nu} v_1^\alpha v_2^\beta v_1^\mu v_2^\nu}{(g_{\alpha\mu} g_{\beta\nu} - g_{\alpha\nu} g_{\beta\mu}) v_1^\alpha v_2^\beta v_1^\mu v_2^\nu}. \tag{4.2.18}$$

This is the Gaussian curvature of a two-dimensional submanifold formed by all the geodesics through P such that their tangent vector at P is in the tangent plane Σ_p.

From the antisymmetry of $R_{\alpha\beta\mu\nu}$ and of $(g_{\alpha\mu} g_{\beta\nu} - g_{\alpha\nu} g_{\beta\mu})$ with respect to $\alpha\beta$ and with respect to $\mu\nu$, the sectional curvature has the same value for any two independent vectors that are linear combinations of $\boldsymbol{v}_1$ and $\boldsymbol{v}_2$, that is, it has the same value for any two independent vectors in Σ_p, and in fact it defines the curvature of the manifold at P associated with the orientation determined by Σ_p. In the case of a two-dimensional manifold (a surface), the sectional curvature (4.2.18) turns out to be the ordinary **Gaussian curvature**[70] of that surface (see the mathematical appendix).

From the definition of isotropy, that is, formal invariance of the metric for infinitesimal rotations, it follows that by a rotation about a point P we can transform from one local orientation Σ_p, determined by $\boldsymbol{v}_1$ and $\boldsymbol{v}_2$, to any other local orientation Σ'_p, while keeping unchanged the values of $R_{\alpha\beta\mu\nu}$ and $(g_{\alpha\mu} g_{\beta\nu} - g_{\alpha\nu} g_{\beta\mu})$ at P. Therefore, for an isotropic manifold the sectional curvature (4.2.18) is independent of Σ_p; that is, the sectional curvature (4.2.18) at P has the same value for every choice of $\boldsymbol{v}_1$, $\boldsymbol{v}_2$ in T_p. If a quantity $X_{\alpha\beta\mu\nu} v_1^\alpha v_1^\mu v_2^\beta v_2^\nu$ is zero for every choice of v_1^α and v_2^β, we must have

$$X_{\alpha\beta\mu\nu} + X_{\alpha\nu\mu\beta} + X_{\mu\beta\alpha\nu} + X_{\mu\nu\alpha\beta} = 0. \tag{4.2.19}$$

Indeed, $Y_{\alpha\beta} w^\alpha w^\beta = 0$ for every w^α implies $Y_{\alpha\beta} + Y_{\beta\alpha} = 0$. Furthermore, if $X_{\alpha\beta\mu\nu}$ has the same symmetry properties of the Riemann tensor

$$\begin{cases} R_{\alpha\beta\mu\nu} = -R_{\beta\alpha\mu\nu}; & R_{\alpha\beta\mu\nu} = -R_{\alpha\beta\nu\mu} \\ R_{\alpha[\beta\mu\nu]} = 0 \quad \text{and} & R_{\alpha\beta\mu\nu} = R_{\mu\nu\alpha\beta} \end{cases} \tag{4.2.20}$$

we have

$$X_{\alpha\beta\mu\nu} = 0, \tag{4.2.21}$$

and, since the sectional curvature (4.2.18) has the same value for every $\boldsymbol{v}_1$, $\boldsymbol{v}_2$, for $X_{\alpha\beta\mu\nu} \equiv R_{\alpha\beta\mu\nu} - K_p(g_{\alpha\mu} g_{\beta\nu} - g_{\alpha\nu} g_{\beta\mu})$, we have:

$$R_{\alpha\beta\mu\nu} = K_p(g_{\alpha\mu} g_{\beta\nu} - g_{\alpha\nu} g_{\beta\mu}). \tag{4.2.22}$$

From this expression, we then have for the Ricci curvature of an n-dimensional manifold:

$$R^\alpha{}_\beta \equiv R^{\sigma\alpha}{}_{\sigma\beta} = (n-1) K_p \delta^\alpha{}_\beta \tag{4.2.23}$$

$$R \equiv R^\alpha{}_\alpha = n(n-1) K_p, \tag{4.2.24}$$

and for the Einstein curvature tensor

$$G^{\alpha}{}_{\beta} \equiv R^{\alpha}{}_{\beta} - \frac{1}{2}\delta^{\alpha}{}_{\beta}R = (n-1)(2-n)\frac{K_p}{2}\delta^{\alpha}{}_{\beta}, \tag{4.2.25}$$

and finally from the Bianchi identities

$$G^{\sigma}{}_{\beta;\sigma} = 0 \tag{4.2.26}$$

for $n > 2$, we conclude that

$$K_{p,\beta} = 0 \qquad \text{or} \qquad K_p = K = \text{constant}. \tag{4.2.27}$$

In other words, the curvature is constant over the manifold, that is, is independent of the location. Such a space is called a manifold of constant Riemannian curvature. Then, the manifold is homogeneous, locally, and isotropic (homogeneity defined by the transitive group of isometries, see after formula 4.2.9).[71]

We therefore have the *Schur theorem.*[72] If a Riemannian manifold is *isotropic about every point*, then it is also *homogeneous.*

From the Friedmann-Robertson-Walker metric (4.2.16) and from expression (4.2.18) for K, using the definition of Ricci curvature and equation (4.2.24), we conclude that the Ricci scalar curvature, ${}^{(3)}R$, of the hypersurfaces t = constant is given by

$${}^{(3)}R = 6K = \frac{6k}{R^2(t)}. \tag{4.2.28}$$

Then, the spatial hypersurfaces t = constant, of the Friedmann-Robertson-Walker metric (4.2.16) are hypersurfaces of positive constant spatial curvature if k is $+1$, spatially flat if $k = 0$, and of negative constant spatial curvature if k is -1 (see fig. 4.5).

The spatially homogeneous and isotropic metric (4.2.16) is sometimes written, with the coordinate transformation $r = \frac{\rho}{(1+\frac{1}{4}k\rho^2)}$,

$$\begin{aligned} ds^2 &= -dt^2 + R^2(t)\frac{(d\rho^2 + \rho^2 d\theta^2 + \rho^2 \sin^2\theta\, d\phi^2)}{(1+\frac{1}{4}k\rho^2)^2} \\ &= -dt^2 + R^2(t)\frac{(dx^2+dy^2+dz^2)}{[1+\frac{1}{4}k(x^2+y^2+z^2)]^2}. \end{aligned} \tag{4.2.29}$$

Another useful form for the Friedmann metric (4.2.16) is expression (4.2.15), given by the coordinate transformations

$$\begin{aligned} r &= \sin\chi && \text{for} \quad k = +1 \\ r &= \chi && \text{for} \quad k = 0 \\ r &= \sinh\chi && \text{for} \quad k = -1; \end{aligned} \tag{4.2.30}$$

that is,

$$
\begin{aligned}
ds^2 &= -dt^2 + R^2(t)[d\chi^2 + \sin^2\chi d\Omega^2], && \text{for} \quad k = +1\\
ds^2 &= -dt^2 + R^2(t)[d\chi^2 + \chi^2 d\Omega^2], && \text{for} \quad k = 0 \qquad (4.2.31)\\
ds^2 &= -dt^2 + R^2(t)[d\chi^2 + \sinh^2\chi d\Omega^2], && \text{for} \quad k = -1
\end{aligned}
$$

where, as usual, $d\Omega^2 \equiv d\theta^2 + \sin^2\theta d\phi^2$

Now, if the universe is spatially homogeneous and isotropic, the four-velocity vector of what we can call the cosmological fluid must have zero Lie derivative for any vector field $\boldsymbol{\xi}$ representing infinitesimal spatial translations and infinitesimal spatial rotations about an arbitrary point, that is, u^α must be formally invariant for spatial translations and rotations. Therefore, by using the definition (4.2.1) of Lie derivative of a vector, with Killing vectors taken to represent rotations and translations, one can conclude from the metric (4.2.16)

$$u^0 = 1, u^k = 0, \qquad (4.2.32)$$

that is, the particles of the cosmological fluid are at rest with respect to the coordinate system of expression (4.2.16). Thus, the fluid particles themselves define the spatial coordinate grid. Such a coordinate system is said to comove with the matter.

For every Friedmann model universe, even though the pressure is in general different from zero, $p \neq 0$, from expression (4.2.16) we have that the particles of the cosmological perfect fluid follow geodesic motion

$$\frac{Du^\alpha}{ds} = \frac{du^\alpha}{ds} + \Gamma^\alpha_{00} = \frac{du^\alpha}{ds} = 0. \qquad (4.2.33)$$

Furthermore, using the spacelike Killing vectors admitted by the Friedmann-Robertson-Walker metric and using the Killing equation for a scalar ϕ: $\phi_{,\alpha}\,\xi^\alpha = 0$, we conclude that ε and p must be functions of t only: $\varepsilon(t)$ and $p(t)$.

We observe that in the Friedmann model universes the rotation pseudovector ω^α and the shear tensor $\sigma^{\alpha\beta}$ (see § 4.5) of the cosmological perfect fluid are null:

$$\omega^\alpha \equiv \frac{1}{2}\epsilon^{\alpha\beta\mu\nu}u_\beta u_{\mu;\nu} = 0 \qquad (4.2.34)$$

and

$$\sigma_{\alpha\beta} \equiv \left(u_{(\mu;\nu)} - \frac{1}{3}h_{\mu\nu}u^\rho{}_{;\rho}\right)h^\mu{}_\alpha h^\nu{}_\beta = 0 \qquad (4.2.35)$$

where $h_{\alpha\beta} \equiv g_{\alpha\beta} + u_\alpha u_\beta$ (see § 4.5) and $\epsilon^{\alpha\beta\mu\nu}$ is the Levi-Civita pseudotensor (defined in § 2.8); that is, the Friedmann models are nonrotating and free of shear.

Redshift Versus Distance Relation

If a photon propagates radially from the point r_e, θ_e, ϕ_e, with initial conditions: $\dot{\theta}_e = \dot{\phi}_e = 0$, from the geodesic equation $\ddot{\theta}_e = \ddot{\phi}_e = 0$, we have that $\dot{\theta} = \dot{\phi} = 0$ all along its world line. Thus, the photon will continue to propagate radially with constant θ and ϕ. From $ds^2 = 0$ we then have

$$dt^2 = \frac{R^2(t)dr^2}{(1 - kr^2)}. \tag{4.2.36}$$

If a wave crest is emitted at time t_e by one observer comoving with the cosmological fluid at the point r_e and received at time $t_\circ$ by another comoving observer at $r_\circ = 0$, we have from (4.2.36)

$$\int_{t_e}^{t_\circ} \frac{dt}{R(t)} = \int_0^{r_e} \frac{dr}{\sqrt{1 - kr^2}}, \tag{4.2.37}$$

and since r is constant in time for the two comoving observers, corresponding to the next wave crest emitted at time $t_e + \delta t_e$ from r_e, we have

$$\int_{t_e+\delta t_e}^{t_\circ+\delta t_\circ} \frac{dt}{R(t)} = \int_0^{r_e} \frac{dr}{\sqrt{1 - kr^2}} = \int_{t_e}^{t_\circ} \frac{dt}{R(t)}. \tag{4.2.38}$$

Therefore, for small δt:

$$\begin{aligned} &\frac{\delta t_\circ}{R_\circ} = \frac{\delta t_e}{R_e}, \text{ and} \\ &\frac{R_\circ}{R_e} = \frac{\delta t_\circ}{\delta t_e} = \frac{\nu_e}{\nu_\circ} = \frac{\lambda_\circ}{\lambda_e} = \frac{\lambda_e + \Delta\lambda}{\lambda_e} = 1 + z \end{aligned} \tag{4.2.39}$$

where $z \equiv \frac{\Delta\lambda}{\lambda_e} \equiv \frac{\lambda_\circ - \lambda_e}{\lambda_e}$ is the redshift parameter, and ν_e, λ_e and $\nu_\circ, \lambda_\circ$ are frequency and wavelength of the electromagnetic wave at the point of emission and at the point of observation. Therefore, if the scale factor $R(t)$ is changing in time, the observed frequency of an electromagnetic wave emitted from a distant galaxy will be different from the emission frequency, redshifted if $R(t)$ is increasing and the universe is expanding, or blueshifted if $R(t)$ is decreasing and the universe is contracting.

A **redshift** in the spectral lines **of extragalactic objects** was observed as early as 1910 by Slipher[73] and the *velocity versus distance relation* and the *expansion of the universe*, theoretically predicted by Friedmann in 1922, were in fact confirmed with the famous 1929 result by Hubble,[74] based on various observations, that the redshift of "local" galaxies is proportional to their distance from us: $z = H_\circ d$, where z is the redshift parameter of a galaxy, d is its distance from our galaxy measured via the luminosity method (see below), and $H_\circ$ is the **Hubble constant**, today estimated[75,76,205] to be about $[\frac{3\times 10^{17}\text{ s}}{h}]^{-1}$, or about $[\frac{10^{10}\text{ yr}}{h}]^{-1}$ where $h\in\{0.3, 1\}$. This is the famous **Hubble law**, theoretically well explained, for small z, by the expansion of the Friedmann model universe.

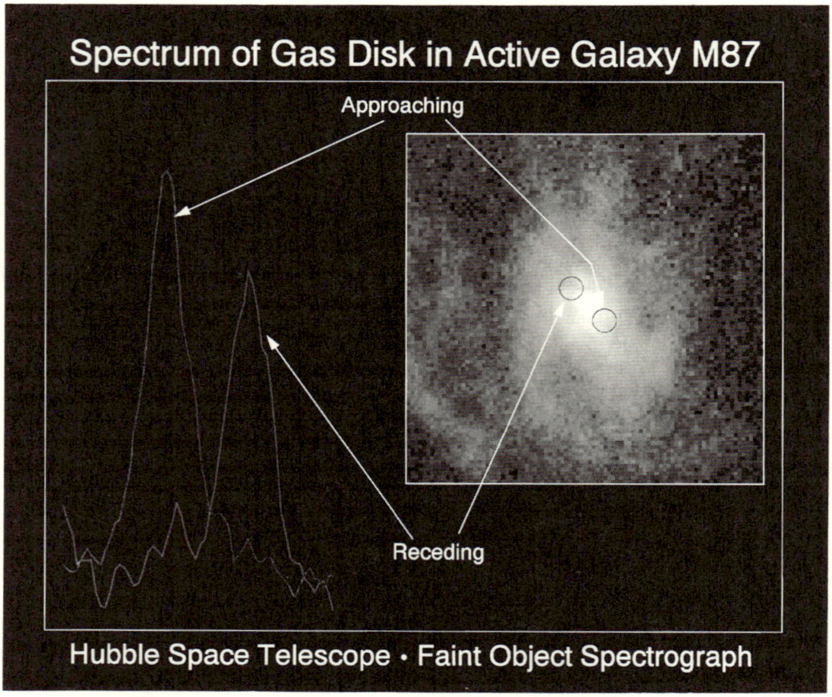

PICTURE 4.5. A schematic representation of the observations of a rotating disk of hot gas, in the core of the active elliptical galaxy M87 (see also picture 6.4 of chap. 6), made with the *Hubble Space Telescope* Faint Object Spectrograph. The gas on one side of the disk is approaching us at 500 ± 50 km s^{-1} and on the other side is receding from us at 500 ± 50 km s^{-1}. The corresponding disk rotation velocity, considering the disk inclination, has been estimated $\cong 750$ km s^{-1}, thus the total mass of the nucleus has been calculated $\sim 2.4 \pm 0.7 \times 10^9\ M_\odot$, within 18 pc of the nucleus, and the mass-to-light ratio $\left(\frac{M}{M_\odot} / \frac{L}{L_\odot}\right) \cong 170$. This observation provides a further experimental evidence for a supermassive black hole in the core of M87, source of the strong gravitational field that keeps the gas orbiting at high velocity[227,228] (courtesy of NASA 1994).

Thus the observations confirmed experimentally the dynamical character of the universe that was predicted, on a purely theoretical ground, by Friedmann in 1922,[21,22] using general relativity (however, we have not yet used the Einstein field equation and so far our results are valid in any metric theory of gravity (without prior geometry); § 3.1).

Let us now derive the **redshift versus distance relation** in a Friedmann model universe.

There are various methods to measure the distance of an astronomical object (see, for example, Weinberg).[46] We can determine the distance of a "near" astronomical object with laser ranging and radar delay measurements (see §§ 3.4.2 and 6.7); with proper motion, by measuring the change in the angular position of an object in the sky due to its proper motion, if we can estimate its true velocity; or with parallax, by measuring the apparent change in the angular position of an object in the sky due to the revolution of Earth around the Sun. However, for a distant object, if we can estimate its true dimensions, we can determine its distance from us by measuring its angular diameter and, for a very distant object, by measuring its apparent luminosity, if we can estimate its absolute luminosity, which may be done by using some standard astronomical "candle." The distance measured with this method is called the *luminosity distance* D_L. If L is the absolute luminosity of an object and l its measured apparent luminosity, one may *define*

$$L \equiv 4\pi D_L^2 l \qquad \text{or} \qquad D_L = \left(\frac{L}{4\pi l}\right)^{\frac{1}{2}}. \tag{4.2.40}$$

We observe that, according to the Friedmann-Robertson-Walker metric (4.2.16), the radial proper distance between an observer at $r = 0$ and an object at r is given by:

$$D_p = R(t)\int_{\circ}^{r} \frac{dr}{\sqrt{(1 - kr^2)}} \tag{4.2.41}$$

and therefore

$$\begin{aligned} D_p(k = +1) &= R(t)\sin^{-1} r \\ D_p(k = 0) &= R(t) r \\ D_p(k = -1) &= R(t)\sinh^{-1} r. \end{aligned} \tag{4.2.42}$$

Now, according to the Friedmann-Robertson-Walker metric, the photons emitted at an event $r = r_e$, at time of emission t_e, and observed at $r = 0$, at time $t_\circ$, cross a surface of total area $4\pi R_\circ^2 r_e^2$ at time $t_\circ$ (integral over θ and ϕ). The energy $h\nu$ of each photon is redshifted by a factor $\frac{1}{1+z}$ due to the expansion of the universe, as shown by formula (4.2.39); furthermore, corresponding to an interval Δt, the number of photons per unit time received by the observer at $r = 0$ is also decreased by a factor $\frac{1}{1+z}$, by reason of the same time-stretching factor. Therefore, for a source of total luminous power output L, the apparent luminosity is, at $r = 0$, and time $t_\circ$:

$$l = \frac{L}{4\pi R_\circ^2 r_e^2}\frac{1}{(1+z)^2}, \tag{4.2.43}$$

and from expression (4.2.39)

$$l = \frac{L R_e^2}{4\pi R_\circ^4 r_e^2}. \tag{4.2.44}$$

For the luminosity distance defined by expression (4.2.40) we deduce the result

$$D_L = r_e \frac{R_\circ^2}{R_e} = r_e R_\circ (1+z). \tag{4.2.45}$$

Now, from expression (4.2.39), $\frac{R_\circ}{R_e} \equiv \frac{R_e + \Delta R}{R_e} = (1+z)$, we have for small z, $\Delta R = R_\circ(z - z^2 + \cdots)$, and with a Taylor expansion for small $\Delta t \equiv (t_\circ - t_e)$ and for small $\Delta R \equiv (R_\circ - R_e)$, we get

$$\begin{aligned}\int_{t_e}^{t_\circ} \frac{dt}{R(t)} &= \int_{R_e}^{R_\circ} \frac{dR}{R(t)\dot{R}(t)} = -\int_{R_\circ}^{R_\circ - \Delta R} \frac{dR}{R(t)\dot{R}(t)} \\ &\cong \frac{\Delta R}{R_\circ \dot{R}_\circ} + \frac{1}{2}\frac{\left(R_\circ \ddot{R}_\circ + \dot{R}_\circ^2\right)}{R_\circ^2 \dot{R}_\circ^3}(\Delta R)^2 + \cdots \\ &\cong \frac{1}{\dot{R}_\circ} z + \frac{1}{2}\frac{\left(R_\circ \ddot{R}_\circ - \dot{R}_\circ^2\right)}{\dot{R}_\circ^3} z^2 + \cdots ;\end{aligned} \tag{4.2.46}$$

with a Taylor expansion for small r_e,

$$\int_0^{r_e} \frac{dr}{\sqrt{1-kr^2}} \cong r_e + O\left(r_e^3\right);$$

therefore, from equation (4.2.37), we get

$$r_e = \frac{1}{\dot{R}_\circ} z + \frac{1}{2}\frac{\left(R_\circ \ddot{R}_\circ - \dot{R}_\circ^2\right)}{\dot{R}_\circ^3} z^2 + \cdots = \frac{1}{R_\circ H_\circ} z - \frac{1}{2}\frac{1}{R_\circ H_\circ}(1+q_\circ)z^2 + \cdots \tag{4.2.47}$$

where $H_\circ \equiv \dot{R}_\circ / R_\circ$ and $q_\circ \equiv -\ddot{R}_\circ R_\circ / \dot{R}_\circ^2$ are the so-called **Hubble constant** at time $t_\circ$ and **deceleration parameter** at time $t_\circ$. Finally, from expression (4.2.45) we get the luminosity distance as a function of the redshift for small z:

$$D_L = H_\circ^{-1} z + \frac{1}{2} H_\circ^{-1}(1 - q_\circ) z^2 + \cdots, \tag{4.2.48}$$

and inverting this expansion, we get z as a function of the luminosity distance for small D_L:

$$z \cong H_\circ D_L - \frac{1}{2} H_\circ^2 (1 - q_\circ) D_L^2, \tag{4.2.49}$$

and at the lowest order we have the Hubble law: $z \cong H_\circ D_L$. Since for $v \ll 1$ we have $\frac{\nu_e}{\nu_\circ} \equiv 1 + z \cong 1 + v$, at the lowest order we also get $v \cong H_\circ D_L$. We observe that for the *Friedmann-Robertson-Walker metric*, for *any velocity* $v_p \equiv \dot{D}_p$, from expression (4.2.41) we get $\frac{v_p}{D_p} = \frac{\dot{R}}{R} \equiv H$ and therefore we get the *velocity versus distance law*:

$$v_p = H D_p. \tag{4.2.50}$$

Hubble constant $H_\circ$ and deceleration parameter $q_\circ$ are crucial parameters in order to determine the evolution of a Friedmann model universe. In principle, they may be observationally measured by using the luminosity distance versus redshift relation (4.2.48), that is, if the absolute luminosities of some astronomical objects are known, by measuring their apparent luminosity and their redshift and by using expressions (4.2.40) and (4.2.48). Unfortunately a number of systematic errors affect these measurements, such as errors in the estimation of the absolute luminosity of the standard "candles" and "proper," noncosmological, motions of the astronomical objects used. At present, the value of the Hubble time $H_\circ^{-1}$ is estimated to be between $\cong 10 \times 10^9$ yr and $\cong 30 \times 10^9$ yr.[75,76,205]

In the near future we may have a better determination of the Hubble time using the **Hubble Space Telescope**[77–79] (see also ref. 229). The Space Telescope, a large-aperture, long-term optical and ultraviolet space observatory, was launched into orbit in 1990. It carries a 2.4 m aperture mirror; the optical design is the Ritchey-Chrétien variation of the Cassegrain configuration. It covers the range of wavelengths from far-ultraviolet to far-infrared. Since Hubble is orbiting Earth at an altitude of about 500 km its reception is essentially unaffected by distortion and absorbing effects of the atmosphere. This advantage increases the angular resolution of the Space Telescope with respect to the best ground telescopes. Among the various important contributions of Hubble in astronomy, astrophysics, and cosmology, which include the study of the solar system, stars and stellar evolution, interstellar medium, nebulae, galaxies, active galactic nuclei, quasars, and evolution of the universe, especially relevant for general relativity are the observations of active galactic nuclei and quasars, which may have central black holes as possible engines (see picture 4.5 and § 6.3). Moreover, use of distant Cepheid variables and supernovae as standard candles, through the redshift versus distance relation, may allow a better determination of the Hubble constant, which, as we have just observed, is at present regarded as undetermined by a factor of about 2–3, between about 30–50 km s^{-1} Mpc^{-1} and about 100 km s^{-1} Mpc^{-1}, corresponding to a Hubble time between about 30×10^9 yr and about 10×10^9 yr. Indeed, new observations of Cepheid variable stars in the spiral galaxy M100 in the Virgo cluster, using the Hubble Space Telescope[230] (see also ref. 231), have yielded an accurate estimate of the distance to M100, and have thus provided a new "locally" *estimated* value of the Hubble constant, $H_\circ \cong 80 \pm 17$ km s^{-1} Mpc^{-1}. The Space Telescope may also help in determining a more accurate value of the deceleration parameter $q_\circ$, a crucial parameter, in a Friedmann model, to test the universe for closure in time (see below).

Dynamics of Spatially Homogeneous and Isotropic Models

Let us now study the dynamics of a spatially homogeneous and isotropic model universe. We generalize ($\eta_{\alpha\beta} \rightarrow g_{\alpha\beta}$) the energy-momentum tensor of special relativity to rewrite the expression of the energy-momentum tensor $T_{\alpha\beta}$ of a fluid in curved spacetime in the form:[80,194,195]

$$T_{\alpha\beta} = (\varepsilon + p)u_\alpha u_\beta + (q_\alpha u_\beta + u_\alpha q_\beta) + p g_{\alpha\beta} + \pi_{\alpha\beta}. \tag{4.2.51}$$

Where ε is the total energy density of the fluid, u_α its four-velocity, q_α the energy flux (heat flow), p the isotropic pressure, and $\pi_{\alpha\beta}$ the viscosity contribution to the energy-momentum tensor (see § 2.3). For a perfect fluid, by definition, $\pi_{\alpha\beta} = q_\alpha \equiv 0$, and therefore

$$T_{\alpha\beta} = (\varepsilon + p)u_\alpha u_\beta + p g_{\alpha\beta}, \tag{4.2.52}$$

and for a pressureless perfect fluid or dust,

$$T_{\alpha\beta} = \varepsilon u_\alpha u_\beta. \tag{4.2.53}$$

By definition, every metric theory of gravity (§ 3.1) satisfies the "conservation law":

$$T^{\alpha\beta}{}_{;\beta} = 0. \tag{4.2.54}$$

Using expression (4.2.52) for the energy-momentum tensor of a perfect fluid and the Friedmann-Robertson-Walker metric (4.2.16), written in comoving coordinates, we then have from the $\alpha = 0$ component of equation (4.2.54):

$$T^{0\beta}{}_{;\beta} = T^{00}{}_{,0} + T^{ik}\Gamma^0_{ik} + T^{00}\Gamma^k_{k0} = \dot{\varepsilon} + 3(\varepsilon + p)\frac{\dot{R}}{R} = 0 \tag{4.2.55}$$

(understandable in terms of work done in expansion, as if against a piston) and for a pressureless perfect fluid (cloud of dust):

$$[\varepsilon R^3(t)]_{,t} = 0 \tag{4.2.56}$$

or

$$\varepsilon(t) = \frac{\text{constant}}{R^3(t)} \tag{4.2.57}$$

as the model universe evolves, expands, or contracts.

So far, with the hypotheses of spatial homogeneity and isotropy about one point, we have derived and discussed results which are common to any metric theory of gravity (§ 3.1), that is, common to any theory of gravity whose spacetime is a Lorentzian manifold and such that the equivalence principle in the medium strong form (§ 2.1) and the conservation law (4.2.54) are satisfied.

However, in order to specify the dynamics of the universe, that is, the **dynamics of the scale factor**, $R(t)$, of the space geometry, we have to specify the *field equations* and the *equations of state* for matter and fields. Therefore, let us now turn to the standard and well-tested metric theory of gravity: Einstein general relativity. We assume here that the field equations are the general relativistic equations without the *cosmological term* Λ : $\boldsymbol{G} = \chi \boldsymbol{T}$. It is well known that the term $\Lambda \boldsymbol{g}$, on the left-hand side of the field equation, was introduced by Einstein in 1917 to allow the equation to admit a static model universe as a solution. However, as we have observed, in 1929 Hubble discovered the red-shift versus distance relation for galaxies,[74] known as the Hubble law, which is explained by an overall expansion of the universe. Then, in 1930 Einstein abandoned the cosmological constant Λ in the field equation and concluded that the cosmological term was the "biggest blunder" of his life[81] (however, see the discussion on the inflationary models at the end of this section). Quantum and particle cosmology seems to suggest a nonzero primordial cosmological constant. However, there are strong observational limits on its "present" value, of the order of $|\Lambda| \lesssim 3 \times 10^{-52}\ \text{m}^{-2}$. Therefore, for simplicity, we assume here $\Lambda = 0$; furthermore, we assume that the matter in the universe may on average be described by the energy-momentum tensor (4.2.52) of a perfect fluid.

From the field equation, using the Friedmann-Robertson-Walker metric, after some calculations, we thus have

$$G_{00} = 8\pi T_{00} \longrightarrow \dot{R}^2 = -k + \frac{8\pi}{3}\varepsilon R^2 \tag{4.2.58}$$

$$G_{ij} = 8\pi T_{ij} \xrightarrow{i=j} \ddot{R}R + \frac{\dot{R}^2}{2} = -\frac{k}{2} - 4\pi p R^2 \tag{4.2.59}$$

$$G_{\alpha\beta} = 8\pi T_{\alpha\beta} \xrightarrow{\alpha\neq\beta} \text{identically zero.}$$

From the "conservation law" $T^{\alpha\beta}{}_{;\beta} = 0$, we have already obtained

$$\dot{\varepsilon} + 3(\varepsilon + p)\frac{\dot{R}}{R} = 0 \tag{4.2.60}$$

(reduction of energy density by work done in expansion).

Let us now study the dynamics of spatially homogeneous and isotropic cosmological models in Einstein theory, in the, so-called, **matter-dominated** phase, when $\varepsilon_M \gg \varepsilon_R$, that is, in the epoch when the energy density of non-relativistic matter in the universe is much larger than the energy density of radiation and relativistic particles. We thus assume $p \cong 0$ and from the conservation law (4.2.56) we obtain $\varepsilon R^3(t) = \text{constant}$, which substituted in the evolution equation (4.2.58) gives

$$\dot{R}^2 = -k + \frac{E}{R}; \qquad \text{where} \quad E \equiv \frac{8}{3}\pi\varepsilon R^3 = \text{constant.} \tag{4.2.61}$$

This is the so-called Friedmann equation.

We now have three possibilities:

(1) $k = 0$, "**flat**" **Friedmann model**, corresponding to a spatially flat model universe (see also next section)

$$\dot{R}^2 = \frac{E}{R} \tag{4.2.62}$$

with the simple solution

$$R(t) = \left(\frac{9}{4}E\right)^{\frac{1}{3}} t^{\frac{2}{3}}. \tag{4.2.63}$$

This is the so-called Einstein–de Sitter cosmological model,[82–84] a **spatially flat model universe** expanding forever (see fig. 4.2).

(2) $k = +1$, **"closed" Friedmann model**, corresponding to a model universe with **positive spatial curvature** (see also next section)

$$\dot{R}^2 = -1 + \frac{E}{R} \tag{4.2.64}$$

with solution

$$t = E\left(\sin^{-1}\sqrt{\frac{R}{E}} - \sqrt{\frac{R}{E} - \left(\frac{R}{E}\right)^2}\right), \tag{4.2.65}$$

which one can rewrite

$$R = \frac{E}{2}(1 - \cos\varphi); \qquad t = \frac{E}{2}(\varphi - \sin\varphi). \tag{4.2.66}$$

This is the equation of a cycloid: the model universe expands and then recollapses (see fig. 4.2).

(3) $k = -1$, **"open" Friedmann model**, corresponding to a model universe with **negative spatial curvature** (see also next section)

$$\dot{R}^2 = 1 + \frac{E}{R} \tag{4.2.67}$$

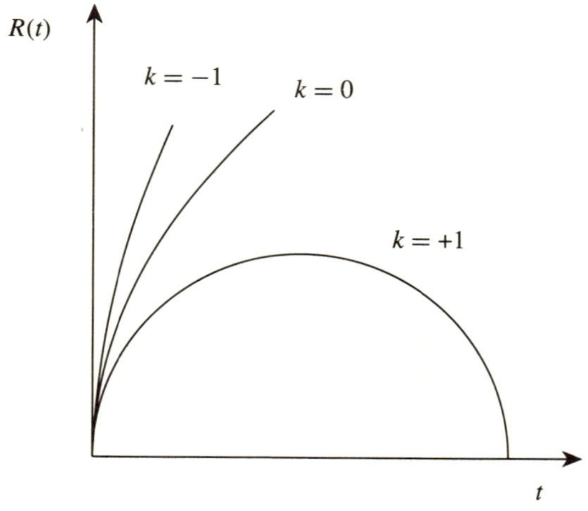

FIGURE 4.2. The standard Friedmann models of the universe.

with solution

$$t = E\left(-\sinh^{-1}\sqrt{\frac{R}{E}} + \sqrt{\frac{R}{E} + \left(\frac{R}{E}\right)^2}\right), \tag{4.2.68}$$

which one can rewrite

$$R = \frac{E}{2}(\cosh\varphi - 1); \qquad t = \frac{E}{2}(\sinh\varphi - \varphi). \tag{4.2.69}$$

In this model as t increases the scale factor R increases monotonically (see fig. 4.2) and the model universe expands forever.

Let us now study how closure in time and spatial curvature are related to mass-energy density in the Friedmann spatially homogeneous and isotropic cosmological models.

From the $\alpha = \beta = 0$ component (4.2.58) of the Einstein field equation, at time $t_\circ$ ("now"), for the curvature index k we have the relation

$$k = \left[\frac{8\pi\varepsilon_\circ}{3} - H_\circ^2\right]R_\circ^2, \tag{4.2.70}$$

and therefore the criterion for "open" versus "closed" model universe reduces to

$$k > 0, = 0, \text{ or } < 0 \iff \varepsilon_\circ > \varepsilon_c, = \varepsilon_c, \text{ or } < \varepsilon_c \equiv \frac{3H_\circ^2}{8\pi} \tag{4.2.71}$$

where ε_c is the **critical density**, $\varepsilon_c = 1.88 \times 10^{-29} h_\circ^2$ g/cm^3, and, as previously defined $h_\circ$ is the actual value of the Hubble "constant" today, relative to one presumptive value, $H_\circ = 100$ km/(s Mpc), of this expansion parameter (see eqs. (4.2.47) and (4.2.50)). From the $\alpha = \beta = i$ components of the field equation, at $t = t_\circ$, we then have for the curvature index

$$k = -8\pi p_\circ R_\circ^2 + H_\circ^2 R_\circ^2 (2q_\circ - 1). \tag{4.2.72}$$

By assuming that the universe is **matter-dominated** at the present time $t_\circ$, $\varepsilon_\circ \gg p_\circ$, from $p_\circ \cong 0$, we have

$$k \cong H_\circ^2 R_\circ^2 (2q_\circ - 1), \tag{4.2.73}$$

and therefore

$$k > 0, = 0, \text{ or } < 0 \iff q_\circ > \frac{1}{2}, = \frac{1}{2}, \text{ or } < \frac{1}{2}. \tag{4.2.74}$$

These relations show the importance of determining the present values of the cosmological density $\varepsilon_\circ$ of mass-energy, of the Hubble constant $H_\circ$, and of the deceleration parameter $q_\circ$.

If one assumes (1) that the classical Einstein field equation continues to hold even under the most extreme initial conditions with energy density tending to infinity, (2) that the spacetime is a spatially homogeneous and isotropic, inextendible (see § 2.9), Lorentzian manifold, and (3) that the universe is expanding, in agreement with the observations, from equations (4.2.63), (4.2.66), and (4.2.69) (see fig. 4.2), we have that the "scale factor" $R(t)$ must have been zero at some initial time t_i in the past, which we may define $t_i \equiv 0$. This is the initial singularity, $R(0) = 0$, or the so-called "big bang." (For a discussion of the initial states of the universe and inflationary cosmologies, see below.)

From figures 4.2 and 4.3, by drawing a tangent to any of the three curves (dashed line of fig. 4.3), since $H_\circ^{-1} = \frac{R_\circ}{\dot{R}_\circ}$, we see that the "age of a Friedmann model universe" $t_\circ$ must be less than the **Hubble time** $H_\circ^{-1}$ in any of the three Friedmann cosmological models.

Suitable observations of the relations between luminosity distance and redshift,[46,85–91] make it possible to infer a tentative value of the deceleration parameter $q_\circ$. In particular, using VLBI, from observations of the angular size of compact radio sources[206] (associated with active galaxies and quasars) and their redshift (see also refs. 207 and 208) some researchers have inferred a tentative value of $q_\circ \approx \frac{1}{2}$. For a matter-dominated Friedmann model universe, from expression (4.2.70), a value of $q_\circ \approx \frac{1}{2}$ corresponds to a value of the density parameter $\Omega_\circ \equiv \frac{\varepsilon_\circ}{\varepsilon_c}$ near one, that is, to a present energy density $\varepsilon_\circ$ of about $\varepsilon_c = 1.88 \times 10^{-29} h_\circ^2$ g/cm^3, where $h_\circ$ probably lies between 0.3 and 1. However, it is important to observe that uncertainties in galactic evolution may bias the value of $q_\circ$, as inferred from the correlation between luminosity distance and redshift. Indeed, we do not precisely know the luminosity of

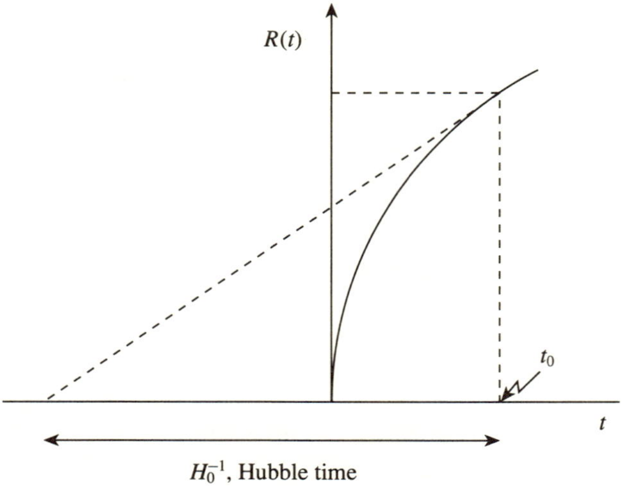

FIGURE 4.3. The Hubble time exceeds the age of the universe in every expanding Friedmann model universe.

TABLE 4.1 Curvature, density of mass-energy, and deceleration parameter in "matter-dominated" Friedmann cosmologies

Curvature Index, k		*Density, $\varepsilon_\circ$, of Mass-Energy*		*Deceleration Parameter, $q_\circ$*
$k = +1$	$\Longleftrightarrow$	$\varepsilon_\circ > \varepsilon_c$	$\Longleftrightarrow$	$q_\circ > \frac{1}{2}$
$k = 0$	$\Longleftrightarrow$	$\varepsilon_\circ = \varepsilon_c$	$\Longleftrightarrow$	$q_\circ = \frac{1}{2}$
$k = -1$	$\Longleftrightarrow$	$\varepsilon_\circ < \varepsilon_c$	$\Longleftrightarrow$	$q_\circ < \frac{1}{2}$

Note: Here ε_c is the critical density $1.88 \times 10^{-29} h_\circ^2$ g/cm^3 of equation (4.2.71).

faraway young galaxies. On the contrary, estimating the mass of clusters and galactic halos with dynamical methods[53] and assuming that the mass in the universe is mainly concentrated in clusters and galactic halos, one gets[210] an average energy density in the universe not higher than $\varepsilon_\circ \cong 0.2\varepsilon_c$. Whereas, the energy density of the luminous matter is less than $10^{-2}\varepsilon_c$. One gets even smaller densities by considering other standard forms of energy, such as radiation. This is usually called the **cosmological "missing mass problem,"** or cosmological "dark matter problem."[232] The standard and well-tested theory of the production of the elements by nuclear reactions in the early days of the universe would give results incompatible with the observed abundances of the elements if more than approximately 15% of the mass of the universe were baryonic in character. Therefore the signs point to a not-yet-identified source of mass which does not have the characteristics of matter. Thus, this puzzle is better called "the mystery of the missing mass" or the "missing mass problem." Confronted by this mystery, one finds it difficult not to recognize that gravitational radiation carries zero baryon number, yet possesses the power to coalesce into black holes of significant mass. The question is still open at this writing whether it is a defensible thesis to speak of the "missing mass as collapsed gravitational waves."[217]

Moreover, the **galactic "missing mass problem"** consists in the riddle that the mass of galaxies and clusters of galaxies inferred with dynamical methods is larger than the mass of the luminous matter by approximately one order of magnitude.[53] Therefore, many astrophysicists hypothesize the existence of some form of dark mass. As we have already observed, some researchers have suggested explaining the "missing mass problem" with massive neutrinos[7] or with some exotic particles,[58] such as "axions," "photinos," "selectrons," "gravitinos," etc., or both. These hypothetical particles are also known as WIMPs, *Weakly Interacting Massive Particles*. There have also been some attempts to explain the galactic missing mass problem using gravitational theories with a non-Newtonian limit in some regime.[204] A possible explanation of the galactic missing mass problem has been suggested as unseen matter in the form of dark bodies, or *MACHOs*, that is, *MAssive Compact Halo Objects*. These dark bodies might be brown dwarfs, "jupiters" (bodies not massive enough to produce energy by fusion), neutron stars, old white dwarfs, or black holes (however, see references 233, 234 and 235). In the halo of our Galaxy these objects might amplify the apparent brightness of stars in nearby galaxies via the gravitational lensing effect mentioned in section 3.4.1.[218] Some preliminary reports of possible gravitational microlensing due to stars with masses of a fraction of a solar mass have been presented after monitoring the brightness of millions of stars in the Large Magellanic Cloud for observation times of about 1 and 3 yr.[219,220] However, at this writing cosmological and galactic missing mass problems are still open problems in physics.

If one assumes that the evolution of the Friedmann model universe is, at some stage, essentially dominated by radiation, in the sense that the energy density of radiation and relativistic particles is much larger than the energy density of nonrelativistic matter, from the equation of state for radiation $p_r = \frac{\varepsilon_r}{3}$, and from the "conservation law" (4.2.55), we have

$$\dot{\varepsilon}_r + 4\varepsilon_r \frac{\dot{R}}{R} = 0 \tag{4.2.75}$$

and

$$(\varepsilon_r R^4(t))_{,t} = 0 \qquad \text{or} \qquad \varepsilon_r \propto \frac{1}{R^4(t)}. \tag{4.2.76}$$

From the $\alpha = \beta = 0$ and $\alpha = \beta = i$ components of the Einstein field equation, equations (4.2.58) and (4.2.59), in the case of a **"radiation-dominated"** universe, one then gets

$$k = 0 \Longleftrightarrow \varepsilon_\circ = \varepsilon_c \Longleftrightarrow q_\circ = 1, \tag{4.2.77}$$

the characteristic features of a *"radiation-dominated" Friedmann model universe*.

Particle Horizon, Event Horizon, and Inflationary Models.

Finally, we introduce the useful concepts of **"particle horizon"** and **"event horizon"**[92,93] (see fig. 4.4). In some cosmological models there exists a "particle horizon": a surface that divides the particles in the universe that we have already observed from the ones that we have not yet seen because their "light" has not yet reached us. In some cosmological models there also exists an "event horizon": a hypersurface that divides the events of the spacetime into those that we will be able to observe and those that we can never see because no light ray emitted from these events will ever reach us.

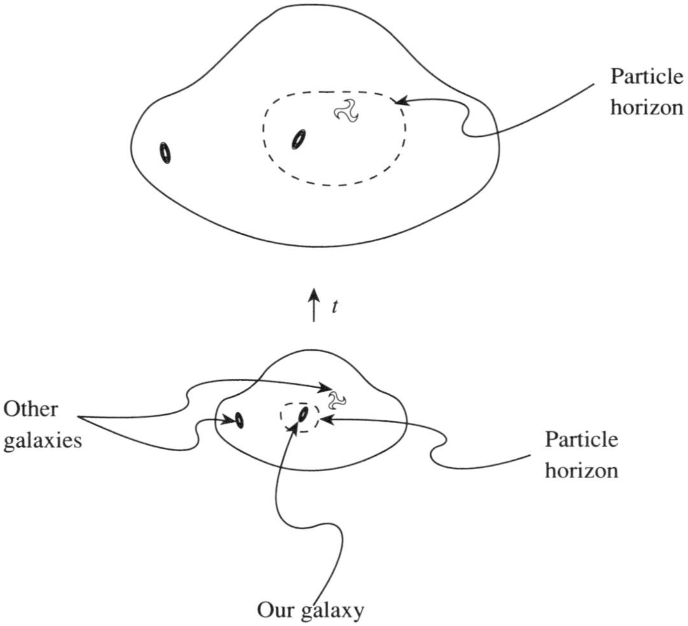

FIGURE 4.4. Particle horizon depicted in an expanding, spatially compact, model universe at two different cosmological times. We, in our galaxy, see more (expanding particle horizon surface corresponds to dashed line in the diagram) the longer we wait (increasing t).

In the case of a Friedmann model universe, for a radial photon emitted at r_e, at time t_e, and received at $r_\circ$, at time $t_\circ$, we have (see eq. 4.2.37)

$$\int_{r_\circ}^{r_e} \frac{dr}{(1-kr^2)^{\frac{1}{2}}} = \int_{t_e}^{t_\circ} \frac{dt}{R(t)}. \tag{4.2.78}$$

By this formula one can obtain the coordinate r_{PH} of the **particle horizon**, at time $t_\circ$, by setting $r_e = r_{\text{PH}}$ and our radial coordinate $r_\circ = 0$. The time of emission t_e is naturally defined to be zero for any model expanding from an initial singularity (however, in some models one has $t_e \longrightarrow -\infty$). Thus, we have

$$\int_{0}^{r_{\text{PH}}} \frac{dr}{(1-kr^2)^{\frac{1}{2}}} = \int_{0}^{t_\circ} \frac{dt}{R(t)}, \tag{4.2.79}$$

that is

$$\sin^{-1} r_{\mathrm{PH}} = \int_0^{t_\circ} \frac{dt}{R(t)}, \qquad \text{for} \quad k = +1$$

$$r_{\mathrm{PH}} = \int_0^{t_\circ} \frac{dt}{R(t)}, \qquad \text{for} \quad k = 0 \tag{4.2.80}$$

$$\sinh^{-1} r_{\mathrm{PH}} = \int_0^{t_\circ} \frac{dt}{R(t)}, \qquad \text{for} \quad k = -1.$$

The particle horizon, $r_{\mathrm{PH}}(t_\circ)$, divides the particles we, at $r = 0$, have already observed, at time $t_\circ$, from the ones we have not yet seen. The condition for the existence of the particle horizon is $\int_0^{t_\circ} \frac{dt}{R(t)} < \infty$.

We then introduce the **event horizon**. Given some event (r_e, t_e), from equation (4.2.78) we have that as time of observation $t_\circ$ increases and eventually, in some models, goes to infinity, $r_\circ$ may or may not go to zero. We shall observe the event r_e, t_e, if $r_\circ = 0$, at some time $t_\circ$; otherwise, if it is always $r_\circ \neq 0$, even for $t_\circ \longrightarrow \infty$, we shall never observe the event t_e, r_e. If, for *every* event r_e, t_e, $r_\circ = 0$ at some time $t_\circ$, then there is no event horizon. However, if there is an event horizon, corresponding to a time of emission t_e, the r_{EH} coordinate of the event horizon is given, for $t_\circ \longrightarrow \infty$, or for $t_\circ \longrightarrow t_{\mathrm{F}}$, time corresponding to the final singularity, by

$$\sin^{-1} r_{\mathrm{EH}} = \int_{t_e}^{\infty} \frac{dt}{R(t)}, \qquad \text{for} \quad k = +1$$

$$r_{\mathrm{EH}} = \int_{t_e}^{\infty} \frac{dt}{R(t)}, \qquad \text{for} \quad k = 0 \tag{4.2.81}$$

$$\sinh^{-1} r_{\mathrm{EH}} = \int_{t_e}^{t_{\mathrm{F}}} \frac{dt}{R(t)}, \qquad \text{for} \quad k = -1.$$

The condition for the existence of the event horizon is thus $\int_t^{\infty,\ \mathrm{or}\ t_{\mathrm{F}}} \frac{dt}{R(t)} < \infty$, for some t.

In special relativity there is no event horizon, since one can observe every event by waiting a sufficiently long time. However, in general relativity the expansion of the universe may prevent some photons from ever reaching us, even if we wait an infinite time, $t_\circ \longrightarrow \infty$. In other words, in some cosmological models the proper distance of some galaxies may increase faster than the speed of light. This phenomenon does not contradict causality and the special relativistic constraint, $v \leq 1$, valid for "causally connected" motions, in fact the motions in question are not causally connected; the validity of special relativity is restricted to *local* Minkowskian frames. In general relativity there are solutions with both particle and event horizons, solutions with one of the two only, and solutions with no horizons.

Before we turn to a discussion on closure in time and the spatial compactness characteristic of some cosmological models, it is interesting to describe briefly the so-called **inflationary models.**[94–105].

These models offer an explanation for the very large scale average homogeneity and the substantial level of isotropy of the universe. They also explain why we do not observe the large production of magnetic monopoles predicted by Grand Unified Theories in the initial phases of the universe.

As we have seen, the cosmic background microwave radiation is isotropic to a level of $\frac{\Delta T}{T}$ of a few parts in 10^5. Large angular scale anisotropies have been observed by COBE only at a level of $\approx 10^{-5}$ (§ 4.1). According to the theory of the big bang, this radiation was last scattered when the universe became transparent to radiation, some $\sim 10^5$ years after the big bang. Therefore, in *standard cosmological models*, the cosmic microwave radiation that we now observe to be isotropic to a level of a few parts in 10^5 was emitted also from regions whose distance from one another was greater than the particle horizon distance[92,93] (see fig. 4.4). In other words the cosmic microwave radiation that we now receive substantially isotropically from every part of the sky was, according to standard cosmological models, emitted from regions that, at time $\sim 10^5$ years, were causally disconnected, that is, regions that were outside the particle horizon of each other. Here we understand the particle horizon distance to be the maximum distance that a signal traveling at the speed of light could have covered since the initial big bang. In standard Friedmann model universes there is not sufficient time before $\sim 10^5$ years for any information, propagating at the speed of light since the big bang, to homogenize the plasma. Consequently, there is no explanation for the observed homogeneity of the universe and for the observed level of isotropy of the cosmic radiation, unless one assumes it as given a priori, as an initial condition. This is the so-called "*horizon problem*" (Rindler 1956).[92]

Another problem presses: what was the *origin of initial density perturbations* of the right amplitude to form the large structures we now observe in the universe:[1,7–11] galaxies, clusters of galaxies, clusters of clusters, etc.? And what was the origin of the large angular scale fluctuations observed by COBE?

Some cosmologists point to a further issue: the "*flatness problem*," the apparent lack of explanation of why the value of the density parameter, Ω, is near the critical value one, as inferred from some observations of the expansion of the universe.[207,208]. A value of $\Omega = 1$ would correspond to a spatially flat Friedmann model universe.[83]

In addition, according to Grand Unified Theories (GUTs) of elementary interactions, in the very early state of the universe an enormous production of *magnetic monopoles* should have occurred, with enormously high energy density, that should have led to a fast recollapse of the universe, in strong disagreement with the observations.

Inflationary models attempt to respond to some of these problems. The original inflationary model (Guth 1981; see also Gliner et al.)[94–100] assumed the validity of some Grand Unified Theories of strong and electro-weak interactions and, at the same time, the validity of the classical Einstein field equation with appropriate energy-momentum tensor. According to this model, the very early universe was in a metastable "false vacuum" state, characterized by symmetry of the elementary interactions. It had a temperature T larger than a critical temperature T_c at which a first-order phase transition could take place. The "false vacuum" was a minimum of the energy and was characterized by an enormously large energy density. In the original inflationary model this energy density has been calculated to be comparable to that of a massive star squeezed to the size of a proton! The classical Einstein field equation has been assumed to be valid under these extreme conditions. The equation of state of the "false vacuum" has been taken to be $p_f = -\varepsilon_f$, giving a pressure p_f equal in magnitude but opposite in sign to the energy density ε_f. The Einstein field equation (4.2.58)–(4.2.59) predicts the expansion rate

$$\ddot{R} = -\frac{4\pi}{3}(\varepsilon_f + 3p_f)R = \frac{8\pi}{3}\varepsilon_f R \tag{4.2.82}$$

and therefore an exponentially growing scale factor $R(t)$: $R \propto e^{\alpha t}$, with the very short time scale $\alpha^{-1} = \left(\frac{3}{8\pi\varepsilon_f}\right)^{1/2}$. This very fast exponential expansion has been called inflation. It resembles the exponential expansion of the spatially flat, $k = 0$, de Sitter model universe[82–84] with a positive cosmological constant, $\Lambda > 0$.

According to the inflationary model this very fast expansion supercooled the universe to a temperature $T < T_c$, and then a phase transition took place from the symmetric "false vacuum" to an asymmetric, stable, true vacuum, a global minimum of the energy, with breakdown of symmetry between the strong and electro-weak interactions. From this time, at about 10^{-30} s, the standard Friedmann model replaced the phase of inflation. During the phase transition, the enormous energy difference between the two states was released and the particle production occurred.

This original inflationary model, sometimes called the old inflationary model, gives an explanation for the so-called horizon problem. During the inflationary era, the "radius" of the universe was much smaller than the "radius" of a standard Friedmann model universe at the corresponding time, therefore, in the inflationary models the whole universe is within the distance of the particle horizon, that is, the whole universe is within a causally connected region. Inflation may explain the very large-scale homogeneity of the universe, may solve the "monopole problem," and may explain the so-called "flatness problem," that is, the value of the density parameter near 1. However, the old inflationary model does not produce the density perturbations necessary for the formation of galaxies and clusters. On the contrary, among the problems of this model, it predicts very large inhomogeneities with production of "bubbles," that is, regions with energy mainly concentrated on their surface and nothing inside, in strong disagreement with our observations of the universe.

To avoid the problem of these large inhomogeneities other inflationary models have been developed, with different effective potentials for the vacuum field, (Linde 1982; Albrecht and Steinhardt 1982).[101–104] These models have been called "new inflationary models." However, some of these models give rise to initial density perturbations that are too large. Others give initial density perturbations that are sufficiently small and of the right size to explain the formation of the galaxies and of the clusters that we observe; however, these models do not explain the so-called horizon problem, since they may not produce any inflation at all, or a too late inflation. "*Extended*" *inflation*, based on Jordan-Brans-Dicke theory of gravity, and "*hyperextended*" *inflation*, with Brans-Dicke parameter ω changing with time, have also been proposed to avoid the production of large bubbles and to give suitable, initial, cosmological density perturbations (the 1993 limit to the present value of the Brans-Dicke parameter is about $\omega \gtrsim 620$; see § 3.2.5).

Finally, in order to solve these horizon and bubble problems the so-called "*Chaotic Inflationary Scenario*" has been developed.[105] The Chaotic Inflationary Scenario predicts the generation of different regions with different types of inflation; these regions have different physical properties, dimensions, etc., and have been called "child-" or "baby-universes." One of these child-universes is conjectured to be the four-dimensional universe where we live and that we now observe. For a detailed review of these various inflationary models see, for example, Linde 1987;[101] for creation of perturbations in these models see, for example, Lukash and Novikov 1991.[102]

According to some cosmologists, during the phase transition between false vacuum and true vacuum, very thin, practically one-dimensional, long objects, with very high energy density, called *cosmic strings*,[106] might have formed. For example, for an energy scale of symmetry breaking of $\sim 10^{16}$ GeV, it has been calculated[106] that the strings have a thickness of $\sim 10^{-30}$ cm and a linear mass density of $\sim 10^{22}$ g/cm! In some gauge theories the cosmic strings have no ends and are either infinitely long or closed.

Cosmic strings have been invoked to explain the formation of the large structures we observe in universe, galaxies, clusters,[107] etc. However, some of the effects of cosmic strings should be observable,[37] such as possible production of gravitational lensing of distant quasars, anisotropies in the cosmic microwave background radiation, and low-frequency gravitational waves.

An idea[108,109] proposed to explain the apparent homogeneity of the universe, which agrees with the views of Einstein and Wheeler on spatial compactness (§ 4.8), is the "*small universe*" hypothesis. The universe, which on a very large scale and on the average is observed to be homogeneous and isotropic to a substantial level, might be a spatially compact "small universe" with some nontrivial spatial topology obtained by suitable identifications of regions of a three-manifold, for example with the spatial topology of a three-torus (see § 4.3). Therefore, we might have observed many times round the universe since decoupling, and we might see many images of each galaxy due to the particular topology of the universe (for example, a $k = 0$ Friedmann model may have the spatial topology of a *flat* three-torus; see § 4.3).

4.3 [CLOSURE IN TIME VERSUS SPATIAL COMPACTNESS]

As we have seen in the previous section, one can write the Riemann tensor of a homogeneous and isotropic n-dimensional hypersurface, "hypersurface of constant curvature," in the form

$$^{(n)}R_{abcd} = K(g_{ac}g_{bd} - g_{ad}g_{bc}) \tag{4.3.1}$$

from which we immediately get the Ricci scalar of curvature

$$^{(n)}R = {}^{(n)}R^{a}{}_{a} = n(n-1)K. \tag{4.3.2}$$

For the homogeneous and isotropic, three-dimensional, spacelike hypersurfaces described by the Friedmann-Robertson-Walker metric (4.2.16) with $t =$ constant, we then have the expression (4.2.28)

$$^{(3)}R = 6K = \frac{6k}{R^2(t)}. \tag{4.3.3}$$

K positive, zero, or negative, and Friedmann-Robertson-Walker metrics with $k = +1$, $k = 0$, or $k = -1$, correspond to a manifold with positive, zero, or negative curvature (see fig. 4.5).

Let us study further the three-dimensional spacelike manifolds described by the Friedmann-Robertson-Walker spatial metric with $k = +1$, $k = 0$, or $k = -1$, corresponding to $K > 0, = 0$, or < 0.

In the case $k = +1$ one may locally find a map between the Friedmann-Robertson-Walker hypersurface of homogeneity and a three-space embedded in a four-dimensional Euclidean manifold (with coordinates x^0, x^1, x^2, x^3). The map is[110]

$$\begin{aligned} x^0 &= R(1-r^2)^{\frac{1}{2}} \\ x^1 &= Rr\sin\theta\cos\phi \\ x^2 &= Rr\sin\theta\sin\phi \\ x^3 &= Rr\cos\theta. \end{aligned} \tag{4.3.4}$$

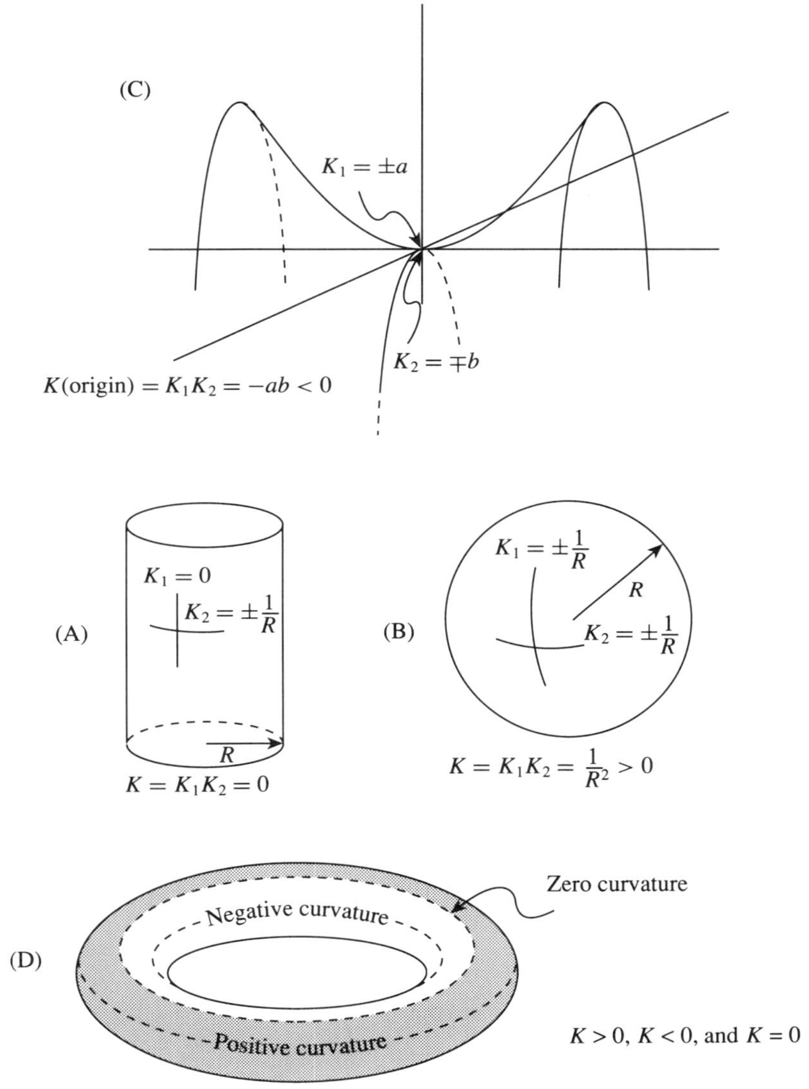

FIGURE 4.5. Examples of two-surfaces with different curvature.[61] (A) *Zero curvature surface*. Cylinder, Gaussian curvature: $K = 0$. (B) *Positive curvature surface*. Sphere, Gaussian curvature: $K = K_1 K_2 = \frac{1}{R^2}$. (C) *Negative curvature surface*. Hyperbolic paraboloid, Gaussian curvature at origin: $K(0, 0, 0) = K_1 K_2 = -ab < 0$. (D) *Standard two-torus*. *Positive curvature* (elliptic points) on the outer part, *negative curvature* (hyperbolic points) on the inner part, and *zero curvature* (planar) on the upper and lower circumferences (see also fig. 4.6).

The metric of the hypersurface of homogeneity Σ is then

$$\begin{aligned} d\Sigma^2 &= R^2(t)\left[\frac{1}{(1-r^2)}dr^2 + r^2 d\theta^2 + r^2\sin^2\theta d\phi^2\right] \\ &= (dx^0)^2 + (dx^1)^2 + (dx^2)^2 + (dx^3)^2 = \delta_{\alpha\beta}dx^\alpha dx^\beta \end{aligned} \tag{4.3.5}$$

and the hypersurface of homogeneity Σ satisfies:

$$(x^0)^2 + (x^1)^2 + (x^2)^2 + (x^3)^2 = \delta_{\alpha\beta}x^\alpha x^\beta = R^2(t), \tag{4.3.6}$$

that is, the equation of a *three-sphere, with constant positive curvature*, embedded in a four-dimensional Euclidean space.

In the case $k = 0$, with the coordinate transformation

$$\begin{aligned} x^1 &= Rr\sin\theta\cos\phi \\ x^2 &= Rr\sin\theta\sin\phi \\ x^3 &= Rr\cos\theta \end{aligned} \tag{4.3.7}$$

from the Friedmann-Robertson-Walker metric (4.2.16), we immediately find the metric of the hypersurface of homogeneity

$$d\Sigma^2 = (dx^1)^2 + (dx^2)^2 + (dx^3)^2, \tag{4.3.8}$$

that is, the metric of a *three-dimensional, flat, Euclidean manifold*, for example $\mathfrak{R}^3$ (however, see below for other possible global topologies).

Finally, in the case $k = -1$ one may find a local map between the $k = -1$ hypersurface of homogeneity and a three-dimensional hypersurface embedded in a four-dimensional Minkowskian manifold. The map is[110]

$$\begin{aligned} x^0 &= R(1+r^2)^{\frac{1}{2}} \\ x^1 &= Rr\sin\theta\cos\phi \\ x^2 &= Rr\sin\theta\sin\phi \\ x^3 &= Rr\cos\theta. \end{aligned} \tag{4.3.9}$$

The metric of the hypersurface of homogeneity Σ is then

$$\begin{aligned} d\Sigma^2 &= R^2(t)\left[\frac{1}{(1+r^2)}dr^2 + r^2 d\theta^2 + r^2\sin^2\theta d\phi^2\right] \\ &= -(dx^0)^2 + (dx^1)^2 + (dx^2)^2 + (dx^3)^2 = \eta_{\alpha\beta}dx^\alpha dx^\beta \end{aligned} \tag{4.3.10}$$

and the hypersurface of homogeneity Σ satisfies

$$-(x^0)^2 + (x^1)^2 + (x^2)^2 + (x^3)^2 = \eta_{\alpha\beta}x^\alpha x^\beta = -R^2(t), \tag{4.3.11}$$

that is, the equation of a *three-dimensional hyperboloid*, with *constant negative curvature*, embedded in a four-dimensional flat Minkowskian manifold.

One can immediately verify the formal invariance of these metrics for rotations and translations from expressions (4.3.5), (4.3.8), and (4.3.10) of these metrics of the hypersurfaces of homogeneity corresponding to $k = +1$, $k = 0$, and $k = -1$. The metric (4.3.5) of a three-sphere, corresponding to the $k = +1$ Friedmann-Robertson-Walker metric, is manifestly formally invariant for rotations in $\Re^4$: $x'^{\alpha} = O^{\alpha'}{}_{\beta} x^{\beta}$, where the $O^{\alpha'}{}_{\beta}$ are constants and satisfy $\delta_{\alpha\beta} = O^{\sigma'}{}_{\alpha} O^{\rho'}{}_{\beta} \delta_{\sigma'\rho'}$. Similarly, the metric (4.3.8) is formally invariant for rotations plus translations in $\Re^3$. The metric (4.3.10) of a three-hyperboloid, corresponding to the $k = -1$ Friedmann-Robertson-Walker metric, is then manifestly formally invariant for rotations in M^4 (homogeneous Lorentz transformations): $x'^{\alpha} = \Lambda^{\alpha'}{}_{\beta} x^{\beta}$, where the $\Lambda^{\alpha'}{}_{\beta}$ are constants and satisfy $\eta_{\alpha\beta} = \Lambda^{\sigma'}{}_{\alpha} \Lambda^{\rho'}{}_{\beta} \eta_{\sigma'\rho'}$. Therefore, one can "locally" perform rotations and translations on the hypersurfaces described by the Friedmann-Robertson-Walker metric (4.2.16) with $t =$ constant, with the transformations $\phi = \psi^{-1} \cdot R \cdot \psi$ where ψ is a suitable continuous map between a neighborhood of the hypersurface of homogeneity and a neighborhood of the corresponding three manifold (three-sphere, $\Re^3$, or three-hyperboloid), with inverse continuous map ψ^{-1}, and R corresponds to a rotation in $\Re^4$ or M^4, and to a rotation or a translation in $\Re^3$. The transformations ϕ are isometries. The metric (4.2.16) is manifestly formally invariant under ϕ:

$$\phi(\mathbf{g}(x)) = \mathbf{g}'(x') = \mathbf{g}(x'). \tag{4.3.12}$$

These isometries show again the formal invariance of the Friedmann-Robertson-Walker metric under the transitive and isotropy (SO(3)) groups of transformations, that is, its spatial homogeneity and isotropy about every point. Furthermore, since we have derived the Friedmann-Robertson-Walker metric under the hypotheses of spatial homogeneity and isotropy about *one* point, we explicitly obtain again, in a three-dimensional case, the result that a manifold homogeneous and isotropic about one point is also isotropic about *every* other point.

In the previous section we have seen that $k = +1$ corresponds to a model universe *closed in time*, that is, expanding and then recollapsing. However, if $k = 0$, or $k = -1$, the model universe is *open in time*, that is, it will expand forever. Furthermore, in the case $k = +1$, one *may* have a homeomorphism between the three-dimensional, $k = +1$, homogeneity hypersurface and a *three-sphere* embedded in a four-dimensional Euclidean manifold; for $k = -1$ one *may* have a homeomorphism between the $k = -1$, homogeneity hypersurface and a *hyperboloid* embedded in a four-dimensional Minkowskian manifold, and finally, for $k = 0$, one *may* have a homeomorphism between the $k = 0$ hypersurface and a three-dimensional *Euclidean manifold* $\Re^3$. We recall that two manifolds are homeomorphic, or topologically equivalent, if between them there exists a homeomorphism, that is, a one-to-one and surjective map, continuous, with continuous inverse map.

Therefore, in these examples, the homogeneity hypersurface is compact for $k = +1$, but it is not compact for $k = 0$ and $k = -1$. In these examples spatial compactness corresponds to closure in time and vice versa; therefore, it is natural to ask an interesting question: *is a model universe that is "closed" in space is also "closed" in time and vice versa*?

The answer is no, in general; however, a special compact spatial topology is a *necessary* condition for a model universe to recollapse and probably, with some additional standard requirements, also sufficient (see below). This question is relevant to our investigation of the origin of inertia in geometrodynamics which, according to the ideas of Einstein and Wheeler, is consistent with a spatially compact universe.

Let us first discuss the simple case of spatially homogeneous and isotropic model universes. The corresponding metric is the Friedmann-Robertson-Walker metric (4.2.16); with the above three choices of global geometry, the hypersurfaces of homogeneity of this metric were topologically globally chosen in such a way that spatial compactness was equivalent to closure in time. However, other global homeomorphisms between the hypersurfaces of homogeneity of the Friedmann-Robertson-Walker metric and other three-manifolds with different global topologies are possible. A known example is the *flat*[61] *three-torus*, T^3. The hypersurfaces of homogeneity of the Friedmann-Robertson-Walker metric corresponding to $k = 0$ may in fact have the topology of the flat three-torus (see fig. 4.6).

We recall that a one-sphere, or circle, is defined as a manifold homeomorphic to the circle S^1 in the Euclidean plane $\Re^2$:

$$S^1 = \{x \in \Re^2 : d(x, 0) = 1\} \tag{4.3.13}$$

where $d(x, y)$ is the usual Pythagorean metric

$$d(x, y) = \left(\sum_{i=1}^{n} (x^i - y^i)^2 \right)^{\frac{1}{2}} .$$

In general, an n-sphere is defined[61]

$$S^n = \{x \in \Re^{n+1} : d(x, 0) = 1\}, \tag{4.3.14}$$

and an n-torus is the product[61] of n copies of S^1:

$$S^1 \times S^1 \times \cdots \times S^1 : \quad n \text{ times.} \tag{4.3.15}$$

In particular, $S^1 \times S^1$ is the usual two-torus, homeomorphic to a subset of $\Re^4$ but also homeomorphic to the standard two-torus subset of $\Re^3$ (the surface of a doughnut; see figs. 4.5 and 4.6).

A two-torus is homeomorphic to the surface of a doughnut in $\Re^3$, with negative curvature on the internal parallel circle and inner region, positive curvature on the external parallel circle and outer region, and zero curvature on the upper

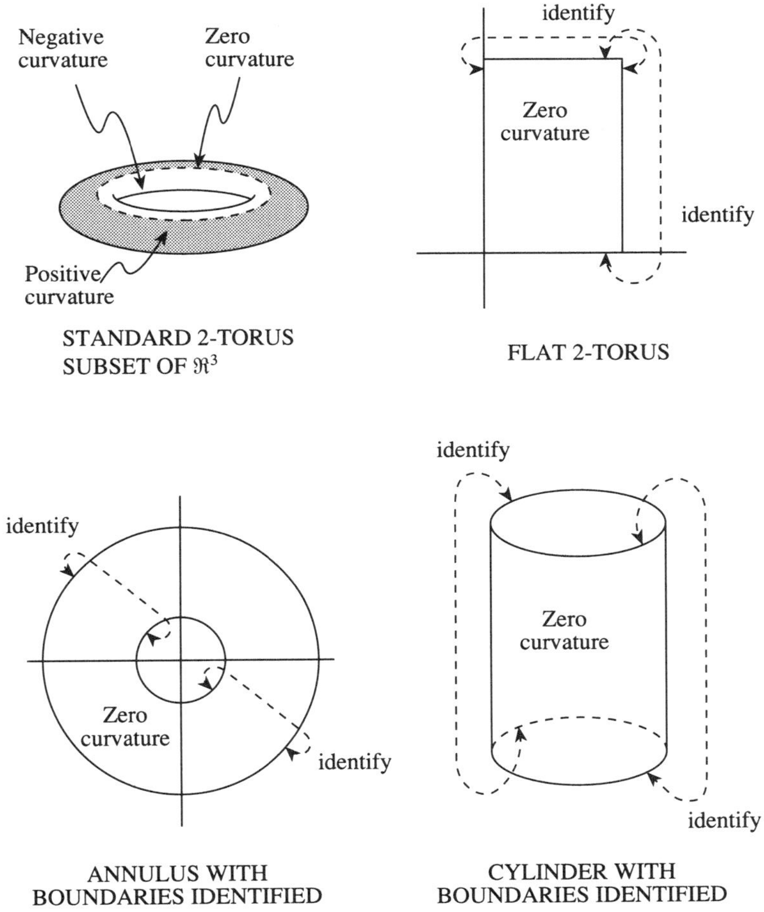

FIGURE 4.6. Four examples of two-torus, three of them flat, the fourth (upper left) with curvature. These two-surfaces are homeomorphic, or topologically equivalent, to a standard doughnut-like two-torus embedded in $\Re^3$ but they have different curvature.[61]

and lower parallel circles. However, a two-torus is also homeomorphic to a subset $S^1 \times S^1$ of $\Re^2 \times \Re^2$ which one can, for example, construct by properly identifying on a real plane $\Re^2$ the points $x = 0$ with the points $x = L$, and the points $y = 0$ with the points $y = L$, where $L \in \mathbb{Z}$ is an integer. This two-dimensional subset of $\Re^4$ has zero curvature and is called "flat-torus"[61] (see fig. 4.6).

The spacelike hypersurfaces of homogeneity described by the Friedmann-Robertson-Walker metric (4.2.16) with $k = 0$ may have the global topology of a *flat three-torus*:

$$S^1 \times S^1 \times S^1 \subset \Re^2 \times \Re^2 \times \Re^2. \qquad (4.3.16)$$

For example, one can identify the opposite faces of a three-cube of edge l by identifying the points $x = 0$ with the points $x = l$, the points $y = 0$ with the points $y = l$, and the points $z = 0$ with the points $z = l$, where $l \in \mathbb{Z}$ is an integer. One then gets a three-torus of finite volume $V = R^3 l^3$ (calculated using the metric $d\Sigma^2 = R^2 dl^2$).

This hypersurface, corresponding to the Friedmann-Robertson-Walker metric with $k = 0$, is flat; however, topologically, it is a closed manifold, compact and without boundary (see mathematical appendix) and, as we have seen in section 4.2, it expands forever, that is, it is open in time. Indeed, in a model universe containing dust, that is, a perfect fluid with zero pressure, from the 0-0 component of the Einstein field equation we have (see chap. 5)

$$\text{(extrinsic curvature)} + \text{(intrinsic curvature)} = 16\pi\,\text{(energy density)}$$

or[110]

$$(\mathrm{Tr}\,\boldsymbol{K})^2 - \mathrm{Tr}(\boldsymbol{K}^2) + {}^{(3)}R = 16\pi\varepsilon. \qquad (4.3.17)$$

For a $k = 0$, Friedmann model universe with spacelike hypersurfaces of homogeneity which have the topology of a flat three-torus with "edge length" $L(t)$, we have ${}^{(3)}R = 0$. Since the *extrinsic curvature tensor* $\boldsymbol{K}$ represents the fractional rate of change and deformation with time of three-geometry (see chap. 5) and for a Friedmann model $\boldsymbol{K} = -\frac{1}{L}\frac{dL}{dt}\mathbf{1}$, calling $\mathcal{M}$ the total mass content of the model universe we have

$$\frac{6}{L^2}\left(\frac{dL}{dt}\right)^2 = 16\pi\,\frac{\mathcal{M}}{L^3} \qquad (4.3.18)$$

or

$$L(t) = (6\pi\mathcal{M}t^2)^{\frac{1}{3}}. \qquad (4.3.19)$$

Therefore, this spatially homogeneous and isotropic model universe is open in time, that is, it expands for ever, as we have previously seen in section 4.2 (eq. 4.2.63).

This example clearly shows that spatial compactness is not equivalent to closure in time, even for spatially homogeneous and isotropic model universes. What about generic model universes? In general (see below), every three-torus model universe[111] does not have a time of maximum expansion, that is, it does not have a maximal hypersurface where the extrinsic curvature, $\mathrm{Tr}\,\boldsymbol{K}$, is zero. It is open in time.

It has been rigorously proved[112] that every model universe closed in time must have the spatial topology of a *three-sphere* S^3 or of a *two sphere times a circle*, $S^2 \times S^1$, or any other topology obtained by connected summation of these two topologies or by some identifications of points of S^3 (see below). No converse general proof is, at this writing, known which shows the recollapse of cosmological models having these spatial topologies, but there are some findings in this direction (see end of this section). Let us describe the interesting result on the global topology required for recollapse.

We first have the important result of Schoen and Yau (1978, 1979)[113–115] that a nonflat, compact, orientable, three-dimensional manifold Σ, admitting a metric with nonnegative Ricci scalar curvature ${}^{(3)}R$, must have the topology:

$$\left(\frac{S^3}{P_1}\right) \# \left(\frac{S^3}{P_2}\right) \# \cdots \# \left(\frac{S^3}{P_n}\right) \# \, k(S^2 \times S^1) \tag{4.3.20}$$

(where the differentiable structure of $\Sigma \times \Re$ is assumed to be non-exotic;[116–119,202,203] see after (4.3.23). Here the symbol # means connected sum of two manifolds that can be essentially obtained by cutting out and throwing away a spherical region of each manifold and then by gluing together the two remaining manifolds on the corresponding boundaries, that is, by identifying with a homeomorphism the corresponding points of the boundaries of the two spherical regions (see fig. 4.7). Two examples of connected sum are the n-holed torus and the sphere with n-handles of fig. 4.8.

In expression (4.3.20) $\frac{S^3}{P_i}$ ("S^3 quotient the subgroup P_i") is the manifold obtained by identifying all the points of a three-sphere S^3 that are carried into one another under the action of a finite subgroup P_i of the three-dimensional rotation group of motions on the three-sphere S^3, and $k(S^2 \times S^1)$ is the connected sum of k copies of $S^2 \times S^1$.

In summary, the Schoen and Yau theorem tells us, if a model universe has a nonflat spacelike hypersurface Σ, with ${}^{(3)}R \geq 0$, the topology of Σ must be a connected sum of copies of $S^2 \times S^1$ and of S^3 quotient a subgroup P_i (it is assumed that $\Sigma \times \Re$ has the standard, nonexotic, differentiable structure).[116–119,202,203]

Turning from space to spacetime let us now assume (1) the Einstein field equation and (2) the weak energy condition, that is, $T_{\alpha\beta}u^\alpha u^\beta \geq 0$, for every timelike vector u^α. From the Einstein field equation, on a spacelike hypersurface Σ, we then have (see chap. 5):[110]

$$(\mathrm{Tr}\,\boldsymbol{K})^2 - \mathrm{Tr}(\boldsymbol{K}^2) + {}^{(3)}R = 16\pi T_{\alpha\beta}n^\alpha n^\beta \tag{4.3.21}$$

where n^α is the unit vector field normal to the hypersurface. We recall that a maximal hypersurface is defined to be a hypersurface of maximum expansion, where the extrinsic curvature, $\mathrm{Tr}\,\boldsymbol{K}$, is equal to zero, corresponding to the time of maximum expansion.

Since $\mathrm{Tr}(\boldsymbol{K}^2) \geq 0$ (indeed in Gaussian normal coordinates we have $\mathrm{Tr}(\boldsymbol{K}^2) = n_{i;k}n^{i;k} \geq 0$), if the model universe has a compact, orientable, nonflat, spacelike hypersurface of maximal expansion Σ where $(\mathrm{Tr}\,\boldsymbol{K})_\Sigma = 0$, from $T_{\alpha\beta}n^\alpha n^\beta \geq 0$

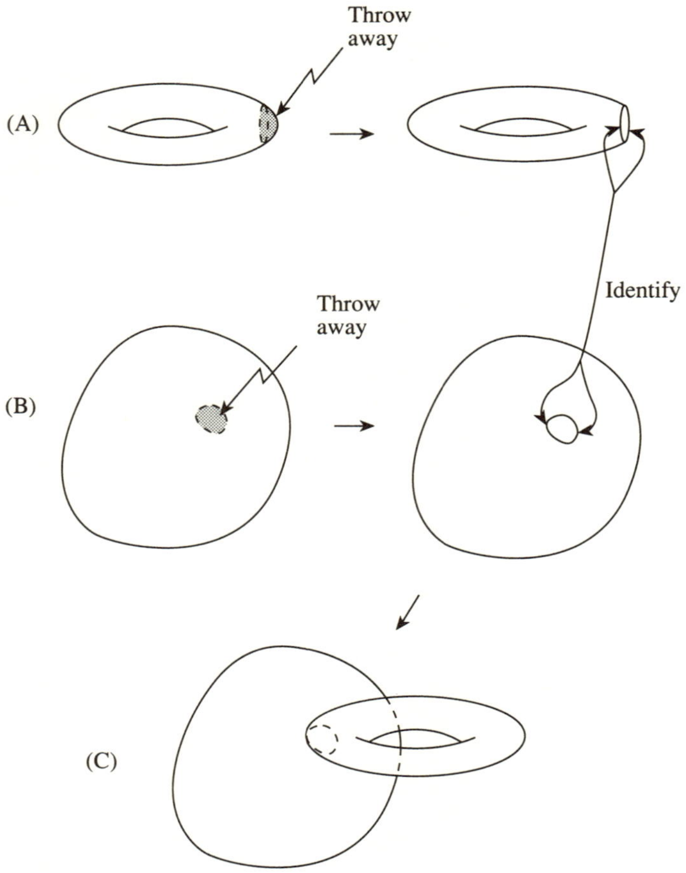

FIGURE 4.7. Handles and spheres with handles. (A) A handle: a torus with a hole. (B) A sphere with a hole. (C) A sphere with a handle (see also fig. 4.8).

in (4.3.21), we must have on Σ:

$$^{(3)}R \geq 0, \tag{4.3.22}$$

and therefore, according to the Schoen and Yau theorem,[113–115] Σ must have the topology (4.3.20).

We then have the following theorem (Barrow and Tipler 1985).[112] If a model universe has a compact, orientable, spacelike, maximal hypersurface Σ, and (1) the Einstein field equation (without cosmological term) is satisfied, (2) the weak energy condition, $T_{\alpha\beta}u^\alpha u^\beta \geq 0$, is satisfied for every timelike vector u^α, (3) the induced metric on Σ is not flat, (4) the differentiable structure on $\Sigma \times \Re$

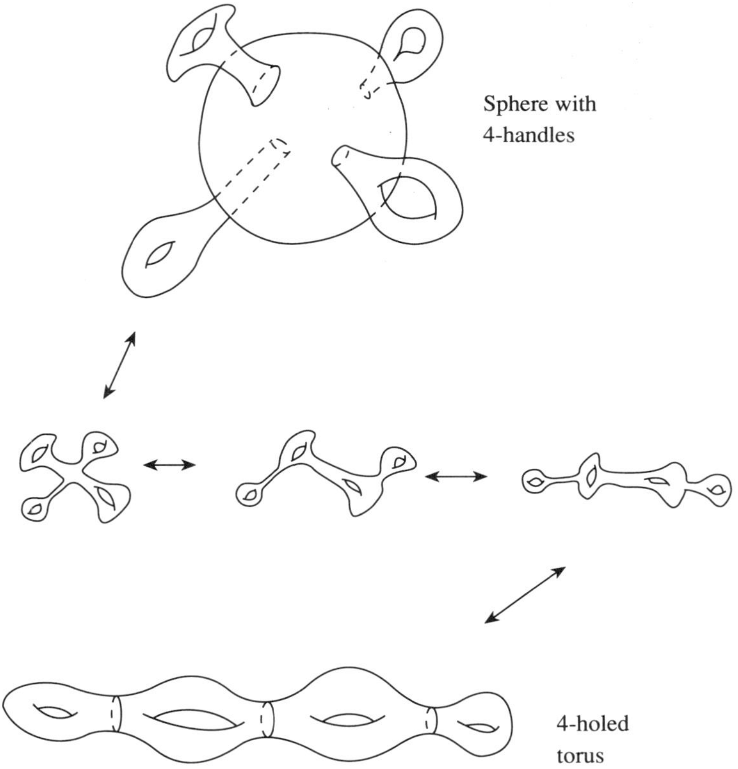

FIGURE 4.8. A sphere with n-handles and an n-holed torus ($n = 4$), two topologically equivalent (homeomorphic) manifolds that in general have different curvatures.

is not exotic, then Σ must have the topology

$$\left(\frac{S^3}{P_1}\right) \# \cdots \# \left(\frac{S^3}{P_n}\right) \# \, k(S^2 \times S^1). \tag{4.3.23}$$

By definition, a nonexotic differentiable structure on $S^3 \times \Re$ is provided by any standard coordinate system on $S^3 \times \Re$. (In 1982–1983, Freedman (1982)[116] and Donaldson (1983)[117] (see also refs. 118, 119, 202, and 203) discovered *exotic differentiable manifolds*, $\Re^4_{\text{FAKE}}$, homeomorphic but not diffeomorphic to $\Re^4$ with the standard differentiable structure $\Re^4_{\text{STD}}$. It is amazing that these exotic manifolds exist in four dimensions only, for $\Re^n$ with $n \neq 4$ any differentiable structure on $\Re^n$ is equivalent to the standard one: $\Re^n_{\text{STD}}$.)

Furthermore, the following theorem has been proved (Marsden and Tipler 1980[120] and Barrow and Tipler 1985)[112]: a model universe has a compact maximal hypersurface if and only if it has initial and final all-encompassing, true curvature singularities (or the length of every timelike curve in the model universe is less than a universal constant T). The proof assumes that $R_{\alpha\beta}u^{\alpha}u^{\beta} \geq 0$, called the timelike convergence condition (§ 2.9), the existence of a compact Cauchy surface (§ 5.2.2), plus a few other mathematical conditions.

In conclusion, in Einstein theory provided some positive energy conditions hold, if a model universe is closed in time, that is, if it expands and then recollapses, then its spatial topology must be a connected sum of copies of a three-sphere (and of a three-sphere quotient some subgroup P_i, S^3/P_i) and of copies of a two-sphere times a circle, $S^2 \times S^1$. Therefore, a model universe with a spacelike Cauchy surface homeomorphic to a three-torus T^3 cannot have a maximal hypersurface and cannot recollapse. For example, from this theorem we immediately have that the $k = 0$, spatially homogeneous and isotropic Friedmann model, spatially homeomorphic to the flat T^3 (described before), will expand forever.

What about the *converse*? Will a model universe with the spatial topology (4.3.20) recollapse?

Not always, depending on the energy-momentum tensor. Indeed, Barrow, Galloway, and Tipler[121] have given examples of ever-expanding Friedmann model universes with compact spatial topology of a three-sphere S^3 (eq. 4.2.16, with $k = +1$) and such that the energy-momentum tensor satisfies the strong energy condition ($(T_{\alpha\beta} - \frac{1}{2}Tg_{\alpha\beta})u^{\alpha}u^{\beta} \geq 0$, for every timelike vector u^{α}), the weak $\left(T_{\alpha\beta}u^{\alpha}u^{\beta} \geq 0\right)$ and dominant (in any orthonormal basis $T_{00} \geq |T_{\alpha\beta}|$, for every $\alpha\beta$) energy conditions, and the generic condition (every nonspacelike geodesic has at least a point at which $K_{[\alpha}R_{\beta]\gamma\delta[\mu}K_{\nu]}K^{\gamma}K^{\delta}$ is not zero, where K^{α} is the tangent vector; see § 2.9). These nonrecollapsing, S^3, Friedmann model universes are characterized by a fluid with pressure p and density ρ satisfying the abnormal equation of state $p = -\frac{1}{3}\rho$; in addition they contain dust with $p = 0$.

However, in the case of the Friedmann models with spatial topology of a three-sphere, it has been proved,[121] within the framework of general relativity, that whenever the positive pressure condition is satisfied, $\sum_{i=1}^{3} p_i \geq 0$ (p_i are the principal pressures), that is, $p \geq 0$ for the Friedmann models, along with the dominant energy condition, along with some standard matter regularity conditions for the energy-momentum tensor, then these S^3 Friedmann model universes will recollapse (and eventually will recollapse to a final singularity).

Furthermore, in the case of nonrotating homogeneous Bianchi IX models, which have spatial topology of a three-sphere[122] (see §§ 4.4 and 4.7), it has been proved[123] that if the positive pressure condition $\sum_{i=1}^{3} p_i \geq 0$ and the dominant energy condition are satisfied, then these Bianchi IX models will recollapse.

This proof has been announced[123] to be extendible to all the Bianchi IX and Kantowski-Sachs models.

Barrow, Galloway, and Tipler have conjectured[121] the following within the framework of general relativity:

Conjecture (1): Every globally hyperbolic (with a Cauchy surface), spatially compact, vacuum, model universe with spatial topology S^3 or $S^2 \times S^1$, or in general with spatial topology (4.3.20), $\left(\frac{S^3}{P_1}\right) \# \left(\frac{S^3}{P_2}\right) \# \cdots \# \left(\frac{S^3}{P_n}\right) \# k\left(S^2 \times S^1\right)$, will recollapse (toward an all-encompassing final singularity).

Conjecture (2): Every globally hyperbolic, spatially homogeneous and compact, model universe with spatial topology S^3 or $S^2 \times S^1$ or in general with spatial topology (4.3.20), and with energy-momentum tensor satisfying (1) the positive pressure condition, (2) the strong energy condition, (3) the dominant energy condition, and (4) standard matter regularity conditions, will recollapse (toward an all-encompassing final singularity).

The three known types of spatially compact and homogeneous model universes with spatial topology S^3 or $S^2 \times S^1$ are: Kantowski-Sachs models ($S^2 \times S^1$), general Bianchi IX (S^3), and Taub (S^3), that is, Bianchi IX with axial symmetry. Since conjecture (2) has been proved[123] for nonrotating Bianchi IX models and has been announced by Lin, Wald, and Burnett to be provable for general Bianchi IX models and for Kantowski-Sachs models,[123] it might be valid in general.

Conjecture (3): Every globally hyperbolic, spatially compact, model universe with spatial topology S^3 or $S^2 \times S^1$, or in general with spatial topology (4.3.20), and with energy-momentum tensor satisfying (1) the positive pressure condition and (2) the strong energy condition, will recollapse (toward an all-encompassing final singularity).

In regard to the relation of these results for the understanding of the origin of inertia in Einstein geometrodynamics, we observe that if, one day, we finally determine the value of some parameter which would imply the time closure of the universe (such as a deceleration parameter $q_\circ > \frac{1}{2}$ in the case of a matter-dominated Friedmann model universe) then, by knowing that the universe is closed in time and by the theorems shown above, we would also know that the spatial topology of the universe is compact, or topologically closed and bounded, in agreement with the ideas of Einstein and Wheeler on the origin of inertia (see chap. 5).

4.4 SPATIALLY HOMOGENEOUS MODELS: THE BIANCHI TYPES

In sections 2.6 and 4.2 we have seen that the symmetries of a geometric object $\boldsymbol{G}$ on a manifold can be characterized by the vector fields $\boldsymbol{\xi}$ for which the **Lie derivative**[59–62] of the geometric object is zero, $\mathcal{L}_{\boldsymbol{\xi}}\boldsymbol{G} = \boldsymbol{0}$.

In particular, the symmetries of a manifold itself[60] can be characterized by the Lie derivative of the metric tensor $\boldsymbol{g}$: $\mathcal{L}_{\xi}\boldsymbol{g}$. If $\mathcal{L}_{\xi}\boldsymbol{g} = \boldsymbol{0}$, $\boldsymbol{\xi}$ is called **Killing vector** and represents some **symmetry of the manifold**

$$\begin{aligned}\left(\mathcal{L}_{\xi}\boldsymbol{g}\right)_{\alpha\beta} &\equiv g_{\alpha\beta,\sigma}\xi^{\sigma} + \xi^{\sigma}{}_{,\alpha}g_{\sigma\beta} + \xi^{\sigma}{}_{,\beta}g_{\alpha\sigma} = \\ &= \xi_{\alpha;\beta} + \xi_{\beta;\alpha} \equiv 2\xi_{(\alpha;\beta)} = 0\end{aligned} \tag{4.4.1}$$

(see expression 4.2.6). The infinitesimal transformation (4.4.1) associated with a Killing vector field is called **infinitesimal isometry** (see §§ 2.6 and 4.2). The group of isometries of a Riemannian manifold (with a finite number of connected components) forms a *Lie group*,[60–62] that is, a group which is itself a smooth manifold, with the group operation xy^{-1} that is a smooth map into the manifold itself (see the mathematical appendix). On a geodesically complete Riemannian manifold the set of all the Killing vector fields, $\boldsymbol{\xi}$, generators of isometries of the manifold, forms the associated *Lie algebra*.[61,62] This set of all the Killing vector fields is in fact a vector space under addition (any linear combination of Killing vector fields with constant coefficients is a Killing vector field), and the operation of product of two Killing vector fields is defined by the Lie derivative, that is, by the commutator of the two relevant vector fields

$$\left(\mathcal{L}_{\xi_A}\xi_B\right)^{\alpha} = \frac{\partial \xi_B^{\alpha}}{\partial x^{\sigma}}\xi_A^{\sigma} - \frac{\partial \xi_A^{\alpha}}{\partial x^{\sigma}}\xi_B^{\sigma} \equiv \left[\boldsymbol{\xi}_A, \boldsymbol{\xi}_B\right]^{\alpha} \tag{4.4.2}$$

where $\left[\boldsymbol{\xi}_A, \boldsymbol{\xi}_B\right]^{\alpha}$ defines the commutator of $\boldsymbol{\xi}_A$ and $\boldsymbol{\xi}_B$. Indeed, if $\boldsymbol{\xi}_A$ and $\boldsymbol{\xi}_B$ are Killing vectors, $\left[\boldsymbol{\xi}_A, \boldsymbol{\xi}_B\right]$ is also a Killing vector

$$\left(\mathcal{L}_{[\xi_A,\xi_B]}\boldsymbol{g}\right)_{\alpha\beta} = \left[\boldsymbol{\xi}_A, \boldsymbol{\xi}_B\right]_{\alpha;\beta} + \left[\boldsymbol{\xi}_A, \boldsymbol{\xi}_B\right]_{\beta;\alpha} = \boldsymbol{0}. \tag{4.4.3}$$

Furthermore, the product operation, $\left[\boldsymbol{\xi}_A, \boldsymbol{\xi}_B\right]$ satisfies

$$\left[\boldsymbol{\xi}_A, \boldsymbol{\xi}_B\right] = -\left[\boldsymbol{\xi}_B, \boldsymbol{\xi}_A\right] \tag{4.4.4}$$

and the cyclic Jacobi identity

$$\left[\boldsymbol{\xi}_A, \left[\boldsymbol{\xi}_B, \boldsymbol{\xi}_C\right]\right] + \left[\boldsymbol{\xi}_B, \left[\boldsymbol{\xi}_C, \boldsymbol{\xi}_A\right]\right] + \left[\boldsymbol{\xi}_C, \left[\boldsymbol{\xi}_A, \boldsymbol{\xi}_B\right]\right] = \boldsymbol{0}. \tag{4.4.5}$$

Therefore the Killing vectors $\boldsymbol{\xi}$ form a Lie algebra.

By choosing a basis for the algebra of the Killing vectors $\boldsymbol{\xi}_{(b)}$ where $b = 1, \ldots, n$, since $\left[\boldsymbol{\xi}_{(a)}, \boldsymbol{\xi}_{(b)}\right]$ is still an element of the algebra, we can write the commutator between any two elements of the basis:

$$\left[\boldsymbol{\xi}_{(a)}, \boldsymbol{\xi}_{(b)}\right] = C^{c}_{ab}\boldsymbol{\xi}_{(c)}. \tag{4.4.6}$$

C^{c}_{ab} are called the **structure constants** of the Lie algebra and by construction are antisymmetric with respect to the two lower indexes a, b.

Let us now define the quantities A^{cd}:

$$A^{cd} \equiv \frac{1}{2}C^{c}_{ab}\bar{\epsilon}^{abd}. \tag{4.4.7}$$

Here, $\bar{\epsilon}^{abc}$ is the completely antisymmetric symbol $\bar{\epsilon}^{abc} = 1$, for even permutations of 123, -1 for odd permutations of 123, and 0 when two or more indexes are repeated. We can then rewrite the quantities A^{ab} as sum of an antisymmetric part and a symmetric one:

$$A^{ab} \equiv N^{(ab)} + \bar{\epsilon}^{abc} A_c. \tag{4.4.8}$$

One can rewrite the Jacobi identities (4.4.5) in terms of the structure constants C^c_{ab} with help of (4.4.6):

$$C^s{}_{[bc} C^d{}_{e]s} = 0. \tag{4.4.9}$$

We substitute $C^c_{ab} = \bar{\epsilon}_{abd} A^{cd}$, multiply by $\bar{\epsilon}^{bce}$, sum over bce (see § 2.8) and find

$$\bar{\epsilon}_{abc} A^{bc} A^{ad} = 0, \tag{4.4.10}$$

and therefore, from (4.4.8), we have

$$N^{(db)} A_b = 0. \tag{4.4.11}$$

One can redefine the basis Killing vectors, $\boldsymbol{\xi}_{(a)}$, by taking linear combinations of the $\boldsymbol{\xi}_{(a)}$ with constant coefficients $\boldsymbol{\xi}'_{(a)} = K^b_{a'} \boldsymbol{\xi}_{(b)}$, where $K^b_{a'}$ are constants. Under such a transformation the $A^{ab} J$ transform as components of a tensor (J is the Jacobian, that is, the determinant of the matrix of the first partial derivatives of the transformation). Therefore, one can choose a basis for the Killing vectors that diagonalizes the symmetric quantities $N^{(ab)} \to n_a \delta_{ab}$. Furthermore, with a rotation we can transform the A_i to be $A_1 \equiv a$ and $A_2 = A_3 = 0$, while keeping $N^{(ab)}$ diagonal. We can then make $a \geq 0$. Since we have the three conditions $N^{(ab)} A_b = 0$, we have $n_1 a = 0$, that is either $a = 0$ or $n_1 = 0$, or both, $a = n_1 = 0$, in the notation of expression (4.4.12) below. Finally, by rescaling the basis Killing vectors we can transform the n_a which are different from zero to be ± 1. However, in general, is not possible to simultaneously rescale both the n_i and a, different from zero, to be ± 1. We then have

$$A^{ab} = \begin{pmatrix} n_1 & 0 & 0 \\ 0 & n_2 & a \\ 0 & -a & n_3 \end{pmatrix} \tag{4.4.12}$$

where $n_i = -1, 0$, or $+1$, and $a n_1 = 0$.

It is now possible to classify all the spatially homogeneous model universes[124–128] by the values of a and n_a. Indeed, these quantities characterize the structure constants and therefore the Killing vectors that describe the symmetries of a manifold, as summarized in table 4.2. The various spatially homogeneous models are called the **Bianchi types.**[130]

TABLE 4.2 The Bianchi homogeneous models[124–130]

Class	*Type*	a	n_1	n_2	n_3
A	I	0	0	0	0
	II	0	1	0	0
	VI_0 (or Bianchi type VI)	0	0	1	-1
	VII_0 (or Bianchi type VII)	0	0	1	1
	VIII	0	1	1	-1
	IX	0	1	1	1
B	V	1	0	0	0
	IV	1	0	0	1
	{ III ($\mathrm{VI}_{h=-1}$)	1	0	1	-1
	{ VI_h	$\sqrt{-h} \neq 0$	0	1	-1
	VII_h	$\sqrt{h} \neq 0$	0	1	1

Note: Luigi Bianchi (1856–1928) first set forth a scheme of classification of spatially homogeneous manifolds in 1897.

4.5 EXPANSION, ROTATION, SHEAR, AND THE RAYCHAUDHURI EQUATION

In order to describe some interesting rotating cosmological models and to investigate the influence of rotating masses on local inertial frames, we shall first analyze what we mean by rotation for a collection of matter, idealized here as a fluid. We shall define some quantities useful to characterize the fluid motion, such as the **volume expansion scalar**, the **shear tensor**, and the **rotation tensor** of a fluid, and we shall establish the connection between these features of the fluid motion and the Ricci curvature of the spacetime, that is, the **Raychaudhuri equation.**[131–133]

Given a fluid described by the four-velocity field of the fluid particles $u^\alpha = \frac{dx^\alpha}{ds}$, we reparametrize the world lines of the fluid particles with the arc-length s (§ 2.4), so that we have $g_{\alpha\beta}u^\alpha u^\beta = -1$.

We can then decompose[134,80] any vector, or "project" any index of a tensor, into a timelike component and a spacelike component by applying a **time-projection tensor**, $-(u_\alpha u_\beta)$, and a **space-projection tensor**, $h_{\alpha\beta} \equiv g_{\alpha\beta} + u_\alpha u_\beta$:

$$P_t(v^\alpha) \equiv -u_\alpha u_\beta v^\beta \tag{4.5.1}$$

$$P_\Sigma(v^\alpha) \equiv h_{\alpha\beta} v^\beta \tag{4.5.2}$$

One can verify that $P_t(v^\alpha)$ and $P_\Sigma(v^\alpha)$ are respectively spacelike and timelike.

We can then rewrite the spacetime distance ds^2 between two events x^α and $x^\alpha + dx^\alpha$

$$ds^2 = h_{\alpha\beta} dx^\alpha dx^\beta - u_\alpha u_\beta dx^\alpha dx^\beta. \tag{4.5.3}$$

Then, an observer in x^α, comoving with the fluid particle with four-velocity u^α, measures between the two events x^α and $x^\alpha + dx^\alpha$ a space distance $dl = (h_{\alpha\beta} dx^\alpha dx^\beta)^{1/2}$ and a time interval $d\tau = (u_\alpha u_\beta dx^\alpha dx^\beta)^{1/2}$.

Therefore, $h_{\alpha\beta}$ is a projection tensor into the space orthogonal to the particle four-velocity u^α, with the following properties:

$$h_{\alpha\beta} u^\beta = 0; \quad h_{\alpha\beta} \dot{u}^\beta \equiv h_{\alpha\beta} u^\beta{}_{;\sigma} u^\sigma \equiv h_{\alpha\beta} a^\beta = a_\alpha \tag{4.5.4}$$

$$h^{\alpha\beta} h_{\beta\gamma} = h^\alpha{}_\gamma; \quad \text{and} \quad h^\alpha{}_\alpha = 3 \tag{4.5.5}$$

where $\dot{u}^\beta \equiv u^\beta{}_{;\sigma} u^\sigma \equiv a^\beta$ is the four-acceleration of the fluid particles. This vector measures the deviations of the particles from the geodesic motion $u^\alpha{}_{;\sigma} u^\sigma = 0$; that is, in metric theories of gravity (§ 3.2), where test particles satisfy $\frac{Du^\alpha}{ds} = 0$, a^β is due to nongravitational forces (such as pressure) acting on the particles.

We now decompose the tensor $u_{\alpha;\beta}$ into various spacelike and timelike components using the projection tensors just defined. We thus have

$$\begin{aligned} u_{\alpha;\beta} &= u_{\mu;\nu} \big(h^\mu{}_\alpha - u^\mu u_\alpha\big)\big(h^\nu{}_\beta - u^\nu u_\beta\big) \\ &= u_{\mu;\nu} h^\mu{}_\alpha h^\nu{}_\beta - u_{\alpha;\nu} u^\nu u_\beta \\ &= \Big\{\Big[\frac{1}{2}(u_{\mu;\nu} + u_{\nu;\mu}) - \frac{1}{3}\Theta h_{\mu\nu}\Big] \\ &\quad + \Big[\frac{1}{2}(u_{\mu;\nu} - u_{\nu;\mu})\Big] + \frac{1}{3}\Theta h_{\mu\nu}\Big\} h^\mu{}_\alpha h^\nu{}_\beta - a_\alpha u_\beta \\ &= \sigma_{\alpha\beta} + \omega_{\alpha\beta} + \frac{1}{3}\Theta h_{\alpha\beta} - a_\alpha u_\beta, \end{aligned} \tag{4.5.6}$$

where

(1) the scalar Θ of volume expansion is defined as

$$\Theta \equiv u^\alpha{}_{;\alpha}, \tag{4.5.7}$$

(2) the symmetric traceless tensor of shear $\sigma_{\alpha\beta}$ is

$$\sigma_{\alpha\beta} \equiv \left[u_{(\mu;\nu)} - \frac{1}{3}\Theta h_{\mu\nu}\right] h^\mu{}_\alpha h^\nu{}_\beta, \tag{4.5.8}$$

(3) the antisymmetric tensor of vorticity $\omega_{\alpha\beta}$ is

$$\omega_{\alpha\beta} \equiv u_{[\mu;\nu]} h^\mu{}_\alpha h^\nu{}_\beta, \tag{4.5.9}$$

and we have

$$a^\beta u_\beta = \sigma_{\alpha\beta} u^\beta = \omega_{\alpha\beta} u^\beta = 0. \tag{4.5.10}$$

Therefore, $\omega_{\alpha\beta}$, $\sigma_{\alpha\beta}$, and Θ represent the decomposition of the space-space projection of the tensor $u_{\alpha;\beta}$ into antisymmetric part, symmetric traceless part, and trace of the symmetric part. These quantities $\omega_{\alpha\beta}$, $\sigma_{\alpha\beta}$, and Θ are respectively called **vorticity or rotation tensor, shear tensor**, and **volume expansion.**[80,133,135,136]

To display the role of the various parts of the field $u_{\alpha;\beta}$ we now examine the vector δx^i that connects a typical fluid particle with a typical *neighboring* particle. When we use comoving coordinates, that is, when we label each particle of the fluid with a y^i, the vector that connects two infinitesimally close particles, $\delta y^i \equiv (y^i + \delta y^i) - y^i$, and joins them at the same proper time ($\delta y^0 \equiv \delta s = 0$), will join the same particles at all times. In a general coordinate system we have $\delta x^\alpha = \frac{\partial x^\alpha}{\partial y^k}\delta y^k$ (see fig. 4.9) and the Lie derivative of δx^α with respect to u^α is zero:

$$\delta x^\alpha{}_{;\sigma} u^\sigma - u^\alpha{}_{;\sigma}\delta x^\sigma = \pounds_{\mathbf{u}}\boldsymbol{\delta x} \equiv \delta x^\alpha{}_{,\sigma} u^\sigma - u^\alpha{}_{,\sigma}\delta x^\sigma$$

$$= \Big(\frac{\partial x^\alpha}{\partial y^\rho}\delta y^\rho\Big)_{,\sigma}\frac{\partial x^\sigma}{\partial s} - \Big(\frac{\partial x^\alpha}{\partial s}\Big)_{,\sigma}\frac{\partial x^\sigma}{\partial y^\rho}\delta y^\rho$$

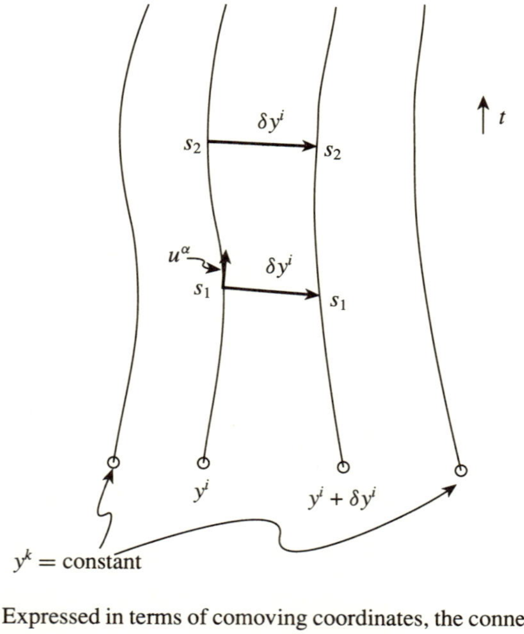

FIGURE 4.9. Expressed in terms of comoving coordinates, the connecting vector δy^i joins at all times the same two neighboring particles labeled y^i and $y^i + \delta y^i$, ($\frac{\partial}{\partial s}\delta y^i = 0$); in this figure δy^i also joins the particles at the same proper time ($\delta y^0 \equiv \delta s = 0$).

$$= \frac{\partial}{\partial s}\left(\frac{\partial x^\alpha}{\partial y^\sigma}\right)\delta y^\sigma - \frac{\partial}{\partial y^\sigma}\left(\frac{\partial x^\alpha}{\partial s}\right)\delta y^\sigma$$

$$= 0 \tag{4.5.11}$$

where we have used

$$\Gamma^\alpha_{[\mu\nu]} = 0, \ \frac{\partial}{\partial s}\delta y^\rho = 0 \qquad \text{and} \qquad \frac{\partial}{\partial s}\left(\frac{\partial x^\alpha}{\partial y^\sigma}\right) = \frac{\partial}{\partial y^\sigma}\left(\frac{\partial x^\alpha}{\partial s}\right).$$

and where u^α is the four-velocity of a particle of fluid with y^i = constant. Then, one can understand the meaning of Θ, $\boldsymbol{\sigma}$, and $\boldsymbol{\omega}$ by imagining a local orthonormal tetrad (or vierbein) $\lambda^\beta_{(\alpha)}$ tied to each particle of the fluid, each tetrad is constructed in such a way that its timelike vector is the four-velocity of the corresponding particle, $\lambda^\alpha_{(0)} \equiv u^\alpha$, and $\lambda^\sigma_{(\alpha)}\lambda_{(\beta)\sigma} \equiv \eta_{\alpha\beta}$, and its spatial vectors $\lambda^\alpha_{(i)}$ are *Fermi-Walker transported* along the particle world line[137,138] (§ 3.4.3). Therefore, locally, the spatial axes $\lambda^\alpha_{(i)}$ of each tetrad may be constructed by three orthogonal gyroscopes in torque-free motion (that is, obeying Fermi-Walker transport):

$$\lambda^\alpha_{(i);\beta}u^\beta = (a^\beta \cdot \lambda_{(i)\beta})u^\alpha. \tag{4.5.12}$$

In this *Fermi frame* the spatial components of the separation vector between neighboring particles, δx^α, are given by

$$\delta x^{(i)} = \delta x^\alpha \lambda^{(i)}_\alpha, \tag{4.5.13}$$

and therefore the rate of change of this separation is

$$\begin{aligned} \delta x^{(i)}{}_{;\sigma}u^\sigma &= \left(\delta x^\alpha \lambda^{(i)}_\alpha\right)_{;\sigma} u^\sigma \\ &= \delta x^\alpha{}_{;\sigma}u^\sigma\lambda^{(i)}_\alpha + \delta x^\alpha\left(\lambda^{(i)}_\alpha\right)_{;\sigma} u^\sigma \\ &= u^\alpha{}_{;\sigma}\delta x^\sigma\lambda^{(i)}_\alpha + \left(a^\sigma\lambda^{(i)}_\sigma\right)u_\alpha\delta x^\alpha = u^\alpha{}_{;\sigma}\delta x^\sigma\lambda^{(i)}_\alpha \end{aligned} \tag{4.5.14}$$

where we have used equation 4.5.11, the definition of Fermi-Walker transport, and the definition of connecting vector, $u_\alpha\delta x^\alpha = 0$. In terms of comoving coordinates a typical particle has four-velocity $u^\alpha = \delta^\alpha{}_0$, and one can redefine the $\lambda^{(i)}_\alpha$ to be $\lambda^{(i)}_\alpha = \delta^i{}_\alpha$, that is, $\lambda^{(\mu)}_\nu = \delta^\mu{}_\nu$. We then have for the rate of separation of particles

$$\frac{D(\delta x^{(i)})}{ds} = (u^{(i)}{}_{;(k)})\delta x^{(k)} \tag{4.5.15}$$

that one can also immediately get by writing equation (4.5.11) in a comoving frame. From equation (4.5.6), the rate of change of the connecting vector is in the Fermi frame

$$\frac{D(\delta x_{(i)})}{ds} = (\sigma_{ik} + \omega_{ik} + \frac{1}{3}\Theta h_{ik})\delta x^{(k)}. \tag{4.5.16}$$

This expression explicitly shows[139] that σ_{ik}, ω_{ik}, and Θ give the rate of change with time of the separation $\delta x^{(i)}$ between neighboring particles, that is, the motions and the rate of change of the dimensions of an infinitesimal volume element of the fluid relative to a local comoving Fermi frame.[80,133,136] In particular, for an infinitesimal spherical surface described by the fluid particles with $\delta x^{(i)}\delta x_{(i)} \equiv \delta r^2$ in the Fermi frame, its deformation with time is given by expression (4.5.16), Θ gives its fractional rate of volume expansion, σ_{ik} its shear (a measure of rate of change of shape), and ω_{ik} its angular velocity relative to the Fermi axes, that is, relative to local gyroscopes (see fig. 4.10).

In general, the scalar of expansion Θ represents the fractional rate of volume expansion of local ensembles of fluid particles, the **shear tensor** $\sigma_{\alpha\beta}$ represents the distortion of the generic local fluid ensemble at constant volume, the **vorticity tensor** $\omega_{\alpha\beta}$ represents the rotation of the generic local ensemble of fluid particles with respect to local gyroscopes, that is, with respect to the spatial axes, $\lambda^{\alpha}_{(i)}$, of a local comoving Fermi frame of reference, that are carried forward in time with Fermi-Walker transport. The pseudovector

$$\omega^{\alpha} \equiv \frac{1}{2}\epsilon^{\alpha\beta\mu\nu}u_{\beta}\omega_{\mu\nu} \tag{4.5.17}$$

(where $\epsilon^{\alpha\beta\mu\nu}$ is the Levi-Civita pseudotensor defined in § 2.8) is called the **vorticity or rotation vector**[140–142] and is built from the vorticity tensor $\omega_{\mu\nu}$

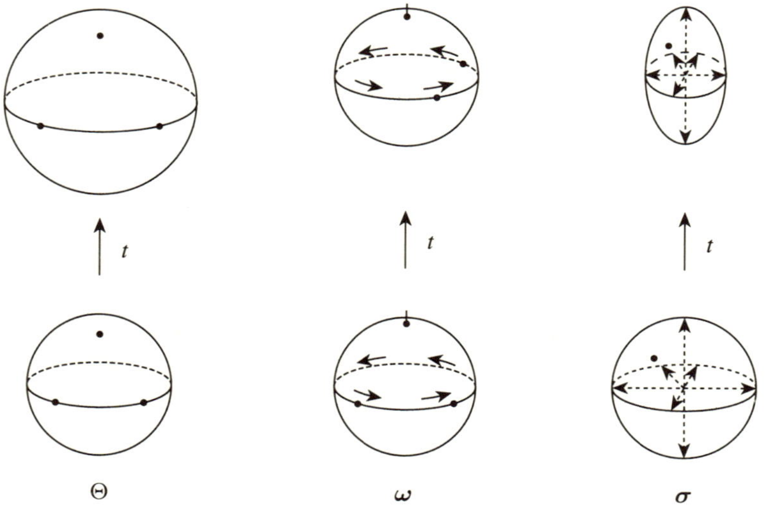

FIGURE 4.10. Examples of expansion Θ, rotation $\boldsymbol{\omega}$, and shear $\boldsymbol{\sigma}$, of a small element of fluid (adapted from Ellis 1971.)[80] In the case of $\Theta = 0$, $\boldsymbol{\omega} = \mathbf{0}$, and $\boldsymbol{\sigma} \neq \mathbf{0}$, volume is constant and the directions of the principal axes of shear do not change but the other directions change.

itself. Let us investigate the meaning of the vorticity vector ω^α. We calculate the components of the vorticity ω^α at an event in a local inertial frame momentarily at rest with respect to the fluid, we have $\overset{(0)}{g}_{\alpha\beta} = \eta_{\alpha\beta}$, $\overset{(0)}{\Gamma}{}^\alpha_{\mu\nu} = 0$, and $u^\alpha = \delta^\alpha{}_0$, but in general we have a nonzero acceleration $u^\alpha{}_{,\beta} \neq 0$, and therefore:

$$\omega^0 = 0 \qquad \text{and} \qquad \omega^i = \frac{1}{2}\epsilon^{i0kl}u_{k,l} = \frac{1}{2}\left(\frac{\partial u_l}{\partial x^k} - \frac{\partial u_k}{\partial x^l}\right) \tag{4.5.18}$$

where $i \neq (k, l)$ and ikl is an even permutation of 123. From this expression (4.5.18) we see that the ω^i are the components of the classical mechanics vector representing the angular velocity of rotation of a fluid:

$$\vec{\omega} = \frac{1}{2}\vec{\nabla} \times \vec{v} \qquad \text{and} \qquad \omega^0 = 0. \tag{4.5.19}$$

Therefore, the vorticity vector ω^α is the relativistic generalization of the rotation vector of a fluid of classical mechanics.

To find the relation between the descriptors of motion of matter, Θ, $\sigma_{\alpha\beta}$, $\omega_{\alpha\beta}$, and the Ricci curvature of the spacetime we use the equation for the commutator of the covariant derivatives of a vector field, given by the Riemann tensor

$$u^\alpha{}_{;\mu;\nu} - u^\alpha{}_{;\nu;\mu} = R^\alpha{}_{\sigma\nu\mu}u^\sigma. \tag{4.5.20}$$

By contracting over α and ν and then by multiplying with u^μ and contracting, we have

$$u^\alpha{}_{;\mu;\alpha}u^\mu - u^\alpha{}_{;\alpha;\mu}u^\mu = R_{\sigma\mu}u^\sigma u^\mu.$$

Inserting the decomposition (4.5.6) of the covariant derivative $u_{\alpha;\beta}$ of the four-velocity we obtain

$$-a^\alpha{}_{;\alpha} + \dot{\Theta} + \frac{1}{3}\Theta^2 + 2(\sigma^2 - \omega^2) = -R_{\sigma\mu}u^\sigma u^\mu \tag{4.5.21}$$

where $\dot{\Theta} = \Theta_{,\alpha}u^\alpha$ is the rate of change of fractional rate of volume expansion and where

$$\sigma^2 \equiv \frac{1}{2}\sigma_{\alpha\beta}\sigma^{\alpha\beta}, \; \omega^2 \equiv \frac{1}{2}\omega_{\alpha\beta}\omega^{\alpha\beta} \tag{4.5.22}$$

are the "squares" of, respectively, shear and rotation tensors.

Equation (4.5.21) is called the **Raychaudhuri equation.**[131–133] By using the Einstein field equation and the expression of the energy-momentum tensor of a perfect fluid, we transform the Raychaudhuri equation to a form that links mass-energy density, pressure, and fluid motion:

$$\begin{aligned} &-a^\alpha{}_{;\alpha} + \dot{\Theta} + \frac{1}{3}\Theta^2 + 2(\sigma^2 - \omega^2) = -R_{\sigma\mu}u^\sigma u^\mu = \\ &-\chi\left(T_{\sigma\mu} - \frac{1}{2}Tg_{\sigma\mu}\right)u^\sigma u^\mu = -4\pi(\varepsilon + 3p). \end{aligned} \tag{4.5.23}$$

The quantity $(\varepsilon + 3p)$, on the right-hand side, acts as the *active gravitational mass density*; once again (see chaps. 2 and 3) this equation shows that in general relativity every form of energy contributes to the total mass, including pressure itself, a general-relativity effect largely responsible for gravitational collapse (see § 2.9). For a static spherical star, from equation (4.5.23) we have $a^{\alpha}{}_{;\alpha} = 4\pi(\varepsilon + 3p)$, whereas in the Newtonian limit, $a^{\alpha}{}_{,\alpha} \cong 4\pi\rho$. Then we see that pressure, trying to keep the star in equilibrium through the pressure gradients, at the same time contributes to the active gravitational mass, that is, to the gravitational attraction, and therefore, at the same time, it promotes stellar collapse (*"self-regeneration" of pressure*; "the harder if fights, the harder it falls").

When the particles follow geodesic motion the four-acceleration is zero, $a^{\alpha} = 0$, and we have

$$\dot{\Theta} + \frac{1}{3}\Theta^2 + 2(\sigma^2 - \omega^2) = -4\pi(\varepsilon + 3p). \tag{4.5.24}$$

From the Friedmann-Robertson-Walker metric (4.2.16) and from the definition of Θ we get: $\frac{\Theta(t)}{3} = \frac{\dot{R}(t)}{R(t)} \equiv H(t)$, where $H(t)$ is the Hubble "constant" at time t, and $\boldsymbol{\omega} = \boldsymbol{\sigma} = \mathbf{0}$; thus, $\dot{\Theta} + \frac{1}{3}\Theta^2 = 3\frac{\ddot{R}}{R}$. From the Raychaudhuri equation and from $a^{\alpha}{}_{;\alpha} = 0$ we then get

$$3\frac{\ddot{R}}{R} = -4\pi(\varepsilon + 3p).$$

Therefore, independently from the equation of state, provided only that $(\varepsilon + 3p) > 0$, one has $\ddot{R}(t) < 0$ at any time t, which for $R \to 0$ implies a singularity at a finite time, $t_{\circ}$, in the past; furthermore $t_{\circ} < \frac{1}{H_{\circ}}$ (see fig. 4.3), that is, the "age of the universe" is less than the Hubble time.

4.6 [THE GÖDEL MODEL UNIVERSE]

In section 4.2, by considering a manifold that is spatially homogeneous and spatially isotropic about one point, and therefore spatially homogeneous and spatially isotropic about every point, we have derived the Friedmann-Robertson-Walker metric. Then, by assuming the validity of the Einstein field equation, we have studied the time evolution of the Friedmann model universe and of the corresponding scale factor $R(t)$.

In this section we describe a four-dimensional model universe, **homogeneous both in space and time**, which admits the whole four-dimensional simply transitive (for any two points there is *only* one "translation" that carries one point into the other) group of isometries, in other words, a spacetime that admits all four "simple translations" as independent Killing vectors.

The metric $\boldsymbol{g}$ of such a spacetime must admit a timelike Killing vector $\boldsymbol{\xi}_t$ representing a translation in time. Consequently, in some coordinate system one can write $\xi_t^\alpha = (1, 0, 0, 0)$. From the Killing equation (4.4.1), the metric $\boldsymbol{g}$ must be stationary, that is, time-independent in such a coordinate system. For any standard field equation, energy density and pressure of mass-energy should also be time independent in such a model universe. Therefore, any model universe that admits the largest group of simply transitive isometries G_4, or the maximum number of independent Killing vectors representing "simple translations," that is, four, must be stationary, with stationary matter; the scalar of expansion Θ, representing the fractional rate of change of volume in the model universe, should then be zero. The distortion tensor σ_{ik} should also vanish.

Furthermore, from the conservation law (4.2.54), one can prove that a perfect fluid homogeneous both in space and time must pursue geodesic motion; therefore, from the Raychaudhuri equation, we have

$$2\omega^2 = R_{\alpha\beta}u^\alpha u^\beta. \tag{4.6.1}$$

Now, either $R_{\alpha\beta}u^\alpha u^\beta$ and ω^2 are both zero, or, if $R_{\alpha\beta}u^\alpha u^\beta > 0$, we must have $\omega^2 > 0$. In this latter case, the fluid particles in the model universe are rotating relative to the local Fermi frames, that is, galaxies are rotating relative to local gyroscopes.

If we demand that the timelike convergence condition $R_{\alpha\beta}u^\alpha u^\beta \geq 0$ be satisfied for every timelike vector u^α, and if we further assume the validity of the Einstein field equation with the cosmological term $\Lambda g_{\alpha\beta}$ (see § 4.2) on the left-hand side, then from expression (4.2.52) of the energy-momentum tensor of a perfect fluid, we must have

$$4\pi(\varepsilon + 3p) - \Lambda \geq 0 \tag{4.6.2}$$

where Λ is the so-called cosmological constant. The first possibility $R_{\alpha\beta}u^\alpha u^\beta = 0$ corresponds to the Einstein static model universe. Therefore, in any zero-pressure Einstein static model universe, the push of the cosmological constant must balance the pull of the energy density; that is, $\Lambda = 4\pi\varepsilon$. The other possibility $R_{\alpha\beta}u^\alpha u^\beta > 0$ implies $2\omega^2 = 4\pi(\varepsilon + 3p) - \Lambda > 0$.

In 1949 the great mathematician and logician Kurt Gödel[141] discovered a rotating model universe with $\Lambda = -4\pi\varepsilon$, $p = 0$, and $\omega^2 = 4\pi\varepsilon$. In the original 1949 paper, $\Lambda \neq 0$ and $p = 0$; however, the **Gödel rotating solution** also corresponds to a model universe with $\Lambda = 0$, $p = \varepsilon$, and $\omega^2 = 8\pi\varepsilon$. The equation of state[143] $p = \varepsilon$ gives the maximum allowable pressure, corresponding to the speed of sound in the fluid equal to the speed of light, $c \equiv 1$. If we assume a model universe that is homogeneous both in space and time, with $R_{\alpha\beta}u^\alpha u^\beta > 0$, and that satisfies the Einstein field equation, we have that the spacetime geometry is given by the **Gödel metric**

$$ds^2 = \frac{1}{2\omega^2}\left[-(dt + e^x dz)^2 + dx^2 + dy^2 + \frac{1}{2}e^{2x}dz^2\right]. \tag{4.6.3}$$

The **Gödel model universe** has three-dimensional hypersurfaces, $y =$ constant, with null and timelike curves, which admit Killing vectors with structure constants of the Bianchi VIII type admitting the three-dimensional Lorentz group of isometries. However, as we have observed, the whole spacetime is homogeneous, and therefore the Gödel model universe has four independent Killing vectors representing "simple translations." In addition it also has a fifth Killing vector.[136] Indeed, the Gödel metric also has rotational symmetry about the y-axis at every spacetime event. The *five Killing vectors* are the generators of the isometry group G_5 admitted by the Gödel rotating solution, in the coordinate basis they are[136]

$$\xi_1 = \frac{\partial}{\partial t}; \qquad \text{representing a translation along the } t\text{-axis}$$

$$\xi_2 = \frac{\partial}{\partial y}; \qquad \text{representing a translation along the } y\text{-axis}$$

$$\xi_3 = \frac{\partial}{\partial z}; \qquad \text{representing a translation along the } z\text{-axis}$$

$$\xi_4 = \frac{\partial}{\partial x} - z\frac{\partial}{\partial z}; \qquad \text{representing a translation along the } x\text{-axis plus a contraction along the } z\text{-axis}, \tag{4.6.4}$$

plus a fifth Killing vector

$$\xi = -2e^{-x}\frac{\partial}{\partial t} + z\frac{\partial}{\partial x} + \left[e^{-2x} - \frac{1}{2}z^2\right]\frac{\partial}{\partial z}. \tag{4.6.5}$$

Let us summarize the most important properties of the Gödel model universe:[133,136,141,144]

(1) At each point the cosmological fluid appears to be *rotating* with angular velocity $\omega = \sqrt{8\pi\varepsilon} = \sqrt{8\pi p}$ ($\omega = \sqrt{4\pi\varepsilon}$, if $\Lambda \neq 0$ and $p = 0$) relative to the local Fermi frames, $\lambda^{\alpha}_{(\beta)}$. In other words, every nearby cluster of galaxies rotates relative to the local gyroscopes carried by Fermi-Walker transport along the world lines of the observers. Furthermore, the cosmological fluid also rotates globally relative to local gyroscopes. In this sense the Gödel rotating model universe appears to contradict some strong general relativistic interpretation of the Mach principle that is outlined in section 4.8.

(2) The Gödel model universe is homogeneous both in space and time and therefore it is *stationary*. In other words, in this model the cosmological fluid is characterized by zero expansion, $\Theta = 0$, and zero shear, $\sigma_{\alpha\beta} = 0$. Thus the Gödel model runs into difficulty with the expansion of the universe. (For model universes *homogeneous* both *in space and time* but *expanding*, thus with *continuous creation* of matter and with field

equations different from Einstein equation, the so-called "*steady-state*" *model universes*, see Bondi and Gold, and Hoyle).[196–198]

(3) Through each point of the model universe there are *closed timelike curves* that are not geodesics.[145] In other words, observers who travel all around one of such curves return to their starting spacetime event. Such a feature of spacetime appears to disagree with the *principle of causality*. However, Novikov, Thorne, et al.[146–149] (see also ref. 236) explain that the existence of a universe with closed timelike curves is still conceivable. They conjecture that the spacetime might contain closed timelike curves, or "timelike wormholes," which, however, also in regard to the matter present in the universe, have to be part of *self-consistent solutions*, that is, solutions free from self-inconsistent causality violations. This granted, the universe might contain closed timelike curves which are globally self-consistent. To exclude the occurrence of closed timelike curves Hawking[150] has then formulated the "Chronology Protection Conjecture," however, see also Kim 1992[199] and Ori 1993.[221] For a discussion on the existence of closed timelike curves in the Gödel model universe, see, for example, Hawking and Ellis.[144]

(4) The Gödel model universe has the interesting property of being singularity-free, contrary to the dynamical standard Friedmann cosmological models. Indeed, the Gödel model universe is *geodesically complete*; that is, every geodesic can be extended to every value of the affine parameter.

(5) The standard Gödel model universe is *not spatially compact* (however, see § 4.7). Finally, the three-dimensional manifolds y = constant are of Bianchi VIII type,[136] admitting the three-dimensional Lorentz group of isometries, and have the metric of a three-dimensional unit hyperboloid embedded in $\Re^4$.

Ozsváth and Schücking[151] have observed that the previous definition of rotation of a fluid with respect to a local Fermi frame, that is, rotation relative to local gyroscopes (the compass of inertia), or $\omega^i \neq 0$ in the Fermi frames, may not describe a global rotation in the case when the shear $\sigma_{\alpha\beta}$ differs from zero. Suppose that, in classical mechanics, we have a fluid rotating in the xy-plane in circular orbits about the z-axis, with tangential velocity proportional to the inverse of the distance from the z-axis. According to our definition of the rotation vector ω^i, the local angular velocity of the fluid is zero. Nevertheless, the fluid undergoes a global rotation about the z-axis. Therefore, in general, global rotation seems not to be well defined by ω^i, unless the shear $\sigma_{\alpha\beta}$ is equal to zero, as it is in the Gödel model universe.

After the Gödel discovery of rotating model universes, several other rotating cosmological models were discovered with interesting properties different from the Gödel model universe. In the next section we shall describe some of these cosmological solutions relevant to the interpretation of the origin of local inertia in general relativity.

4.7 [BIANCHI IX ROTATING COSMOLOGICAL MODELS]

In the previous section we have described the Gödel model universe, which is characterized by a cosmological fluid rotating with respect to the local compass of inertia, that is, relative to the gyroscopes of the local Fermi frames. As we have observed, this interesting model universe has, however, some features generally considered to be unphysical, in particular: (1) It contains closed timelike curves (however, see the discussions of Novikov et al.[146–150,199,219] on the existence of self-consistent solutions with closed timelike curves); (2) it is stationary, that is, nonexpanding, $\Theta = 0$, in disagreement with the observed cosmological redshift (for other, nonconventional, explanations of the cosmological redshift see for example refs. 152–154; for some observational evidence against these nonconventional explanations see for example ref. 155); and (3) it is spatially noncompact (several authors, among which Einstein[156] and Wheeler,[157–159] have conjectured that only in a spatially compact model universe we have a satisfactory explanation of the origin of local inertia: see chap. 5 and discussion at the end of this chapter). However, spatial compactness is a theoretical assumption not required by any present observational evidence.

To avoid the problem of existence of closed timelike curves and the problem of spatial noncompactness, **Gödel**, in 1950,[142] announced the existence of **other rotating cosmological models**, with an arbitrary value of Λ, including zero, that have no closed timelike curves and that expand but are spatially homogeneous and compact.

The metric corresponding to some of these rotating cosmological models, of **Bianchi IX type**, was explicitly given by Ozsváth and Schücking in 1962.[160] Later, in 1969,[151] they discussed and gave detailed proofs of the rotation in these models, of the nonexistence of closed timelike curves, and of the spatial compactness of these homogeneous model universes. However, the *Ozsváth-Schücking model universes* are stationary and do not expand, and thus disagree with the observations. They demand a cosmological constant different from zero.

With regard to rotating models with a null cosmological constant Λ (as in standard Einstein general relativity), Maitra found in 1966[161] a rotating, nonhomogeneous, cylindrically symmetric solution with $\Lambda = 0$, dust-filled, with shear $\sigma_{\alpha\beta}$ different from zero, without closed timelike curves, geodesically complete, and with null volume expansion $\Theta = 0$. Again, this model disagrees with the observations, because inhomogeneous and stationary.

All these models apparently disagree (see below) with the idea that the compass of inertia, that is, the local gyroscopes, should follow the average distribution of mass-energy in the universe; however, all these models are generally considered to be "unphysical" primarily because they disagree with the observed cosmological expansion.

Nevertheless, Shepley,[162] in his 1965 Ph.D. thesis with Wheeler, and Matzner, Shepley, and Warren in 1970,[122] explicitly studied cosmological models with compact, spacelike, three-dimensional hypersurfaces of homogeneity admitting the transitive group of isometries SO(3, $\Re$) (the group of unit determinant, orthogonal, 3×3 matrices with real elements; a realization consisting in the group of the three-dimensional rotations), that is, cosmological models of *Bianchi IX type*.[136,163–166] They analyzed the Bianchi IX models filled with a pressureless perfect fluid, or dust, in standard general relativity with $\Lambda = 0$. These model universes may expand, reach a time of maximum expansion, and then recollapse, and they have regions of infinite density, that is, singularities (as in the standard Friedmann models). These models are anisotropic generalizations of the isotropic Friedmann model universes.

In particular, in a subset of these models the cosmological fluid is *rotating*, that is, $\omega^i \neq 0$.

As we have just observed, the Bianchi IX models have compact, homogeneous, spacelike three-dimensional hypersurfaces, with SO(3, $\Re$) as transitive group of isometries. It is possible to choose a one-parameter family of these homogeneous spacelike hypersurfaces filling these model universes; such a slicing has been called a foliation of the spacetime. Thus, one can label each hypersurface by using a parameter. In every spacetime one can choose a spacelike hypersurface and at each point take timelike geodesics orthogonal to the hypersurface. Then, one can assign spatial coordinates x^i = constant to each of these geodesics and choose a coordinate time equal to the proper time measured by clocks on each of these geodesics, with the same initial time assigned on the initial hypersurface (see also § 4.5). Then, clocks at different points on each of these hypersurfaces are synchronized.[136,167] This coordinate system is called the *synchronous system*, and the coordinates *Gaussian normal coordinates*, the corresponding metric is given by expression (4.7.1) below.

The metric of the Bianchi IX model universes in the synchronous system is then:[122,136]

$$ds^2 = -\boldsymbol{dt}^2 + g_{ij}\boldsymbol{\chi}^i\boldsymbol{\chi}^j \tag{4.7.1}$$

where the g_{ij} are functions of t only and the three $\boldsymbol{\chi}^i$ are the differential forms

$$\begin{aligned}
\boldsymbol{\chi}^1 &= -\sin x^3 \boldsymbol{dx}^1 + \sin x^1 \cos x^3 \boldsymbol{dx}^2 \\
\boldsymbol{\chi}^2 &= \cos x^3 \boldsymbol{dx}^1 + \sin x^1 \sin x^3 \boldsymbol{dx}^2 \\
\boldsymbol{\chi}^3 &= \cos x^1 \boldsymbol{dx}^2 + \boldsymbol{dx}^3.
\end{aligned} \tag{4.7.2}$$

In terms of an orthonormal tetrad, one can write the metric (4.7.1)

$$ds^2 = -dt^2 + (\boldsymbol{\lambda}^{(1)})^2 + (\boldsymbol{\lambda}^{(2)})^2 + (\boldsymbol{\lambda}^{(3)})^2 \qquad (4.7.3)$$

where

$$\boldsymbol{\lambda}^{(i)} = b_{is}\boldsymbol{\chi}^{s},$$

and

$$b_{il}b_{lj} = g_{ij} \qquad \text{and} \qquad b_{[ij]} = 0.$$

These models expand or contract (§ 4.3) $\Theta \equiv u^{\alpha}{}_{;\alpha} \neq 0$, and may rotate $\omega^i \neq 0$. In the synchronous system, in the orthonormal basis (4.7.3), with the help of the geodesic equation, one can write Θ and ω_{ik}

$$\Theta = \frac{l_{kl}u_k u_l}{u_0} - u_0 l_{kk} \qquad (4.7.4)$$

and

$$\begin{aligned} \omega_{0i} &= -\omega_{i0} = -\frac{1}{2}\frac{u_k u_l d^k_{li}}{u_0} \\ \omega_{ij} &= -\omega_{ji} = -\frac{1}{2}u_k d^k_{ij} \end{aligned} \qquad (4.7.5)$$

where u^{α} is the four-velocity of the cosmological fluid and

$$d^i_{jk} = b_{il}\bar{\epsilon}_{lmn}b^{-1}_{mj}b^{-1}_{nk}$$

$$l_{ij} = k_{(ij)} = \frac{1}{2}(k_{ij} + k_{ji})$$

and

$$k_{ij} = \dot{b}_{il}b^{-1}_{lj}$$

where, as usual, dot means time derivative, b^{-1}_{ij} is the inverse of the element b_{ij}, and $\bar{\epsilon}_{lmn}$ is the completely antisymmetric symbol with $\bar{\epsilon}_{123} \equiv 1$.

The Bianchi IX models can be subdivided into three subclasses:[122] (1) nonrotating, (2) rotating with rotation axis fixed in one direction, describable as nontumbling models, and (3) general models. If the four-velocity of the cosmological fluid is orthogonal to the t = constant, synchronous, spacelike hypersurfaces, we have $u^0 = 1$ and $u^i = 0$, and therefore $\omega^i = 0$; these are the nonrotating models of the first subclass.

A model belonging to the first subclass is the Friedmann model universe closed in time. However, in the models of the second subclass the four-velocity u^{α} of the cosmological fluid is not orthogonal to the synchronous spacelike hypersurfaces, that is, $u^i \neq 0$. It can be shown that the spacelike component of the four-velocity is along a fixed direction that can be chosen as one of the three spatial directions, for example the direction labeled "three." Therefore $u_1 = u_2 = 0$, and from the field equation $b_{13} = b_{23} = l_{13} = l_{23} = 0$ at any time. From the definition (4.5.17) of rotation vector ω^α and from the expressions (4.7.5), we thus have $\omega^1 = \omega^2 = 0$ and $\omega^0 \neq 0$, $\omega^3 \neq 0$. Therefore, such rotating models have a fixed axis of rotation at each point (fig. 4.11). This axis is orthogonal to the four-velocity of the cosmological fluid (see definition 4.5.17).

Summarizing:

(1) The Matzner-Shepley-Warren models are *Bianchi IX models* with compact, homogeneous, spacelike hypersurfaces admitting SO(3, $\mathfrak{R}$) as transitive group of isometries. A particular case in which the hypersurfaces are also isotropic at each point is the Friedmann closed model universe. Therefore, as we have already observed, these models may be considered as generalizations of the Friedmann closed model.

(2) These model universes are *nonstationary*, that is, in general: $\Theta \neq 0$. Furthermore they are closed in time, that is, after a stage of maximum expansion they recollapse. Therefore, they are compatible with the observed cosmological redshift.

(3) In general, these models have *rotation* $\boldsymbol{\omega}$ and *shear* $\boldsymbol{\sigma}$ different from zero.

(4) The spacetime does *not* have *closed timelike curves* (which, instead, exist in the original Gödel model).

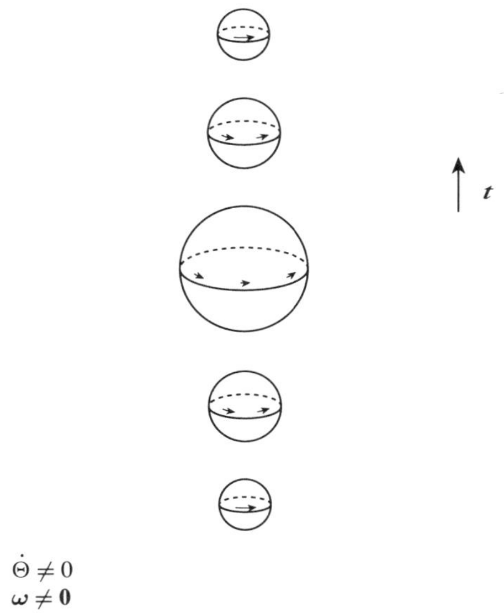

FIGURE 4.11. An element of cosmological fluid in an expanding and recollapsing, rotating cosmological model (for simplicity the shear $\boldsymbol{\sigma}$ is not represented here).

(5) Their metric is a solution of the standard Einstein field equation with *null cosmological constant*, $\Lambda = 0$.

In conclusion, the Bianchi IX cosmological models studied by Matzner, Shepley, and Warren seem to contradict a conceivable idea that the compass of inertia, that is, the local gyroscopes, should follow the average distribution of mass-energy in the universe in every "physical" cosmological model (however, see the following discussion regarding some alternative interpretations). The Matzner-Shepley-Warren models are in fact nonstationary, expanding, and then recollapsing, spatially compact, with no closed timelike curves, with $\Lambda = 0$, and yet the matter, in the subclasses 2 and 3, is rotating relative to the local compass of inertia.

However, Dehnen and Hönl[168,169] and Wheeler[158,170] proposed an explanation and an interpretation of the rotation of matter relative to local gyroscopes in these cosmological models. Their interpretation satisfies the idea that the "compass of inertia" should follow an average flux of energy in the universe. Dehnen and Hönl showed that the metric of the Ozsváth-Schücking model universe can be interpreted as the metric of the Einstein static model universe (background), plus a time-independent deformation of the Einstein static background, plus a cosmological circulating gravitational wave. This gravitational wave circulates around the model universe in a direction opposite to the rotation of matter relative to the local compass of inertia. They explained that the energy flux of this cosmological gravitational wave has to drag the local compass of inertia, as does the energy flux of matter; using the words of Dehnen and Hönl, the energy flux of the cosmological circulating gravitational wave has to produce Coriolis-type forces. Indeed, as we saw in section 2.10, gravitational waves, as electromagnetic waves, have an associated effective energy-momentum pseudotensor, which is properly defined over a certain "macroscopic" region of several wavelengths size (of course there is no contradiction with the equivalence principle which holds locally). In other words, since gravitational waves carry energy and momentum, as do electromagnetic waves, and since in general relativity any form of energy equally contributes to the inertial mass and to the gravitational mass (see § 3.2.5), it follows that energy and momentum of gravitational waves, that is, their energy-momentum pseudotensor, must contribute to the large-scale background curvature (see chap. 2 and Misner, Thorne, Wheeler 1973).[110] Furthermore, as any other flux of energy, gravitational waves must also contribute to the "dragging of inertial frames," that is, they must drag local gyroscopes, as matter currents do. The model universes corresponding to the time-dependent metric (4.7.1) may be interpreted as having cosmological circulating gravitational waves, which carry energy and momentum.[171] Consequently, in these rotating models the local compass of inertia must be influenced by both the flux of rotating matter and the flux of circulating gravitational waves. Therefore, the energy flux of the rotating gravitational waves may compensate the energy flux of the rotating matter, or the drag of inertial frames due to the rotating gravitational waves may compensate their drag due to the flux of the oppositely rotating matter. The local Fermi observers should then see the matter in the universe rotating, relative to their local gyroscopes, in a direction opposite to the rotation of the cosmological circulating gravitational waves.

In agreement with some strong general relativistic interpretation of the Mach principle, if one also includes the energy flux of cosmological gravitational waves, the local compass of inertia, that is, the local gyroscopes, should then be at rest relative to an "average global flux of matter and energy" in the universe. Such an interpretation of the origin of "local inertia" in rotating models has been confirmed by King (1990) for model universes with three-sphere spatial topology.[171]

4.8 [COSMOLOGY AND ORIGIN OF INERTIA]

The equivalence principle (see chaps. 2 and 3), at the foundations of Einstein geometrodynamics, has been the subject of many discussions[172] and also criticisms[137] over the years. In the end, a convenient way to proceed has been to determine and distinguish among various formulations of the equivalence principle, more or less strong (chap. 2), and then leave the last word to the experiments (chap. 3).

Similarly, with regard to the origin of inertia, we *try* to do the same in this section (and in this book): to determine and distinguish among some formulations and interpretations of the origin of inertia in Einstein geometrodynamics, in other metric theories, and in classical mechanics, and come up with experiments which might test these different interpretations (see table 4.3).

Classical Mechanics.

In classical mechanics, according to the ideas of Newton[173,174] (chap. 7), there exists an "**absolute space**" with properties completely independent of the mass-energy content in the universe. Inertia, in the sense of the local inertial forces experienced by the observers accelerating or rotating relative to the "compass of inertia," originates from the acceleration of the observers relative to this absolute space (see the Newton bucket experiment, chap. 7). In other words, the inertial frames are all the frames at rest or uniformly moving with respect to this hypothetical absolute space and the "compass of inertia" (gyroscopes) has a fixed direction in space, forever, and is in fact one and the same thing as the absolute space.

Einstein Geometrodynamics.

In Einstein geometrodynamics there is no place for an absolute space, since the properties of the spacetime geometry and of the metric $\boldsymbol{g}$ are influenced, and at least in part determined, through the field equation (see below and chap. 5) by the distribution of mass-energy and mass-energy currents in the universe.

TABLE 4.3 Inertia in Classical Mechanics and in Einstein Geometrodynamics

Interpretation of Origin of Inertia	*Main Characteristics*	*Possible Tests*
	Classical Mechanics "ABSOLUTE SPACE"	
• Inertial forces generated by accelerations and rotations with respect to an absolute space, or absolute frame, independent of the mass-energy in the universe.	• Inertial frames fixed (or in uniform motion) with respect to an "**Absolute Space**" and independent of the mass-energy distribution and currents in the universe.	• Null result for the Lense-Thirring effect.
	Einstein Geometrodynamics "MASS-ENERGY THERE RULES INERTIA HERE"	
• Local inertial forces generated by accelerations and rotations with respect to the local inertial frames which are *influenced* by distribution and currents of mass-energy.	• **Gravitomagnetism and dragging of inertial frames**: the local inertial frames are *dragged* by the currents of mass-energy.	• Test of the Lense-Thirring frame dragging effect (LAGEOS III, GPB, etc.).
	Einstein Geometrodynamics with *additional cosmological requirements*	
• Local inertial frames completely *determined* by the distribution and by the currents of mass-energy in the universe.	• **Universe** with **compact spacelike** Cauchy surface (universe *spatially* "*closed*").	• Measurement of the average density of mass-energy in the universe, ε, or of some other critical parameter, such as the deceleration parameter, $q_\circ$, and, if the universe may be described on some very large scale by a Friedmann model: • Definitive confirmation of very large scale spatial homogeneity and spatial isotropy of the universe, that is, of its very large scale Friedmann character. (If $\varepsilon > \varepsilon_c$, or $q_\circ > \frac{1}{2} \Rightarrow$ universe closed in time $\Rightarrow$ universe spatially compact with topology of a three-sphere, or of a two-sphere times a circle.)
	• **Universe "closed" in time** (recollapsing).	

TABLE 4.3 (*continued*)

Interpretation of Origin of Inertia	*Main Characteristics*	*Possible Tests*
Einstein Geometrodynamics with *additional cosmological requirements*		
• Local inertial frames completely *determined* and *dragged* by the distribution and currents of mass-energy in the universe (including gravitational waves).	• Average distribution of mass-energy in the **universe nonrotating** relative to local inertial frames (apart from local dragging effects).	• Test of nonrotation of the universe via test of very large scale spatial homogeneity and spatial isotropy about us ⇒ spatial isotropy about every point ⇒ nonrotation of the universe relative to local inertial frames (that is, relative to gyroscopes); (for example, one can put limits to the rotation of the universe using upper limits to the large scale anisotropy of the cosmic blackbody radiation, see § 4.8). • Test of nonrotation of the universe via test of nonrotation of local gyroscopes (apart from Thomas, de Sitter, and Lense-Thirring effects) relative to distant quasars (for example, using VLBI and local gyroscopes).

Then, the spacetime metric $\boldsymbol{g}$ determines the local inertial Fermi frames (the compass of inertia) at each point. These local inertial frames satisfy $\overset{(i)}{g}_{\alpha\beta} \cong \eta_{\alpha\beta}$ and $\frac{D\lambda^{\alpha}_{(\beta)}}{ds} = \lambda^{\alpha}_{(\beta);\sigma} u^{\sigma} = \left(a_{\sigma}\lambda^{\sigma}_{(\beta)}\right) u^{\alpha} = 0$ (that is, the local inertial tetrad is parallel transported).

This interpretation of the origin of the local inertial forces is always valid in Einstein general relativity, no matter which cosmological model is considered; furthermore it is also qualitatively valid, but quantitatively different, in other known metric theories of gravity; therefore one may call it a *weak form of "interpretation of the origin of inertia."* According to this general relativistic interpretation of the origin of the local inertial forces, there is *no* Newtonian absolute space and in fact the compass of inertia is not determined by an absolute space but is influenced, and at least in part determined (see below), by the distribution and by the currents of mass-energy in the universe. This weak "interpretation of the origin of inertia" has its physical manifestation in the so-called **dragging of inertial frames**[175] (chaps. 3 and 6) and will be experimentally tested by one of the experiments for the detection of the **gravitomagnetic field**[176,177] and of the Lense-Thirring[178,179] effect described in chapter 6 (LAGEOS III,[180] GPB,[181] etc.).

However, even though in general relativity the spacetime geometry, that is, the metric $\boldsymbol{g}$ (and therefore the local inertial frames), is dynamical and affected by the distribution and by the currents of mass-energy in the universe via the Einstein field equation, there are some different conceivabilities, corresponding to a more or less strong influence of the mass-energy in the universe in determining the local inertia.

(1) The spacetime metric and the local inertial frames are influenced, and at least in part determined, by the distribution and currents of mass-energy in the universe—dragging of inertial frames (weak "interpretation of the origin of inertia"); however, the universe may be spatially noncompact and part of the cosmological metric may be, a priori, unaffected by the distribution and currents of mass-energy (see below). In other words, the boundary conditions at infinity may be mostly responsible for determining the local inertial frames.

(2) The universe is **spatially compact**. The spacetime metric is completely determined by the distribution and currents of mass-energy in the universe, or is fully dynamical, that is, the local inertial frames are fully determined by the mass-energy in the universe. One may call this interpretation a *medium form of "interpretation of the origin of inertia"* (see below).

(3) Not only the spacetime metric and the local inertial frames are fully determined by the mass-energy in the universe (that is, no boundary conditions in solving the initial value problem,[182,183] or compact space), but (a) the universe is also **closed in time**, that is, expanding and then recollapsing, and furthermore (b) the *local compass of inertia* (*gyroscopes*) is *at rest relative to the average flux of energy in the universe*. One may call this interpretation a *strong form of "interpretation of the origin of inertia"* (see below).

Corresponding to the first possibility, the cosmological metric may have some kind of prior geometry in it,[110] that is, a part of the spacetime geometry that is nondynamical (see chap. 5). This may correspond to either alternative metric theories with some type of prior geometry or to general relativity with an asymptotically nondynamical geometry. In this case, to solve the Cauchy problem, one gives asymptotical conditions which are independent of the content of mass-energy in the universe. An example is given by the condition $\boldsymbol{g} \xrightarrow{\infty} \boldsymbol{\eta}$, that is, that the universe be asymptotically flat: $\boldsymbol{\eta}$ is nondynamical and independent from the energy in the universe, $\boldsymbol{\eta}$ may therefore be thought of as some kind of prior geometry. Examples of these solutions are the Schwarzschild (2.6.35) and Kerr-Neumann (2.6.36) geometries. If the universe is asymptotically flat, since the spacetime metric $\boldsymbol{g}$ is in part determined by the boundary condition $\boldsymbol{g} \xrightarrow{\infty} \boldsymbol{\eta}$, then the local inertial frames (where $\overset{(0)}{\boldsymbol{g}} \longrightarrow \overset{(0)}{\boldsymbol{\eta}}$) are in part determined by some matching with the asymptotical frames where $g_{\alpha\beta} \xrightarrow{\infty} (-1, +1, +1, +1)$, and they are also in part determined by the mass-energy distribution and currents which drag them. In other words the local inertial forces are in part generated by accelerations and rotations with respect to the asymptotical frames where $\boldsymbol{g} \xrightarrow{\infty} \boldsymbol{\eta}$. However, $\boldsymbol{\eta}$ is unaffected by the mass-energy in the universe, therefore the local inertial forces are in part generated by accelerations and rotations with respect to some kind of absolute frames where $\boldsymbol{g} \xrightarrow{\infty} \boldsymbol{\eta}$. To rule out this possibility some physicists have conjectured a spatially compact universe, one might call this a medium form of "interpretation of the origin of inertia." Of course, one may also have a spatially noncompact model universe, without asymptotically flat geometry $\boldsymbol{\eta}$, for example a Friedmann model universe with hyperbolic spatial curvature.

General relativity is defined not only by a curved dynamical Lorentzian manifold plus the field equation which determine this geometry but also by proper initial and boundary conditions to solve the field equation. Then, according to Einstein, Wheeler, and others, to avoid any boundary condition and to avoid the existence of any kind of prior geometry and the asymptotic flatness of the metric $\boldsymbol{g} : \boldsymbol{g} \xrightarrow{\infty} \boldsymbol{\eta}$, the universe should be spatially compact (closed and bounded set). Therefore, one might assume the validity of *a medium form of "interpretation of the origin of inertia"*: (1) the weak form is satisfied (that is, there is no absolute space and there is *dragging of inertial frames* by currents of mass-energy) plus (2) the spacetime has a *compact* spacelike Cauchy surface and therefore one does not need any boundary condition to solve the field equation, that is, the spacetime geometry and therefore the local inertial frames are completely determined by the mass-energy content in the universe.

However, even assuming that the universe is spatially compact, we still have two possibilities: (1) model universes with compact space topologies which are closed in time, that is, expanding and then recollapsing, such as some model universes with space topologies homeomorphic (topologically equivalent) to connected sums of copies of a three-sphere, S^3, and of a two-sphere times a circle, $S^2 \times S^1$ (this is a necessary condition for closure in time[112–115] and probably, under standard additional requirements, also sufficient[121,123] (see § 4.3); and (2) model universes with compact space topologies which are open in time, that is, expanding forever, such as model universes with space topologies homeomorphic to a three torus T^3 (see § 4.3). Therefore, as for the case of the spatially noncompact Friedmann model universes, so for $t \to \infty$ the clusters of galaxies in these model universes will be infinitely apart from each other and then, according to the views of some physicists, they should have negligible influence on the dynamics of the local geometry and of the local inertial frames. Following a similar view, Dicke[184] and others have conjectured that, as the universe expands, G should decrease (see chap. 3).

Furthermore, in section 4.7 we have studied some examples of spatially compact Bianchi IX model universes which have matter rotating with respect to the local compass of inertia, with no evidence to drag or influence it (however, see the interpretation given at the end of § 4.7). To rule out these possibilities some physicists have conjectured that a model universe should also satisfy a *strong form of "interpretation of the origin of inertia,"* that one might also call "a general relativistic interpretation of the Mach Principle," or make three demands: (1) the weak form (*dragging of inertial frames*) is satisfied, (2) the *space is compact*, and (3) the universe is *closed in time*, that is, expanding and then recollapsing (or at least not expanding forever), and *the axes of the local inertial frames (gyroscopes)*, that is, the local compasses of inertia, *follow the average flux of energy in the universe*.

If we live in a universe of Friedmann type, to prove experimentally the closure in time one should first confirm the global Friedmann character of the universe, that is, definitively confirm its very large scale homogeneity and isotropy (see § 4.2), and then determine the average density of mass-energy ε, or some equivalent parameter such as the deceleration parameter $q_\circ$. As we have seen in section 4.2, if $\varepsilon > \varepsilon_c$ the universe must recollapse, at least in general relativity. If the universe does not satisfy the cosmological principle of spatial homogeneity and spatial isotropy, even on a very large scale, if one could however verify that the global spatial topology of the universe, is of the type $S^3 \# \cdots \# S^2 \times S^1$, and if one could prove rigorously the conjectures discussed in section 4.3, then one would know, at least according to general relativity, that the universe is closed in time and would eventually recollapse.

Concerning the rotation of matter relative to the local compass of inertia, the result of King,[171] mentioned at the end of the previous section, has shown that, in model universes that are spatially compact and possess a three-sphere topology, the compasses of inertia, that is, the local gyroscopes, follow an average flux of energy in the universe, if one includes with the flux of matter the effective flux of energy carried by the rotating cosmological gravitational waves.

We can put some *experimental limits on the rotation of the universe*. Hawking and Collins[185,186] studied, in 1973, homogeneous and anisotropic perturbations of the Friedmann model universes. For the $k = +1$, spatially compact model, the perturbed model universe is of Bianchi IX type. In this case, by considering the observational limits to the anisotropy of the cosmological background radiation (Collins and Hawking assumed an upper limit to the anisotropies of about 10^{-3}), they placed limits to the present rotation, ω, of the universe between less than 3×10^{-11} and less than 2×10^{-14} arcsec/century, depending on the time of last scattering of the radiation now observed (they assumed a last scattering of radiation at a redshift z between $\cong 7$ and 1000). However, according to other unpublished calculations,[187] more realistic limits for the Bianchi IX models are of the order of 10^{-4} arcsec/century. In the case of the $k = 0$ model, the perturbed model universe is of Bianchi I or Bianchi VII_0 type (for Bianchi I the rotation is zero), and the upper limits to the rotation that they obtained are $\approx 2.5 \times 10^{-5}$ and $\approx 1.5 \times 10^{-7}$ arcsec/century respectively for $z = 7$ and $z = 1000$. For the $k = -1$ model, the perturbed model universe is of Bianchi V or Bianchi VII_h type, and the upper limit to the rotation that Collins and Hawking obtained is $\approx 8 \times 10^{-5}$ arcsec/century. These upper limits to the rotation of the universe may be improved by considering the present (1993) limits to the anisotropy of the cosmic radiation of the order of 10^{-5}; *however*, large angular scale anisotropies in the cosmic microwave background radiation with a quadrupole amplitude of $\approx 2 \times 10^{-6}$ have been observed by COBE (see § 4.1).[42–44,188,226]

Planetary observations set a limit[189] of about 4 milliarcsec/yr to the rotation of local inertial frames in the solar system with respect to the distant stars. A proposed planetary experiment with a lander on Phobos (moon of Mars) should then reach an accuracy[190] of a fraction of milliarcsec/yr in measuring any rotation of local inertial frames with respect to quasars.

We finally observe that some limits to the rotation of the universe can be obtained by a simple interpretation of the test of the de Sitter effect by Bertotti, Ciufolini, and Bender (1987)[191] (see §§ 3.4.3 and 6.10, and refs. 192, 211, and 212). General relativity predicts a precession of gyroscopes orbiting the Sun, at the Earth-Sun distance, of 19.2 milliarcsec/yr with respect to an asymptotic inertial frame. Bertotti, Ciufolini, and Bender obtained a test of the de Sitter precession of the Moon perigee of 19.2 milliarcsec/yr with an accuracy of about 10%, that is, of about 2 milliarcsec/yr. This test was based on the comparison between Lunar Laser Ranging (LLR) and Very Long Baseline Interferometry (VLBI) data, that is, on the comparison between local, LLR measurements, and nonlocal, VLBI measurements relative to a frame defined by the position of the distant quasars. Therefore, assuming the validity of general relativity, this test shows that the frame defined by these distant stars is an asymptotic quasi-inertial frame with an accuracy of about 2 milliarcsec/yr, that is, the frame marked by the distant quasars does not rotate with respect to an asymptotic quasi-inertial frame with an upper limit to a conceivable rotation of about 2 milliarcsec/yr. In other words, apart from the de Sitter effect and other local gravitomagnetic perturbations, the frame defined by distant stars does not rotate with respect to the local inertial frames, that is, with respect to local gyroscopes.

REFERENCES CHAPTER 4

1. P.J.E. Peebles, *The Large-Scale Structure of the Universe* (Princeton University Press, Princeton, 1980).
2. M. Jôeveer, J. Einasto, and E. Tago, Spatial distribution of galaxies and of clusters of galaxies in the southern galactic hemisphere, *Mon. Not. Roy. Astr. Soc.* 185:357–69 (1978).
3. M. Tarenghi, W.G. Tifft, G. Chincarini, H.J. Rood, and L.A. Thompson, The Hercules Supercluster. I. Basic Data, *Astrophys. J.* 234:793–801 (1979).
4. J. Einasto, M. Jôeveer, and E. Saar, Structure of superclusters and supercluster formation, *Mon. Not. Roy. Astron. Soc.* 193:353–75 (1980).
5. S.A. Gregory, L.A. Thompson, and W.G. Tifft, The Perseus supercluster, *Astrophys. J.* 243:411–26 (1981).
6. R.B. Tully, The local supercluster, *Astrophys. J.* 257:389–422 (1982).
7. For a review of observations and theories see Y.B. Zel'dovich, J. Einasto, and S.F. Shandarin, Giant voids in the universe, *Nature* 300:407–13 (1982); see also refs. 8–11.
8. J.H. Oort, Superclusters, *Ann. Rev. Astron. Astrophys.* 21:373–428 (1983).
9. P.R. Shapiro, Pancakes and galaxy formation, in *Clusters and Groups of Galaxies*, ed. F. Mardirossian, G. Giuricin, and M. Mezzetti (Reidel, Dordrecht, 1984), 447–78.
10. P.J.E. Peebles. The origin of galaxies and clusters of galaxies, *Science* 224:1385–91 (1984).
11. M.J. Rees, The emergence of structure in the universe: galaxy formation and "dark matter," in *300 Years of Gravitation*, ed. S.W. Hawking and W. Israel (Cambridge University Press, Cambridge, 1987), 459–98.
12. R.P. Kirshner, A. Oemler, P.L. Schechter, and S.A. Shectman, A million cubic megaparsec void in Boötes?, *Astrophys. J. Lett.* 248:L57–L60 (1981).
13. For a review see H.J. Rood, Voids, *Ann Rev. Astr. and Astrophys* 26:245–94 (1988).
14. R. Giovanelli and M.P. Haynes, The Lynx-Ursa major supercluster, *Astron. J.* 87:1355–63 (1982).
15. V. de Lapparent, M.J. Geller, and J.P. Huchra, A slice of the universe, *Astrophys. J. Lett.* 302:L1–L5 (1986).
16. R.P. Kirshner, A. Oemler, P.L. Schechter, and S.A. Shectman, The texture of the universe, in *11th Texas Symposium on Relativistic Astrophysics, Ann. NY Acad. Sci.* 422:91–94 (1984).
17. S.M. Faber, J.R. Gott, A.L. Melott, and M. Dickinson, The sponge-like topology of large-scale structure in the universe, *Astrophys. J.* 306:341–57 (1986).
18. D. Lynden-Bell, S.M. Faber, D. Burstein, R.L. Davies, A. Dressler, R.J. Terlevich, and G. Wegner, Spectroscopy and photometry of elliptical galaxies. V. Galaxy streaming toward the new supergalactic center, *Astrophys. J.* 326:19–49 (1988).

19. A. Dressler, S.M. Faber, D. Burstein, R.L. Davies, D. Lynden-Bell, R.J. Terlevich, and G. Wegner, Spectroscopy and photometry of elliptical galaxies: A large-scale streaming motion in the local universe, *Astrophys. J. Lett.* 313:L37–L42 (1987).
20. A. Dressler, The supergalactic plane redshift survey: A candidate for the great attractor, *Astrophys. J.* 329:519–26 (1988).
21. A. Friedmann, Über die Krümmung des Raumes, *Z. Phys.* 10:377–86 (1922).
22. A. Friedmann, Über die Möglichkeit einer Welt mit konstanter negativer Krümmung des Raumes, *Z. Phys.* 21:326–32 (1924).
23. L.M. Golden, Observational selection in the identification of quasars and claims for anisotropy, *Observatory* 94:122–26 (1974).
24. A.R. Gillespie, Investigations into reported anisotropies in radio source counts and spectra at 1421 MHz, *Mon. Not. Roy. Astron. Soc.* 170:541–49 (1975).
25. A. Webster, The clustering of radio sources. I. The theory of power-spectrum analysis, *Mon. Not. Roy. Astron. Soc.* 175:61–70 (1976).
26. A. Webster, The clustering of radio sources. II. The 4C, GB and MC_1 surveys, *Mon. Not. Roy. Astron. Soc.* 175:71–83 (1976).
27. J. Machalski, A new statistical investigation of the problem of isotropy in radio source population at 1400 MHz. I. Spatial density, source counts and spectra-index distributions in a new GB sky survey, *Astron. Astrophys.* 56:53–57 (1977).
28. M. Seldner and P.J.E. Peebles, Statistical analysis of catalogs of extragalactic objects. X. Clustering of 4C radio sources, *Astrophys. J.* 225:7–20 (1978).
29. A.M. Wolfe, New limits on the shear and rotation of the universe from the x-ray background, *Astrophys. J. Lett.* 159:L61–L67 (1970).
30. A.M. Wolfe and G.R. Burbidge, Can the lumpy distribution of galaxies be reconciled with the smooth x-ray background?, *Nature* 228:1170–74 (1970).
31. J. Silk, Diffuse cosmic x and gamma radiation: The isotropic component, *Space Sci. Rev.* 11:671–708 (1970).
32. A.C. Fabian, Analysis of x-ray Background Fluctuations, *Nature Phys. Sci.* 237:19–21 (1972).
33. D.A. Schwartz, The isotropy of the diffuse cosmic x-rays determined by OSO-III, *Astrophys. J.* 162:439–44 (1970).
34. D.A. Schwartz, S.S. Murray, and H. Gursky, A measurement of fluctuations in the x-ray background by UHURU, *Astrophys. J.* 204:315–21 (1976).
35. R.H. Dicke, P.J.E. Peebles, P.G. Roll, and D.T. Wilkinson, Cosmic-black-body radiation, *Astrophys. J.* 142:414–19 (1965).
36. A.A. Penzias and R.W. Wilson, A measurement of excess antenna temperature at 4080 Mc/s, *Astrophys. J.* 142:419–21 (1965).
37. R.B. Partridge, The cosmic microwave background, in *Proc. IAU Symp. on Observational Cosmology*, ed. A. Hewitt et al. (1987), 31–53.
38. D.T. Wilkinson, Measurement of the microwave background radiation, *Phil. Trans. Roy. Soc. Lond. A* 320:595–607 (1986).
39. N. Kaiser and J.I. Silk, Cosmic microwave background anisotropy, *Nature* 324:529–37 (1986).

40. C.L. Bennet and G.F. Smoot, The COBE cosmic 3K anisotropy experiment: A gravity wave and cosmic string probe, in *Relativistic Gravitational Experiments in Space*, ed. R.W. Hellings (NASA Conf. Pub. 3046), 114–17 (1989).

41. G.F. Smoot, COBE (Cosmic Background Explorer Satellite) measurements, in *Proc. 6th Marcel Grossmann Meeting on General Relativity*, Kyoto, 1991, ed. H. Sato and T. Nakamura (World Scientific, Singapore, 1992), 283–304.

42. G.F. Smoot et al., Structure in the COBE DMR first-year maps, *Astrophys. J. Lett.* 396:L1–L5 (1992).

43. E.L. Wright et al., Interpretation of the CMBR anisotropy detected by the COBE DMR, *Astrophys. J. Lett.* 396:L13–L18 (1992).

44. C.L. Bennet et al., Preliminary separation of galactic and cosmic microwave emission for the COBE DMR, *Astrophys. J. Lett.* 396:L7–L12 (1992).

45. Ya. B. Zel'dovich and I.D. Novikov, *Relativistic Astrophysics*, vol. 2., *The Structure and Evolution of the Universe* (University of Chicago Press, Chicago, 1983).

46. S. Weinberg, *Gravitation and Cosmology: Principles and Applications of the General Theory of Relativity* (Wiley, New York, 1972).

47. G. Gamow, Expanding universe and the origin of elements, *Phys. Rev.* 70:572–73 (1946).

48. R.A. Alpher, H. Bethe, and G. Gamow, The origin of chemical elements, *Phys. Rev.* 73:803–4 (1948).

49. R.A. Alpher and R.C. Herman, Theory of the origin and relative-abundance distribution of the elements, *Rev. Mod. Phys.* 22:153–213 (1950).

50. H.P. Robertson, Kinematics and world structure, *Astrophys. J.* 82:248–301 (1935).

51. H.P. Robertson, Kinematics and world structure, *Astrophys. J.* 83:187–201, 257–71 (1936).

52. A.G. Walker, On Milne's theory of world-structure, *Proc. Lond. Math. Soc.* 42:90–127 (1936).

53. P.J.E. Peebles, *Physical Cosmology* (Princeton University Press, Princeton, 1971).

54. A.G. Doroshkevich, S.F. Shandarin, and Ya. B. Zel'dovich, Three-dimensional structure of the universe and regions devoid of galaxies, *Comm. Astrophys. Space Phys.* 9:265–73 (1982).

55. J. Silk, A.S. Szalay, and Ya. B. Zel'dovich, The large-scale structure of the universe, *Sci. Am.* 249 (October): 56–64 (1983).

56. S.F. Shandarin and Ya. B. Zel'dovich, The large-scale structure of the universe: Turbulence, intermittency, structures in a self-gravitating medium, *Rev. Mod. Phys.* 61:185–220 (1989).

57. G. Lemaître, Evolution of expanding universe, *Proc. Nat. Acad. Sci. USA* 20:12–17 (1934).

58. H.R. Pagels, Microcosmology: New particles and cosmology, *Ann. NY Acad. Sci* 442:15–32 (1984).

59. K. Yano, *The Theory of Lie Derivatives and Its Applications* (North-Holland, Amsterdam, 1955).

60. S. Helgason, *Differential Geometry and Symmetric Spaces* (Academic Press, New York, 1962).
61. M. Spivak, *A Comprehensive Introduction to Differential Geometry* (Publish or Perish, Boston, 1970).
62. See also B. Schutz, *Geometrical Methods of Mathematical Physics* (Cambridge University Press, Cambridge, 1980).
63. For applications in cosmology see M.P. Ryan, Jr., and L.C. Shepley, *Homogeneous Relativistic Cosmologies* (Princeton University Press, Princeton, 1975).
64. W. Killing, Über die Grundlagen der Geometrie, *J. Reine Angew. Math.* 109:121–86 (1892).
65. R.C. Tolman, *Relativity, Thermodynamics and Cosmology* (Clarendon Press, Oxford, 1934).
66. A.K. Raychaudhuri, *Theoretical Cosmology* (Clarendon Press, Oxford, 1979).
67. R.A. Matzner, Almost Symmetric Spaces and Gravitational Radiation, *J. Math. Phys.* 9:1657–68 (1968).
68. G.F.R. Ellis, Relativistic cosmology, its nature, aims and problems, in *Proc. 10th Int. Conf. on General Relativity and Gravitation*, Padova, 3–8 July 1983, ed. B. Bertotti et al. (Reidel, Dordrecht, 1984), 215–88.
69. G.F.R. Ellis and W. Stoeger, The "fitting problem" in cosmology, *Class. Quantum Grav.* 4:1697–1729 (1987).
70. L.P. Eisenhart, *Riemannian Geometry* (Princeton University Press, Princeton, 1926).
71. D. Kramer, H. Stephani, M. MacCallum, and E. Herlt, *Exact Solutions of Einstein Field Equations* (VEB Deutscher Verlag der Wissenschaften, Berlin, 1980).
72. F. Schur, Über den Zusammenhang der Räume konstanten Krümmungsmasses mit den projektiven Räumen, *Math. Ann.* 27:537–67 (1886).
73. In A.S. Eddington, *The Mathematical Theory of Relativity*, 2d ed. (Cambridge University Press, London, 1924), 162.
74. E.P. Hubble, A relation between distance and radial velocity among extragalactic nebulae, *Proc. Nat. Acad. Sci USA* 15:169–73 (1929).
75. P.W. Hodge, The extragalactic distance scale, *Ann. Rev. Astr. Astrophys.* 19:357–72 (1981).
76. M. Rowan-Robinson, *The Cosmic Distance Scale* (Freeman, New York, 1985).
77. *The Space Telescope Observatory*, ed. D.N.B. Hall (NASA Conf. Pub. 2244, 1982).
78. R. Giacconi, Space telescope and cosmology, in *Highlights of Modern Astrophysics, Concepts and Controversies*, ed. S.T. Shapiro and S.A. Teukolsky (John Wiley and Sons, New York, 1986), 331–55.
79. C.R. O' Dell, The Hubble Space Telescope Observatory, *Phys. Today* (April): 32–38 (1990).
80. G.F.R. Ellis, Relativistic cosmology, in *General Relativity and Cosmology, Proc. Course 47 of the Int. School of Physics "Enrico Fermi,"* ed. R.K. Sachs (Academic Press, New York, 1971), 104–82.
81. G. Gamow, *My World Line* (Viking Press, New York, 1970), 149.

82. W. de Sitter, On the relativity of inertia: Remarks concerning Einstein's latest hypothesis, *Proc. Kon. Ned. Akad. Wet.* 19:1217–25 (1917).

83. W. de Sitter, On the curvature of space, *Proc. Kon. Ned. Akad. Wet.* 20:229–43 (1918).

84. W. de Sitter, Further remarks on the solutions of the field-equations of Einstein's theory of gravitation, *Proc. Kon. Ned. Akad. Wet.* 20:1309–12 (1918).

85. A. Sandage, The redshift-distance relation. I. Angular diameter of first-ranked cluster galaxies as a function of red-shift: The aperture correction to magnitudes, *Astrophys. J.* 173:485–99 (1972).

86. A. Sandage, The redshift-distance relation. II. The Hubble diagram and its scatter for first-ranked cluster galaxies: A formal value for q_0, *Astrophys. J.* 178:1–24 (1972).

87. A. Sandage, The redshift-distance relation. III. Photometry and the Hubble diagram for radio sources and the possible turn-on time for QSOs, *Astrophys. J.* 178:25–44 (1972).

88. A. Sandage, The redshift-distance relation. IV. The composite nature of N galaxies, their Hubble diagram, and the validity of measured redshifts as distance indicators, *Astrophys. J.* 180:687–97 (1973).

89. A. Sandage, The redshift-distance relation. V. Galaxy colors as functions of galactic latitude and redshift: observed colors compared with predicted distributions for various world models, *Astrophys. J.* 183:711–30 (1973).

90. A. Sandage, The redshift-distance relation. VI. The Hubble diagram from S20 photometry for rich clusters and sparse groups: A study of residuals, *Astrophys. J.* 183:731–42 (1973).

91. A. Sandage and E. Hardy, The redshift-distance relation. VII. Absolute magnitudes of the first three ranked cluster galaxies as functions of cluster richness and Bautz-Morgan cluster type: The effect on q_0, *Astrophys. J.* 183:743–57 (1973).

92. W. Rindler, Visual horizons in world-models, *Mon. Not. Roy. Astron. Soc.* 116:662–67 (1956).

93. W. Rindler, *Essential Relativity* (Springer-Verlag, New York, 1977).

94. A.H. Guth, Inflationary universe: A possible solution to the horizon and flatness problems, *Phys. Rev. D* 23:347–56 (1981).

95. E.B. Gliner, Algebraic properties of the energy-momentum tensor and vacuum-like states of matter, *J. Exp. Teor. Fiz.* 49:542–48 (1965); *JETP Lett.* 22:378 (1966).

96. E.B. Gliner, The vacuum-like state of a medium and Friedman cosmology, *Dokl. Akad. Nauk. SSSR* 192:771–74 (1970); *Sov. Phys. Dokl.* 15:559 (1970).

97. E.B. Gliner and I.G. Dymnikova, Non singular Friedmann cosmology, *Pis'ma V Astron. Zh.* 1:7–9 (1975); *Sov. Astr. Lett.* 1:93 (1975).

98. L.E. Gurevich, On the origin of the metagalaxy, *Astrophys. Space Sci.* 38:67–78 (1975).

99. Ya. B. Zel'dovich, The cosmological constant and the theory of elementary particles, *Usp. Fiz. Nauk.* 95:209–30 (1968); *Sov. Phys. Uspekhi* 11:381 (1968).

100. For a review see S.K. Blau and A.H. Guth, Inflationary cosmology, in *300 Years of Gravitation*, ed. S.W. Hawking and W. Israel (Cambridge University Press, Cambridge, 1987), 524–603; see also ref. 101.

101. A. Linde, Inflation and quantum cosmology, in *300 Years of Gravitation*, ed. S.W. Hawking and W. Israel (Cambridge University Press, Cambridge, 1987), 604–30 (1987).

102. V.N. Lukash and I.D. Novikov, Lectures of the very early universe, in *Observational and Theoretical Cosmology*, ed. R. Rebolo (Cambridge University Press, Cambridge, 1991), 5–45.

103. A.D. Linde, A new inflationary universe scenario: a possible solution of the horizon, flatness, homogeneity, isotropy and primordial monopole problems, *Phys. Lett. B* 108:389–93 (1982).

104. A. Albrecht and P.J. Steinhardt, Cosmology for grand unified theories with radiatively induced symmetry breaking, *Phys. Rev. Lett.* 48:1220–23 (1982).

105. A.D. Linde, Chaotic inflation, *Phys. Lett. B* 129:177–81 (1983); see also review in ref. 87.

106. For a review see A. Vilenkin, Cosmic strings and domain walls, *Phys. Rep.* 121:263–315 (1985).

107. For simulations of interactions of cosmic strings, see R. Matzner, Interaction of U(1) cosmic strings: Numerical intercommutation, *Computers in Physics* 2 (5): 51–64 (1988).

108. G.F.R. Ellis Topology and Cosmology, *Gen. Rel. and Grav.* 2:7–21 (1971).

109. G.F.R. Ellis and G. Schreiber, Observational and dynamical properties of small universes, *Phys. Lett. A* 115:97–107 (1986).

110. C.W. Misner, K.S. Thorne, and J.A. Wheeler, *Gravitation* (Freeman, San Francisco, 1973).

111. D.R. Brill, Maximal surfaces in closed and open spacetimes, in *First Marcel Grossmann Meeting on General Relativity*, ed. R. Ruffini (North-Holland, Amsterdam, 1977), 193–206.

112. J.D. Barrow and F.J. Tipler, Closed universes: Their future evolution and final state, *Mon. Not. Roy. Astron. Soc.* 216:395–402 (1985).

113. R. Schoen and S.-T. Yau, Incompressible minimal surfaces, three-dimensional manifolds with non-negative scalar curvature, and the positive mass conjecture in general relativity, *Proc. Nat. Acad. Sci.* 75:2567 (1978).

114. R. Schoen and S.-T. Yau, Existence of incompressible minimal surfaces and the topology of three dimensional manifolds with non-negative scalar curvature, *Ann. Math.* 110:127–42 (1979).

115. R. Schoen and S.T. Yau, On the structure of manifolds with positive scalar curvature, *Manuscripta Math.* 28:159–83 (1979).

116. M.H. Freedman, The topology of four-dimensional manifolds, *J. Diff. Geom.* 17:357–453 (1982).

117. S.K. Donaldson, An application of gauge theory to four dimensional topology, *J. Diff. Geom.* 18:279–315 (1983).

118. R.E. Gompf, Three exotic R^4's and other anomalies, *J. Diff. Geom.* 18:317–28 (1983).

119. D.S. Freed and K.K. Uhlenbeck, *Instantons and Four Manifolds* (Springer-Verlag, New York, 1984).

120. J.E. Marsden and F.J. Tipler, Maximal hypersurfaces and foliations of constant mean curvature in general relativity, *Phys. Rep.* 66:109–39 (1980).

121. J.D. Barrow, G.J. Galloway, and F.J. Tipler, The closed-universe recollapse conjecture, *Mon. Not. Roy. Astron. Soc.* 223:835–44 (1986).

122. R.A. Matzner, L.C. Shepley, and J.B. Warren, Dynamics of SO(3,R)-homogeneous cosmologies, *Ann. Phys.* 57:401–60 (1970).

123. X. Lin and R.M. Wald, Proof of the closed-universe-recollapse conjecture for diagonal Bianchi type-IX cosmologies, *Phys. Rev. D* 40:3280–86 (1989).

124. C.G. Behr, Generalization of the Friedmann world model with positive space curvature, *Z. Astrophys.* 54:268–86 (1962).

125. F.B. Estabrook, H.D. Wahlquist, and C.G. Behr, Dyadic analysis of spatially homogeneous world models, *J. Math. Phys.* 9:497–504 (1968); see also ref. 126.

126. G.F.R. Ellis, and M.A.H. MacCallum, A class of homogeneous cosmological models, *Comm. Math. Phys.* 12:108–41 (1969).

127. M.A.H. MacCallum, Anisotropic and inhomogeneous relativistic cosmologies, in *General Relativity: An Einstein Centenary Survey*, ed. S.W. Hawking and W. Israel (Cambridge University Press, Cambridge, 1979), 533–80.

128. L.D. Landau and E.M. Lifshitz, *The Classical Theory of Fields* (Pergamon Press, New York, 1975).

129. R.T. Jantzen, Perfect fluid sources for spatially homogeneous spacetimes, *Ann. Phys. (NY)* 145:378–426 (1983).

130. L. Bianchi, Sugli spazi a tre dimensioni che ammettono un gruppo continuo di movimenti, *Mem. di Mat. Fis. Soc. Ital. Sci. Ser. III* 11:267–352 (1898).

131. A.K. Raychaudhuri, Relativistic cosmology. I., *Phys. Rev.* 98:1123–26 (1955).

132. A.K. Raychaudhuri, Relativistic and Newtonian cosmology, *Z. Astrophys.* 43:161–64 (1957).

133. A.K. Raychaudhuri, *Theoretical Cosmology* (Clarendon Press, Oxford, 1979).

134. C. Cattaneo, Proiezioni naturali e derivazione trasversa in una varietà a metrica iperbolica normale, *Ann. Mat. Pura ed Appl.* 48:361–86 (1959).

135. C. Cattaneo, *Introduzione alla Teoria Einsteiniana della Gravitazione* (Libreria Eredi Virgilio Veschi, Rome, 1977).

136. M.P. Ryan and L.C. Shepley, *Homogeneous Relativistic Cosmologies* (Princeton University Press, Princeton, 1975).

137. J.L. Synge, *Relativity: The General Theory* (North-Holland, Amsterdam, 1960).

138. E. Fermi, Sopra i fenomeni che avvengono in vicinanza di una linea oraria, *Atti R. Accad. Lincei Rend. Cl. Sci. Fis. Mat. Nat.* 31:21–23, 51–52, 101–3 (1922).

139. G.K. Batchelor, *An Introduction to Fluid Dynamics*, 2d ed. (Cambridge, Cambridge University Press, 1970).

140. J.L. Synge, Relativistic hydrodynamics, *Proc. Lond. Math. Soc.* 43:376–416 (1937).

141. K. Gödel, An example of a new type of cosmological solutions of Einstein's field equations of gravitation, *Rev. Mod. Phys.* 21:447–50 (1949).

142. K. Gödel, Rotating universes in general relativity theory, in *Proc. 1950 Int. Congress of Math* (1950), vol. 1, 175–81.

143. Ya. B Zel'dovich, The equation of state at ultrahigh densities and its relativistic limitations, *Zh. Eksp. Teor. Fiz.* 41:1609–15 (1961); *Sov. Phys.*—JETP 14:1143 (1962).

144. S.W. Hawking and G.F.R. Ellis, *The Large Scale Structure of Space-Time* (Cambridge University Press, Cambridge, 1973).

145. S. Chandrasekhar and J.P. Wright, The geodesics in Gödel's universe, *Proc. Nat. Acad. Sci.* 47:341–47 (1961).

146. M.S. Morris, K.S. Thorne, and U. Yurtsever, Wormholes, time machines, and the weak energy condition, *Phys. Rev. Lett.* 61:1446–49 (1988).

147. U.P. Frolov and I.D. Novikov, Physical effects in wormholes and time machines, *Phys. Rev. D* 42:1057–65 (1990).

148. J. Friedman, M.S. Morris, I.D. Novikov, F. Echeverria, G. Klinkhammer, K.S. Thorne, and U. Yurtsever, Cauchy problem in spacetimes with closed timelike curves, *Phys. Rev. D* 42:1915–30 (1990).

149. I.D. Novikov, Time machine and self-consistent evolution in problems with self-interaction, *Phys. Rev. D* 45:1989–94 (1992).

150. S.W. Hawking, The chronology protection conjecture, in *Proc. 6th Marcel Grossmann Meeting on General Relativity*, Kyoto, June 1991, ed. H. Sato and T. Nakamura (World Scientific, Singapore, 1992), 3–13.

151. I. Ozsváth and E.L. Schücking, The Finite Rotating Universe, *Ann. Phys.* 55:166–204 (1969).

152. H. Arp, Peculiar galaxies and radio sources, *Science* 151:1214–16 (1966).

153. J.C. Pecker, A.P. Roberts, and J.P. Vigier, Non-velocity redshifts and photon-photon interactions, *Nature* 237:227–29 (1972).

154. M.J. Geller and P.J.E. Peebles, Test of the expanding universe postulate, *Astrophys. J.* 174:1–5 (1972). For some observational evidence against these theories see, e.g., ref. 155.

155. J.E. Solheim, T.G. Barnes, and H.J. Smith, Observational evidence against a time variation in Planck's constant, *Astrophys. J.* 209:330–34 (1976).

156. A. Einstein, *The Meaning of Relativity*, 3d ed. (Princeton University Press, Princeton, 1950).

157. J.A. Wheeler, View that the distribution of mass and energy determines the metric, in *Onzième Conseil de Physique Solvay: La Structure et l'evolution de l'univers* (Éditions Stoops, Brussels, 1959), 96–141.

158. J.A. Wheeler, Mach's principle as boundary condition for Einstein's equations, in *Gravitation and Relativity*, ed. H.-Y. Chiu and W.F. Hoffmann (W. A. Benjamin, New York, 1964), 303–49.

159. J. Isenberg and J.A. Wheeler, Inertia here is fixed by mass-energy there in every W model universe, in *Relativity, Quanta, and Cosmology in the Development of the Scientific Thought of Albert Einstein*, ed. M. Pantaleo and F. de Finis (Johnson Reprint Corp., New York, 1979), vol. 1, 267–93.

160. I. Ozsváth and E.L. Schücking, Finite rotating universe, *Nature* 193:1168–69 (1962).

161. S.C. Maitra, Stationary dust-filled cosmological solution with $\Lambda = 0$ and without closed timelike lines, *J. Math. Phys.* 7:1025–30 (1966).

162. L.C. Shepley, SO (3,R)–Homogeneous cosmologies, Ph.D. dissertation., Princeton University (1965).

163. C.W. Misner, Mixmaster universe, *Phys. Rev. Lett.* 22:1071–74 (1969).

164. V.A. Belinski, E.M. Lifshitz, and I.M. Khalatnikov, Oscillatory approach to the singular point in relativistic cosmology, *Uspekhi Fiz. Nauk* 102:463–500 (1971); *Sov. Phys.—Uspekhi* 13:745 (1971).

165. V.A. Belinski, E.M. Lifshitz, and I.M. Khalatnikov, Oscillatory mode of approach to a singularity in homogeneous cosmological models with rotating axes, *Zh. Eksp. Teor. Fiz.* 60:1969–79 (1971); *Sov. Phys.—JETP* 33:1061–66 (1971).

166. M.P. Ryan, Jr., Hamiltonian cosmology, *Lecture Notes in Physics*, vol. 13 (Springer-Verlag, Berlin, 1972).

167. E.M. Lifshitz and I.M. Khalatnikov, Investigations in relativistic cosmology, *Adv. Phys.* 12:185–249 (1963).

168. H. Dehnen and H. Hönl, Finite universe and Mach's principle, *Nature* 196:362–63 (1962).

169. H. Hönl and H. Dehnen, Zur Deutung einer von I. Ozsváth und E. Schücking gefundenen strengen Lösung der Feldgleichungen der Gravitation, *Z. Phys.* 171:178–88 (1963).

170. J.A. Wheeler, A few specializations of the generic local field in electromagnetism and gravitation, *Trans. NY Acad. Sci., Ser. II* 38:219–43 (1977).

171. D.H. King, Mach's principle and rotating universes, Ph.D. dissertation, University of Texas at Austin (1990).

172. See, e.g., V. Fock, *The Theory of Space, Time and Gravitation* (Moscow, 1961); 2d rev. ed., trans. N. Kemmer (Pergamon Press, Oxford, 1966). See also ref. 137.

173. I. Newton, *Philosophiae naturalis principia mathematica* (Streater, London, 1687); trans. A. Motte (1729) and revised by F. Cajori as *Sir Isaac Newton's Mathematical Principles of Natural Philosophy and His System of the World* (University of California Press, Berkeley and Los Angeles, 1934; paperback, 1962). See related discussions in ref. 174.

174. E. Mach, *Die Mechanik in ihrer Entwicklung historisch-kritisch dargestellt* (Brockhaus, Leipzig, 1912); trans. T.J. McCormack with an introduction by Karl Menger as *The Science of Mechanics* (Open Court, La Salle, IL, 1960).

175. A. Einstein, letter to E. Mach, Zurich, 25 June 1913. See, e.g., ref. 110, p. 544.

176. K.S. Thorne, Multipole expansions of gravitational radiation, *Rev. Mod. Phys.* 52:299–339 (1980).

177. K. S Thorne, R.H. Price, and D.A. MacDonald, eds., *Black Holes, the Membrane Paradigm* (Yale University Press, New Haven, 1986).

178. J. Lense and H. Thirring, Über den Einfluss der Eigenrotation der Zentralkörper auf die Bewegung der Planeten und Monde nach der Einsteinschen Gravitationstheorie, *Phys. Z.* 19:156–63 (1918).

179. J. Lense and H. Thirring, On the gravitational effects of rotating masses: The Thirring-Lense papers, trans. B. Mashhoon, F.W. Hehl, and D.S. Theiss, *Gen. Rel. Grav.* 16:711–50 (1984).

180. I. Ciufolini, A comprehensive introduction to the LAGEOS gravitomagnetic experiment: from the importance of the gravitomagnetic field in physics to preliminary error analysis and error budget, *Int. J. Mod. Phys. A* 4:3083–3145 (1989).

181. Papers on the Stanford Relativity Gyroscope Experiment, in *Proc. SPIE*, vol. 619, *Cryogenic Optical Systems and Instruments II* (Society of Photo-Optical Instr. Eng., Los Angeles, 1986), 29–165.

182. J.W. York, Jr., Kinematics and dynamics of general relativity, in *Sources of Gravitational Radiation*, ed. L. Smarr (Cambridge University Press, Cambridge, 1979), 83–126.

183. Y. Choquet-Bruhat and J.W. York, Jr., The Cauchy problem, in *General Relativity and Gravitation*, ed. A. Held (Plenum, New York, 1980), 99–172.

184. R.H. Dicke, *The Theoretical Significance of Experimental Relativity* (Gordon and Breach, New York, 1964).

185. C.B. Collins and S.W. Hawking, The rotation and distortion of the universe, *Mon. Not. Roy. Astron. Soc.* 162:307–20 (1973).

186. S.W. Hawking, in *The Anisotropy of the universe at large times*, *Confrontation of Cosmological Theories with Observational Data*, ed. M.S. Longair (IAU), 283–86 (1974).

187. R.A. Matzner, private communication.

188. J.C. Mather et al., Early results from the Cosmic Background Explorer (COBE), Goddard Space Flight Center, preprint 90-03 (1990).

189. L.I. Schiff, Observational basis of Mach's principle, *Rev. Mod. Phys.* 36:510–11 (1964).

190. J.M. Davidson, Assessment of Phobos VLBI Frame Tie, JPL interoffice memo 335.3-88-81 (1981).

191. B. Bertotti, I. Ciufolini, and P.L. Bender, New test of general relativity: Measurement of de Sitter geodetic precession rate for lunar perigee, *Phys. Rev. Lett.* 58:1062–65 (1987).

192. I.I. Shapiro, R.D. Reasenberg, J.F. Chandler, and R.W. Babcock, Measurement of the de Sitter precession of the Moon: A relativistic three-body effect, *Phys. Rev. Lett.* 61:2643–46 (1988).

193. P.C. Gregory and J.J. Condon, The 87GB catalog of radio sources covering $0° < \delta < +75°$ at 4.85 GHz, *Astrophys. J. Suppl.* 75:1011–1291 (1991).

194. J. Ehlers, Contributions to the relativistic mechanics of continuous media, *Akad. Wiss. Lit. Mainz Abh. Math. Nat. Kl.* 793–837 (1961).

195. J. Ehlers, General relativity and kinetic theory, in *General Relativity and Cosmology, Proc. Course 47 of the Int. School of Physics "Enrico Fermi,"* ed. R.K. Sachs (Academic Press, New York, 1971), 1–70.

196. H. Bondi and T. Gold, The steady-state theory of the expanding universe, *Mon. Not. Roy. Astron. Soc.* 108:252–70 (1948).

197. F. Hoyle, A new model for the expanding universe, *Mon. Not. Roy. Astron. Soc.* 108:372–82 (1948).

198. F. Hoyle, Mach's principle and the creation of matter, in *Proc. Int. Conf. on Mach's Principle, from Newton's Bucket to Quantum Gravity*, Tübingen, Germany, July 1993, ed. J. Barbour and H. Pfister (Birkhäuser, Boston, 1995).

199. S.-W. Kim, Quantum effects of Lorentzian wormhole, in *Proc. 6th Marcel Grossmann Meeting on General Relativity*, Kyoto, June 1991, ed. H. Sato and T. Nakamura (World Scientific, Singapore, 1992), 501–3.

200. J.J. Broadhurst, R.S. Ellis, D.C. Koo, and A.S. Szalay, Large-scale distribution of galaxies at the Galactic poles, *Nature* 343:726–28 (1990).

201. M.J. Geller and J.P. Huchra, Mapping the universe, *Science* 246:897–903 (1989).

202. R.C. Kirby, The topology of 4-manifolds, in *Lecture Notes in Mathematics* (Springer-Verlag, Berlin, 1989), 1374.

203. C.H. Brans and D. Randall, Exotic differentiable structures and general relativity, *Gen. Rel. Grav.* 25: 205–21 (1993).

204. See, e.g., J.D. Bekenstein, New gravitational theories as alternatives to dark matter, in *Proc. 6th Marcel Grossmann Meeting on General Relativity*, Kyoto, June 1991, ed. H. Sato and T. Nakamura (World Scientific, Singapore, 1992), 905–24.

205. For a determination of H_0 see also A. Sandage, $H_0 = 43 \pm 11$ km s^{-1} Mpc^{-1} based on angular diameters of high-luminosity field spiral galaxies, *Astrophys. J.* 402: 3–14 (1993).

206. K.I. Kellermann, The cosmological deceleration parameter estimated from the angular-size/redshift relation for compact radio sources, *Nature* 361:134–36 (1993).

207. N. Kaiser, G. Efstathiou, R. Ellis, C. Frenk, A. Lawrence, M. Rowan-Robinson, and W. Saunders, The large-scale distribution of *IRAS* galaxies and the predicted peculiar velocity field, *Mon. Not. Roy. Astron. Soc.* 252:1–12 (1991).

208. A.J.S. Hamilton, Ω from the anisotropy of the redshift correlation function in the *IRAS* 2 Jansky survey, *Astrophys. J. Lett.* 406:L47–L50 (1993).

209. D.S. Mathewson, V.L. Ford, and M. Buchhorn, No back-side infall into the Great Attractor, *Astroph. J. Lett.* 389:L5–L8 1992.

210. C.W. Allen, *Astrophysical Quantities*, 3rd ed. with corrections (Athlone, London, 1976).

211. J.O. Dickey, X.X. Newhall, and J.G. Williams, Investigating relativity using lunar laser ranging: Geodetic precession and the Nordtvedt effect, *Adv. Space Res.* 9:75–78 (1989); see also *Relativistic Gravitation Symp.* 15 *of the COSPAR 27th Plenary Meeting*, Espoo, Finland, 18–29 July 1988, and ref. 279 of chap. 3.

212. J. Müller, M. Schneider, M. Soffel, and H. Ruder, Testing Einstein's theory of gravity by analyzing lunar laser ranging data, *Astrophys. J. Lett.* 382:L101–L103 (1991).

213. A Gould, An estimate of the *COBE* quadrupole, *Astroph. J. Lett.* 403:L51–L54 (1993).

214. P.B. Stark, Uncertainty of the *Cosmic Background Explorer* quadrupole detection, *Astrophys. J. Lett.* 408:L73–L76 (1993).

215. J.C. Mather et al., A preliminary measurement of the cosmic microwave background spectrum by the *Cosmic Background Explorer* (*COBE*) satellite, *Astrophys. J. Lett.* 354:L37–L40 (1990).

216. J.C. Mather et al., Measurement of the cosmic microwave background spectrum by the *COBE* FIRAS instrument, *Astrophys. J.* 420: 439–44 (1994).

217. D.E. Holz, W.A. Miller, M. Wakano, and J.A. Wheeler, Coalescence of primal gravity waves to make cosmological mass without matter, in *Directions in General Relativity, Proc. of the* 1993 *Int. Symposium, Maryland*, ed. B.L. Hu and T.A. Jacobson (Cambridge University Press, Cambridge, 1993), vol. 2, 339–58.

218. B. Paczyński, Gravitational microlensing by the galactic halo, *Astrophys. J.* 304:1–5 (1986).

219. C. Alcock et al., Possible gravitational microlensing of a star in the Large Magellanic Cloud, *Nature* 365:621–23 (1993).

220. E. Aubourg et al., Evidence for gravitational microlensing by dark objects in the Galactic halo, *Nature* 365:623–25 (1993).

221. A. Ori, Must time-machine construction violate the weak energy condition?, *Phys. Rev. Lett.* 71:2517–20 (1993).

222. R. Crittenden, J.R. Bond, R.L. Davis, G. Efstathiou, and P.J. Steinhardt, Imprint of gravitational waves on the cosmic microwave background, *Phys. Rev. Lett.* 71:324–27 (1993).

223. L. Nicolaci da Costa et al., A complete southern sky redshift survey, *Astrophys. J. Lett.* 424:L1–L4 (1994).

224. G. Smoot and K. Davidson, *Wrinkles in Time* (Morrow, New York, 1994).

225. S. Hancock et al., Direct observation of structure in the cosmic microwave background, *Nature* 367:333–338 (1994).

226. C. Bennet et al., to be published (1994).

227. H.C. Ford et al., Narrowband *HST* images of M87: Evidence for a disk of ionized gas around a massive black hole, *Astrophys. J. Lett.* 435:L27–L30 (1994).

228. R.J. Harms et al., *HST* FOS spectroscopy of M87: Evidence for a disk of ionized gas around a massive black hole, *Astrophys. J. Lett.* 435:L35–L38 (1994).

229. Special issue of the *Astrophysical Journal Letters* dedicated to the *Hubble Space Telescope*, 435:L1–L78 (1994).

230. W.L. Freedman et al., Distance to the Virgo cluster galaxy M100 from Hubble Space Telescope observations of Cepheid, *Nature* 371:757–762 (1994).

231. M.J. Pierce et al., The Hubble constant and Virgo cluster distance from observations of Cepheid variables, *Nature* 371:385–389 (1994).

232. D.W. Sciama, *Modern Cosmology and the Dark Matter Problem* (Cambridge University Press, New York, 1994).

233. K.C. Sahu, Stars within the Large Magellanic Cloud as potential lenses for observed microlensing events, *Nature* 370:275–6 (1994).

234. E.M. Hu, J.-S. Huang, G. Gilmore, and L.L. Cowie, An upper limit on the density of low-mass stars in the Galactic halo, *Naure* 371:493–5 (1994).

235. J.N. Bahcall, C. Flynn, A. Gould, and S. Kirhakos, M dwarfs, microlensing, and the mass budget of the galaxy, *Astrophys. J. Lett.* 435:L51–L54 (1994).

236. K.S. Thorne, *Black Holes and Time Warps* (Norton, New York, 1994).

5

The Initial-Value Problem in Einstein Geometrodynamics

In this chapter we describe the physics and the mathematics of the initial-value problem in Einstein geometrodynamics. We further investigate the relations between the interpretation of the origin of inertia in Einstein general relativity and the initial-value formulation.

5.1 [FROM THE INITIAL-VALUE PROBLEM TO THE ORIGIN OF INERTIA IN EINSTEIN GEOMETRODYNAMICS]

To understand the dynamics of spacetime geometry it is not enough, we know, to listen to Einstein's 1915 and still-standard classical theory, reciting from the pulpit the standard classical creed. "Give me a full statement of conditions now, and I will give you the means to reckon conditions everywhere and for all time." A promise that is, of course, a great promise, but we cannot capitalize on this promised dynamic engine to go around until we have a starter and a steering wheel: starter, to specify what we mean by "now," and steering wheel,[1,2] to set at our will the initial conditions.

We already possess the needed mathematics.[3–8] The classical physics interpretation of this mathematics is, however, what we seek to lay out here for its bearing on quantum gravity and the origin of inertia.

Origin of inertia? Mach's principle?

Mach's principle as Mach thought of it? "[The] investigator must feel the need of . . . knowledge of the immediate connections, say, of the masses of the universe. There will hover before him as an ideal an insight into the principles of the whole matter, from which accelerated and inertial motions will result in the same way."[9]

Mach's principle as Einstein thought of it[10,11]—he who named it so in appreciation of the inspiration he had received from Mach? Einstein wrote to him on 25 June, 1913, well before his final November 1915 formulation of the field equation: "it . . . turns out that inertia originates in a kind of interaction between bodies, quite in the sense of your considerations on Newton's pail . . .

experiment. . . . If one rotates [a heavy shell of matter] relative to the fixed stars about an axis going through its center, a Coriolis force arises in the interior of the shell; that is, the plane of a Foucault pendulum is dragged around (with a practically unmeasurably small angular velocity)"[9] (see § 6.1).

Mach's principle as subsequent workers sought to define it? Few chapters in the history of ideas are richer in proposals and counterproposals, agreements and disagreements, examples and counterexamples. Excellent reviews of this history—of which space forbids citing more than a few[12–19]—make instructive reading, not least because they set forth such varied outlooks (see also ref. 149).

There are some, the literature reveals, who regard the very term "Mach's principle" as vague, deprived of all mathematical content, and mystical or theological or worse. For that reason the present treatment of inertia, outgrowth of a long evolution,[20–28,1] foregoes the explanation demanding phrase, interpretation of Mach's principle, in favor of the neutral term, *interpretation of the origin of inertia, or "energy ruling inertia"*[1] (see chap. 4).

The geometrodynamic "interpretation of the origin of inertia," loosely put, declares: "*Mass there governs spacetime geometry here.*" Restated more fully, but still only in broad terms, it says: (1) Spacetime geometry steers fields and matter. (2) In turn, the combined field-plus-matter momentum-energy steers the geometry.

Let us put the "interpretation of the origin of inertia" in still more concrete terms before we proceed to details: *The specification of the relevant features of a three-geometry and its time rate of change on a closed (compact and without boundary manifold), initial value, spacelike hypersurface, together with the energy density and density of energy flow (conformal) on that hypersurface and together with the equation of state of mass-energy, determines the entire spacetime geometry, the local inertial frames, and hence the inertial properties of every test particle and every field everywhere and for all time.*

Every term in this statement requires clarification: compact space, time (§§ 5.2.1, 5.2.2, and mathematical appendix; details in § 5.2.7); "topology of spacetime"—bounded versus unbounded—as related to the topology of space (§ 5.2.3); conformal part of the three-geometry (§ 5.2.4; details in § 5.2.7); density of this and that specified, not directly, but by the giving of certain well-defined conformal densities (§ 5.2.4; detail on density of mass-energy, § 5.2.7; on density of flow of mass-energy, § 5.2.7; on distortion tensor or gravitomagnetic field, §§ 5.2.6 and 5.2.7); determined: determined by the solution of two coupled partial differential equations on the three-manifold in question, one of them for a conformal scale factor ψ, the other for a three-vector gravitomagnetic potential, $\boldsymbol{W}$ (§ 5.3).

5.2 [THE INITIAL-VALUE PROBLEM AND THE INTERPRETATION OF THE ORIGIN OF INERTIA IN GEOMETRODYNAMICS]

5.2.1. [Compact Space and Open Space Viewed as Enveloped in a Larger Compact Space]

Einstein, discussing in light of his field equation Mach's considerations connecting inertia here with mass there, argued,[29] as others have since,[21–25,30–32] that the only natural boundary condition is no boundary: "In my opinion the general theory of relativity can only solve this problem [of inertia] satisfactorily if it regards the world as spatially self-enclosed."[29]

It is interesting to read further what the person who developed special and general relativity thought about the origin and the meaning of inertia in his theory and about its relations with the Mach principle (see also chap. 6). Indeed, as is well known, the idea that the local inertial "properties"—here—must be fixed by the distribution and motion of mass-energy in the universe—there—helped inspire Einstein's formulation of general relativity.

In his classical book on general relativity, *The Meaning of Relativity*, Einstein wrote: "Our previous considerations, based upon the field equation (96), had for a foundation the conception that space on the whole is Galilean-Euclidean, and that this character is disturbed only by masses embedded in it. This conception was certainly justified as long as we were dealing with spaces of the order of magnitude of those that astronomy has mostly to do with. But whether portions of the universe, however large they may be, are quasi-Euclidean, is a wholly different question. We can make this clear by using an example from the theory of surfaces which we have employed many times. If a certain portion of a surface appears to be practically plane, it does not at all follow that the whole surface has the form of a plane; the surface might just as well be a sphere of sufficiently large radius. The question as to whether the universe as a whole is non-Euclidean was much discussed from the geometrical point of view before the development of the theory of relativity. But with the theory of relativity, this problem has entered upon a new stage, for according to this theory the geometrical properties of bodies are not independent, but depend upon the distribution of masses.

"If the universe were quasi-Euclidean, then Mach was wholly wrong in his thought that inertia, as well as gravitation, depends upon a kind of mutual action between bodies. For in this case, with a suitably selected system of co-ordinates, the $g_{\mu\nu}$ would be constant at infinity, as they are in the special theory of relativity, while within finite regions the $g_{\mu\nu}$ would differ from these constant values by small amounts only, for a suitable choice of co-ordinates, as a result of the influence of the masses in finite regions. The physical properties of space would not then be wholly independent, that is, uninfluenced by matter, but in the main they would be, and only in small measure, conditioned by matter. Such a dualistic conception is even in itself not satisfactory; there are, however, some important physical arguments against it, which we shall consider.

"The hypothesis that the universe is infinite and Euclidean at infinity, is, from the relativistic point of view, a complicated hypothesis. In the language of the general theory of relativity it demands that the Riemann tensor of the fourth rank, R_{iklm}, shall vanish at infinity, which furnishes twenty independent conditions, while only ten curvature components, $R_{\mu\nu}$, enter into the laws of the gravitational field. It is certainly unsatisfactory to postulate such a far-reaching limitation without any physical basis for it.

"If we think these ideas consistently through to the end we must expect the whole inertia, that is, the whole $g_{\mu\nu}$-field, to be determined by the matter of the universe, and not mainly by the boundary conditions at infinity.

"The possibility seems to be particularly satisfying that the universe is spatially bounded and thus, in accordance with our assumption of the constancy of σ [mass-energy density], is of constant curvature, being either spherical or elliptical; for then the boundary conditions at infinity which are so inconvenient from the standpoint of the general theory of relativity, may be replaced by the much more natural conditions for a closed surface.[*]

"According to the second of Equations (123) the radius, a, of the universe is determined in terms of the total mass, M, of matter, by the equation

$$a = \frac{MK}{4\pi^2} \tag{124}$$

The complete dependence of the geometrical upon the physical properties becomes clearly apparent by means of this equation.

"Thus we may present the following arguments against the conception of a space-infinite, and for the conception of a space-bounded, universe:

1. From the standpoint of the theory of relativity, the condition for a closed surface is very much simpler than the corresponding boundary condition at infinity of the quasi-Euclidean structure of the universe.
2. The idea that Mach expressed, that inertia depends upon the mutual action of bodies, is contained, to a first approximation, in the equations of the theory of relativity; it follows from these equations that inertia depends, at least in part, upon mutual actions between masses. As it is an unsatisfactory assumption to make that inertia depends in part upon mutual actions, and in part upon an independent property of space, Mach's idea gains in probability. But this idea of Mach's corresponds only to a finite universe, bounded in space, and not to a quasi-Euclidean, infinite universe. From the standpoint of epistemology it is more satisfying to have the mechan-

[*However, see chap. 4:§§4.2 and 4.3.]

ical properties of space completely determined by matter, and this is the case only in a space-bounded universe.
3. An infinite universe is possible only if the mean density of matter in the universe vanishes.[*] Although such an assumption is logically possible, it is less probable than the assumption that there is a finite mean density of matter in the universe."[11]

How shall we assess the alternative postulate that space geometry shall be asymptotically flat? This postulate poses two problems of principle:

First, it imposes flatness from on high. To do so is totally to repudiate the three inspirations:[33] (M) Mach's vision of inertia, (E) Einstein's own principle that local gravity vanishes in every free-float frame of reference, and (R) Riemann's conception of geometry, that led, Einstein tells us, to his geometrodynamics. Of none of the (MER) trio did he speak more warmly[34] than of Riemann's argument—out of the principle of action and reaction—that geometry cannot be a God-given perfection standing high above the battles of matter and energy. It is instead a participant, an actor, an object on equal terms with the other fields of physics. On equal terms with them close in but far out flatter and flatter, more and more sanctified, more and more God-given? No! Geometry can not be part of physics in some regions and not a part of physics in others.

Second, we have come to realize that quantum fluctuations rule the geometry of space in the small, at all time and everywhere.[35–37] Physics tells us, as forcefully as it can, that asymptotic flatness—except in the classical approximation and in a local domain like a neighborhood of Earth, Sun, or other astrophysical object—is an improbable idea.

No natural escape has ever presented itself from these two difficulties of principle except to say that **space** in the large must be **compact**.

No one will deny that spacetime approaches flatness well out from many a localized center of attraction. However, nothing, anywhere, in any finding of the astrophysics of our day makes it unattractive to treat every such nearly flat region, or even totally flat region, "not as infinite, but as part of a closed universe."[38]

"For a two-dimensional analog, fill a rubber balloon with water and set it on a glass tabletop and look at it from underneath. The part of the universe that is curved acquires its curvature by reason of its actual content of mass-energy or—if animated only by gravitational waves—by reason of its effective content of mass-energy. This mass-energy, real or effective, is to be viewed as responsible—via field equation and initial-value formulation—for the inertial properties of the test particle that at first sight looked all alone in the universe."[39]

It is easy to see the role of spatial closure in the context of a simple problem: find the magnetic field associated with a stationary system of electric currents. To the well-known obvious solution we have only to add any magnetic field

[*However, see chap. 4: §§4.2 and 4.3.]

that is free of curl and divergence—and which therefore goes on and on to infinity with its twisty, wavy lines of force—to obtain another solution. Demand, however, that the background magnetic field or space geometry have a suitably closed topology, then the solution reduces to one or another linear combination of a finite number of harmonic vector fields. That number is exactly zero for some topologies. In any case, closure reduces the wild to the tame.

The solution is tame, for example, when the three-geometry has S^3 topology supplemented by one or more wormholes. In this example, as in others, a simple way exists to reduce a magnetic field, or any other vector field, from the tame to the unique. For this purpose we have only to specify for each wormhole its "charge"—the total of the flux through that wormhole.[39] But for the tensor field of gravity, as Unruh has shown,[40] there is no such thing as charge in the sense of flux through a wormhole. More precisely, he has shown that in geometrodynamics, "ordinary homologies with real coefficients" prohibit any adjustments of anything at all like a geometrodynamic analogue of electric charge.

Is the solution still unique when the compact three-geometry in question is endowed with one or more wormholes? No and yes. No for the magnetic field or any vector field until we restore uniqueness by specifying the flux through each wormhole.

Alternatives to the simple condition that space be closed are more complicated boundary conditions that accomplish an equivalent purpose when the space is regarded as asymptotically flat. What those conditions mean, how to formulate them, and how thus to guarantee existence and uniqueness of a solution,[25,41] we see best by regarding the "asymptotically flat" region as not being truly flat, after all, and at still greater distances, not infinite, but—as above—part of a closed universe; not manifestly but "implicitly closed."

By demanding that a solution to the initial-value problem shall exist and be unique we achieve no small purpose. We determine **spacetime here**—and therefore what we mean by **inertia here**—in terms of **mass-energy there**. In this book we have stated and expounded this thesis, its foundations and its consequences.

This determination of the geometry takes place on a hypersurface that is everywhere spacelike. No signaling! But, as Ellis and Sciama reminded us,[42] "the Coulomb field of a charged particle that lies outside our particle horizon is still in principle detectable today. We can express this situation by saying that although we cannot see a charge outside our light cone, we can certainly feel it." Adopting this language, we can declare that spacetime and inertia here do not see mass-energy there; they feel it.

Not see, but feel? The distinction does not demand curved spacetime or a horizon. It arises in electrodynamics and other field theories. Fermi, making a similar point in his treatise on quantum electrodynamics,[43] nevertheless shook some readers. There he related electric fields on the present-time hypersurface to the present-time location of the several electric charges. Were not those fields in reality generated by those charges at their past locations? By way of answer, we have only to point to Fermi's treatment of the additional and ostensibly free part of the total field. It takes care of the difference. Moreover, we have no alternative once we have discarded a world line or other spacetime description and embarked on an analysis of the electromagnetic field that begins with initial-value data and only initial-value data. With that choice of question granted, the answer to it is inescapable. The appropriate Green's function is not $1/r_{\text{ret}}$. It is $1/r$. The field here—as expressed in the standard elliptic equation of the standard initial-value problem of electrodynamics, $\Delta\Phi = -4\pi\rho$, with the standard uniqueness-giving condition on asymptotic behavior—does not see the charge there. It feels it.

Obvious contributors to space closure are matter and electromagnetic and other radiations, including the effective energy of gravitational waves themselves (see § 2.10).[22,25,44–46] What do we say about a conceivable additional contribution to closure brought about through the action of a so-called cosmological constant? Einstein invented the idea in 1917,[47] we know, as a desperate device to secure a universe static in the large. Then, however, came Hubble's 1929 evidence that the universe is not static in the large (see § 4.2) and later—regarding the very idea of a cosmological constant—Einstein's exclamation to George Gamow: "the biggest blunder of my life."[48] Many and interesting are the theories as to why there should or should not be a cosmological constant. Quantum and particle cosmology seems to suggest a nonzero primordial cosmological constant (see § 4.2). Nevertheless, not the slightest battle-tested astrophysical evidence has ever been adduced for a nonzero value for such a quantity. The observational limit on its "present" value is of the order of $|\Lambda| \lesssim 3 \times 10^{-52}\ \text{m}^{-2}$. We shall take it to be zero.

In the same spirit of sticking to Einstein's 1915 and still-standard and unclouded geometrodynamics (see chap. 3), we forego any attempt to analyze the initial-value equation of Hoyle-Narlikar theory,[49] of Jordan-Brans-Dicke theory,[50] and any of those deviants from standard general relativity which are so numerous (§ 3.7).[51]

Furthermore, for the reasons explained above we demand that the **spacetime manifold** shall be **spatially closed**, that is spatially compact and spatially without boundary.[52] To formulate space compactness properly requires a further statement about space topology. Space topologies that are compact and natural to consider fall into two classes according to whether they do or do not lead to a model universe of limited duration; for this teleological distinction between space topologies of the two kinds, see section 4.3. We now turn to a global parameter suited for specifying time and for analyzing the **geometrodynamic initial-value problem**.

5.2.2. [York Time as the Time Parameter Most Useful in Applications to Black Hole Physics and Spatially Closed Cosmologies]

Time? Three-geometry itself is carrier of information about time.[53–55] Thus, any spacelike slice that wriggles, weaves, and waves its way through any already known and generic spacetime—compact in space, finite or infinite in time, and solution of Einstein's field equation—automatically reveals to us, by its very three-geometry, *where* it is in that four-manifold and therefore all that we need to know about *time* on that hypersurface.

In physical conditions on a spacelike hypersurface so wavy, however, we are normally as little interested as we would be in the state of the electromagnetic field on a comparably wavy hypersurface slicing through flat spacetime. Only for purposes of advanced theory like Tomonaga's bubble-time–derivative formulation of quantum electrodynamics[56] or the superspace version of quantum geometrodynamics[35,57] do we ordinarily need to consider this concept of many-fingered time. In contrast, we foliate spacetime—we slice through it with an all-encompassing monotonic continuous sequence of nonintersecting spacelike hypersurfaces[58,59]—in every working analysis of the evolution of geometry with time, whether in astrophysics,[60,61] gravitational-wave physics,[62] or cosmology.[63] Then the parameter that distinguishes one slice from another provides the appropriate measure of time.

Of all tools to specify a spacelike slice, none has proved more useful than *York time*, T. For a point on a spacelike slice through spacetime, we define T as the fractional rate of decrease of volume per unit advance of proper time,

$$\begin{aligned} T &= \text{York time} = \operatorname{Tr} \boldsymbol{K} = (\text{trace of the extrinsic curvature tensor}) \\ &= (\text{fractional rate of expansion or contraction of the three-volume per unit} \\ &\quad \text{advance in proper time (cm) normal to the hypersurface}). \end{aligned} \tag{5.2.1}$$

We recall that the **extrinsic curvature tensor** or **second fundamental form** K_{ik} of a spacelike hypersurface, represents the fractional rate of change and deformation with time of the three-geometry of the spacelike hypersurface. Using the timelike unit vector n^α normal to the spacelike hypersurface, we may express the components of $\boldsymbol{K}$ as

$$K_{\alpha\beta} = -\frac{1}{2} h^\sigma{}_\alpha h^\rho{}_\beta (n_{\sigma;\rho} + n_{\rho;\sigma}) \tag{5.2.2}$$

where $h^\sigma{}_\alpha$ is the space projection tensor of section 4.5, $h^\alpha{}_\beta = \delta^\alpha{}_\beta + n^\alpha n_\beta$, and $n^\alpha n_\alpha = -1$. If the coordinates are such that $x^0 =$ constant on the spacelike hypersurface, we have

$$K_{ij} = -\frac{1}{2}(n_{i;j} + n_{j;i}) \equiv -n_{(i;j)}. \tag{5.2.3}$$

For example, for the Friedmann model universe with the space topology of a three-sphere of radius $R(t)$, we have

$$T = \operatorname{Tr}\boldsymbol{K} = -\frac{3}{R}\frac{dR}{dt} \tag{5.2.4}$$

where $\frac{dR}{dt}$ is the rate of expansion or contraction. In this language, $T = -\frac{3}{R}\frac{dR}{dt} = 0$ is the York time of maximum expansion of this Friedmann model universe.

An infinite sequence of zero T hypersurfaces, distinguished one from another by some parameter additional to T, was used by André Lichnerowicz and Yvonne Choquet-Bruhat in their beautiful and pioneering investigations[64–66] (see also Georges Darmois, 1923–1924)[67,68] of the dynamics of geometry. Well adapted as this scheme proved to be for spacetimes that are asymptotically flat, it turned out not to succeed with spatially closed cosmologies. Moreover, even the Schwarzschild four-geometry, epitome of the black hole, asymptotic though it is to flat space, turned out[69,70] not to permit zero T foliations that approach the singularity closer than a critical boundary, $r = (3/2)M$. The way out of this impasse came with York's 1972 discovery[3,4] that T itself, not some additional parameter, should be taken as the foliation parameter for any history of geometry that shows bang or crunch or both—hence the name—"York time" employed here as synonym for T. Therefore, to bring into simple evidence "how mass there influences spacetime—and therefore inertia—here," we analyze the dynamics of geometry via York's time T.

"Proper time and cosmological time deal with quite different aspects of physics. Proper time is an attribute of a particle. York time is an attribute of a cosmos."[71]

The slicing of a cosmological model by a T-indexed sequence of spacelike hypersurfaces gives a unique foliation. The proof of this uniqueness[58,59] assumes that the model universe—whether spatially open or spatially closed is globally hyperbolic,[72] not everywhere flat, and endowed with an energy momentum tensor that satisfies everywhere certain natural local conditions.[72] We recall that the condition of global hyperbolicity is equivalent to the existence of a Cauchy surface, that is, a submanifold of the spacetime such that each causal path without endpoints intersects it once and only once.

Does there exist any slow change with cosmic time in the so-called dimensionless constants of physics or in the degree of time-asymmetry in K-meson decay? For any such variation there is up to now not a single piece of battle-tested observational evidence (see § 3.2.3). However, if a detecting scheme of improved sensitivity or cleverness does ever succeed in establishing one or another such effect, the time T alone would seem to have the requisite cosmos-wide standing to orchestrate this change.[71]

Will York time, negative today, rise to $T = 0$ and go on to infinite positive values? In other words, will the world attain a state of maximum expansion and then collapse? It is difficult to cite any issue of large-scale physics on which there is today more diversity of views;[24,73–81] in chapter 4 (§ 4.3) we have discussed the relations between closure in time and spatial compactness.

5.2.3. [The Topology of a Space That Is Compact Falls into One or Another of Two Classes, According to Whether Its Dynamics Does or Does Not Lead to Recollapse of the Universe]

The unanswered question about closure in time connects mathematically in a striking way with the unanswered question about spatial compactness. The mathematics of the linkage again assumes global hyperbolicity,[72] a zero cosmological constant, and a local energy-momentum tensor that fulfills certain suitable and natural requirements[72] of section 4.3. With these granted, the necessary, and in some cases also sufficient (see § 4.3), condition for the model universe to begin with a big bang and end with a big crunch, that is, possess time closure, is to endow the three-dimensional spacelike initial-value hypersurface with a **"recollapse-topology."** "Recollapse-topologies" are compact topologies.

Every "recollapse-topology" is compact, but not every compact three-manifold is endowed with "recollapse-topology." As we have seen in section 4.3 a three-torus, T^3, model universe (we recall that the n-torus is defined as $S^1 \times \cdots \times S^1$, with n factors, where S^1 is the circle or one-dimensional sphere), for example, is spatially compact but—expanding forever from a big bang—it does not give recollapse.[82–84] However, any manifold with the topology of the three-sphere, S^3, may give recollapse and may give a big bang to big crunch history. So too may a one-wormhole topology, $S^2 \times S^1$—or a three-sphere topology with three attached wormholes, $S^3 \# 3(S^2 \times S^1)$. More generally, an initial-value spacelike hypersurface may give time closure, may give recollapse, only if its topology is of the form

$$(S^3/P_1) \# (S^3/P_2) \# \cdots \# (S^3/P_n) \# k(S^2 \times S^1), \qquad (5.2.5)$$

where P_i is a finite subgroup of SO(3), # denotes connected sum, and $k(S^2 \times S^1)$ means the connected sum of k copies of $(S^2 \times S^1)$ (chap. 4, § 3).[84,85]

The York[5,8] and Choquet-Bruhat and York[6] initial-value analysis is the heart of this chapter, where we assume space compactness, and where we do not discuss the question whether the space topology does or does not fall into the "recollapsing class." Nevertheless, the classification of topology, coming out of work in mathematics, is an important issue for physics which we have described in section 4.3.

5.2.4. [Specification of Conformal Three-Geometry]

Within a given three-manifold of a specified topology, the quantity to specify is not three-geometry itself for that would be too much and would ordinarily give to the spacelike hypersurface a wiggly, wavy course through spacetime, incompatible with our demand for simplicity, T = constant. The proper quantity to specify is the **conformal** part of the **three-geometry**, a quantity which we have symbolized as ${}^{(3)}\overset{(c)}{\mathbf{g}}$.

We recall that the three-geometry gives the distance between any two nearby points on a hypersurface, $g_{mn}dx^m dx^n = \psi^4(x, y, z)\overset{(c)}{g}_{mn}dx^m dx^n$, where ψ^4 is the conformal factor or scale of the three-geometry, and $\overset{(c)}{g}_{mn}$ represents the conformal three-geometry and gives the distance between any two nearby points up to the scale factor, that is, $\overset{(c)}{g}_{mn}$ gives at a point the ratio between any two "local" distances.

Let three angle-like coordinates, x, y, z, distinguish points on the three-manifold in question. Employing those coordinates, we do not specify the three-metric g_{mn}. Instead, we give only its conformal part, $\overset{(c)}{g}_{mn}$. The three-geometry itself is totally unknown at this stage, unknown before we turn to any analysis of the initial-value problem, unknown until we solve the initial-value equation (5.3.9) below, which gives the **conformal factor**, ψ—the factor that connects the actual three-geometry with the conformal three-geometry via the equation

$$g_{mn} = \psi^4 \overset{(c)}{g}_{mn}. \tag{5.2.6}$$

5.2.5. [The Larger View: The Variation Principle Itself—as Ratified by the Quantum Principle—Tells the Quantities to Be Fixed]

Central Role of Phase-fixing Factor in Feynman's Sum over Histories

Classical dynamics, whether the dynamics of a geometry or a particle or any other entity, does not spring full blown, we know, out of the head of Minerva. It takes its origin in quantum physics. It arises—in Feynman's attractive formulation[86,87]—from the principle of the democratic equality of all histories. The probability amplitude,

$$\langle \text{final configuration} \mid \text{initial configuration} \rangle, \tag{5.2.7}$$

to transit from the specified initial configuration to one or other final configuration is given by a sum of contributions of identical magnitude but diverse phase,

$$(\text{normalizing factor}) \cdot \exp[(2\pi i/h) \cdot (\text{action integral for history} H)], \tag{5.2.8}$$

over all conceivable histories H that lead from the given initial configuration to the chosen final configuration.

The phase-fixing factor in equation (5.2.8), the action, is the natural focus of attention, whatever the system under consideration. Demand constructive interference? Or require that the phase shall be an extremum,

$$\text{(action integral)} = \text{(extremum)}, \tag{5.2.9}$$

an extremum against every local variation in the history that keeps the end points fixed? To issue this command and trace out its consequences—with or without knowledge of the quantum rationale underlying it—is to arrive, as is well known, at a second-order equation for the dynamics—an equation that is totally standard, totally classical.

Hilbert Choice of Action Principle Supplies Natural Fixer of Phase for Geometrodynamics

The classical equation for the dynamics of geometry won its standard formulation in 1915. At that time any quantum background for any variation principle was, of course, unknown. Nevertheless, David Hilbert, who had followed Einstein's labors with collegial enthusiasm, had the insight to recognize that an action principle which used the four-dimensional scalar curvature invariant ${}^{(4)}R(t, x, y, z)$, offered the natural—and simple—way to get for the first time a truly reasonable field equation (see § 2.3).[88] This is the equation accepted by Hilbert, and subsequently by everyone, as the standard Einstein field equation.

The story of the variational principle has been told and retold in many a treatise on gravity theory. Where in any of them, however, was anything to be seen of the physics of what to fix at limits? The dynamic engine was there, whether cloaked or laid out to clear view; but where and what were the handles for the steering of that engine?

The Search for Handles by Which to Steer: Via Differential Equations? Or Via Variational Principle?

No one except Élie Cartan—in his mathematical publications[89] and in his correspondence with Einstein[90]—seemed to realize that this vital part of the machinery was missing. Thus began the differential-equation investigation of the initial-value problem. It led from Georges Darmois, Cartan, and André Lichnerowicz to Yvonne Choquet-Bruhat and James W. York, Jr.

A totally different orientation toward the initial-value problem presents itself in quantum mechanics. There we are required on occasions to fold together two propagators, the head of one to the tail of the other, to obtain a third:

$$\langle \text{final} \mid \text{initial} \rangle = \sum_{\substack{\text{all intermediate}\\ \text{configurations}}} \langle \text{final} \mid \text{intermediate} \rangle \cdot \langle \text{intermediate} \mid \text{initial} \rangle. \tag{5.2.10}$$

In this sum over intermediate configurations, what do we fix and what do we vary? This information the variation principle itself reveals, whether in the elementary example of mechanics,

$$(\text{action}) = \int [(m/2)(dx/dt)^2 - V(x,t)]\, dt, \tag{5.2.11}$$

or in gravity theory (§ 2.3)

$$(\text{action}) = \iiiint \left[\left({}^{(4)}R/16\pi\right) + \text{Lagrangian}_{\text{matter and other fields}}\right] \times \left(-\det {}^{(4)}g\right)^{\frac{1}{2}}\, dt\, dx\, dy\, dz. \tag{5.2.12}$$

No longer do we center our attention on the variation in the action caused by changes in the course of the history between the initial and intermediate configuration. Instead, we focus on the alteration in the terminal configuration itself and what it does to the action: in mechanics,[93]

$$\delta(\text{action}) = m(dx/dt)\,\delta x - [(m/2)(dx/dt)^2 + V(x,t)]\,\delta t; \tag{5.2.13}$$

and in the simplest version of geometrodynamics, as clarified by Karel Kuchar and James W. York, Jr.,[53,92,93]

$$\delta(\text{action}) = \iiint [(\text{geometrodynamical field momentum}) \cdot \delta({}^{(3)}\mathscr{g}^{(c)}) - (\text{geometrodynamical field Hamiltonian}) \times \delta(\mathrm{Tr}\,\boldsymbol{K})]\, d^3V. \tag{5.2.14}$$

The Folding Together of Two Propagators Calls for the Fixing of the Intermediate Time—and Therefore for Holding Fixed a Clearly Timelike Quantity

Equations (5.2.13) and (5.2.14) for the effect of changing end-point conditions tell us clearly what we must do when we fold together two propagators. In the quantum mechanics of a single particle we hold fixed the time—or set δt equal to zero in equation (5.2.13)—and integrate over x. In quantum gravity we hold fixed the York time, $T = \mathrm{Tr}\,\boldsymbol{K}$, and integrate over all choices of the conformal three-geometry, ${}^{(3)}\mathscr{g}^{(c)}$. This conformal three-geometry thus constitutes the geometrodynamical field coordinate as truly as x serves as particle coordinate in elementary mechanics.

In the folding together of two propagators a difference in character shows itself between time and coordinate, or between T and ${}^{(3)}\overset{(c)}{\mathbf{g}}$. Integrate over time? No. Over coordinate? Yes. Over both? No—unless we are prepared for a nonsensical infinity.

The handles, then, for dynamics, as the variation principle itself gives them to us, are the times at the two chosen limits of the particle dynamics, or the geometrodynamics,[94] and—different in quality—the coordinates at those two times.

If we view conditions at those two boundaries as constituting the two faces of a sandwich (however, see refs. 95 and 96), we may go to the so-called thin sandwich limit: in particle mechanics,

$$[t_2, x_2; x_1, t_1] \longrightarrow [t_0, x_0, (dx/dt)_0], \tag{5.2.15}$$

and in geometrodynamics,

$$[T_2, {}^{(3)}\overset{(c)}{\mathbf{g}}_2; T_1, {}^{(3)}\overset{(c)}{\mathbf{g}}_1] \longrightarrow [T, {}^{(3)}\overset{(c)}{\mathbf{g}}, \text{conformal distortion tensor}]. \tag{5.2.16}$$

Here the distortion is qualitatively of the character $d\,{}^{(3)}\overset{(c)}{\mathbf{g}}/dT$, and is defined more closely in section 5.2.7 (see also § 4.5). When dynamic entities are present in addition to geometry we must, of course, give additional information in expression (5.2.16) to supply the extra handles needed in the initial-value problem. Whatever the details, the variation principle itself tells what the right handles are.

Terminology: Helpful and Not

We forego oceans of talk about covariance with respect to change of coordinates because everyone dealing with relativity nowadays takes covariance for granted. Instead of making much ado about what is whim, that is, coordinates, we deal here with what matters, four-geometry, ${}^{(4)}G$.

We abjure any such term as a "truly four-dimensional Hamiltonian version of general relativity." We know that a ${}^{(4)}G$ has no dynamics. Being the history of change, it does not itself change. Hamiltonian for it? No!

Then what is the dynamic object? It is three-geometry, ${}^{(3)}G$. Thus arises the terminology of geometrodynamics: four-geometry as one among many conceivable histories of a three-geometry changing with time in accordance to Einstein's dynamic law,

$$\begin{pmatrix} \text{dual of} \\ \text{moment of} \\ \text{rotation} \\ \text{associated with} \\ \text{a three-cube} \end{pmatrix} = 8\pi \begin{pmatrix} \text{amount of} \\ \text{enery-momentum} \\ \text{in that} \\ \text{three-cube} \end{pmatrix} \tag{5.2.17}$$

(see § 2.8) and this dynamics taking place in the arena of superspace S.[35,57]

Three-Geometry Occupies a More Favorable Position in Quantum Cosmology

More suited to ${}^{(3)}G$ machinery than classical geometrodynamics is quantum cosmology. There is much interest[97,30,31] in the concept of a unique state functional, $\psi({}^{(3)}G)$—or, rather, $\psi({}^{(3)}G$, other fields)—out of which more than one investigator hopes eventually to derive all of cosmology[98–101] from a big bang start out of nothingness.

How can the present straightforward classical initial-value analysis be envisaged to fit into so ambitious a future quantum scheme? First, if $\psi({}^{(3)}G$, other fields) is unique, then $\psi(T, {}^{(3)}\mathbf{g}^{(c)}$, other fields) is also unique. Second, mere simple quantum fluctuations arising out of that unique start gave rise—according to some—to all the fantastic complexity of the world as we see it today. Nevertheless . . . .

For initial-value data in relativistic astrophysics, forego the single ${}^{(3)}G$ scheme of quantum cosmology and specify York's time T plus conformal three-geometry ${}^{(3)}\mathbf{g}^{(c)}$ and its rate of change plus conformal distribution of mass-energy and conformal currents of mass-energy.

Nevertheless, no one in his right mind will go back to the beginning of things when he wants to investigate, via general relativity, an astrophysical process of the here and now. Let two black holes collide off-center; or let the core of a rotating star collapse with the emission of a powerful gravitational wave; or let a gravitational wave hit a star teetering on the verge of collapse. For any such scenario, and a myriad of others, the whole course of events follows from a statement of conditions at one value of York's T-time:

$$\begin{aligned}\{T,\ {}^{(3)}\mathbf{g}^{(c)},\ &\text{rate of change of } {}^{(3)}\mathbf{g}^{(c)},\ \text{conformal distribution and} \\ &\qquad\qquad \text{conformal currents of mass-energy}\}.\end{aligned} \tag{5.2.18}$$

That is the deeper rationale for steering as laid out in this chapter.

5.2.6. [Specifying Conformal Scalar, Vector, and Tensor rather than Actual Physical Scalar, Vector, and Tensor: The Why and the How]

By giving T and the conformal or base metric is to prepare the way for the rest of the initial-value data. Part of this required information describes the present conformal flow and present conformal density of material mass. Another part specifies the "free part" of the conformal distortion tensor, a feature of the geometry. The conformal part of the three-geometry itself differs as much from the free part of the conformal distortion tensor as does a magnetic field from an electric field, or a general continuum-field coordinate from the conjugate continuum-field momentum. Yet another part of these data provides analogous information about field coordinate and field momentum for the electromagnetic field itself and for other fields. In the interest of simplicity we omit in this chapter anything but dustlike matter and gravitation. The treatment of additional entities—and the pressures and other stresses which they bring with them—follows a pattern almost obvious from the scheme of treatment of mass and geometry.

All of the information just mentioned, as required to be specified by York,[5,8] and by Choquet-Bruhat and York,[6] updating the work of Lichnerowicz and Choquet-Bruhat, follow the plan of the conformal three-geometry itself. Their treatment demands, in every case, not the entity in question in its precise magnitude, but expressed always as a conformal quantity of the same tensorial character, multiplied by the appropriate power of the **conformal factor** ψ. Only by the right choice of power in each case do the initial-value equations separate in a form favorable to solution.

What we therefore specify for each quantity—three-geometry, distribution of flow and of density of dust-like matter, gravity-wave–induced distortion—is not the final physically meaningful scalar or vector or tensor magnitude itself, but the corresponding conformal quantity. Only at the end of our solution of the initial-value problem—just prior to any calculation we may wish to do of the evolution of all this physics for any desired reach of time into the future—only then, when we have in hand the three-vector **"gravitomagnetic" potential $\boldsymbol{W}$**, and an analogous conformal scale factor ψ, do we have the means, the simple means, to get every relevant physical quantity at every point over the entire constant-time T initial-value spacelike hypersurface.

We can restate the nature of this initialization procedure in slightly different language. First we set up trial initial-value data. Then—by solving for ψ and $\boldsymbol{W}$—we adjust these data for consistency with the conditions imposed by the 00 and $0i$ components of the Einstein equation. Finally, we find out what are the actual initial-value data that we have set up.

What is it then, that we are doing as we solve initial-value problem? We are bringing about mutual adjustments, jiggling what we have into concurrence, harmonizing the data and making the physics shape up.

Decide, therefore, the decidable! Only then discover all the initial-value information which that decision carried in its train! No other way do we know that will give us either here, in geometrodynamics, or in electrodynamics or in elastomechanics the desired consistency of the data between region and region over the entire initial-value spacelike hypersurface, consistency demanded and enforced by differential equations of elliptic character. However, having once obtained consistent initial-value data by solving the elliptic equations for the conformal expansion factor ψ and the gravitomagnetic vector potential $\boldsymbol{W}$, we can, if we wish, check ourselves: reinstall the just-obtained physical quantities as if they were newly invented conformal quantities and solve the elliptic equations anew. This time the solution exists and is unique.[4–8,102] It has to be

$\psi = 1$, $\boldsymbol{W}_{\text{new}} = \boldsymbol{W}_{\text{previous}}$, for otherwise, there is some error in the computer arithmetic!

5.2.7. [The Freely Specifiable Conformal Quantities Named and Examined]

Time

Specify an everywhere constant value of the York time, $T = \operatorname{Tr}\boldsymbol{K}$. For illustrative example, specify $T = 0$ (stage of maximum expansion). Alternatively make a specification that corresponds more closely to conditions today $T \cong -3H$ (Hubble "constant," see § 4.2) $\approx -3 \times 75$ km/s Mpc^{-1} ($\sim$ today!) $= -(3 \times 75 \times 10^5$ cm/s)/(3×10^{10} cm/s $\times\ 3.1 \times 10^{24}$ cm); that is, $T \approx -2.4 \times 10^{-28}$ cm^{-1}. The negative value of T today indicates a time prior to the attainment of maximum expansion.

Conformal geometry

Specify for the spacelike initial-value hypersurface a suitable closed (compact and without boundary) three-dimensional differentiable manifold (Cauchy surface). On this manifold we can give the positive-definite-signature **conformal metric** $\overset{(c)}{g}_{mn}(x, y, z)$ as function of three angle-like coordinates x, y, z, with the understanding that the metric of the three-geometry itself is to be obtained from equation (5.2.6). For an illustrative example, specify the geometry of a three-sphere of radius R in terms of the three hyperspherical coordinates χ, θ, ϕ (as in eq. 4.2.31a), $\overset{(c)}{g}_{mn} dx^m dx^n = [d\chi^2 + \sin^2\chi\,(d\theta^2 + \sin^2\theta\, d\phi^2)]$.

Density

Specify the everywhere nonnegative conformal value of the **energy density** of matter over the three-manifold. Let it be expressed in geometric units, say cm of mass-energy per cm^3 of space. This quantity is connected to the actual matter density by the appropriate power of the conformal factor. The specification of the conformal density fixes the relevant term $\overset{(c)}{Q}$ that will appear in the initial-value equation itself:

$$\overset{(c)}{Q} = 16\pi \overset{(c)}{\rho}(x, y, z) = 16\pi\, \psi^8 \rho \text{ (nonnegative).} \tag{5.2.19}$$

For the sake of brevity we are giving here a stripped-down version of the Choquet-Bruhat and York analysis. We are simplifying, excluding from the initial-value problem that we pose, every force, every field and, most specifically, every stress-energy that is of elastic or electromagnetic or other nongravitational origin. In brief, for simplicity, we are idealizing here matter to be dustlike. However, we shall include under this heading of "dustlike matter" not only atoms and molecules but also entities of whatever larger size that does not get into problems of superposability, and, under the same condition, admit neutrinos and photons. Nevertheless, one may in general consider a fluid with stresses, for example with pressure p and with an equation of state $p = p(\rho)$. Indeed, it is assumed[5] that the stresses do not enter the initial value equations, that is, do not constrain the initial values of ρ and $\boldsymbol{j}$.

Flow

Specify the **current** of this dustlike matter (say, in cm of mass per cm^3 of volume times cm of distance moved per cm of time elapsed), again not directly, but up to a well-determined power of the **conformal** scale factor,

$$\boldsymbol{j} = \psi^{-10}\overset{(c)}{\boldsymbol{j}}(x, y, z). \tag{5.2.20}$$

We demand that this current shall not transmit energy faster than light. In other words, we require that the four-vector of density and current shall be everywhere timelike or at most lightlike, that is, the quantity

$$\overset{(c)}{\rho}{}^2 - \overset{(c)}{g}_{mn}\overset{(c)}{\boldsymbol{j}}{}^m\overset{(c)}{\boldsymbol{j}}{}^n \tag{5.2.21}$$

shall be everywhere nonnegative ("*dominance of energy*" *condition*). The power 10 in equation (5.2.20)—along with the exponent 8 in equation (5.2.19) and the arithmetical identity $2 \times 8 = -4 + 2 \times 10$—ensures that this condition (5.2.21) "nothing faster than light," shall preserve itself,[5,8] whatever dependence on position the conformal scale factor ψ may later turn out to have.

The Distortion Tensor

One can decompose a symmetric two-covariant tensor S_{ij} on a proper Riemannian three-manifold, with metric h_{ij}, by the splitting technique of York:[103,104]

$$S_{ij} = S_{\text{free}\,ij} + \left[l_{\boldsymbol{h}}(\boldsymbol{V})\right]_{ij} + \frac{1}{3}h_{ij}\,\mathrm{Tr}\,\boldsymbol{S} \tag{5.2.22}$$

where V^i is some vector field; $\left[l_{\boldsymbol{h}}(\boldsymbol{V})\right]_{ij} = V_{i;j} + V_{j;i} - \frac{2}{3}h_{ij}V^k{}_{;k}$; and where $\mathrm{Tr}\,\boldsymbol{S}_{\text{free}} = S_{\text{free}\,ij}h^{ij} = 0$, that is, $\boldsymbol{S}_{\text{free}}$ is trace free, and $S^{ik}_{\text{free};k} = 0$, that is, $\boldsymbol{S}_{\text{free}}$ is covariant divergence free.

One can therefore write the extrinsic curvature tensor or second fundamental form $\boldsymbol{K}$, of equation (5.2.3) on the compact three-manifold x^0 = constant as

$$K_{ij} = A_{\text{total}\,ij} + \frac{1}{3}g_{ij}\,\mathrm{Tr}\,\boldsymbol{K} \tag{5.2.23}$$

where

$$A_{\text{total}\,ij} = A_{\text{free}\,ij} + \left[l_{\boldsymbol{g}}(\boldsymbol{W})\right]_{ij} \tag{5.2.24}$$

and $A^{i}_{\text{total}\,i} = A^{i}_{\text{free}\,i} = 0$ (traceless), and $A^{ik}_{\text{free};k} = 0$ (covariant divergence free), and $[l_{\boldsymbol{g}}(\boldsymbol{W})]_{ij} = W_{i;j} + W_{j;i} - \frac{2}{3} g_{ij} W^{k}{}_{;k}$.

For a fluid with four-velocity u^{α}, in a region of a spacelike hypersurface where $u^{\alpha} = n^{\alpha}$, we have that $A_{\text{total}\,ij}$ is equal to the negative of the spatial components of the shear tensor defined in section 4.5.

Then, give everywhere on the three-manifold the conformal value of the free part of the extrinsic curvature tensor $\boldsymbol{K}$, that is, the **conformal** value of the **free** part of the **distortion tensor** just defined or in other words a second-order, 3×3, traceless, transverse, symmetric tensor, $\overset{(c)}{A}_{\text{free}\,ij}(x, y, z)$, related to the rate of deformation of the conformal three-geometry $\overset{(c)}{\boldsymbol{g}}$.

When we write *transverse* and *traceless*, we mean with respect to the conformal three-geometry $\overset{(c)}{\boldsymbol{g}}$, that is: *traceless*,

$$\overset{(c)}{A}{}^{i}_{\text{free}\,i} \equiv \overset{(c)}{g}_{ij} \overset{(c)}{A}{}^{ij}_{\text{free}} = 0 \tag{5.2.25}$$

and *transverse*, with zero covariant divergence with respect to $\overset{(c)}{\boldsymbol{g}}$,

$$\overset{(c)}{\nabla}_{j} \overset{(c)}{A}{}^{ij}_{\text{free}} \equiv \overset{(c)}{A}{}^{ij}_{\text{free},j} + \overset{(c)}{\Gamma}{}^{i}_{jk} \overset{(c)}{A}{}^{kj}_{\text{free}} + \overset{(c)}{\Gamma}{}^{j}_{jk} \overset{(c)}{A}{}^{ik}_{\text{free}} = 0 \tag{5.2.26}$$

where we define the covariant derivative $\overset{(c)}{\nabla}_{j}$ and the Christoffel symbols $\overset{(c)}{\Gamma}{}^{i}_{jk}$ with respect to $\overset{(c)}{\boldsymbol{g}}$ in the usual way:

$$\overset{(c)}{\nabla}_{k} T^{i}{}_{j} = T^{i}{}_{j,k} + \overset{(c)}{\Gamma}{}^{i}_{kl} T^{l}{}_{j} - \overset{(c)}{\Gamma}{}^{l}_{jk} T^{i}{}_{l} \tag{5.2.27}$$

and

$$\overset{(c)}{\Gamma}{}^{i}_{jk} \equiv \frac{\overset{(c)}{g}{}^{il}}{2} (\overset{(c)}{g}_{jl,k} + \overset{(c)}{g}_{lk,j} - \overset{(c)}{g}_{kj,l}). \tag{5.2.28}$$

An Interpolation: The Analogy with Electrodynamics

The mathematics for setting up such a 3×3 tensor field[103,104] has many points in common[105] with the method well known in electrodynamics of constructing a three-vector field which shall be free of divergence at every point of the initial-value three-manifold. Thus the total electric field on that hypersurface is the sum of two parts. The first part is fixed by the charges. The second, the free part of the electric field, or otherwise known as the field momentum, we require to be everywhere divergence-free (as is also the magnetic field or conjugate field coordinate). Other than being divergence-free, we specify the free vector field arbitrarily, in accordance with the initial-value physical situation that we wish to describe. Being divergence-free, it has—in a well-known way of counting[25,106]—not three degrees of freedom per space point, but only two. They specify what are often called "the two momentum *degrees of freedom*, per space point," of the electromagnetic field.

In the familiar flat-space Fourier decomposition of the electromagnetic field, there exist for each wave vector two independent modes of oscillation, distinguished by their polarizations. Each of these modes has, like the harmonic oscillator itself, one coordinate degree of freedom and one momentum degree of freedom. For the individual wave vector there are therefore two independently specifiable field coordinates and two independently specifiable field momenta. Consequently it is not surprising that the space description—as contrasted to the wave-number description—of the electric part of the field gives us two degrees of freedom per space point. In the same way, the magnetic field, or, to use the language of dynamics, the "field coordinate," has two degrees of freedom per space point.

In geometrodynamics, when we fix $\overset{(c)}{g}_{mn}$ we do not fix the value of the distance between any two points on a hypersurface, but at any point the ratio between any two "local" distances. In other words we fix g_{mn} up to a scale factor ψ^4. Therefore, we may arbitrarily choose the scale factor ψ^4 to satisfy, for example, the condition

$$\det \overset{(c)}{g}_{mn} \equiv \overset{(c)}{g} \equiv 1. \tag{5.2.29}$$

In addition, with three arbitrary smooth transformations of the coordinates x^i on the hypersurface: $x'^i = x^i + \delta x^i(x^1, x^2, x^3)$, the metric will change to

$$\overset{(c)}{g}{}'_{mn} = \overset{(c)}{g}_{mn} + \delta g_{mn}, \tag{5.2.30}$$

and we may think the values of three of the components of the metric $\overset{(c)}{g}_{mn}$ equivalent to a choice of the spatial coordinates on the hypersurface.

Therefore, in geometrodynamics the dynamic field coordinate, the conformal metric $\overset{(c)}{g}_{mn}$, has this real freedom of specification: 6 (component count for a symmetric three-tensor) -1 (irrelevant scale factor, normalized out, for example, by the condition that the determinant of this conformal metric shall be unity) -3 (number of coordinates whose choice is irrelevant to the physics) $= 2$ meaningful field-strength–like components per space point.

The two conjugate momentum-like degrees of freedom display themselves in the dynamic field momentum, a quantity related to the gravitomagnetic field. It is a geometric entity. It is the conformal free distortion tensor $\overset{(c)}{A}_{\text{free}\,ab}(x, y, z)$. That tensor field, like $\overset{(c)}{g}_{mn}$, has two meaningful components per space point: 6 (ostensibly free components) -1 (condition of zero trace) -0 (no remaining freedom in choice of coordinates) -3 (the effective number of equations per space point implied by the condition of transversality) $= 2$ truly physical or nongaugelike quantities subject to adjustment at each point on the initial-value hypersurface.

Behind the mathematics for both fields, vector and tensor, lies an idea central to modern geometry: deformation of structure.[107] A magnetic field lets itself be deformed continuously from one configuration to a totally different configuration by the moving about of knitting needles, in the shape of ion clouds that slide along but never transverse to the magnetic lines of force.[25,108] A working view of the gravitomagnetic field of comparable insight and ease of application is starting to develop as more problems are formulated and solved in the domain of gravity waves,[62] astrophysics,[60,61] and cosmology[63] (see also ref. 150).

5.3 [THE SOLUTION OF THE INITIAL-VALUE EQUATIONS]

5.3.1. [The Laplacian Equation for the Gravitomagnetic Vector Potential *W*, in Terms of the Current *j* of Mass-Energy]

Insert the already specified conformal density of current $\overset{(c)}{\boldsymbol{j}}$ as source term on the right-hand side of the **vector initial-value equation,**[103,104] obtained from the Einstein equation, for an initial-data hypersurface of constant mean extrinsic curvature, $\nabla_i(\mathrm{Tr}\,\boldsymbol{K}) = \overset{(c)}{\nabla}_i(\mathrm{Tr}\,\boldsymbol{K}) = (\mathrm{Tr}\,\boldsymbol{K})_{,i} = 0$; we then have

$$\overset{(c)}{\Delta}{}^* W_i = \frac{2}{3}\psi^6 \overset{(c)}{\nabla}_i(\mathrm{Tr}\,\boldsymbol{K}) + 8\pi \overset{(c)}{j}_i = 8\pi \overset{(c)}{j}_i(x, y, z). \tag{5.3.1}$$

Now solve for the **gravitomagnetic vector potential *W***. There are at least three Laplace operators, all different, that take vectors to vectors. In equation (5.3.1) there appears the **conformal vector Laplacian**, $\overset{(c)}{\Delta}{}^*$ (as defined in ref. 103) as specialized to three-space,

$$\overset{(c)}{\Delta}{}^* W^i \equiv \overset{(c)}{\nabla}_j \left(\overset{(c)}{\nabla}{}^i W_j + \overset{(c)}{\nabla}_j W^i - 2/3 \overset{(c)}{g}{}^{ij} \overset{(c)}{\nabla}_k W^k \right). \tag{5.3.2}$$

In this formula $\overset{(c)}{\nabla}_j$ is the standard covariant derivative as defined on three-space in terms of the conformal metric $\overset{(c)}{g}_{mn}(x, y, z)$ and its first derivatives by formula (5.2.27).

5.3.2. [The Current of Mass-Energy Must Be Confinable]

Now comes a caution, a condition not yet stated, a requirement which the current of mass-energy—that is to say, the source term in equation (5.3.1)—must satisfy in a closed space if there is to exist a solution of that Laplacian-like equation for the gravitomagnetic potential $\boldsymbol{W}$ in that space. This condition? We cannot allow any source of field that would drive the field wild. No unconfinable source!

The distinction between a confinable and an unconfinable source shows most simply in everyday electrostatics in the familiar equation,

$$\Delta\Phi = -4\pi\rho(x, y, z) \tag{5.3.3}$$

that connects electrostatic potential Φ with density of electric charge ρ. Multiply both sides of this equation by the element of volume (determinant of ${}^{(3)}g_{mn})^{\frac{1}{2}}\,dx\,dy\,dz$, and integrate over the entire volume of the closed space in question. The left-hand side vanishes, and therefore the right-hand side, the total charge, must also vanish. What if it did not? There would then be electric lines of force left over with no place to go. The source would be unconfinable.

Any physically acceptable distribution of electric charge over a closed manifold must satisfy the condition of confinability,

$$\iiint \rho(x, y, z)(\det{}^{(3)}g_{mn})^{\frac{1}{2}}\,dx\,dy\,dz = 0. \tag{5.3.4}$$

Out of what deeper insight could we have foreseen the existence of such a condition? Why not out of this, that an oscillator driven at resonance goes wild? The driving force may have in its Fourier decomposition the most fantastic variety of frequencies without causing any difficulty for the motion. However, if one of those frequencies agrees with the natural frequency of the oscillator, then the displacement grows without limit. In other words, any driving mode is acceptable in the force if it is orthogonal to the natural mode—the mode of force-free motion—of the driven system; otherwise not. The key to confinability is orthogonality.

How does orthogonality show itself in electrostatics in a closed three-space? The source of drive is the charge density, $\rho(x, y, z)$. The driven object is the electrostatic potential, $\Phi(x, y, z)$. The law of drive expresses itself in equation (5.3.3). We annul the right-hand side of this equation of drive to find the condition for a natural mode. In a closed space with three-sphere topology there exists a solution of one and only one type.

$$\Phi(x, y, z) = \text{constant}. \tag{5.3.5}$$

To demand that this natural mode of the source-free potential shall be orthogonal to the driving source, $\rho(x, y, z)$, is to arrive in a single step at equation (5.3.4), the condition of confinability.

For the gravitomagnetic potential $\boldsymbol{W}$, we have in equation (5.3.1) the equation of drive. When we annul its right-hand side we have the equation for what we may call a normal mode of the gravitomagnetic potential. This equation is familiar in the domain of differential geometry under another name, the equation for a *conformal Killing vector*[109] (Wilhelm Killing, 1847–1923): $\overset{(c)}{\nabla}_i\xi_j + \overset{(c)}{\nabla}_j\xi_i - 2/3\overset{(c)}{g}_{ij}\overset{(c)}{\nabla}_k\xi^k = 0$. A Killing vector, as we have explained in sections 2.6, 4.2 and 4.4, for our present purpose is a vector $\boldsymbol{\xi}$ such that an infinitesimal vector displacement, $\boldsymbol{\delta x}(x, y, z) = \varepsilon\boldsymbol{\xi}(x, y, z)$, of the points of the three-geometry results in a new three-geometry totally indistinguishable from the original one. An ideal three-sphere admits six linearly independent Killing vector fields. $\boldsymbol{\xi}_A (A = 1, 2, \ldots, 6)$; the generic three-geometry, none; and of three-geometries with an intermediate number, s, of symmetries there are many examples[110] (see § 4.4). However, a three-space of constant curvature admits a maximum number of ten conformal Killing vectors, that is, four more.

The **condition of confinability** for the gravitomagnetic potential $\boldsymbol{W}$ can be satisfied when and only when[102] the source term for this field, the conformal current $\overset{(c)}{\boldsymbol{j}}$ of mass-energy, is orthogonal to every one of the conformal Killing vector fields $\overset{(c)}{\boldsymbol{\xi}}_A$, of the conformal three-geometry in question,

$$\iiint (\overset{(c)}{\boldsymbol{j}}(x, y, z) \cdot \overset{(c)}{\boldsymbol{\xi}}_A(x, y, z))(\det{}^{(3)}\overset{(c)}{g}_{mn})^{\frac{1}{2}}\, dx\, dy\, dz = 0 \qquad (5.3.6)$$

where $A = (1, 2, \ldots, s)$, provided that the three-geometry has any symmetry. In the generic case it does not. Then no such supplementary condition has to be imposed on the current of mass-energy to keep it from driving the gravitomagnetic potential wild. Whatever the current is, it will not.

Confinability being secured, the equation for the gravitomagnetic potential is guaranteed to possess one and only one solution, $\boldsymbol{W}(x, y, z)$. To have solved equation (5.3.1) for it, is to have accomplished a major part of the initial-value problem enterprise: to get a self-consistent setting for the handles that steer the geometrodynamics, and thus to find how *mass and mass flow there influences geometry—and inertia—here.*

5.3.3. [From Gravitomagnetic Potential and Free Part of Conformal Distortion Tensor to Total Conformal Distortion, and from It to Conformal Density of Gravitational-Wave Effective Kinetic Energy]

A Laplacian equation is not the only guide we find in electrodynamics to the plan of geometrodynamics. The electrostatic field, $-\vec{\nabla}\Phi(x, y, z)$ generated by the source charges, where they are now,[105] we have to supplement by the so-called free field to have in hand the total initial-value electric field. Similarly, taking the conformal distortion generated by the present conformal currents of mass-energy (through the intermediary of the gravitomagnetic vector potential, $\boldsymbol{W}$), we have to add the free part of the conformal distortion tensor in order to obtain the **total conformal distortion tensor**,[5–8] thus

$$\begin{aligned}\overset{(c)}{A}_{\text{total}\,ab}(x, y, z) &= \overset{(c)}{A}_{\text{free}\,ab}(x, y, z)\\ &+ \overset{(c)}{\nabla}_a W_b + \overset{(c)}{\nabla}_b W_a - 2/3\overset{(c)}{g}_{ab}\overset{(c)}{\nabla}_m W^m.\end{aligned} \qquad (5.3.7)$$

In electrodynamics we never have to square the total electric field to obtain a source of electromagnetic field—never, because that field theory is linear. Geometrodynamics, however, is not linear. From this nonlinearity the field equation itself leaves us no escape. The Einstein theory of gravity forces us to insert into the quasi-Laplacian equation for our remaining unknown field quantity (the conformal factor ψ) a nonlinear source term $\overset{(c)}{M}$. This term is exactly quadratic. It is the "square" of the conformal distortion tensor:

$$\overset{(c)}{M}(x, y, z) = \overset{(c)}{A}{}^{ab}_{\text{total}} \overset{(c)}{A}_{\text{total}\, ab}. \tag{5.3.8}$$

We might call $\overset{(c)}{M}$ the conformal density of gravitational-wave effective kinetic energy.

5.3.4. [Finding the Last Unknown, the Conformal Scale Factor ψ]

We come now to the last differential equation that has to be solved to complete the mutual adjustment of the handles that steer the geometry: the equation of Lichnerowicz and Choquet-Bruhat, as updated by York, by O'Murchadha and York and by Choquet-Bruhat and York, obtained from the Einstein equation

$$8\overset{(c)}{\Delta}\psi - {}^{(3)}\overset{(c)}{R}\psi + \overset{(c)}{M}\psi^{-7} + \overset{(c)}{Q}\psi^{-3} - (2/3)T^2\psi^5 = 0. \tag{5.3.9}$$

Here $\overset{(c)}{\Delta}$ is the familiar **Laplacian**, as defined, however, on the curved three-geometry with the metric $\overset{(c)}{g}_{ij}$: $\overset{(c)}{\Delta} \equiv \overset{(c)}{g}{}^{ij}\overset{(c)}{\nabla}_i\overset{(c)}{\nabla}_j$, that is, $\overset{(c)}{\Delta}\psi = \overset{(c)}{g}{}^{-1/2}\frac{\partial}{\partial x^a}\left[\overset{(c)}{g}{}^{1/2}\overset{(c)}{g}{}^{ab}\left(\frac{\partial}{\partial x^b}\right)\right]\psi$ and ${}^{(3)}\overset{(c)}{R}$ is the scalar curvature invariant, or trace of the Ricci curvature tensor of three-geometry, for the same conformal metric. The first two terms in equation (5.3.9) transform individually in a slightly complicated way on change of the *conformal* factor. However, the combination of the two transforms as a simple power of ψ. The other factors in equation (5.3.9) we have already met: $\overset{(c)}{M}$, the conformal density of gravitational-wave effective kinetic energy; $\overset{(c)}{Q} = 16\pi\overset{(c)}{\rho}$, a measure of the conformal density of ordinary mass-energy; and T, the York time.

Under suitable mathematical conditions there always exists an everywhere-positive solution, $\psi(x, y, z)$, of the **scalar initial-value equation** (5.3.9), and it is unique.[6–8,102] The generic case of unique existence is when ($T \neq 0$) there is somewhere at least a bit of $\overset{(c)}{M}$ and, or, a bit of $\overset{(c)}{Q}$. A little "kinetic gravitational-wave energy" $\overset{(c)}{M}$ or a little ordinary mass-energy $\overset{(c)}{Q}$ (no matter how little), or both, suffice.

5.4 [THE FINALLY ADJUSTED INITIAL-VALUE DATA FOR THE DYNAMICS OF GEOMETRY]

Once we have found the conformal scale factor $\psi(x, y, z)$, we have in hand every item that we need to fit together consistently all the parts of the initial-value data. We have all we need to say "done" to the task of adjusting the initial-value data into conformity. We have all we need to assure ourselves that we have got all of the starting physics to fall into place, side by side, harmoniously. Specifically, we get every one of the handles that we need to steer the dynamics of the geometry:

(1) The actual **three-metric** itself—and therefore the geometrodynamic field coordinate and electric-like part of the gravity field—by multiplying the conformal metric by ψ^4 (eq. 5.2.6);

(2) The actual **density of mass-energy** from the initially prescribed conformal density (or from $\overset{(c)}{Q}/16\pi$) by multiplication by ψ^{-8} (eq. 5.2.19);

(3) The actual **current of mass-energy** from the initially prescribed conformal current by multiplication by ψ^{-10} (eq. 5.2.20);

(4) The actual **total distortion**—the descriptor of the gravitomagnetic field, and itself the measure of the geometrodynamic field momentum—by scaling the value of the total conformal distortion as it appears in expressions (5.3.7), and (5.3.8), thus,

$$A^{ab}_{\text{total}} = \psi^{-10} \overset{(c)}{A}{}^{ab}_{\text{total}}. \tag{5.4.1}$$

(5) The **extrinsic curvature** of the $T = \operatorname{Tr} \boldsymbol{K} =$ constant initial-value hypersurface, telling how that hypersurface is curved with respect to, and is to be embedded into, the totally unique (unique apart from mere coordinate transformations) but yet-to-be-constructed, enveloping four-geometry;

$$K_{ij} = A_{\text{total}\, ij} + (g_{ij}/3) \operatorname{Tr} \boldsymbol{K}. \tag{5.4.2}$$

What we have done can be put into other terms. Over the entire spacelike hypersurface we now stand in possession of a consistent set of values for the four fields, $\rho, \boldsymbol{j}, \boldsymbol{g}$, and $\boldsymbol{K}$, one of them scalar in character, one a vector field, and two tensorial. The **evolution equations,** consequence of the Einstein field equation, are ready to take hold at this point. They give us four more fields, of the same tensorial orders, qualitatively of the character of $d\rho/dt$, $d\boldsymbol{j}/dt$, $d\boldsymbol{g}/dt$, and $d\boldsymbol{K}/dt$—where by "t" we mean, of course, the York time T. From them we evaluate the four fields themselves at the next time slice, and so on, far into the future. These evolution equations and the metric between two three-dimensional spacelike hypersurfaces may be usefully written[27,94] in terms of a **lapse function**, $\alpha(t, x, y, z) = (-g^{00})^{-\frac{1}{2}}$, determining the lapse of proper time between the two hypersurfaces, the three-metric g_{ij}, determining distances on the initial hypersurface, and a **shift function**, $\beta_i = g_{0i}$:

$$ds^2 = -\alpha^2 dt^2 + g_{ij}(dx^i + \beta^i dt)(dx^j + \beta^j dt).$$

5.5 [THE DYNAMICS OF GEOMETRY]

A few of the words in the foregoing account, and ways of describing the physics, may be new, but to steer and to follow the dynamics of geometry is not new. We have only to look at the physics on any $\mathrm{Tr}\,\boldsymbol{K} =$ constant slice through any exact solution of Einstein's field equation to see a standing demonstration of handles already harmonized to steer!

Moreover, any general relativity calculation of the dynamics of geometry, whether it deals with astrophysics or gravitational waves or classical cosmology, and whether it translates the field equation into difference equations or Regge calculus[111] or null-strut calculus, has to start off with initial-value data.[112] Furthermore, those data—if they make sense at all—can, must, and do satisfy (whether we recognize it or not) the four initial-value equations—equations which in simplest language tell us that "**mass there rules geometry here**."

In the numerical integration each step forward in time leads to a new spacelike hypersurface and to what can be described as new initial-value data. To check that these data in their turn satisfy the initial-value equations, that density and current as well as intrinsic and extrinsic geometry are again in total accord, is a test of consistency by now so frequently applied as to have become almost obligatory.

The books edited by Centrella, LeBlanc and Bowers,[61] by Smarr,[62] and by Centrella[63] present more than a half-dozen beautiful examples of the geometrodynamic engine, steered by mutually adjusted initial-value data, marching ahead, steering-checked computer time step by steering-checked computer time step, to predictions of real physical interest.

In summary, the scalar equation for the conformal scale factor ψ, plus the associated vector equation for the three-vector shift field $\boldsymbol{W}$, constitute the steering machinery.

What are we doing when we solve these *initial-value equations*? We are adjusting initial-value data into harmony. We are allowing for the influence of mass there on geometry here. We are securing a consistent starting point so that we can predict everywhere and for all time to come the evolution of every field and the motion of every particle. By so doing, we are also automatically defining all that we mean by inertia, all that we can mean by inertia within the framework of Einstein's standard geometrodynamics. This we do in the very act of figuring the **spacetime geometry** in its entirety: to know the geometry is to know the **geodesics, and the local inertial frames** along them, and to know the geodesics and the local inertial frames is to know the essence of **inertia**.

5.6 [FURTHER PERSPECTIVES ON THE CONNECTION BETWEEN MASS-ENERGY THERE AND INERTIA HERE]

5.6.1. [The Effective Mass-Energy of a Gravitational Wave]

Energy of a gravitational wave? Of course, in standard general relativity there is no covariant way whatsoever to define a truly local density of gravitational-wave energy (see §§ 2.7 and 2.10). The reasoning is simple. We make whatever choice of place we please, and whatever choice of local free-float frame of reference at that point, nevertheless, because of the equivalence principle, we will find at that point and in that frame no such thing as gravity. A local density of gravitational-wave energy? No.

A nonlocal effective density of gravitational-wave energy? Yes. We know straight out of at least five deductions what that concept means and how it works from standard Einstein geometrodynamics (see § 2.10).

(1) The predicted response of a detector to a gravitational wave.[113,114]
(2) The observations by Joseph Taylor and his collaborators of the famous binary pulsar (Hulse and Taylor 1975; see § 3.5.1 and also § 2.10).[115] They show with ever greater precision that the system loses energy at the rate predicted[116] by outward transport through gravitational waves.[51]
(3) The gravitational geon[117] is a special example of a gravitational wave (§ 2.10). It holds itself together by its own gravitational attraction for a time long compared with any natural period of the system. It manifests externally all the attraction exerted by any everyday mass despite the fact that nowhere within or outside it is any "real" mass to be seen.
(4) The concept of an effective density of gravitational-wave energy makes sense[44,45,118,148] over regions small compared to the total extent of the wave yet large compared to a single wavelength (§ 2.10).
(5) The head-on encounter of two gravitons creates matter out of the emptiness of space by the standard Ivanenko process.[119]

$$2h\nu_{\text{graviton}} \longrightarrow e^{+} + e^{-}. \tag{5.6.1}$$

The system has mass-energy after the encounter. Consequently it must have possessed mass-energy before the encounter. The gravitons here—and, more generally, all gravitational waves—carry effective energy.

How are we to take account of the pulling power of this effective gravitational-wave energy in predicting what will happen to the system under study as time goes on? Do we make special provision for it, add a source term to Einstein's field equation, include this effective density in the quantity $\rho(x, y, z)$ of equation (5.2.19)? No. To do so would be a total mistake. It would be attempting to count again—and to count wrong—what has already been counted right: the effective energy of all gravitational waves. The field-coordinate part of those waves, the analogue of the magnetic part of an electromagnetic wave, has its full effect already included in the three-geometry. Their field-momentum part, the analogue of the electric part of an electromagnetic wave, has its full effect included in the distortion tensor and in the square of that tensor as it appears in the initial-value equation (5.3.9) for the scalar $\psi(x, y, z)$.

Considerations of physics and mathematics alike require that the effective mass-energy of gravitational waves must make itself felt on the spacetime geometry—and therefore on the gyro-defined local inertial frame of reference—on the same level as matter itself. Therefore, nothing would do more to clarify the inertia-influencing effect of a gravitational wave on the local frame than to derive a measure for this effect that is comparable in its ease of interpretability to the measures we have in expressions (5.7.2) and (5.7.3) for the effect of localized mass and localized angular momentum.

5.6.2. [No Inertia in the Environs of a Point Mass Located in a Model Universe That Contains No Other Mass?]

There have been many misunderstandings over the years of the idea which we have abbreviated as "Mass there rules inertia here." As far back as 1917 Kretschmann proposed a flat-space account of inertia.[120] He—and before him Boltzmann in the mid-1890s[121]—supporting the position of Leibniz, Berkeley, and Mach (see chap. 7)—took as basic notions a set of point masses and the network of Euclidean distances between those masses. Only one mass? Then no distances are available to define its location and consequently there is no inertia.

The idea that, "no other masses, then no inertia" transformed itself in the hands of others[16,122] to the view that no solution of Einstein's field equation can give a reasonable account of inertia in a model universe in which the geometry nowhere displays any identifiable center of mass. If there are no centers present other than the test particle whose motion is at issue, then there is no way to measure distances and therefore no way to define inertia!

Today we have outgrown the view that mass points are the be-all and end-all of physics. Fields are taken to be the primordial entities, and ranking high among them, geometry. Spacetime geometry, in and by itself, defines the location of the test mass, defines its velocity four-vector, its acceleration, the local inertial frames, and therefore its inertia—without search for "material masses" to serve for markers.

Gravitational waves alone suffice to curve up one and another model universe into closure,[123–125] an effect exhibited in simplest form in various mixmaster or Bianchi type IX models (see chap. 4),[126,127] of which the Taub model universe[128] is the simplest. There is nothing about the inertia of a test mass in such a model universe that differs from inertia in a universe occupied by stars, planets, and atoms.

The effective mass-energy of a gravitational wave, there, ranks no lower in the operation of the initial-value equations than any other source of *mass energy, there*, in fixing *spacetime geometry, here*, and therefore *inertia, here*.

To think of material mass as the ultimate primordial entity and spacetime geometry as totally derivative from that mass is completely wrong.

5.6.3. [An Integral Equation to Give Spacetime Geometry, and Therefore Inertia, Here and Now, in Terms of the Density and Flow of Material Mass, There and Then?]

We can best probe the very meaning of the question by stating the corresponding question in electrodynamics, that ever-useful guide to problems in geometrodynamics. Express the electromagnetic field, here and now, in terms of the world lines of the charges, there and then? Lienard and Wiechert have told us how. Subsequently we learned how to put their result into extraordinarily simple language.[129]

Do we make everyday working use of this direct-action-at-a-distance formulation of electrodynamics? No, neither in analyzing wave guides, radio waves, nor particle accelerators. We treat an electromagnetic wave on the same footing of democratic equality whether it does or does not have an assignable origin in the identifiable charges. We do not ask whether it bubbles out from a Lienard-Wiechert action-at-a-distance potential or from "outside," from ectogenesis. It would create immense difficulties for electrodynamics to impose for every wave a distinction of parentage between legitimate and illegitimate.

Likewise for geometrodynamics. Past efforts[130–133] could not establish any appealing physical distinction between gravitational waves—and entire cosmologies—that do and that do not trace their parentage to matter. Let two counter-current gravitational waves come to a common focus, collide, and produce matter. Let the gravitational field of this matter be examined. What then is spacetime curvature of legitimate and illegitimate origin?

Even if we put aside this difficulty of principle, we face in geometrodynamics a difficult proposition in working with any proposed integral equation for spacetime geometry here and now in terms of the energy-momentum tensor associated with identifiable mass-energy, and only with it.

Integral, but integral over what? Over the past light cone is the simple answer in the case of a geometry as simple as Friedmann's. However, the generic model universe, according to Belinski, Khalatnikov, and Lifshitz,[134–140] is an ocean of heaving waves, of local oscillations in the geometry, oscillations of which the amplitudes, the phases, and the directions of the principal axes vary chaotically from point to point. To imagine what becomes of an influence function in such a geometry, it is enough to listen to the claps, the reverberations, and the rolling of the thunder from a single lightning stroke. Or to try to read back from the record of an underwater sound detector the nature of the source when the intervening sea is a maelstrom of gradients of temperature and salinity! Enough supplementary information, theory, and computer power will extort something, reliable or unreliable, out of the record of the thunder or the output of the sonar. However, no one will choose that route to knowledge who has in initial-value data and dynamic equations a simpler way to follow what is going on—simpler because the initial-value equations operate on a spacelike hypersurface rather than on the past light "cone."

We forego here any integral over the past light "cone" in favor of **initial-value data on a spacelike hypersurface plus evolution by dynamic equation.**

5.7 [POOR MAN'S ACCOUNT OF INERTIAL FRAME]

A prescription. A prescription that is always definite and sometimes simple. This prescription—provided by the initial-value equations, vector plus scalar—let us figure spacetime here, and therefore inertia here, in terms of mass-energy there. It is definite because the equations are elliptic, definite because they shake down and harmonize any acceptable preliminary data on geometry and distribution of mass-energy, definite because they yield one and only one final shaken-down self-consistent set of data. Physics at each point depends in principle, however, on conditions at every other point on the initial value hypersurface. Correct physics, yes: but complicated physics, too.

Quicker, simpler and more illuminating than this global analysis is what we can only describe as inertial frame via a "quasi-local" system.

A "quasi-local" system makes sense when we deal with the bending of light by the mass of the Sun. The "quasi-local" system is a region large enough for the passing light to have settled down to a well-defined deflection angle but not so large that bending by other stars or the general field of the galaxy comes into play. It is a region, therefore, which counts as asymptotically flat at the relevant distances. On a far grander scale there must, of course, be some very small effective average overall curvature if the cosmos is to close. This global curvature we neglect in the scheme of the "quasi-local" system.

This quasi-local analysis does not make sense in the predicted final phase of collapse of the generic big bang to big crunch model universe. At that time, Belinski, Khalatnikov, and Lifshitz tell us,[134–140] space geometry must be loaded with a wild chaos of gravitational waves. This "quasi-local" scheme does make sense as a way to treat the bending of light by the Sun, the delay of light in passing through the gravity field of the Sun and the motion of Earth in orbit around the Sun; like "quasi-local" considerations apply to the effect of Earth on a satellite and of a black hole on the surrounding accretion disk. In any problem so simple, no one will bother with the full-blown formalism of the elliptic equations of the initial-value problem. The sensible approach is well known. We adopt the standard expression[141] for the weak field geometry far from a rotating object of mass M and of angular momentum J, expression (2.6.39) with $Q = 0$, in Boyer-Lindquist coordinates:

$$ds^2 = -\left(1 - 2\frac{M}{r}\right) dt^2 + \left(1 + 2\frac{M}{r}\right) dr^2 + r^2(d\theta^2 + \sin^2\theta d\phi^2) - 4\frac{J}{r}\sin^2\theta d\phi\, dt. \tag{5.7.1}$$

In equation (5.7.1) we see the metric, therefore the spacetime geometry, therefore the foundation for all that we can know and say about inertia. Cleanly separated in this expression are the Minkowski geometry of the outlying regions of the "quasi-local" system and the inertia-determining potential of the rotating mass in question. Moreover, the linearity of expression (5.7.1) in the mass M assures us that we can add like terms to expression (5.7.1) to evaluate the geometry and the geodesics and—in the same weak field approximation—the inertia as influenced by a number of masses M_i.

We can summarize these considerations in a "poor man's prescription" for evaluating the inertia-determining power of a number of slowly moving and rotating masses. In the "quasi-local" system we assign to each mass M_i at its distance r_i the "voting power" or effectiveness, or the inertial-frame–determining potential given by the dimensionless ratio M_i/r_i. By this phraseology we mean that every geodesic "deviates" from Minkowskian straightness by an amount given by the gradient of the Sciama sum for inertia,[14,16]

$$(\text{"sum for inertia"}) = \sum_i M_i/r_i. \tag{5.7.2}$$

Moreover, the local gyro triad turns with respect to an asymptotic inertial frame at a rate given—to the same level of approximation—by the sum of *Thirring-Lense contributions* (see § 6.1)[142,143] from all the individual concentrations of angular momentum: thus,

$$\begin{aligned}(\text{angular velocity of gyro triad}) &= \sum \nabla \times \{(\boldsymbol{J} \times \boldsymbol{r})/r^3\} \\ &= \sum \{-\boldsymbol{J} + 3(\boldsymbol{J}\cdot\boldsymbol{r})\boldsymbol{r}/r^2\}/r^3.\end{aligned} \tag{5.7.3}$$

In the interest of simplicity we have omitted the subscript "i" that should appear on each symbol in formula (5.7.3). Normally no more than one angular momentum J contributes to the precession.

The gyro-frame dragging can alternatively be viewed as an aspect of the Fermi-Walker transport. The metric of the spacetime in which this transport takes place differs from that seen in equation (5.7.1) only in this, that the departure from ideal Minkowskian geometry arises, not from a single mass and angular momentum at a single distance, but from many masses and many angular momenta at a variety of distances.

5.8 [A SUMMARY OF ENERGY THERE RULING INERTIA HERE]

One of Einstein's great contributions to the understanding of the universe, following an intuition by Riemann, was to show that the spacetime geometry is not flat, as in special relativity, and that its curvature is determined, via the field equation, by the distribution and motion of mass-energy in the universe. For this reason, we have called Einstein general relativity "geometrodynamics."[22]

If we give a strong interpretation to this idea of the geometry of the universe being dynamical, we may conjecture that the whole spacetime geometry be dynamical and be determined *only* by the distribution and motion of mass-energy in the universe; that is, we may require that in the universe there be no part of the metric $\boldsymbol{g}$ that is unaffected by mass-energy. In other words, the metric $\boldsymbol{g}$ does not contain "absolute geometric elements" that are independent of the distribution and motion of mass-energy in the universe. An equivalent way of stating this requirement is to assume that there is no prior geometry in the universe, recalling that prior geometry is defined as that part of the spacetime geometry that is not dynamical.[51] However, the boundary condition that the universe be asymptotically flat, that is, $\boldsymbol{g} \underset{\infty}{\longrightarrow} \boldsymbol{\eta}$, gives a part of $\boldsymbol{g}$, the asymptotic Minkowski metric $\boldsymbol{\eta}$, that is nondynamical and *in a sense* is a kind of prior geometry independent of the energy content in the universe. Therefore, if we wish to avoid the existence of every kind of prior geometry we should also avoid the boundary condition of asymptotic flatness of the metric $\boldsymbol{g}$. We might then conjecture that the **universe** is **spatially compact**. Of course, we may also have a spatially noncompact model universe, without asymptotically flat geometry $\boldsymbol{\eta}$, for example a Friedmann model universe with hyperbolic spatial curvature.

To solve for the spacetime geometry in a spatially closed universe, we need to give proper initial conditions on a closed spacelike Cauchy surface Σ and then solve the "**Cauchy problem**"[5–8] *in Einstein geometrodynamics* (see fig. 5.1).

Nevertheless, it is necessary to observe that, in order to solve the **initial-value problem** in geometrodynamics, the spacelike hypersurface Σ does not necessarily need to be compact, but for example can be homeomorphic, that is, topologically equivalent, to $\Re^3$ and asymptotically flat; however, as we just observed, such model universes have some kind of prior geometry, $\boldsymbol{\eta}$. They do

(3) FIND: The spacetime geometry $g_{\alpha\beta}$, the evolution of matter and fields and the local inertial frames (where $g_{\alpha\beta} \to \eta_{\alpha\beta}$).

(2) SOLVE: The initial value equations and the dynamical equations

"York" time $\equiv \mathrm{Tr}\, K_{ij}$

Closed spacelike Cauchy hypersurface S, defined by:

$\mathrm{Tr}\, K_{ij} = $ constant

(1) GIVEN on S

$\overset{(c)}{g}_{ik} = $ conformal spatial metric
$\overset{(c)}{A}{}^{ik}_{\mathrm{free}} = $ conformal, transverse part of distortion tensor
$\overset{(c)}{\rho} = $ conformal mass-energy density
$\overset{(c)}{j}{}^{i} = $ conformal mass-energy current
plus equations of state

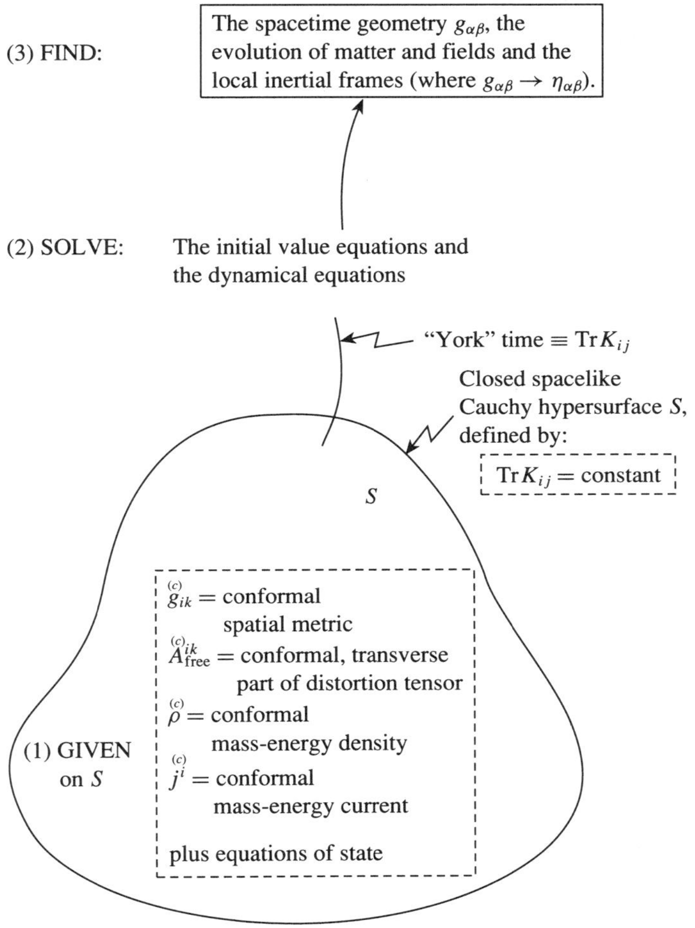

FIGURE 5.1. The initial-value problem in Einstein geometrodynamics (case of closed Cauchy surface).

not agree with the views of Einstein and others[11,21,26] on the origin of inertia in geometrodynamics.

Part of the initial conditions to be fixed on Σ are conditions on the distribution and the motion of mass-energy on Σ, **"mass-energy there."** Therefore, once the mass-energy distribution and motion on Σ are given, and once the initial-value and Cauchy problems of geometrodynamics are solved, we fully know the **spacetime geometry g** of the universe.

Once the geometry is known, the **local inertial frames** are determined at each event of the spacetime, thereby determining **"inertia here."** These are the

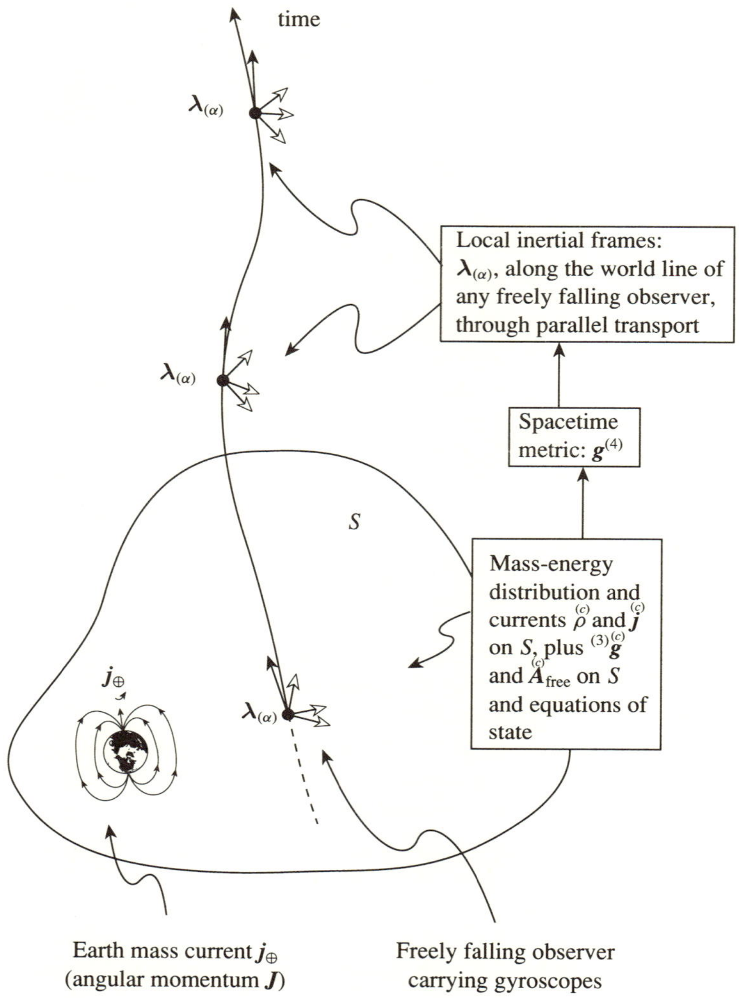

FIGURE 5.2. Initial-value problem, gravitomagnetism, and dragging of inertial frames.

frames where the metric g becomes at the event the Minkowski metric η. The *nonrotating local inertial frames* are determined all along the world line of a freely falling test particle, once the metric g is known, through the Fermi-Walker[144] transport, that is, for a freely falling test particle, through parallel transport of their spatial axes.

Summarizing, we may distinguish between two possibilities.

If one conjectures that the whole spacetime geometry is fully dynamical, one should avoid the boundary condition of asymptotic flatness of $\boldsymbol{g}$. A way to satisfy this requirement is to consider spatially compact model universes. In chapter 4 we have called this conjecture of a spatially compact universe, and the corresponding interpretation of the initial-value problem, the medium strong "interpretation of the origin of inertia" in general relativity.

The other possibility is that the universe is spatially noncompact, and yet the axes of the local inertial frames ("local inertia") are influenced by the mass-energy currents in the universe, $j^k = \varepsilon u^k$, in agreement with the Einstein field equation. We may call this description of the origin of "local inertia," depending on the mass-energy currents in agreement with the initial-value formulation, the weak "interpretation of the origin of inertia" in general relativity.

In both cases the proper initial conditions on a spacelike Cauchy surface Σ and the initial conditions on the mass-energy distribution and motion (conformal) on Σ, "*mass-energy there*," plus the Einstein field equation, plus the equations of state for matter and fields, completely determine[5–8] the geometry of the spacetime $\boldsymbol{g}$, and then $\boldsymbol{g}$ determines the local inertial frames at each event of the spacetime "*inertia here*."

Therefore, the "*dragging of inertial frames*," that is, the dragging of the spatial axes of the local inertial frames due to mass-energy currents is a manifestation of the way "local inertia" arises in Einstein geometrodynamics, that is, evidences that the local inertial frames are determined (in the case of spatially compact universes) or at least partially determined and influenced (in the case of quasi-Minkowskian universes) by the distribution and motion of mass and energy in the universe: "**inertia here is ruled by mass there**."

In the next chapter we shall describe in detail dragging of inertial frames and *gravitomagnetism*, which may be thought of as a manifestation of some weak general relativistic interpretation of the Mach principle; their measurement would provide experimental foundation for this general relativistic interpretation of the origin of inertia.

5.9 [A SUMMARY OF THE INITIAL-VALUE PROBLEM AND DRAGGING OF INERTIAL FRAMES]

Let us briefly summarize the initial-value problem, which we have described in the previous sections, and its relations with the dragging of inertial frames.

The dragging of inertial frames may be thought of as a consequence of the way local inertia (local inertial frames) arises in Einstein geometrodynamics, influenced not only by the mass-energy distribution ρ, but also by the mass-energy currents j^i in the universe.

In this chapter we have assumed a model universe that is **spatially compact** and *spatially without boundary* (*spatially closed*). However, as we have observed, the initial-value and Cauchy problems may also be solved in a model universe that is spatially noncompact, or asymptotically flat. In this case, local inertia is still influenced not only by the mass-energy distribution ρ but also by the mass-energy currents j^i in the universe. Nevertheless, in the asymptotically flat case, there is a part of the geometry, the asymptotic Minkowski metric $\eta_{\alpha\beta}$, that is unaffected by the distribution and motion of mass-energy in the universe and not dynamical in origin.

We have assumed that the spacetime M, in which we solve the Cauchy problem to find the spacetime geometry, is a Lorentzian manifold[145,146] with metric $g_{\alpha\beta}$ and that the spacetime M has embedded in it a closed spacelike Cauchy surface S, that is, a closed submanifold such that each causal path without endpoints intersects it once and only once. Then:

(1) In M there are no closed (or "almost closed") causal paths (no causality violation, such as there is in the Gödel universe with closed timelike curves; see § 4.6).
(2) The Cauchy surface S can be described by a universal time function (a function with gradient everywhere timelike).
(3) The spacetime is topologically equivalent (homeomorphic) to a product $S \times \mathfrak{R}$ ($\mathfrak{R}$ is the real line).

For simplicity, to decouple the initial-value Einstein field equation,[5] we have assumed that the closed spacelike Cauchy surface S is determined by the condition $\mathrm{Tr}\,\boldsymbol{K} \equiv$ "*York time*" $=$ constant, that is, "**trace of the extrinsic curvature**" $=$ constant, or the hypersurface S has constant mean extrinsic curvature. The extrinsic curvature tensor K_{ik} represents the fractional rate of change and deformation with time of the three-geometry of a spacelike hypersurface. Using the timelike unit vector n^i normal to a spacelike hypersurface, if $x^0 =$ constant on it, one can write the components of $\boldsymbol{K}$ as $K_{ik} = -n_{(i;k)}$ where ";" is the covariant derivative relative to the four-geometry.

In order to solve the Cauchy problem for the Einstein field equation, we take the quantities to be fixed on S, to find the spacetime geometry $\boldsymbol{g}$, to be the following:

(1) The **conformal three-geometry** of the hypersurface S:$\overset{(c)}{g}_{ij}(x, y, z)$, that is, the metric $g_{ij}(x, y, z)$ of S up to a scale factor $\psi^{-4}(x, y, z)$:

$$\overset{(c)}{g}_{ij} = \psi^{-4} g_{ij}. \tag{5.9.1}$$

(2) A second-order, symmetric, traceless, transverse (null covariant divergence with respect to $\overset{(c)}{g}_{ij}$) tensor: $\overset{(c)}{A}{}^{ab}_{\text{free}}(x, y, z)$ on S, called the **conformal free distortion tensor** (§ 5.2.7). This tensor may be obtained

by decomposing, with the technique[103] of section 5.2.7, any second-order, symmetric tensor into a combination of tensor, vector, and scalar quantities. $\overset{(c)}{A}{}^{ab}_{\text{free}}$ is related to[1] the "rate of deformation with time of the conformal three-geometry."

(3) For a fluid, the **"conformal" mass-energy density** on S: $\overset{(c)}{\rho}(x, y, z)$, that is, the mass-energy density $\rho(x, y, z)$ on S up to a power of the conformal factor $\psi(x, y, z)$. We write the conformal energy density:

$$\overset{(c)}{\rho} = \psi^8 \rho. \tag{5.9.2}$$

(4) For a fluid, the **"conformal" momentum density** or **"conformal" current density** vector $\overset{(c)}{j}{}^i(x, y, z)$, that is, the momentum density $j^i(x, y, z)$ on S up to a power of the conformal factor $\psi(x, y, z)$. We write the conformal current density vector:

$$\overset{(c)}{j}{}^i = \psi^{10} j^i. \tag{5.9.3}$$

These four quantities: $\overset{(c)}{g}_{ij}$, $\overset{(c)}{A}{}^{ij}_{\text{free}}$, $\overset{(c)}{\rho}$ and $\overset{(c)}{j}{}^i$, to be specified on S are, under suitable mathematical conditions (see § 5.3),[5–8] necessary and sufficient in order to solve for $g_{ij}(x, y, z)$ and $K_{ij}(x, y, z)$ on S.

The solution proceeds as follows. Consider the special case, $\text{Tr}\,\boldsymbol{K}$ = constant, from four of the ten components of the Einstein field equation: $G_{\alpha\beta} \equiv R_{\alpha\beta} - \frac{1}{2} R g_{\alpha\beta} = 8\pi T_{\alpha\beta}$, $(G \equiv c \equiv 1)$, we get the **initial-value equations** to be solved (eqs. 5.3.1 and 5.3.9)

$$\overset{(c)}{\Delta}{}^* W_k = 8\pi \overset{(c)}{j}_k \tag{5.9.4}$$

and

$$8\overset{(c)}{\Delta}\psi - {}^{(3)}\overset{(c)}{R}\psi + \overset{(c)}{M}\psi^{-7} + \overset{(c)}{Q}\psi^{-3} - \frac{2}{3}(\text{Tr}\,\boldsymbol{K})^2\psi^5 = 0 \tag{5.9.5}$$

where $\overset{(c)}{\Delta}{}^* W_k \equiv \overset{(c)}{\nabla}_m \big(\overset{(c)}{\nabla}_k W^m + \overset{(c)}{\nabla}{}^m W_k - \frac{2}{3}\delta^m{}_k \overset{(c)}{\nabla}_n W^n\big)$; where $\overset{(c)}{\nabla}_m$ is the covariant derivative with respect to $\overset{(c)}{g}{}^{ij}$; where $\overset{(c)}{\Delta}\psi = \overset{(c)}{g}{}^{-1/2} \frac{\partial}{\partial x^a}\big[\overset{(c)}{g}{}^{1/2}\, \overset{(c)}{g}{}^{ab} \frac{\partial}{\partial x^b}\big]\psi$; where ${}^{(3)}\overset{(c)}{R}$ = conformal Ricci scalar three-curvature of S; and where $\overset{(c)}{Q} = 16\pi\overset{(c)}{\rho}$ and $\overset{(c)}{M}$ is the "square" of the total conformal distortion tensor $\overset{(c)}{A}{}^{ij}_{\text{total}}$ (see formulae 5.3.7 and 5.3.8): $\overset{(c)}{M} = \overset{(c)}{A}_{\text{total}\,ij} \overset{(c)}{A}{}^{ij}_{\text{total}}$.

We first solve equation (5.9.4) for W_k (the solution exists and is unique, up to conformal Killing vectors, when the conditions 5.3.6 of § 5.3.2 are satisfied). Once we have W_k, we find the second-order, symmetric, traceless tensor $\overset{(c)}{A}{}^{ab}_{\text{total}}$, the total conformal distortion tensor, defined by the formula

$$\overset{(c)}{A}{}^{ab}_{\text{total}} = \overset{(c)}{A}{}^{ab}_{\text{free}} + \overset{(c)}{\nabla}{}^a W^b + \overset{(c)}{\nabla}{}^b W^a - \frac{2}{3}\overset{(c)}{g}{}^{ab}\overset{(c)}{\nabla}_m W^m. \tag{5.9.6}$$

$\overset{(c)}{M}$ is now known by "squaring" $\overset{(c)}{A}{}^{ab}_{\text{total}}$.

We can then solve the initial-value (eq. 5.9.5) for the conformal factor ψ (the solution exists and is unique[5–8] when the conditions explained in § 5.3.4 are satisfied).

Knowing $\overset{(c)}{g}{}^{ab}$, $\overset{(c)}{A}{}^{ab}_{\text{total}}$ and ψ we have $A^{ab}_{\text{total}} = \psi^{-10}\overset{(c)}{A}{}^{ab}_{\text{total}}$, $g^{ab} = \psi^{-4}\overset{(c)}{g}{}^{ab}$ and the extrinsic curvature $K^{ij}(x, y, z)$ on S:

$$K^{ij} = A^{ij}_{\text{total}} + \frac{1}{3} g^{ij}\, \text{Tr}\, \boldsymbol{K}. \tag{5.9.7}$$

Then, we have a consistent set $\{g^{ij}, K^{ij}, \rho, j^i\}$ on S, and we can solve the remaining *dynamical* components of the Einstein field *equation and the equations of motion* $T^{\alpha\beta}{}_{;\beta} = 0$ (plus equations of state for matter and fields), representing the "time derivatives" of g^{ij}, K^{ij}, ρ, and j^i, and finally find the evolution of matter and fields and the spacetime geometry $g_{\alpha\beta}$. Existence, uniqueness, and stability of the solution of this Cauchy problem have been proved under suitable mathematical regularity conditions.[5–8] For a discussion on the relations between observable physical quantities and initial data on a Cauchy surface see reference 147.

In summary, once specified the initial data on a Cauchy spacelike surface, among which the mass-energy currents j^i, we can solve for the **spacetime geometry** $g_{\alpha\beta}$ (fig. 5.1) and therefore determine the nonrotating **local inertial frames** (in which $g_{\alpha\beta} \longrightarrow \eta_{\alpha\beta}$) along the world line of any freely falling observer in the universe through parallel transport (fig. 5.2).

The axes of the local inertial frames (gyroscopes) are therefore influenced and in part determined by the mass-energy currents in the universe. Formulae (6.1.28) and (6.1.33) in the next chapter describe the gravitomagnetic dragging of inertial frames and display a special case of this concept, in which the mass-energy current is due to the rotation of a central body (fig. 5.2). If a model universe is spatially closed, the whole spacetime geometry $g_{\alpha\beta}$ is dynamical in the sense that there is no part of $g_{\alpha\beta}$ independent of the distribution and currents of the mass-energy in the universe (in contrast with the nondynamical Minkowski metric $\eta_{\alpha\beta}$ in asymptotically flat model universes).

REFERENCES CHAPTER 5

1. J. Isenberg and J.A. Wheeler, Inertia here is fixed by mass-energy there in every W model universe, in *Relativity, Quanta, and Cosmology in the Development of the Scientific Thought of Albert Einstein*, ed. M. Pantaleo and F. de Finis (Johnson Reprint Corporation, New York, 1979), vol. 1, 267–93.
2. J. A. Isenberg, Wheeler-Einstein-Mach space-times, *Phys. Rev. D* 24:251–56 (1981).

3. J.W. York, Jr., Role of conformal three-geometry in the dynamics of gravitation, *Phys. Rev. Lett.* 28:1082–85 (1972).
4. N. O'Murchadha and J.W. York, Jr., Existence and uniqueness of solutions of the Hamiltonian constraint of general relativity, *J. Math. Phys.* 14:1551–57 (1973).
5. J.W. York, Jr., Kinematics and dynamics of general relativity, in *Sources of Gravitational Radiation: Proc. Battelle Seattle Workshop*, 24 July–4 August 1978, ed. L.L. Smarr (Cambridge University Press, Cambridge, 1979), 83–126.
6. Y. Choquet-Bruhat and J.W. York, Jr., The Cauchy problem, in *General Relativity and Gravitation*, ed. A. Held (Plenum, New York, 1980), 99–172.
7. N.O. Murchadha and J.W. York, Jr., The initial-value problem of general relativity. I. *Phys. Rev. D* 10:428–36 (1974).
8. J.W. York, Jr., The initial-value problem and dynamics, in *Gravitational Radiation*, Les Houches, 1982, ed. N. Deruelle and T. Piran (North-Holland, Amsterdam, 1983), 175–201.
9. E. Mach, *Die Mechanik in ihrer Entwicklung historisch-kritisch dargestellt* (Brockhaus, Leipzig, 1912); trans. T.J. McCormack with an introduction by Karl Menger as *The Science of Mechanics* (Open Court, La Salle, IL, 1960).
10. A. Einstein, letter to Ernst Mach, Zurich, 25 June 1913, in ref. 27, 544–45.
11. A. Einstein, *The Meaning of Relativity*, 3d ed. (Princeton University Press, Princeton, 1950).
12. H. Weyl, *Was ist Materie?* (Springer, Berlin, 1924).
13. H. Bondi, The problem of inertia, in *Cosmology* (Cambridge University Press, Cambridge, 1952), 27–33.
14. D.W. Sciama, Inertia, *Sci. Am.* 196:99–109 (1957).
15. R.H. Dicke, *The Theoretical Significance of Experimental Relativity* (Gordon and Breach, New York, 1964); see also Experimental relativity, in *Relativity, Groups and Topology*, ed. C. DeWitt and B.S. DeWitt (Gordon and Breach, New York, 1964), 165–316.
16. D.W. Sciama, *Physical Foundations of General Relativity* (Doubleday, London, 1969).
17. H. Goenner, Mach's principle and Einstein's theory of gravitation, in *Boston Studies in the Philosophy of Science* (Reidel, Dordrecht, 1970), vol. 6, 200–215.
18. M. Reinhardt, Mach's principle: A critical review, *Z. Naturforsch.* 28(a): 529–37 (1972).
19. D.J. Raine and M. Heller, *The Science of Space-Time* (Pachart Publishing House, Tucson, AZ, 1981).
20. J.E. Marsden, The initial-value problem and the dynamics of gravitational fields, in *Proc. 9th Int. Conf. on General Relativity and Gravitation*, ed. E. Schmützer (VEB Deutscher Verlag der Wissenschaften, Berlin, 1983), 115–26.
21. J.A. Wheeler, View that the distribution of mass and energy determines the metric, in *Onzième Conseil de Physique Solvay: La Structure et l'évolution de l'univers* (Éditions Stoops, Brussels, 1959), 97–141.

22. J. A. Wheeler, *Geometrodynamics, Topics of Modern Physics* (Academic Press, New York, 1962), vol. 1; expanded from *Rendiconti della Scuola Internazionale di Fisica Enrico Fermi XI Corso*, July 1959 (Zanichelli, Bologna 1960).
23. H. Hönl and H. Dehnen, Über Machsche und anti-Machsche Lösungen der Feldgleichungen der Gravitation. I., *Ann. Physik* 11:201–15 (1963).
24. H. Hönl and H. Dehnen, Über Machsche und anti-Machsche Lösungen der Feldgleichungen der Gravitation. II., *Ann. Physik* 14:271–95 (1964).
25. J.A. Wheeler, Geometrodynamics and the issue of the final state, in *Relativity, Groups and Topology*, 1963 Les Houches Lectures, ed. C. DeWitt and B. DeWitt (Gordon and Breach, New York, 1964), 315–520.
26. J.A. Wheeler, Mach's principle as boundary condition for Einstein's equations, in *Gravitation and Relativity*, ed. H. Y. Chiu and W. F. Hoffman (Benjamin, New York, 1964), 65–89 and 303–49.
27. C.W. Misner, K.S. Thorne, and J.A. Wheeler, *Gravitation* (W. H. Freeman, San Francisco, 1973), chap. 21.
28. J.A. Isenberg, The construction of spacetimes from initial data, Ph. D. dissertation, University of Maryland, 1979.
29. A. Einstein, *Essays in Science* (Philosophical Library, New York, 1934), 54; trans. from *Mein Weltbild* (Querido, Amsterdam, 1933).
30. S.W. Hawking, Quantum cosmology, in *Relativity, Groups and Topology II*, ed. B. DeWitt and R. Stora (Elsevier, Amsterdam, 1984), 333–79.
31. J.B. Hartle, Quantum cosmology, in *High Energy Physics 1985: Proc. Yale Theoretical Advanced Study Institute*, ed. M.J. Bowick and F. Gursey (World Scientific, Singapore, 1986), 471–566.
32. M. Carfora, Initial data sets and the topology of closed three-manifolds in general relativity, *Nuovo Cimento B* 77:143–61 (1983).
33. J. A. Wheeler, From relativity to mutability, in *The Physicist's Conception of Nature*, ed. J. Mehra (Reidel, Dordrecht, 1973), 202–47 and fig. 1, p. 205.
34. See ref. 29, p. 68.
35. J. A. Wheeler, Superspace and the nature of quantum geometrodynamics, in *Battelle Rencontres: 1967 Lectures in Mathematics and Physics*, ed. C.M. De Witt and J.A. Wheeler (Benjamin, New York, 1968), 242–307; also appeared as Le superspace et la nature de la géometrodynamique quantique, in *Fluides et Champ Gravitationnel en Relativité Générale, No. 170, Colloques Internationaux* (Éditions de Centre National de La Recherche Scientifique, Paris, 1969), 257–322, see also ref. 36.
36. J.A. Wheeler, *Einstein's Vision: Wie steht es heute mit Einsteins Vision, alles als Geometrie aufzufassen* (Springer, Berlin, 1968).
37. J.A. Wheeler, On the nature of quantum geometrodynamics, *Ann. Phys.* 2:604–14 (1957).
38. R.A. Sussman, Conformal structure of a Schwarzschild black hole immersed in a Friedmann universe, *Gen. Rel. Grav.* 17:251–91 (1985).
39. C.W. Misner and J.A. Wheeler, Classical physics as geometry, *Ann. Phys.* 2:525–603 (1957); reprinted in ref. 21. The subject of harmonic fields is reviewed, and references are given to the literature.

40. W.G. Unruh, Dirac particles and geometrodynamical charge in curved geometries, Ph. D. dissertation, Princeton University, 1971.

41. See ref. 22, § 10.5.

42. G.F.R. Ellis and D.W. Sciama, Global and non-global problems in cosmology, in *General Relativity: Papers in Honour of J. Synge*, ed. L. O'Raifeartaigh (Oxford University Press, 1972), 35–59.

43. E. Fermi, Quantum theory of radiation, *Rev. Mod. Phys.* 4:87–132 (1932).

44. R.A. Isaacson, Gravitational radiation in the limit of high frequency. I. The linear approximation and geometrical optics, *Phys. Rev.* 166:1623–71 (1968).

45. R.A. Isaacson, Gravitational radiation in the limit of high frequency. II. Nonlinear terms and the effective stress tensor, *Phys. Rev.* 166:1272–80 (1968).

46. T.A. Barnebey, Gravitational waves: The nonlinearized theory, *Phys. Rev. D* 10:1741–48 (1974).

47. A. Einstein, Kosmologische Betrachtungen zur allgemeinen Relativitätstheorie, *Preuss. Akad. Wiss. Berlin, Sitzber.* 142–52 (1917).

48. G. Gamow, *My World Line* (Viking Press, New York, 1970), 44.

49. J. Narlikar, *The Structure of the Universe* (Oxford University Press, 1977); the Hoyle-Narlikar theory is reviewed herein. See also ref. 149.

50. C. Brans and R.H. Dicke, Mach's principle and a relativistic theory of gravitation, *Phys. Rev.* 124:925–35 (1961).

51. C.M. Will, *Theory and Experiment in Gravitational Physics*, rev. ed. (Cambridge University Press, 1993); the PPN formalism is reviewed herein.

52. H. Seifert and W. Threlfall, *Lehrbuch der Topologie* (Teubner, Leipzig, 1934); reprint (Chelsea, New York, 1947).

53. K. Kuchar, A bubble-time canonical formalism for geometrodynamics, *J. Math. Phys.* 13:768–81 (1972).

54. R.F. Baierlein, D.H. Sharp, and J.A. Wheeler, Three-dimensional geometry as carrier of information about time, *Phys. Rev.* 126:1864–1965 (1962).

55. J.A. Wheeler, Three-dimensional geometry as carrier of information about time, in *The Nature of Time*, ed. T. Gold (Cornell University Press, Ithaca, NY, 1967), 90–107.

56. S. Tomonaga, On a relativistically invariant formulation of the quantum theory of wave fields, *Prog. Theor. Phys.* 1:27–42 (1946).

57. J.A. Wheeler, Superspace, in *Analytic Methods in Mathematical Physics*, ed. R.P. Gilbert and R.G. Newton (Gordon and Breach, New York, 1970), 335–78.

58. F.J. Tipler and J.E. Marsden, Maximal hypersurfaces and foliations of constant mean curvature in general relativity, *Phys. Reports* 66:109–39 (1980).

59. F.J. Tipler and J.D. Barrow, Closed universes: Their future evolution and final state, *Mon. Not. Roy. Astron. Soc.* 216:395–402 (1985).

60. S.L. Shapiro and S.A. Teukolsky, Gravitational collapse to neutron stars and black holes: computer generation of spherical spacetimes, *Astrophys. J.* 235:199–215 (1980).

61. C.R. Evans, A method for numerical simulation of gravitational collapse and gravitational radiation generation, in *Numerical Astrophysics*, ed. J.M. Centrella, J.M. LeBlanc, and R.L. Bowers (Jones and Bartlett, Boston, 1985), 216–56.

62. L.L. Smarr, ed., *Sources of Gravitational Radiation, Proc. Battelle Seattle Workshop*, 24 July–4 August 1978 (Cambridge University Press, Cambridge, 1979).

63. J.M. Centrella, ed., *Dynamical Spacetimes and Numerical Relativity*, Proc. Workshop held at Drexel University, 7–11 October 1985 (Cambridge University Press, Cambridge, 1986).

64. A. Lichnerowicz, L'intégration des équations de la gravitation relativisté et le problème des n corps, *J. Math. Pures Appl.* 23:37–63 (1944).

65. Y. Choquet-Bruhat, Cauchy problem, in *Gravitation: An Introduction to Current Research*, ed. L. Witten (Wiley, New York, 1962), 138–68.

66. A. Lichnerowicz, *Relativistic Hydrodynamics and Magnetohydrodynamics: Lectures on the Existence of Solutions* (W. A. Benjamin, New York, 1967).

67. G. Darmois, Sur l'intégration locale des équation d'Einstein, *Comp. Rend. Acad. Sci. (Paris)* 176:646–48 (1923).

68. G. Darmois, Éléments de géométrie des espaces: Introduction aux théories de la relativité générale, *Ann. Physique* 1:5–87 (1924).

69. D.R. Brill, J.M. Cavallo, and J.A. Isenberg, K-surfaces in the Schwarzschild spacetime and the construction of lattice cosmologies, *J. Math. Phys.* 21 (12): 2789–96 (1980).

70. D.M. Eardley and L. Smarr, Time functions in numerical relativity: marginally bound dust collapse, *Phys. Rev. D* 19:2239–59 (1979).

71. A. Qadir and J.A. Wheeler, York's cosmic time versus proper time as relevant to changes in the dimensionless "constants," K-meson decay, and the unity of black hole and big crunch, in *From SU(3) to Gravity*, ed. E. Gotsman and G. Tauber (Cambridge University Press, Cambridge, 1985), 383–94.

72. J.D. Barrow and F.J. Tipler, *The Anthropic Cosmological Principle* (Oxford University Press, New York, 1986), chap. 10, section on global hyperbolicity and stress-energy.

73. See ref. 24, and R.W. Lindquist and J.A. Wheeler, Dynamics of a lattice universe by the Schwarzschild-cell method, *Rev. Mod. Phys.* 29:432–43 (1959).

74. G.M. Patton and J.A. Wheeler, Is physics legislated by cosmogony?, in *Quantum Gravity: An Oxford Symposium*, ed. C.J. Isham, R. Penrose, and D.W. Sciama (Clarendon Press, Oxford, 1975), 538–605.

75. J.A. Wheeler, World as system self-synthesized by quantum networking, *IBM J. Res. Dev.* 32:4–15 (1988).

76. J.D. Barrow and F.J. Tipler, Action principles in nature, *Nature* 331:31–34 (1988).

77. A. Guth, Inflationary universe: A possible solution to the horizon and flatness problems, *Phys. Rev. D* 23:347–56 (1981).

78. A. Linde, A new inflationary universe scenario: A possible solution of the horizon, flatness, homogeneity, isotropy and primordial monopole problems, *Phys. Lett. B* 114:393–431 (1982).

79. A. Linde, Coleman-Weinberg theory and the new inflationary universe scenario, *Phys. Lett. B* 114:431–35 (1982).

80. F.J. Dyson, Time without end: Physics and biology in an open universe, *Rev. Mod. Phys.* 51:447–60 (1979).

81. S. Frautschi, Entropy in an expanding universe, *Science* 217:593–99 (1982).

82. D. Brill, Maximal surfaces in closed and open spacetimes, in *Proc. 1st Marcel Grossmann Meeting on General Relativity*, Trieste, Italy ed. R. Ruffini (North-Holland, Amsterdam, 1977), 193–206.

83. See ref. 72, p. 678, and refs. 27 and 34.

84. F.J. Tipler, Do closed universes recollapse?, in *Proc. 13th Texas Symposium on Relativistic Astrophysics*, Chicago, 14–19 December 1986, ed. M.P. Ulmer (World Scientific, Singapore, 1987), 122–26.

85. See ref. 72, p. 678.

86. R.P. Feynman, The principle of least action in quantum mechanics, Ph.D. dissertation, Princeton University, 1942.

87. R.P. Feynman and A.R. Hibbs, *Quantum Mechanics and Path Integrals* (McGraw-Hill, New York, 1965).

88. D. Hilbert, Die Grundlagen der Physik, *Königl. Gesell. Wiss., Göttingen, Nachr., Math.-Phys. Kl.* 395–407 (1915); *Zweite Mitteilung* 53–76 (1917); English translation of key passages of both communications in ref. 27, 433–34.

89. É. Cartan, Sur les équations de la gravitation de Einstein, *J. Math. Pures Appl.* 1:141–203 (1922).

90. R. Debever, ed., *Élie Cartan–Albert Einstein Letters on Absolute Parallelism* (Princeton University Press, Princeton, 1979).

91. J.S. Ames and F.D. Murnaghan, *Theoretical Mechanics* (Ginn, Boston, 1929); reprint (Dover, New York, 1958).

92. K. Kuchar and J.W. York, Jr., Exercise 21.24: The extremal action associated with the Hilbert action principle depends on conformal three-geometry and extrinsic time, in C.W. Misner, K.S. Thorne, and J.A. Wheeler, *Gravitation* (W. H. Freeman, New York, 1973), 551.

93. J.W. York, Jr., Boundary Terms in the Action Principles of General Relativity, *Found. Phys.* 16:249–57 (1986).

94. R. Arnowitt, S. Deser, and C.W. Misner, The dynamics of general relativity, in *Gravitation*, ed. L. Witten (Wiley, New York, 1962), 227–65.

95. E.P. Belasco and H.C. O'Hanian, Initial conditions in general relativity: Lapse and shift formulation, *J. Math. Phys.* 10:1503–7 (1969).

96. D. Christodoulou and M. Francaviglia, Remarks about the thin sandwich conjecture, *Rep. Math. Phys.* 11:377–83 (1977).

97. J.A. Wheeler, Assessment of Everett's "relative state" formulation of quantum theory, *Rev. Mod. Phys.* 29:463–65 (1957).

98. J.B. Hartle and S.W. Hawking, Wave function of the universe, *Phys. Rev. D* 28:2960–75 (1983).

99. A. Vilenkin, Classical and spectrum cosmology of the Starobinsky inflationary model, *Phys. Rev. D* 32:2511–21 (1985).

100. A. Linde, Eternally existing self-reproducing inflationary universe, *Physica Scripta T* 15:169–75 (1987).

101. J.B. Hartle, Quantum cosmology, in *Proc. Goa, India, Int. Conf. on General Relativity*, December 1987 (Cambridge University Press, Cambridge, 1988), 144–55.

102. N. O'Murchadha and J.W. York, Jr., The initial value problem of general relativity. II. *Phys. Rev. D* 10:437–46 (1974).

103. J.W. York, Jr., Conformally invariant orthogonal decomposition of symmetric tensors on Riemannian manifolds and the initial-value problem of general relativity, *J. Math. Phys.* 14:456–64 (1973).

104. J.W. York, Jr., Covariant decompositions of symmetric tensors in the theory of gravitation, *Ann. Inst. Henri Poincaré* 21:319–32 (1974).

105. J.D. Jackson, *Classical Electrodynamics*, 2d ed. (Wiley, New York, 1975).

106. See ref. 27, § 25.2.

107. H. Goldschmidt and D. Spencer, On the Non-linear Cohomology of Lie Equations, *Acta Math.* 136:103–239 (1976).

108. See ref. 27, diagram on p. 530.

109. W. Killing; Über die Grundlagen der Geometrie, *J. Reine Angew. Math.* 109:121–86 (1892).

110. L. Bianchi, *Lezioni sulla teoria dei gruppi continui finiti di trasformazioni* (Spoerri, Pisa, 1918).

111. T. Regge, General relativity without coordinates, *Nuovo Cimento* 19:558–71 (1961).

112. See, e.g., J. Bower, J. Rauber, and J.W. York, Jr., Two black holes with axisymmetric parallel spins: Initial data, *Class. Quantum Grav.* 1:591–610 (1984).

113. See ref. 27, chap. 37.

114. K. Thorne, Gravitational Radiation, in *300 Years of Gravitation*, ed. S. W. Hawking and W. Israel (Cambridge University Press, Cambridge, 1987), 330–458.

115. R.A. Hulse and J.H. Taylor, Discovery of a pulsar in a binary system, *Astrophys. J. Lett.* 195:L51–L53 (1975).

116. See ref. 27, chap. 36.

117. J.A. Wheeler, Geons, *Phys. Rev.* 97:511–36 (1955).

118. See ref. 25, pp. 410–27.

119. D. Ivanenko and A. Sokolov, *Klassische Feldtheorie* (Berlin, 1953); section on 2 graviton $\rightarrow$ electron pair.

120. E. Kretschmann, Über den physikalischen Sinn der Relativitätspostulate. A. Einsteins neue und seine ursprungliche Relativitätstheorie, *Ann. Physik* 53:575–614 (1917).

121. L. Boltzmann, Über die Grundprinzipien und Grundgleichungen der Mechanik, Clark University 1889–1899, Decennial Celebration Worchester Massachusetts (1899), 261–309. Translation in *Ludwig Boltzmann: Theoretical Physics and*

Philosophical Problem: Selected Writings, ed. B. McGuinness (Reidel, Dordrecht, 1974). On p. 102, "absolute space is nowhere accessible to our experience, which only ever gives us the relative change in the position of bodies."

122. D.J. Raine, Mach's principle and space-time structure, *Rep. Prog. Phys.* 44:1152–95 (1981).
123. J.A. Wheeler, ref. 26, Appendix B, 344–46.
124. L.P. Grishchuk, A.G. Doroshkevich, and V.M. Yudin, Long gravitational waves in a closed universe, *Sov. Phys.–JETP* 42:943–49 (1975).
125. J.A. Wheeler, The beam and stay of the Taub Universe, in *Essays in General Relativity: A Festschrift for Abraham Taub*, ed. F.J. Tipler (Academic Press, New York, 1980), 59–70.
126. M.P. Ryan, Jr., and L.C. Shepley, *Homogeneous Relativistic Cosmologies* (Princeton University Press, Princeton, 1975).
127. R.A. Matzner, L.C. Shepley, and J.B. Warren, Dynamics of SO(3,R)-homogeneous cosmologies, *Ann. Pys.* 57:401–60 (1970).
128. A.H. Taub, Empty space-times admitting a three parameter group of motions, *Ann. Math.* 53:472–90 (1951).
129. J.A. Wheeler and R.P. Feynman, Classical electrodynamics in terms of direct interparticle action, *Rev. Mod. Phys.* 21:425–33 (1949); this elucidates and gives references to the literature.
130. B.L. Altshuler, Mach's principle. Part 1. Initial state of the universe, *Int. J. Theor. Phys.* 24:99–118 (1985).
131. D. Lynden-Bell, On the origins of space-time and inertia, *Mon. Not. Roy. Astron. Soc.* 135:413–28 (1967).
132. D.J. Raine, Mach's principle in general relativity, *Mon. Not. Roy. Astron. Soc.* 171:507–28 (1975).
133. V.K. Mat'tsev and M.A. Markov, On integral formulation of Mach principle in conformally flat space, Paper E 2-9722, Institute of Nuclear Research AS USSR, Moscow, 1976.
134. E.M. Lifshitz and I.M. Khalatnikov, Investigations in relativistic cosmology; trans. J.L. Beeby in *Adv. Phys.* 12:185–249 (1963).
135. V.A. Belinski and I.M. Khalatnikov, On the nature of the singularities in the general solution of the gravitational equations, *Zh. Eksp. Teor. Fiz.* 56:1700–1712 (1960); translation in *Sov. Phys.–JETP* 29:911–17 (1960).
136. V.A. Belinski and I.M. Khalatnikov, General solution of the gravitational equations with a physical singularity, *Zh. Eksp. Teor. Fiz.* 57:2163–75 (1969); translation in *Sov. Phys.–JETP* 30:1174–80 (1970).
137. V.A. Belinski and I.M. Khalatnikov, General solution of the gravitational equations with a physical oscillatory singularity, *Zh. Eksp. Teor. Fiz.* 59:314–21 (1970); translation in *Sov. Phys.–JETP* 32:169–72.
138. V.A. Belinski and I.M. Khalatnikov and E.M. Lifshitz, Oscillatory approach to a singular point in the relativistic cosmology, *Usp. Fiz. Nauk.* 102:463–500 (1970); translation in *Adv. Phys.* 19:525–73 (1970).

139. I.M. Khalatnikov and E.M. Lifshitz, General cosmological solutions of the gravitational equations with a singularity in time, *Phys. Rev. Lett.* 24:76–79 (1970).

140. V.A. Belinski, E.M. Lifshitz, and I.M. Khalatnikov, Oscillatory mode of approach to a singularity in homogeneous cosmological models with rotating axes, *Zh. Eksp. Teor. Fiz.* 60:1969–79 (1971); translation in *Sov. Phys.–JEPT* 33:1061–66 (1971).

141. See ref. 27, chap. 33.

142. J. Lense and H. Thirring, Über den Einfluss der Eigenrotation der Zentralkörper auf die Bewegung der Planeten und Monde nach der Einsteinschen Gravitationstheorie, *Phys. Z.* 19:156–63 (1918).

143. B. Mashhoon, F.W. Hehl, and D.S. Theiss, On the gravitational effects of rotating masses: The Thirring-Lense papers, *Gen. Rel. Grav.* 16:711–50 (1984).

144. E. Fermi, Sopra i fenomeni che avvengono in vicinanza di una linea oraria, *Atti R. Accad. Lincei Rend. Cl. Sci. Fis. Mat. Nat.* 31 (I): 21–23, 51–52, 101–3 (1922).

145. S.W. Hawking and G.F.R. Ellis, *The Large Scale Structure of Space-Time* (Cambridge University Press, Cambridge, 1973).

146. M. Spivak, *A Comprehensive Introduction to Differential Geometry* (Publish or Perish, Boston 1970).

147. R.A. Coleman and H. Korté, The relation between the measurement and Cauchy problems of GTR, in *Proc. 6th Marcel Grossmann Meeting on General Relativity*, Kyoto, 1991, ed. H. Sato and T. Nakamura (World Scientific, Singapore, 1992), 97–119.

148. H. Bondi, Plane gravitational waves in general relativity, *Nature* 179:1072–73 (1957).

149. See also, *Proc. Int. Conf. on Mach's Principle, from Newton's Bucket to Quantum Gravity*, Tübingen, Germany, July 1993, ed. J. Barbour and H. Pfister (Birkhäuser, Boston, 1995).

150. R.A. Matzner, Computation of gravitational wave production from black-hole collisions: status report, in *Proc. Int. Conf. on Phenomenology of Unification from Present to Future*, Rome, March 1994, ed. G. Diambrini-Palazzi, in press (1994).

6

The Gravitomagnetic Field and Its Measurement

We understand today both in broad outline and in precise equations the connection between mass-energy there and inertia here. We nevertheless also know that there remain dark areas yet to be lighted up. Among them, nothing would do more to demonstrate the inertia-influencing effect of mass in motion than to detect and measure the Einstein-Thirring-Lense–predicted *gravitomagnetism* of this rotating Earth.

Hans Christian Oersted proved in 1820 that electric currents produce a magnetic field.[1] According to Einstein geometrodynamics,[2–7] mass currents produce a field called, by analogy, the gravitomagnetic field:[8,9] a force of nature, predicted in 1896–1918, yet even now (1994) not detected and measured. Indeed, gravitomagnetism and gravitational waves are two main predictions of Einstein geometrodynamics that still await direct measurement.

To verify the existence of gravitomagnetism will prove that mass-energy currents influence the spacetime geometry and create curvature (§§ 6.1 and 6.11), and to measure it is to test the ideas of Einstein and others on the origin of inertia in geometrodynamics,[2,10–13] that is, the concept that the local inertial frames are determined by the distribution and currents of mass-energy throughout the universe (§§ 6.1 and 6.2). Furthermore this field has a key role in the theory of jet alignment and production from a rapidly rotating star or black hole (see § 6.3 on high-energy astrophysics).[9]

In this chapter, after a description of the gravitomagnetic field in Einstein theory and of its importance in physics (§§ 6.1, 6.2, and 6.3), and a list of the main experiments proposed to measure it, we describe in particular two space experiments, LAGEOS III and Gravity Probe B, and a few other proposed Earth-laboratory experiments.

6.1 THE GRAVITOMAGNETIC FIELD AND THE MAGNETIC FIELD

As magnetism is to electricity, so gravitomagnetism is to everyday gravity. The difference is that Oersted,[1] by July 1820, had both detected and measured magnetism, but no one has yet detected and measured **gravitomagnetism**: a "new" force of nature, predicted almost one century ago and yet still directly unmeasured.

Before we turn to schemes to detect and measure magnetogravitation we can look more closely into the frame-dragging effect of Earth—that silent whirl of matter, that great array of parallel circulating currents of mass. Sometimes we hear the phrase "gravitomagnetic effect of one elementary sector of a current of mass." However, no one can forget Ampère's 1825 mathematical analysis[14] of the magnetic effect of a small sector of a current-bearing wire. It is a mistake, he showed, to expect a unique mathematical formula for the effect of a single sector. Whittaker, among others, demonstrated the same point in more detail later.[15] The law of conservation of charge makes it impossible to treat the effect of one sector without regard to the current supplied by the sector on its left and taken up by the sector on its right. In consequence there is a far-reaching ambiguity associated with the very terminology, "effect of a single sector," an ambiguity as inescapable in gravitomagnetism as in electromagnetism. Every such indeterminism disappears when we deal with the totalized effect of an entire closed circuit of charge or mass in motion. From this circumstance the Thirring-Lense formula[16–17] (expression 6.1.33 below) gains its unshaken status (see § 6.11 for a complete discussion on gravitomagnetism and its analogies and differences with magnetism).

The analogy of a massive sphere in Einstein geometrodynamics[2–7] with an electrically charged sphere in electrodynamics is interesting. In electrodynamics, in the frame in which an electrically charged sphere is at rest, we have an electric field. If we then rotate the sphere we observe a magnetic field, and the strength of this field depends on the angular velocity. Similarly, in geometrodynamics, a nonrotating, massive sphere produces the standard Schwarzschild field. If we then rotate the sphere, we have a gravitomagnetic field,[8,9] given, in the weak field and slow motion approximation, by the expression (6.1.25) below. The strength of this gravitomagnetic field is proportional to the angular velocity.

Briefly, in classical Galilei-Newton mechanics the external field of a rotating uncharged, idealized spherical body is only given by its mass; in Einstein geometrodynamics the external field of such a body is instead given by its mass and by its angular momentum. In the weak field and slow motion limit of the Kerr-Newman metric (2.6.36), with $Q = 0$, the metric is given, in Boyer-Lindquist coordinates, by[18–20]

$$ds^2 \cong -\left(1 - \frac{2M}{r}\right) dt^2 + \left(1 - \frac{2M}{r}\right)^{-1} dr^2 + r^2(d\theta^2 + \sin^2\theta\, d\phi^2) - \frac{4J}{r} \sin^2\theta\, d\phi\, dt. \tag{6.1.1}$$

The $g_{\phi t}$ component of the metric is the gravitomagnetic potential.[8,9]

The weak field and slow motion formal analogy of the gravitomagnetic field in general relativity with the magnetic field in electromagnetism[21] is similar to the weak field formal analogy of gravitational waves with electromagnetic waves.[22,23]

In electrodynamics,[21] the wave equation describing electromagnetic waves in vacuum is, in the Lorentz gauge, $A^{\alpha}{}_{,\alpha} = 0$:

$$\Box A^{\alpha} = 0 \tag{6.1.2}$$

where $\Box = \eta^{\alpha\beta} \frac{\partial^2}{\partial x^{\alpha} \partial x^{\beta}}$ and A^{α} is the four-vector potential.

Similarly, in general relativity, in the weak field limit, the wave equation describing gravitational waves in vacuum is equation (2.10.11)

$$\Box h_{\alpha\beta} = 0 \tag{6.1.3}$$

where in the weak field limit $h_{\alpha\beta} \cong g_{\alpha\beta} - \eta_{\alpha\beta}$ ($g_{\alpha\beta}$ is the spacetime metric tensor and $\eta_{\alpha\beta}$ is the Minkowski metric tensor).

A similar analogy is valid for the gravitomagnetic field.[9] In electrodynamics,[21] from the Maxwell equations (2.8.43) and (2.8.44) and in particular from the equation for the magnetic induction, $\boldsymbol{B}$, expressing the absence of free magnetic monopoles, $\boldsymbol{\nabla} \cdot \boldsymbol{B} = \boldsymbol{0}$, one can write $\boldsymbol{B} = \boldsymbol{\nabla} \times \boldsymbol{A}$, where $\boldsymbol{A}$ is the vector potential. From Ampere's law for a stationary current distribution: $\boldsymbol{\nabla} \times \boldsymbol{B} = \frac{4\pi}{c}\boldsymbol{j}$, where $\boldsymbol{j}$ is the current density, one has then:

$$\boldsymbol{\nabla} \times (\boldsymbol{\nabla} \times \boldsymbol{A}) = \boldsymbol{\nabla}(\boldsymbol{\nabla} \cdot \boldsymbol{A}) - \Delta \boldsymbol{A} = \frac{4\pi}{c}\boldsymbol{j}, \tag{6.1.4}$$

and in the Coulomb gauge, $\boldsymbol{\nabla} \cdot \boldsymbol{A} \equiv A^{i}{}_{,i} = 0$,

$$\Delta \boldsymbol{A} = -\frac{4\pi}{c}\boldsymbol{j}. \tag{6.1.5}$$

The solution of this Poisson equation can be written

$$\boldsymbol{A}(\boldsymbol{x}) = \frac{1}{c} \int \frac{\boldsymbol{j}(\boldsymbol{x}')}{|\boldsymbol{x} - \boldsymbol{x}'|} \, d^3x'. \tag{6.1.6}$$

One can then define the magnetic moment $\boldsymbol{m}$ of a current distribution:

$$\boldsymbol{m} \equiv \frac{1}{2c} \int \boldsymbol{x} \times \boldsymbol{j}(\boldsymbol{x}) \, d^3x. \tag{6.1.7}$$

For example the magnetic moment generated by a plane loop of electric current I, with area A, is $m = \frac{I}{c} A$.

For a localized, stationary current distribution $\boldsymbol{j}$, far from the current one can expand equation (6.1.6). The lowest nonvanishing term of $\boldsymbol{A}(\boldsymbol{x})$, the magnetic dipole vector potential, can then be written

$$\boldsymbol{A}(\boldsymbol{x}) \cong \frac{\boldsymbol{m} \times \boldsymbol{x}}{|\boldsymbol{x}|^3}, \tag{6.1.8}$$

and therefore, for a localized current distribution, the lowest nonvanishing term of $\boldsymbol{B}$ is the field of a magnetic dipole with magnetic moment given by (6.1.7):

$$\boldsymbol{B} = \nabla \times \boldsymbol{A} \cong \frac{3\hat{\boldsymbol{x}}(\hat{\boldsymbol{x}} \cdot \boldsymbol{m}) - \boldsymbol{m}}{|\boldsymbol{x}|^3}. \tag{6.1.9}$$

For a system of charged particles, with charges q_i, masses m_i, and velocities $\boldsymbol{v}_i$, the magnetic moment (6.1.7) can be written as a function of the orbital angular momenta $\boldsymbol{l}_i$ of the particles

$$\boldsymbol{m} = \frac{1}{2c} \sum_i q_i (\boldsymbol{x}_i \times \boldsymbol{v}_i) = \frac{1}{2c} \sum_i \frac{q_i \boldsymbol{l}_i}{m_i}; \tag{6.1.10}$$

if all the particles have the same charge to mass ratio $\frac{q_i}{m_i} \equiv \frac{q}{m}$, we then have

$$\boldsymbol{m} = \frac{q}{2mc} \sum_i \boldsymbol{l}_i = \frac{q \boldsymbol{l}_{\text{Tot}}}{2mc}. \tag{6.1.11}$$

In electrodynamics, the equation of motion of a particle of mass m and charge q, subjected to an electric field $\boldsymbol{E}$ and to a magnetic induction $\boldsymbol{B}$, is the Lorentz equation:

$$m \frac{d^2 \boldsymbol{x}}{dt^2} = q(\boldsymbol{E} + \frac{1}{c} \frac{d\boldsymbol{x}}{dt} \times \boldsymbol{B}). \tag{6.1.12}$$

Therefore, given a current distribution with current density $\boldsymbol{j}(\boldsymbol{x})$, the total force on the current distribution due to $\boldsymbol{B}$ is

$$\boldsymbol{F} = \frac{1}{c} \int \boldsymbol{j}(\boldsymbol{x}) \times \boldsymbol{B}(\boldsymbol{x})\, d^3x, \tag{6.1.13}$$

and the torque on it, due to $\boldsymbol{B}$, is

$$\boldsymbol{\tau} = \frac{1}{c} \int \boldsymbol{x} \times (\boldsymbol{j} \times \boldsymbol{B})\, d^3x. \tag{6.1.14}$$

For a localized current distribution $\boldsymbol{j}(\boldsymbol{x})$, with magnetic moment $\boldsymbol{m}'$, in the field of an external magnetic induction $\boldsymbol{B}(\boldsymbol{x})$ that varies slowly over the region of the current, one can expand $\boldsymbol{B}(\boldsymbol{x})$ about a suitable origin around the localized current distribution: $\boldsymbol{B}(\boldsymbol{x}) = \boldsymbol{B}(\boldsymbol{0}) + (\boldsymbol{x} \cdot \nabla)\boldsymbol{B}|_0 + \cdots$. We then find the lowest nonvanishing term of the force (6.1.13)

$$\begin{aligned} \boldsymbol{F} = (\boldsymbol{m}' \times \nabla) \times \boldsymbol{B}|_0 &= \nabla(\boldsymbol{m}' \cdot \boldsymbol{B})|_0 - \boldsymbol{m}'(\nabla \cdot \boldsymbol{B})|_0 \\ &= \nabla(\boldsymbol{m}' \cdot \boldsymbol{B})|_0, \end{aligned} \tag{6.1.15}$$

and for a stationary field, outside the source, $\nabla \times \boldsymbol{B} = \boldsymbol{0}$, we have $\boldsymbol{F} = (\boldsymbol{m}' \cdot \nabla)\boldsymbol{B}|_0$. From expression (6.1.14), the lowest nonvanishing term of the torque on a localized, stationary current distribution is then the torque on a magnetic dipole with magnetic moment given by (6.1.7):

$$\boldsymbol{\tau} = \boldsymbol{m}' \times \boldsymbol{B}(\boldsymbol{0}). \tag{6.1.16}$$

In geometrodynamics in the weak field and slow motion approximation (see § 3.7), for a stationary, localized ($g \xrightarrow{\infty} \eta$), mass-energy distribution, the $(0i)$ components of the Einstein field equation can be written in the Lorentz gauge (see §§ 2.10 and 3.7):

$$\Delta h_{0i} \cong 16\pi\rho v^i \tag{6.1.17}$$

(compare with eq. 6.1.5 in electrodynamics, $\Delta A^i = -4\pi\rho_e v^i$), with solution

$$h_{0i}(\boldsymbol{x}) \cong -4\int \frac{\rho(\boldsymbol{x}')v^i(\boldsymbol{x}')}{|\boldsymbol{x}-\boldsymbol{x}'|}\, d^3x'. \tag{6.1.18}$$

From the classical definition of **angular momentum**

$$\boldsymbol{J} = \int \boldsymbol{x} \times (\rho\boldsymbol{v})\, d^3x \tag{6.1.19}$$

for a spheroidal distribution of matter rotating about an axis with angular velocity $\dot{\boldsymbol{\alpha}}$, in the weak field and slow motion approximation, we may substitute $\boldsymbol{v} = \dot{\boldsymbol{\alpha}} \times \boldsymbol{x}$, in expression (6.1.19) and by a rotation of the spatial axes, to make $\dot{\boldsymbol{\alpha}} = (0, 0, \dot{\alpha}(r))$, we then have

$$\begin{aligned} \boldsymbol{J} &= \int \boldsymbol{x} \times (\dot{\boldsymbol{\alpha}} \times \boldsymbol{x})\rho r^2\, d\Omega\, dr \\ &= \frac{8\pi}{3}\int \dot{\boldsymbol{\alpha}}(r)\rho(r)r^4\, dr \end{aligned} \tag{6.1.20}$$

where $d\Omega \equiv \sin\theta\, d\theta\, d\phi$. Then, from formula (6.1.18):

$$\begin{aligned} \boldsymbol{h}(\boldsymbol{x}) &= -4\int \frac{\rho(r')\dot{\boldsymbol{\alpha}}(r') \times \boldsymbol{x}'}{|\boldsymbol{x}-\boldsymbol{x}'|} r'^2\, d\Omega'\, dr' \\ &= \frac{16\pi}{3r^3}\left(\boldsymbol{x} \times \int \dot{\boldsymbol{\alpha}}(r')\rho(r')r'^4\, dr'\right) \end{aligned} \tag{6.1.21}$$

where $\boldsymbol{x}' \equiv r'\hat{\boldsymbol{x}}'$, and where the integral over $d\Omega'$ has been simply evaluated by rotating the spatial axes so that $\boldsymbol{x} = (0, 0, r)$, by integrating and then by rotating back the axes to the original position, to get the general expression for any $\boldsymbol{x}$, that is:

$$\int \frac{\boldsymbol{x}'}{|\boldsymbol{x}-\boldsymbol{x}'|}\, d\Omega' =$$

$$\int_0^{\pi} \sin\theta' d\theta' \int_0^{2\pi} \frac{(r'\sin\theta'\cos\phi',\, r'\sin\theta'\sin\phi',\, r'\cos\theta')}{\sqrt{r'^2 - 2rr'\cos\theta' + r^2}}\, d\phi' =$$

$$= \begin{bmatrix} \left(0, 0, \dfrac{4\pi r'^2}{3r^2}\right) = \dfrac{4\pi r'^2}{3r^3}\boldsymbol{x}, \quad \text{when} \quad r > r' \qquad (6.1.22) \\ \text{or} \\ \left(0, 0, \dfrac{4\pi}{3}\dfrac{r}{r'}\right) = \dfrac{4\pi}{3r'}\boldsymbol{x}, \quad \text{when} \quad r < r'. \qquad (6.1.22') \end{bmatrix}$$

Therefore, for a spheroidal distribution of matter,[3] or far from the stationary source,[16–17] $\boldsymbol{h} \equiv (h_{01}, h_{02}, h_{03})$ can be rewritten

$$\boldsymbol{h}(\boldsymbol{x}) \cong -2\frac{\boldsymbol{J} \times \boldsymbol{x}}{|\boldsymbol{x}|^3}. \qquad (6.1.23)$$

For a slowly rotating sphere with angular momentum $\boldsymbol{J} \equiv (0, 0, J)$, we have in spherical coordinates

$$h_{0\phi} \cong -\frac{2J}{r}\sin^2\theta, \qquad (6.1.24)$$

that is, the $g_{0\phi}$ component of the Kerr metric[18] (2.6.36, with $Q = 0$) in the weak field and slow motion approximation (eq. 6.1.1). $\boldsymbol{h}$ is called the gravitomagnetic potential.

One can then define a **gravitomagnetic field:**[8,9] $\boldsymbol{H} = \boldsymbol{\nabla} \times \boldsymbol{h}$ (see fig. 1.2),

$$\boldsymbol{H} = \boldsymbol{\nabla} \times \boldsymbol{h} \cong 2\left[\frac{\boldsymbol{J} - 3(\boldsymbol{J} \cdot \hat{\boldsymbol{x}})\hat{\boldsymbol{x}}}{|\boldsymbol{x}|^3}\right]. \qquad (6.1.25)$$

From these equations we see that, in general relativity, in the weak field and slow motion limit, the angular momentum $\boldsymbol{J}$ of a stationary, localized, mass-energy current has a role similar to the magnetic dipole moment $\boldsymbol{m}$ of a stationary, localized, charge current in electrodynamics (the difference between electromagnetism and weak field general relativity shows in an extra factor -4 in general relativity).

By using the geodesic equation $\frac{Du^\alpha}{ds} \equiv \frac{d^2x^\alpha}{ds^2} + \Gamma^\alpha_{\beta\gamma}\frac{dx^\beta}{ds}\frac{dx^\gamma}{ds} = 0$, in the weak field and slow motion limit, one has then

$$m\frac{d^2\boldsymbol{x}}{dt^2} \cong m\left(\boldsymbol{G} + \frac{d\boldsymbol{x}}{dt} \times \boldsymbol{H}\right) \qquad (6.1.26)$$

where $\boldsymbol{G} \cong -\frac{M}{|\boldsymbol{x}|^2}\hat{\boldsymbol{x}}$ is the standard Newtonian acceleration and $\boldsymbol{H}$ is the gravitomagnetic field (6.1.25).

Furthermore, as for electromagnetism, so for general relativity[9] the "**torque**" acting **on a gyroscope** with angular momentum $\boldsymbol{S}$, in the weak field and slow motion approximation, is

$$\boldsymbol{\tau} \cong \frac{1}{2}\boldsymbol{S} \times \boldsymbol{H} = \frac{d\boldsymbol{S}}{dt} \equiv \dot{\boldsymbol{\Omega}} \times \boldsymbol{S}. \qquad (6.1.27)$$

Therefore, the gyroscope precesses with respect to an asymptotic inertial frame with angular velocity (eq. 3.4.38):

$$\dot{\boldsymbol{\Omega}} = -\frac{1}{2}\boldsymbol{H} = \frac{-\boldsymbol{J} + 3(\boldsymbol{J}\cdot\hat{\boldsymbol{x}})\hat{\boldsymbol{x}}}{|\boldsymbol{x}|^3} \tag{6.1.28}$$

where $\boldsymbol{J}$ is the angular momentum of the central object. This phenomenon is the "**dragging of a gyroscope**" or "**dragging of an inertial frame**," of which the gyroscope define an axis.

As for electromagnetism, the "**force**" exerted **on the gyroscope** by the gravitomagnetic field $\boldsymbol{H}$ is (see §6.10)

$$\boldsymbol{F} = \left(\frac{1}{2}\boldsymbol{S}\cdot\nabla\right)\boldsymbol{H}. \tag{6.1.29}$$

Finally, a central object with angular momentum $\boldsymbol{J}$ via the second term in the "force" (6.1.26), drags the orbital plane (and the orbital angular momentum) of a test particle (which can be thought of as an enormous gyroscope) in the sense of rotation of the central body. This dragging of the whole orbital plane is described by a formula for the rate of change of the longitude of the nodes, discovered by Lense and Thirring in 1918.[16,17] For a test particle with orbital angular momentum $\boldsymbol{l} = m(\boldsymbol{x}\times\boldsymbol{v})$, orbiting in the field of a central body with angular momentum $\boldsymbol{J}$, from equation (6.1.26), the torque on its *orbit* due to $\boldsymbol{J}$ is $\boldsymbol{\tau} = \boldsymbol{x}\times\left(m\frac{d\boldsymbol{x}}{dt}\times\boldsymbol{H}\right)$, where $\boldsymbol{H}$ is given by equation (6.1.25). We then have $\boldsymbol{\tau} = \frac{d\boldsymbol{l}}{dt} = \dot{\boldsymbol{\Omega}}\times\boldsymbol{l}$ and therefore

$$\dot{\boldsymbol{\Omega}}\times\left(\boldsymbol{x}\times\frac{d\boldsymbol{x}}{dt}\right) = \boldsymbol{x}\times\left(\frac{d\boldsymbol{x}}{dt}\times\boldsymbol{H}\right). \tag{6.1.30}$$

By integrating this equation over an orbital period, in the simple case of a circular polar orbit (orbital plane containing the vector $\boldsymbol{J}$), we immediately find $\dot{\boldsymbol{\Omega}} = \frac{2\boldsymbol{J}}{r^3}$.

For a general elliptic orbit, by a rotation of the spatial axes, we may first set $\boldsymbol{J} = (0, 0, J)$, and for a test particle in the field of a central body, we may then write:[24]

$$\begin{cases} x = r(\cos u\cos\Omega - \sin u\cos I\sin\Omega) \equiv r\cos u \\ y = r(\cos u\sin\Omega + \sin u\cos I\cos\Omega) \equiv r\sin u\cos I \\ z = r\sin u\sin I \end{cases} \tag{6.1.31}$$

where I is the *orbital inclination*, that is, the angle between the orbital plane of the test particle and the equatorial plane of the central body; u is the *argument of latitude*, that is, the angle on the orbital plane measuring the departure of the test particle from the equatorial plane; and Ω is the *nodal longitude*, that is, the angle on the equatorial plane measuring the orientation of the nodal line (intersection between the equatorial and the orbital plane). In equation (6.1.31) we have redefined the origin of Ω, so that $\cos\Omega \equiv 1$ and $\sin\Omega \equiv 0$. Indeed, for a motion under a central force, when we can neglect any other perturbation, including relativistic effects, we classically have $\dot{\Omega} = 0$ and $\dot{I} = 0$. The argument of the latitude can then be written $u = \omega + f$. Here ω is the *argument of pericenter*, that is, the angle on the orbital plane measuring the departure of the pericenter from the equatorial plane, and f is the *true anomaly*, that is, the angle on the orbital plane measuring the departure of the test particle from the pericenter. For a motion under a central force $\propto 1/r^2$ we have classically $\dot{\omega} = 0$, and for an elliptic orbit $r \cos f = a(\cos E - e)$, where a is the *orbital semimajor axis* of the test particle, e its *orbital eccentricity*, and E the *eccentric anomaly*, the angle defining the position of the test particle with respect to the circle circumscribed about the ellipse.

By inserting equation (6.1.31) in equation (6.1.30) for $\dot{\mathbf{\Omega}}$ and by integrating over one orbital period, also using the stated orbital relations, and evaluating *to first order* the relativistic effect, we thus find

$$\int \dot{\mathbf{\Omega}}\, dt = \int \frac{2\boldsymbol{J}}{|\boldsymbol{x}|^3}\, dt. \tag{6.1.32}$$

By writing $|\boldsymbol{x}| = r = a(1 - e\cos E)$ and by using the relation $dt = \frac{P}{2\pi}(1 - e\cos E)\, dE$, where P is the *orbital period*, we finally get, to first order, the formulae discovered by **Lense and Thirring** in 1918,[16,17] for the secular **rate of change of the longitude of the nodes** (intersection between the orbital plane of the test particle and the equatorial plane of the central object)

$$\dot{\mathbf{\Omega}}^{\text{Lense–Thirring}} = \frac{2\boldsymbol{J}}{a^3(1 - e^2)^{3/2}} \tag{6.1.33}$$

where a is the test particle semimajor axis, e its orbital eccentricity, and $\boldsymbol{J}$ the angular momentum of the central body.

Similarly, by integrating the equation of motion (6.1.26) of the test particle, one can find the formulae derived by Lense and Thirring[16,17] for the **secular rates of change of the longitude of the pericenter,**[25] $\dot{\tilde{\boldsymbol{\omega}}}$ (determined by the Runge-Lenz vector) **and of the mean orbital longitude** $\dot{L}_0$

$$\dot{\tilde{\boldsymbol{\omega}}}^{\text{Lense–Thirring}} = \frac{2J}{a^3(1 - e^2)^{\frac{3}{2}}}(\hat{\boldsymbol{J}} - 3\cos I\hat{\boldsymbol{l}}) \tag{6.1.34}$$

where $\cos I \equiv \hat{\boldsymbol{J}} \cdot \hat{\boldsymbol{l}}$ and I is the orbital inclination (angle between the orbital plane and the equatorial plane of the central object). Therefore, the term containing $\cos I$ drags the pericenter in the same sense as $\hat{\boldsymbol{l}}$ when $\hat{\boldsymbol{J}} = -\hat{\boldsymbol{l}}$, and in the opposite sense when $\hat{\boldsymbol{J}} = \hat{\boldsymbol{l}}$, and

$$\dot{L}_0^{\text{Lense–Thirring}} = \frac{2J}{a^3(1 - e^2)^{\frac{3}{2}}}(1 - 3\cos I) \tag{6.1.35}$$

In section 2.6 we have seen that inside a hollow, static, spherically symmetric distribution of matter, in vacuum, we have the flat metric $\eta_{\alpha\beta}$. Therefore, in the weak field and slow motion limit, the **metric inside a slowly rotating massive shell** may be written $g_{\alpha\beta} \cong \eta_{\alpha\beta} + h_{\alpha\beta}$, and we may then apply the results (6.1.17) and (6.1.18). Inside a thin shell of total mass M and radius R, rotating with angular velocity $\dot{\boldsymbol{\alpha}}$, we thus have the field (6.1.18); furthermore, by integrating expression (6.1.18) inside the shell,[17,26,27] with $\boldsymbol{v} = \dot{\boldsymbol{\alpha}} \times \boldsymbol{x}$, by rotating the spatial axes so that $\dot{\boldsymbol{\alpha}} = (0, 0, \dot{\alpha})$ and by using the mass density of a thin spherical shell, $\rho(\boldsymbol{x}') = \frac{M}{4\pi R^2}\,\delta(R - r')$, we have:

$$\begin{aligned} \boldsymbol{h} &= -4 \int \frac{\rho(\boldsymbol{x}')(\dot{\boldsymbol{\alpha}} \times \boldsymbol{x}')}{|\boldsymbol{x} - \boldsymbol{x}'|} r'^2 \, d\Omega' \, dr' \\ &= -\frac{M}{\pi} \dot{\boldsymbol{\alpha}} \times \int_0^{\pi} \sin\theta' \, d\theta' \int_0^{2\pi} \frac{R\hat{\boldsymbol{x}}'}{|\boldsymbol{x} - R\hat{\boldsymbol{x}}'|} \, d\phi'. \end{aligned} \tag{6.1.36}$$

The integral over $d\Omega'$ has already been evaluated and is given by formula (6.1.22′). Therefore, for any $\boldsymbol{x}$ inside the shell, from (6.1.36) we have

$$\boldsymbol{h} \equiv (h_{0x}, h_{0y}, h_{0z}) = -\frac{4}{3}\frac{M}{R}\dot{\boldsymbol{\alpha}} \times \boldsymbol{x} = \left(\frac{4M}{3R}\dot{\alpha} y, -\frac{4M}{3R}\dot{\alpha} x, 0\right). \tag{6.1.37}$$

By substituting the components of $h_{\alpha\beta}$ inside the slowly rotating shell in the geodesic equation (6.1.26) we find the 1918 Thirring result[17,26,27] for the **acceleration of a test particle inside a rotating shell,** due to the rotation of the shell:

$$\begin{aligned} \ddot{x} &= -\frac{8}{3}\frac{M}{R}\dot{\alpha}\dot{y} + \frac{4}{15}\frac{M}{R}\dot{\alpha}^2 x, \\ \ddot{y} &= \frac{8}{3}\frac{M}{R}\dot{\alpha}\dot{x} + \frac{4}{15}\frac{M}{R}\dot{\alpha}^2 y, \\ \ddot{z} &= -\frac{8}{15}\frac{M}{R}\dot{\alpha}^2 z, \end{aligned} \tag{6.1.38}$$

where the $\dot{\alpha}^2$ terms are due to the other components of $h_{\alpha\beta}$ and may be thought of as a change in the inertial, and gravitational, mass of the shell due to the velocity $\dot{\alpha}R$. Therefore, due to the rotation of the shell, the test particle is affected by forces *formally* similar to Coriolis and centrifugal forces.

Furthermore, from expressions (6.1.28) and (6.1.36), we conclude that the axes of local inertial frames, that is, **gyroscopes**, are **dragged by the rotating shell** with constant angular velocity $\dot{\boldsymbol{\Omega}}^G$, according to

$$\dot{\boldsymbol{\Omega}}^G \cong -\frac{1}{2}\boldsymbol{H} = -\frac{1}{2}\boldsymbol{\nabla} \times \boldsymbol{h} = \frac{4}{3}\frac{M}{R}\dot{\boldsymbol{\alpha}}. \tag{6.1.39}$$

Indeed, by performing the transformation $\phi' = \phi - \dot{\Omega}^G t$ one gets $h'_{0i} = 0$, and by a suitable redefinition of t and r one gets the flat Minkowski metric inside the shell. For other solutions inside a rotating shell with arbitrary mass M, Brill and Cohen (1966),[28] or to higher order in the angular velocity[29] $\dot{\alpha}$, and for discussions on the interpretation of the forces[29–32] inside a rotating shell, see references 28–34.

We conclude this section by stressing, once more, that despite the beautiful and illuminating analogies between classical electrodynamics and classical geometrodynamics, the two theories are, of course, fundamentally different (see §§ 2.7, 6.11, and 7.1). Their fundamental difference is valid even for weak field, linearized gravity, as discussed in section 6.11 and as unambiguously displayed by the equivalence principle.

6.2 GRAVITOMAGNETISM AND THE ORIGIN OF INERTIA IN EINSTEIN GEOMETRODYNAMICS

Gravitomagnetism may be thought of as a manifestation of the way inertia originates in Einstein geometrodynamics: "*mass-energy there rules inertia here*" (chap 5). The measurement of the gravitomagnetic field will be the experimental evidence of this interpretation of the origin of the local inertial forces, that might be called a weak general relativistic interpretation of the Mach principle. In his book, *The Meaning of Relativity*,[2] Einstein writes (pp. 95–98): "But in the second place, the theory of relativity makes it appear probable that Mach was on the right road in his thought that inertia depends upon a mutual action of matter. For we shall show in the following that, according to our equations, inert masses do act upon each other in the sense of the relativity of inertia, even if only very feebly. What is to be expected along the line of Mach's thought?

1. The inertia of a body must increase when ponderable masses are piled up in its neighbourhood.[*]
2. A body must experience an accelerating force when neighbouring masses are accelerated, and, in fact, the force must be in the same direction as that acceleration.
3. A rotating hollow body must generate inside of itself a 'Coriolis field,' which deflects moving bodies in the sense of the rotation, and a radial centrifugal field as well.

We shall now show that these three effects, which are to be expected in accordance with Mach's ideas, are actually present according to our theory, although their magnitude is so small that confirmation of them by laboratory experiments is not to be thought of. For this purpose we shall go back to the

[*However, regarding this point it has to be stressed that in general relativity, because of the very strong equivalence principle, the inertial mass and the gravitational mass, or G, do not change in the field of other masses. This change may however be present in alternative metric theories[36] (see chap. 3). Indeed, in 1962, Brans had shown that in general relativity this change is a mere coordinate effect.[35]]

equations of motion of a material particle (90), and carry the approximations somewhat further than was done in equation (90a).

First, we consider γ_{44} as small of the first order. The square of the velocity of masses moving under the influence of the gravitational force is of the same order, according to the energy equation. It is therefore logical to regard the velocities of the material particles we are considering, as well as the velocities of the masses which generate the field, as small, of the order $1/2$. We shall now carry out the approximation in the equations that arise from the field equations (101) and the equations of motion (90) so far as to consider terms, in the second member of (90), that are linear in those velocities. Further, we shall not put ds and dl equal to each other, but, corresponding to the higher approximation, we shall put

$$ds = \sqrt{g_{44}}\,dl = \left(1 - \frac{\gamma_{44}}{2}\right) dl$$

From (90) we obtain, at first,

$$\frac{d}{dl}\left[\left(1 + \frac{\gamma_{44}}{2}\right)\frac{dx_\mu}{dl}\right] = -\Gamma^{\mu}_{\alpha\beta}\frac{dx_\alpha}{dl}\frac{dx_\beta}{dl}\left(1 + \frac{\gamma_{44}}{2}\right) \tag{116}$$

From equation (101) we get, to the approximation sought for,

$$\begin{aligned} -\gamma_{11} = -\gamma_{22} = -\gamma_{33} = \gamma_{44} &= \frac{\chi}{4\pi}\int\frac{\sigma\,dV_0}{r} \\ \gamma_{4\alpha} &= -\frac{i\chi}{2\pi}\int\frac{\sigma\frac{dx_\alpha}{ds}\,dV_0}{r} \\ \gamma_{\alpha\beta} &= 0 \end{aligned} \tag{117}$$

in which, in (117), α and β denote the space indices only.

On the right-hand side of (116) we can replace $1 + \frac{\gamma_{44}}{2}$ by 1 and $-\Gamma^{\alpha\beta}_{\mu}$ by $\begin{bmatrix}\alpha\beta\\ \mu\end{bmatrix}$. It is easy to see, in addition, that to this degree of approximation we must put

$$\begin{aligned} \begin{bmatrix}44\\ \mu\end{bmatrix} &= -\frac{1}{2}\frac{\partial\gamma_{44}}{\partial x_\mu} + \frac{\partial\gamma_{4\mu}}{\partial x_4} \\ \begin{bmatrix}\alpha 4\\ \mu\end{bmatrix} &= \frac{1}{2}\left(\frac{\partial\gamma_{4\mu}}{\partial x_\alpha} - \frac{\partial\gamma_{4\alpha}}{\partial x_\mu}\right) \\ \begin{bmatrix}\alpha\beta\\ \mu\end{bmatrix} &= 0 \end{aligned}$$

in which α, β and μ denote space indices. We therefore obtain from (116), in the usual vector notation,

$$\frac{d}{dl}\left[(1+\bar{\sigma})\boldsymbol{\nu}\right] = \boldsymbol{grad}\,\bar{\sigma} + \frac{\partial \boldsymbol{A}}{\partial l} + [\boldsymbol{rot}\,\boldsymbol{A} \times \boldsymbol{\nu}]$$

$$(118) \qquad \bar{\sigma} = \frac{\chi}{8\pi}\int \frac{\sigma\, dV_0}{r}$$

$$\boldsymbol{A} = \frac{\chi}{2\pi}\int \frac{\sigma \frac{dx_\alpha}{dl}\, dV_0}{r}$$

The equations of motion, (118), show now, in fact, that

(1) The inert mass is proportional to $1 + \bar{\sigma}$, and therefore increases when ponderable masses approach the test body.[*]
(2) There is an inductive action of accelerated masses, of the same sign, upon the test body. This is the term $\frac{\partial \boldsymbol{A}}{\partial l}$.
(3) A material particle, moving perpendicularly to the axis of rotation inside a rotating hollow body, is deflected in the sense of the rotation (Coriolis field). The centrifugal action, mentioned above, inside a rotating hollow body, also follows from the theory, as has been shown by Thirring.†

Although all of these effects are inaccessible to experiment, because χ is so small, nevertheless they certainly exist according to the general theory of relativity. We must see in them a strong support for Mach's ideas as to the relativity of all inertial actions. If we think these ideas consistently through to the end we must expect the whole inertia, that is, the whole $g_{\mu\nu}$-field, to be determined by the matter of the universe, and not mainly by the boundary conditions at infinity."

From these 1922 considerations of Einstein a few conclusions are central: In agreement with the idea of Mach that the centrifugal forces are due to rotational motion of a system relative to other matter, Thirring showed in 1918 the existence, inside a rotating hollow sphere, of the gravitational accelerations (6.1.38), *formally* similar to the Coriolis and centrifugal forces[17,25,26] (for the interpretation of these forces see refs. 28–34). Lense and Thirring in 1918[16,17] showed the existence, outside a rotating body, of a gravitational field due to the rotation of the body and later called, by its analogy with the magnetic field of electromagnetism, the gravitomagnetic field (expression 6.1.25). They also derived the Lense-Thirring nodal precession (eq. 6.1.33) of a test particle orbiting a rotating central body.

"†That the centrifugal action must be inseparably connected with the existence of the Coriolis field may be recognized, even without calculation, in the special case of a co-ordinate system rotating uniformly relatively to an inertial system; our general co-variant equations naturally must apply to such a case."

[*However, regarding this point, see our previous note and ref. 35.]

6.3 THE GRAVITOMAGNETIC FIELD IN ASTROPHYSICS

Astrophysics reveals no source of energy more powerful than the quasi-stellar object or "quasar," central object of many an active galactic nucleus and source of jets of particles and intense electromagnetic radiation. Every feature of the *quasar* speaks for its being powered by a supermassive rotating black hole, source of an intense *gravitomagnetic field* (see picture 4.5 of the M87 nucleus in chap. 4). Treatments of the quasar, and in particular theories of energy storage, power generation, jet formation, and jet alignment of quasars and active galactic nuclei, will be found in Thorne, Price, and MacDonald (1986)[9] and Bardeen and Patterson (1975).[37]

The typical quasar or active galactic nucleus usually has a central source of continuum radiation of dimension of the order of the solar system, that is, about 10^{15} cm, and a luminosity that may reach 10^{48} erg/s, that is, approximately 3×10^{14} times the Sun's luminosity. Today's model for this phenomenon envisages an accretion disk around a supermassive black hole. From one side or from opposite sides of the central region usually originate jets, or linear features that may have a length of hundreds of kiloparsecs, with strong emission at radio wavelengths. The orientation of the accretion disk and therefore the orientation of the emitted jets may be explained by the gravitomagnetic field of the central body. The hydrodynamics of the accretion disk is interpreted by the Navier-Stokes equation including the action of the classical gravity field and the gravitomagnetic field of the central object.

Under simplifying hypotheses, neglecting the spherical, Schwarzschild, post-Newtonian terms, since they should have a small effect on the disk orientation, including, however, the gravitomagnetic field as a correction to the classical Newtonian field (the nonspherical effects due to the quadrupole moment of the central body have been calculated to be smaller than the gravitomagnetic effect), from expression (2.3.24) for $T^{\alpha\beta}$ and from $T^{i\beta}{}_{;\beta} = 0$ (see also eq. 6.1.26), the Navier-Stokes equation can be written[9]

$$\rho \left[\frac{\partial u^i}{\partial t} + u^j u^i_{\ |j} \right] \cong \rho \left(G^i + \epsilon^i_{\ jk} u^j H^k \right) - p^{|i} + \left(2\eta\sigma^{ij} \right)_{|j} \tag{6.3.1}$$

where ρ is the density of mass-energy, $u^i = dx^i/dt$, "|" denotes covariant derivative with respect to the spatial metric, G^i is the classical gravitational acceleration, H^i the gravitomagnetic field, p the pressure, η the coefficient of shear viscosity, and σ^{ij} the shear tensor of the fluid. As noted, this equation is the classical Navier-Stokes equation plus classical gravitational and gravitomagnetic contributions of the central body.

From the hydrodynamic equation (6.3.1), assuming $p = \eta = 0$, plus the expression (6.1.25) for the gravitomagnetic field, one can show[38] that a thin ring of matter orbiting the central body will precess about the angular momentum vector $\boldsymbol{J}$ of the central object with angular velocity $\dot{\boldsymbol{\Omega}} \cong \frac{2\boldsymbol{J}}{r^3}$. However, as a result of the joint action of the gravitomagnetic field and of the viscous forces, the accretion disk, at radii $r \lesssim 100M$ (M = central mass), tends to be oriented into the equatorial plane of the central body.[37] The jets should then be ejected normally to the accretion disk, that is, normally to the equatorial plane of the central object and along its angular momentum vector.

Since the angular momentum vector of the compact central body should act as a gyroscope, this mechanism may explain the observed constancy of direction of the jets that may reach a length of millions of light years, corresponding to an emission time of several millions of years (see pictures 6.1–6.4).[9]

6.4 THE PAIL, THE PIROUETTE, AND THE PENDULUM

In every application of expression (6.1.28) with which we now deal, the decisive quantities are the angular momentum of Earth,[39] $J \cong 5.86 \times 10^{40}$ g cm^2/s = 145 cm^2 (as contrasted to Earth mass, 0.4438 cm) and the distance r to the point of observation. The water in the pail lies flat, Newton explained, when pail and water do not rotate with respect to the stars:[40] that is, with respect to what we would today call the *local inertial frame of reference*. Or, as Weinberg puts it, "First stand still, and let your arms hang loose at your sides. Observe that the stars are more or less unmoving, and that your arms hang more or less straight down. Then pirouette. The stars will seem to rotate around the zenith, and at the same time your arms will be drawn upward by centrifugal force."[41] Or let the swinging pendulum of Foucault hang from a support at the South Pole. In the course of a twenty-four hour day the plane of vibration, referred to axes scratched on Earth's surface, will turn through 360°. Referred to the distant stars, it will turn not at all.

Turn not at all? Wrong. A wrong prediction. Wrong because it overweights the voting power of the stars and wrong because it deprives Earth of any vote at all. Overweight the visible? Imagine that we could see no extraterrestrial object except Neptune. The temptation would be great to assign to it the influence that slowly turns the plane of vibration of the polar pendulum relative to Earth-fixed axes. However, better measurement then reveals that the line from Earth to Neptune does not have the authority that was first claimed for it. Relative to that line the direction of oscillation moves around, very slowly indeed, but at a discernible rate, a rate great enough periodically to bring it back to its original orientation roughly once ever 60,188 days. We draw three conclusions. First, Neptune itself is not one of the decisive determiners of the inertial frame. Second, we are getting no sight of those determiners. Third, whatever and wherever those determiners are, relative to the inertial frame itself Neptune swings once around in orbit every 60,188 days.

Replace the one word Neptune by the many words for the many "fixed" stars of the province that we call the Milky Way and end up—apart from changes in the number—with the same conclusion: influences from outside the"province" fix the number-one feature of the metric (6.1.1), its Minkowskian part. How weak by comparison the influences are from within the "province" of our galaxy we see at once from "the sum for inertia," expression (1.1.1). It tells us to attribute to the i-th star an inertia-determining power $\frac{M_i}{r_i}$. We put in for each star a typical mass of the order of one solar mass or, more simply, of the order of 10^5 cm and distance of the order of 30 kpc or about 10^{23} cm and get a measure of the order of 10^{-18} for its inertia-determining power, and for all those Milky Way stars out there, 10^{11} times as much, or an inertia-determining power of only 10^{-7}. A voting voice so tiny gives neither right nor reason for them to be the arbiters of the local inertial frame.

Outward and outward we go in our search for determiners of the inertial frame. In the end we find ourselves appreciating in a new light the image (§ 5.2.1) of the closed universe as a rubber balloon, and the local frame as the almost flat top of a glass table—flat except insofar as curvature is already initiated by marbles and oranges lying on the table and dimpling the rubber.

The minuscule inertia-determining potential of these stars of the Milky Way, nevertheless, by way of its gradient, produces a nonnegligible gravity field. That weak field is enough to hold those stars in orbit—orbits with periods of the order of 200 million years, as compared to the 165 year period of Neptune; Cassiopeia, Sirius, Vega, and all the other bright lights, we recognize, are not "fixed stars." Then how is it that they reveal the asymptotic inertial reference frame? Not by marking it, but by responding to it. Analyze the differential motions of the stars of the Milky Way—so Oort,[42] Clemence, and others[43,44] taught us—to determine the asymptotic Minkowski frame to impressive precision.

Onto that frame can we not at any rate expect the plane of vibration of the pendulum to lock itself? Yes, if it were out in space and even on Earth, if Earth were not rotating; no, when the pendulum hangs at the Pole on this rotating Earth. If rotation of Earth affects the plane of vibration of the pendulum at all, why is its effect not total? As between total rule over inertia and total impotence, where does the indicator lie for any intermediate influence? In the sum for inertia, expression (1.1.1).

The vote of Earth on the course of a geodesic—a vote on a scale on which the asymptotic inertial reference frame counts as unity—is not zero. It expresses itself in the potential of the Newtonian gravity field, the inertia-determining factor:

$$\begin{aligned}\text{(mass of Earth)/(distance, center to Pole)} &\cong 0.444\ \text{cm}/6.36\times 10^{8}\ \text{cm}\\ &\cong 6.98\times 10^{-10}.\end{aligned} \tag{6.4.1}$$

Here we have translated the conventional expression for the mass of Earth, 5.98×10^{27} g, into the required geometric units via the factor $G/c^2 = 0.7421 \times 10^{-28}$ cm/g.

Our concern, however, is not gravity-influence on a geodesic, but gravitomagnetic twisting of a plane. For the rate of this twisting of the local inertial frame what counts is not directly the mass of Earth, 0.444 cm, but its angular momentum, $J_\oplus \cong 145\ \text{cm}^2$ (translation of the accepted figure[39] of $\cong 5.86 \times 10^{40}\ \text{g cm}^2/\text{s}$ into geometric units; however, there is also the de Sitter or geodesic precession, a twist of the local inertial frame due to the motion of Earth around the Sun; see §§ 3.4.3 and 6.10). The twisting potential, a vector, proceeds by a rule distinct from that for the scalar bending potential. The vote-measuring power of a rotating mass, the vector potential, the scalar-product multiplier of $\boldsymbol{dx}dt$ in the metric (6.1.1), is given by expression (6.1.23):

$$-2\boldsymbol{J} \times \boldsymbol{x}/r^3. \tag{6.4.2}$$

The vector potential (6.4.2) of the spinning Earth gives rise to a calculated frame drag at the Pole of the amount

$$\begin{aligned}(\text{angular velocity}) &\cong 2J_\oplus/r^3\\ &\cong 2 \times 145\ \text{cm}^2/(6.36 \times 10^8\ \text{cm})^3\\ &= 1.13 \times 10^{-24}\ \text{radian per cm of time}\\ &= 220\ \text{milliarcsec per year.}\end{aligned} \tag{6.4.3}$$

There exists an unsophisticated way to arrive at a figure of the order of magnitude (6.4.3) without benefit of metric and vector potential. If Earth had complete governance over the plane of vibration of the pendulum, that plane would turn at the same rate as Earth itself, 360° per day, or 473×10^9 milliarcsec per year. However, the scalar voting power of Earth is only $M_{\text{Earth}}/R_{\text{Earth}} = 0.698 \times 10^{-9}$. Therefore expect the actual turning rate to be of the order of $(0.698 \times 10^{-9}) \times (473 \times 10^9) = 330$ milliarcsec/yr, not far from the proper prediction (6.4.3).

6.5 MEASUREMENT OF THE GRAVITOMAGNETIC FIELD

In the previous sections we have remarked the importance, in physics, to measure the gravitomagnetic field. First, because this field can be considered as a never directly measured "new" field of nature, analogous to the magnetic field generated by a magnet in electrodynamics (§§ 6.1 and 6.11), and the direct test of its existence will be a fundamental test of Einstein general relativity. Second, because the measurement of the gravitomagnetic field will supply the direct experimental proof that local inertial frames are influenced and dragged by mass-energy currents relative to other mass (§§ 6.1, 6.2, 6.11 and chap. 5), in agreement with some weak general relativistic interpretation of the Mach principle. Finally, we have noted the importance of the gravitomagnetic field in astrophysics. Theories of energy storage, power generation, jet formation, and jet alignment of quasars and galactic nuclei are based on the existence of the gravitomagnetic field of a supermassive black hole (§ 6.3).[9]

Therefore, since 1896, many experiments have been discussed and proposed to measure the gravitomagnetic field, among them: Benedikt and Immanuel Friedländer (1896)[149] (see below), August Föppl (1904)[150] (see below), de Sitter (1916)[24] (see below), Lense-Thirring (1917)[16,17] (see below), Ginzburg (1959),[45] Yilmaz (1959)[46,47] (see below), Pugh (1959),[48] Schiff (1960),[49] Everitt (1974),[50] Lipa, Fairbank, and Everitt (1974),[51] Van Patten and Everitt (1976)[52,53] (see below), Braginski, Caves, and Thorne (1977),[54] Scully (1979)[55] (see § 6.9), Braginski and Polnarev (1980),[56] Braginski, Polnarev, and Thorne (1984)[57] (see § 6.9), Paik, Mashhoon, and Will (1987)[58] (see § 6.9), and Cerdonio et al. (1989)[59] (see § 6.9). Of special interest for their extensively studied feasibility are the Gravity Probe-B experiment to measure the gravitomagnetic precession (6.1.28) of a gyroscope orbiting Earth (see § 6.6), and the LAGEOS III experiment (see below and §§ 6.7 and 6.8).

In 1896 B. and I. Friedländer[149] tried to measure the dragging effect due to a rapidly rotating, heavy fly-wheel on a torsion balance. I. Friedländer wrote,[151] "In the same way as centrifugal force is acting on a static wheel due to the rotation of the heavy earth and the cosmos, there should, I thought, appear on accordingly smaller scale a centrifugal force action on bodies near moving heavy fly-wheels. Would this phenomenon be detectable . . . " In 1904, A. Föppl[150] tried to measure the dragging effect on a gyroscope due to the rotation of Earth; he reached an accuracy of about 2% of Earth angular velocity. However, the general relativistic dragging effect on a gyroscope at the surface of Earth (at a U.S. or European latitude) is about 2×10^{-10} of its rotation rate! (see §§ 6.4 and 6.9).

In 1916 de Sitter[24] calculated the tiny shift of the perihelion of Mercury due to the rotation of the Sun, a particular case of the shift of the pericenter of an orbiting test particle due to the angular momentum of the central body, equation (6.1.34). This shift of the order of $-0.002''$/century is about 5×10^{-5} times smaller than Mercury's standard general relativistic precession of $\cong 43''$/century and is too small to be measured by today's tools. In 1918 Lense and Thirring[16,17] calculated the gravitomagnetic secular perturbations of the moons of various planets, in particular the fifth moon of Jupiter has a considerable gravitomagnetic secular precession. However the observations do not yet allow separation and measurement of this effect. In 1959 Yilmaz[46,47] proposed using polar satellites to detect the Earth's gravitomagnetic field, avoiding in such a way the effects due to the nonsphericity of Earth's gravity field. Indeed, for a polar satellite the classical nodal precession is zero. In 1976 Van Patten and Everitt[52,53] proposed measuring the Lense-Thirring nodal precession using two drag-free, guided satellites, counterrotating in the same polar plane. The reason for proposing two counterrotating satellites was to avoid the error associated with the determination of the inclination. Section 6.7 describes the new experiment LAGEOS III, proposed by one of us in 1984–1988,[60–68] to detect the gravitomagnetic field by measuring the orbital drag on nonpolar passive laser ranged satellites, and the accuracy achievable with such an experiment.

6.6 [THE STANFORD GYROSCOPE: GRAVITY PROBE-B (GP-B)]

Inspired by independent proposals of G. E. Pugh,[48] and of Leonard Schiff,[49] a Stanford University group consisting of Francis Everitt, William Fairbank, and others have worked for more than twenty-five years to make and fly superconducting gyroscopes in an Earth-orbiting satellite (see figs. 6.1 and 6.2, and picture 6.5).[50,51] They have been preparing and testing equipment with the intention to detect the gravitomagnetism of Earth and measure it to about 1% accuracy.

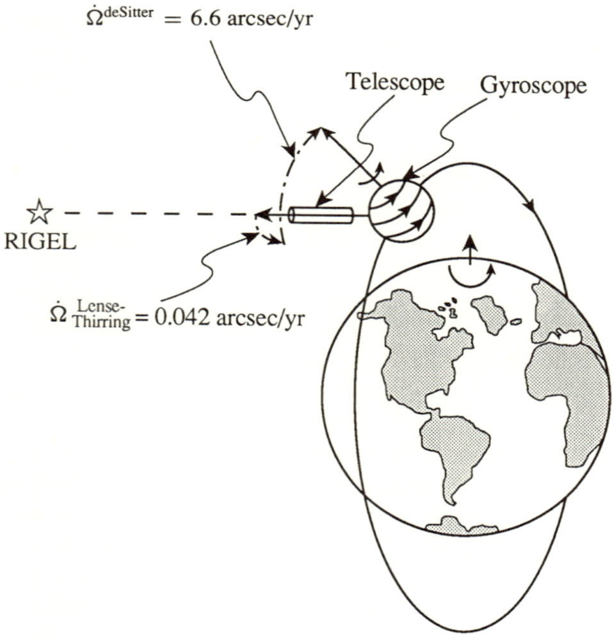

FIGURE 6.1. The GP-B, Gravity Probe-B, Experiment (§ 6.6): $\dot{\Omega}^{\text{de Sitter}}$ is the de Sitter precession; $\dot{\Omega}^{\text{Lense–Thirring}}$ is the dragging of inertial frames by the Earth angular momentum.

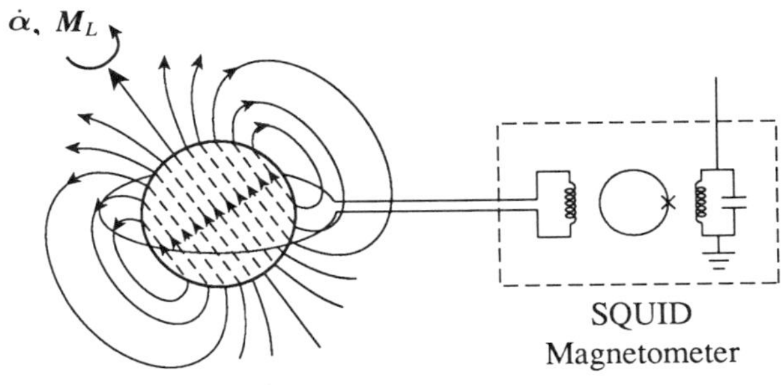

FIGURE 6.2. Readout of GP-B superconducting gyroscope, based on the London magnetic moment $\boldsymbol{M}_L$ and on a Superconducting Quantum Interference Device (SQUID) (adapted from ref. 51).

Flying at an altitude of about 650 km, the axis of a gyroscope is predicted to undergo a **gravitomagnetic frame dragging** of about 42 milliarcsec per year. This precession rate is less than that of the pendulum on Earth by reason of (1) the $1/r^3$ dependence of the frame-dragging field and (2) the averaging over latitudes—some favorable, some unfavorable—experienced by the gyroscope in its polar orbit (see fig. 6.1). The precession rate is given by equations (3.4.38) and (6.1.28):

$$\dot{\boldsymbol{\Omega}}_G = \frac{3}{2}\frac{M_\oplus}{r^2}(\hat{\boldsymbol{r}} \times \boldsymbol{v}) + \frac{3\hat{\boldsymbol{r}}(\boldsymbol{J}_\oplus \cdot \hat{\boldsymbol{r}}) - \boldsymbol{J}_\oplus}{r^3}. \tag{6.6.1}$$

The first term is the de Sitter precession, of about 6.6 arcsec per year for a GP-B gyroscope, and the second term the Lense-Thirring drag, of about 42 milliarcsec per year for GP-B.

The Earth-orbiting satellite will carry four gyroscopes. Each gyro is a quartz sphere 1.9 cm in radius, spherical to one part in a million, and homogeneous to 3×10^{-7} (see fig. 6.2). To reduce friction on that spinning sphere to an acceptable level, the satellite has to operate free of drag to the level of 10^{-10} g or better. To allow magnetic readout of the direction of spin, without unacceptable perturbation, and for this purpose to measure with a SQUID (Superconducting Quantum Interference Device) the so-called *London magnetic moment* (see § 6.9 and refs. 105 and 106) due to the rotation of a thin layer of superconducting niobium on the quartz sphere, a magnetic shield has to cut the ambient magnetic field to 10^{-7} gauss or less. An on-board telescope has to monitor the direction of a reference star to better than half milliarcsec per year—preferably to 0.2 milliarcsec (the angle subtended by the cross section of a hair seen at a distance of 50 km!) per year. Furthermore, the proper motion of the reference star (Rigel) should be known to better than half milliarcsec.

6.7 [LAGEOS III]

The fundamental idea[60–68] (in particular see refs. 62, 65, 67, 68 and 90) (Ciufolini 1984) of the **LAGEOS III experiment** can be broken down into two parts:

(1) Position measurements of laser ranged satellites, of the LAGEOS (1976) type (see below, fig. 6.3, and pictures 6.6 and 6.7), are accurate enough to detect the very tiny effect due to the gravitomagnetic field: the **Lense-Thirring precession** (eq. 6.1.33).
(2) To "cancel out" the enormous perturbations due to the nonsphericity of the Earth gravity field, we need a *new satellite*: *LAGEOS III*, with inclination supplementary to that of LAGEOS, and with the other orbital parameters, a and e, equal to those of LAGEOS.

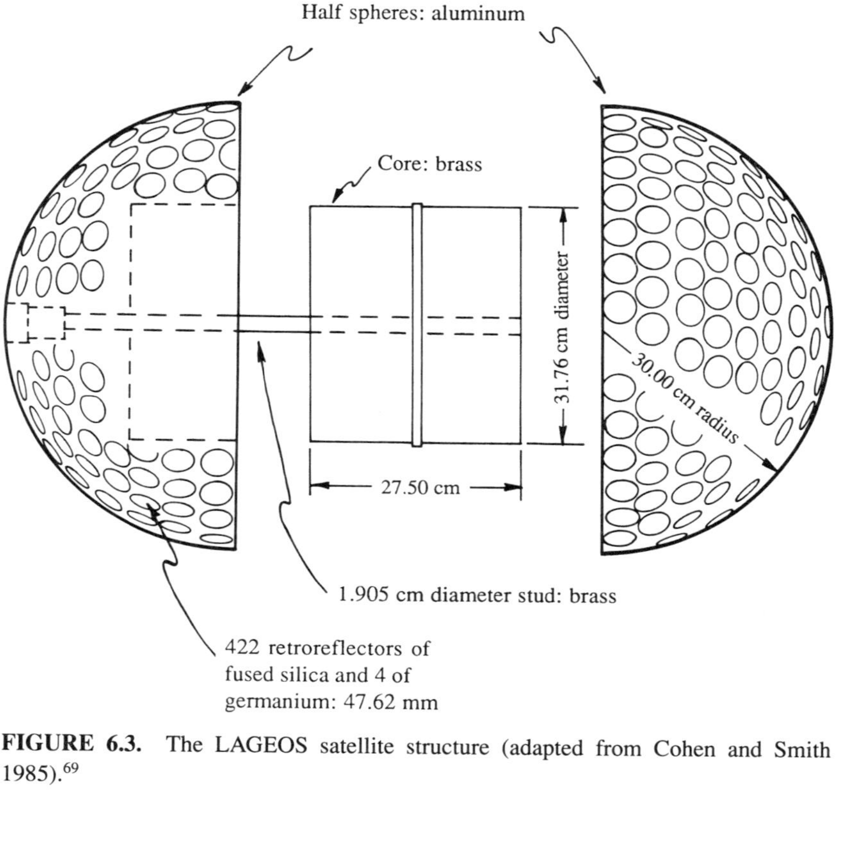

FIGURE 6.3. The LAGEOS satellite structure (adapted from Cohen and Smith 1985).[69]

LAGEOS (LAser GEOdynamics Satellite) (fig. 6.3)[70–72] is a high-altitude, small cross-sectional area-to-mass ratio, spherical, laser ranged satellite. It is made of heavy brass and aluminum, is completely passive and covered with laser retro-reflectors. It acts as a reference target for ground-based laser-tracking systems. LAGEOS was launched in 1976 to measure—via laser ranging—"crustal movements, plate motion, polar motion and Earth rotation." It continues to orbit and to pay scientific dividends. We know the LAGEOS position (and that of few similar laser ranged satellites) better than that of any other object in the sky. The relative accuracy[72] in tracking its position is of the order of 10^{-8} to 10^{-9}, of less than 1 cm over 5900 km of altitude! The LAGEOS[70] semimajor axis is a =12,270 km, the period $P = 3.758$ hr, the eccentricity $e = 0.004$, and the inclination $I = 109.94°$.

In 1977 Rubincam calculated[73] the LAGEOS general relativistic perigee precession: $\cong$ 3.3 arcsec/yr or about 195 m/yr. The problem of measuring the general relativistic perigee shift was then treated by Ashby and Bertotti in 1984.[74] Since the observable quantity is $e \times \dot{\omega}$ and since the LAGEOS orbit is almost circular, that is, $e \ll 1$, even though the relativistic precession of the pericenter $\dot{\omega}$ is significant, the relativistic advance $\dot{\omega}_{\text{rel}}$ is not easily measurable.

However, in 1989 Ciufolini and Matzner[75] showed that from the analysis of the LAGEOS data, it may be possible to have a 20%, or better, measurement of the LAGEOS relativistic perigee precession which can be used to put strong validity limits on some alternative theories of gravity, such as the non-Riemannian, nonsymmetric Moffat theory (see § 3.5.1). In 1993, Ciufolini and Nordtvedt by studying data of LAGEOS and of other laser ranged satellites have set a new limit to a hypothetical spatial anisotropy of the gravitational interaction of about $\frac{|\delta G|}{G} \lesssim 2 \times 10^{-12}$ (see § 3.2.3). Using LAGEOS data the limits to conceivable deviations from the *weak field* inverse square gravity law have also been improved (see § 3.2.1 and fig. 3.2). Furthermore, from the LAGEOS data analysis, it might be possible to get a further test of the very strong equivalence principle and of the Nordtvedt effect (although much less accurate than the Lunar Laser Ranging test, see § 3.2.5), that is, of the effect due to a conceivable violation of the very strong equivalence principle, that is, due to a hypothetical difference in the contributions of the gravitational binding energy of Earth to its inertial and to its gravitational mass. However, as already observed in section 3.2.5, the magnitude of this effect has already strong constraints (of the order of 0.1%) from Lunar Laser Ranging.[76,77,157]

One of the relativistic perturbations best measurable on satellites like LAGEOS, with $e \ll 1$, is the precession of the nodal lines. For LAGEOS, the measured total nodal precession is[70]

$$\begin{aligned}\dot{\Omega}^{\text{exp}}_{\text{LAGEOS}} &= \dot{\Omega}^{\text{various effects}}_{\text{LAGEOS}} + \dot{\Omega}^{\text{de Sitter}}_{\text{LAGEOS}} + \dot{\Omega}^{\text{Lense–Thirring}}_{\text{LAGEOS}} \\ &\cong 126°/\text{yr},\end{aligned} \tag{6.7.1}$$

and the Lense-Thirring precession is[62]

$$\dot{\Omega}^{\text{Lense–Thirring}}_{\text{LAGEOS}} = \frac{2GJ_\oplus}{c^2a^3(1-e^2)^{\frac{3}{2}}} \cong 31 \text{ milliarcsec/yr} \tag{6.7.2}$$

where $J_\oplus \cong 5.9 \times 10^{40}$ g cm^2/s $\cong 145$ cm^2 (in geometrized units) is the angular momentum of Earth.

The total nodal precession can be determined on LAGEOS with an accuracy better than 1 milliarcsec/yr.[72]

Unfortunately, the Lense-Thirring precession cannot be extracted from the experimental value of $\dot{\Omega}^{\text{exp}}_{\text{LAGEOS}}$ because of the uncertainty in the value of the **classical precession**:[78]

$$\dot{\Omega}^{\text{class}}_{\text{LAGEOS}} = -\frac{3}{2}n\left(\frac{R_\oplus}{a}\right)^2 \frac{\cos I}{\left(1-e^2\right)^2}\left\{J_2 + J_4\left[\frac{5}{8}\left(\frac{R_\oplus}{a}\right)^2 \times (7\sin^2 I - 4)\frac{(1+\frac{3}{2}e^2)}{(1-e^2)^2}\right] + \cdots\right\} \tag{6.7.3}$$

where $n = 2\pi/P$ is the orbital mean motion, $R_\oplus$ is Earth's equatorial radius, and the J_{2n} are the even zonal harmonic coefficients. This classical precession is due to the quadrupole and higher multipole mass moments of Earth, measured by the coefficients J_{2n}. The orbital parameters n, a, and e in formula (6.7.3) are determined with sufficient accuracy via the LAGEOS laser ranging and the average inclination angle I can be determined with sufficient accuracy over a long enough period of time (see next section).[67] Any other quantity in equation (6.7.3) can be determined or is known with sufficient accuracy, apart from the J_{2n}. In fact, the largest uncertainty in the classical precession $\dot{\Omega}^{\text{class}}_{\text{LAGEOS}}$ arises from the uncertainty in the coefficients J_{2n}. This uncertainty, relative to J_2, is of the order of[79,80]

$$\frac{\delta J_{2n}}{J_2} \sim 10^{-6}. \tag{6.7.4}$$

For J_2, this corresponds, from formula (6.7.3), to an uncertainty in the nodal precession of about 450 milliarcsec/yr, plus the uncertainty due to the higher J_{2n} coefficients. Therefore, the uncertainty in $\dot{\Omega}^{\text{class}}_{\text{LAGEOS}}$ is several times larger than the Lense-Thirring precession.

A solution would be to orbit several high-altitude, laser ranged satellites, similar to LAGEOS, to measure J_2, J_4, J_6, etc., and one satellite to measure $\dot{\Omega}^{\text{Lense–Thirring}}$.

Another solution would be to orbit polar satellites; in fact, from formula (6.7.3), for polar satellites, since $I = 90°$, $\dot{\Omega}^{\text{class}}$ is equal to zero. As mentioned before, Yilmaz proposed the use of polar satellites in 1959.[46,47] In 1976 Van Patten and Everitt[52,53] proposed a complex and high cost experiment with two drag-free, guided, counterrotating, polar satellites. The reason for proposing *two* counterrotating polar satellites was to avoid inclination measurement errors.

A new solution would be to orbit a second satellite, of LAGEOS type, with the same semimajor axis, the same eccentricity, but the *inclination supplementary* to that of LAGEOS (see fig. 6.4).[60–68] Therefore, "LAGEOS III" should have the following orbital parameters:

$$I^{III} = 180^\circ - I^I \cong 70^\circ, \quad a^{III} = a^I, \quad e^{III} = e^I. \tag{6.7.5}$$

With this choice, since the classical precession $\dot{\Omega}^{\rm class}$ is linearly proportional to $\cos I$, $\dot{\Omega}^{\rm class}$ will be equal and opposite for the two satellites:

$$\dot{\Omega}^{\rm class}_{III} = -\dot{\Omega}^{\rm class}_{I}. \tag{6.7.6}$$

By contrast, since the Lense-Thirring precession $\dot{\Omega}^{\rm Lense\text{–}Thirring}$ is independent of the inclination (eq. 6.7.2), $\dot{\Omega}^{\rm Lense\text{–}Thirring}$ will be the same in magnitude and sign for both satellites:

$$\dot{\Omega}^{\rm Lense\text{–}Thirring}_{III} = \dot{\Omega}^{\rm Lense\text{–}Thirring}_{I}. \tag{6.7.7}$$

"Adding" the measured nodal precessions $\dot{\Omega}^{\rm exp}$, we will get

$$\dot{\Omega}^{\rm exp}_{III} + \dot{\Omega}^{\rm exp}_{I} - \dot{\Omega}^{\rm other\ forces}_{I+III} = 2\dot{\Omega}^{\rm Lense\text{–}Thirring}_{III} = 2\dot{\Omega}^{\rm Lense\text{–}Thirring}_{I}, \tag{6.7.8}$$

where $\dot{\Omega}^{\rm other\ forces}_{I+III}$ is the sum of the nodal precessions of LAGEOS and LAGEOS III due to all the other perturbations and calculable with sufficient accuracy (see next section).[62–68] We observe that this method of measuring the Lense-Thirring

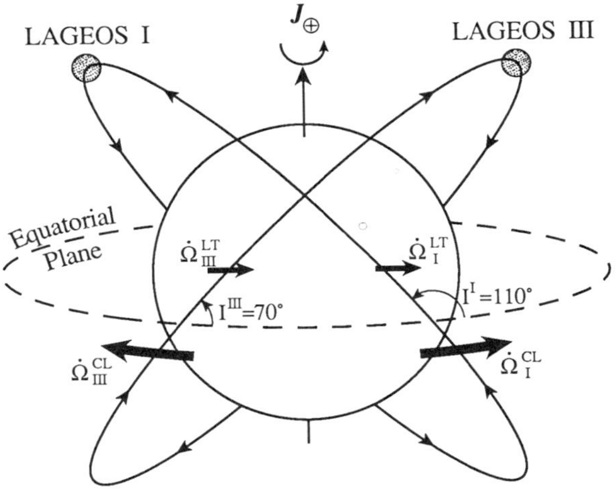

FIGURE 6.4. The LAGEOS and LAGEOS III orbits and their classical and gravitomagnetic nodal precessions. A new configuration to measure the Lense-Thirring effect (adapted from Ciufolini 1989).[60–68]

effect should not be thought of as the subtraction of two large numbers to get a very small number, but as the sum of the small *unmodeled* nodal precession of LAGEOS, $\dot{\Omega}_I^{\text{unmodeled}}$, with the small *unmodeled* nodal precession of LAGEOS III, $\dot{\Omega}_{III}^{\text{unmodeled}}$ (corresponding to the same Earth gravity field solution), to get $2\dot{\Omega}^{LT} = \dot{\Omega}_I^{\text{unmodeled}} + \dot{\Omega}_{III}^{\text{unmodeled}}$.

This idea to orbit a satellite LAGEOS III, to couple to LAGEOS, with the same orbital parameters of LAGEOS but supplementary inclination, can be described in the following way. Since the classical nodal precession is equal and opposite for two satellites, the bisector of the angle between the nodal lines of the two satellites would define *a kind of gyroscope*, in the sense that this line will not be affected by the partially unknown classical precession (6.7.3) but only by the general relativistic "dragging of inertial frames" (6.7.2) and by the de Sitter precession (see fig. 6.5). In other words, one may think of the bisector as the equatorial component of the orbital angular momentum vector of a satellite

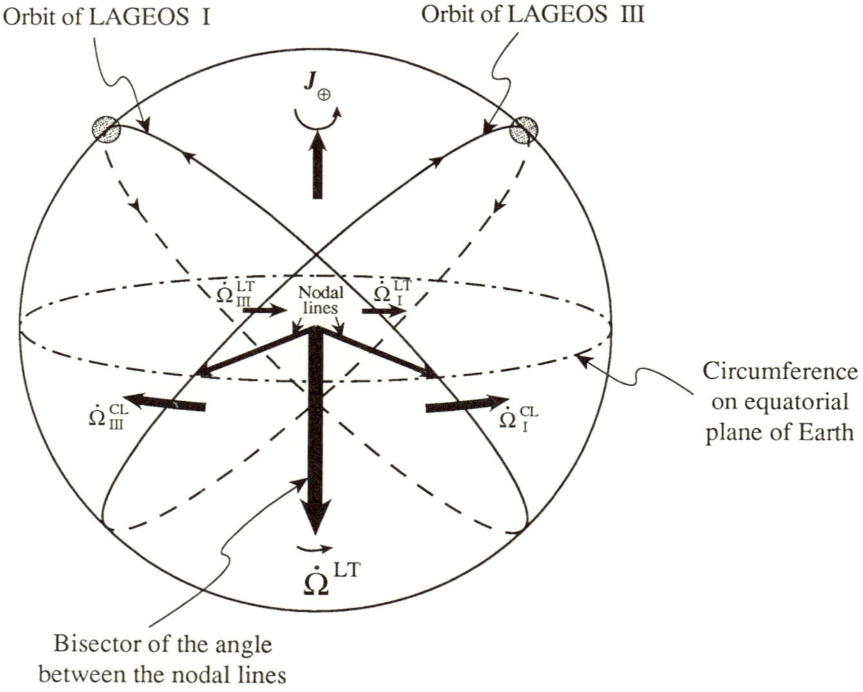

FIGURE 6.5. A kind of enormous gyroscope to measure the gravitomagnetic field:[67] the bisector of the angle between the nodal lines of LAGEOS and LAGEOS III (see § 6.7).

orbiting a central body, perfectly spherically symmetric, with null quadrupole and null higher multipole mass moments.

Since the LAGEOS III experiment requires only a copy of the LAGEOS passive satellite launched in 1976, *no* new technology or instrumentation needs to be developed, and therefore LAGEOS III will be a very low cost space experiment; since it uses *passive* satellites, it is also essentially riskless.

6.8 [ERROR SOURCES AND ERROR BUDGET OF THE LAGEOS III EXPERIMENT]

A fundamental problem of the LAGEOS III experiment, to measure the gravitomagnetic field by laser ranged satellites with supplementary inclinations, is to estimate the accuracy of the measurement. Since some of the error sources that may affect the experiment are common to some other relativistic experiments, it may be interesting to describe and summarize them.[67] These errors may be divided into the following categories.[65,67]

(1) *Errors from gravitational perturbations.* These arise from uncertainties in modeling the LAGEOS (and LAGEOS III) gravitational nodal perturbations. These errors may be subdivided into the following:
   (a) *Errors from orbital injection errors,*[62,67] due to the uncertainties in the knowledge of the static part of the even zonal harmonic coefficients, J_{2n}, of the Earth gravity field. These errors are, a priori, zero for two satellites that orbit at supplementary inclinations. However, any deviation of the orbital parameters of LAGEOS III from the optimal values (6.7.5) will introduce uncertainties which must be evaluated.
   (b) *Errors from other gravitational perturbations:*[65,67] Static odd zonal harmonic perturbations; static nonzonal harmonic perturbations; nonlinear harmonic perturbations; solid and ocean Earth tides; de Sitter (or geodetic) precession; Sun, Moon, and planetary tidal accelerations; nonlinear, n-body, general relativistic effects;[67,147] tiny deviations from geodesic motion of LAGEOS due to the LAGEOS quadrupole moment, to the LAGEOS intrinsic angular momentum, and to the coupling between Earth spin and LAGEOS spin (however, these last three perturbations are *completely* negligible). The main error source[65,67] is due, however, to the uncertainties in modeling the dynamical, time-dependent, part of the Earth gravity field, that is, due to the uncertainties in modeling solid and ocean Earth tides.

(2) *Errors from nongravitational perturbations.*[63–68] These include direct solar radiation pressure, Earth's albedo radiation pressure, satellite eclipses,

anisotropic thermal radiation, Poynting-Robertson effect, Earth's infrared radiation, atmospheric drag, solar wind, interplanetary dust, and Earth's magnetic field. The main error sources[63,67,68] are, however, due to the uncertainties in modeling the Earth's albedo and the satellite anisotropic thermal radiation (see below).

(3) *Errors from uncertainties in the orbital parameters.*[67,68] These errors are due to the uncertainties in the determination of the LAGEOS orbital parameters[62] and particularly to the errors in the determination of the inclination I, and of the LAGEOS nodal longitude Ω, relative to an asymptotic inertial frame.

We observe that LAGEOS (1) has a very small ratio of cross-sectional area to mass: $A/m \cong 0.0069$ cm^2/g, (2) is a high-altitude satellite: $(a - R_{\oplus}) \cong 5900$ km, orbiting well above the dense atmosphere, (3) is essentially spherically symmetric, and (4) has a very small orbital eccentricity, $e = 0.004$. For these reasons, such perturbations as radiation pressure and atmospheric drag are minimized, and it is possible to model these perturbations with enough accuracy to measure the Lense-Thirring effect.

Furthermore, several sources of nodal errors are negligible because they are axially symmetric and therefore equal in magnitude but opposite in sign for two satellites with supplementary inclinations (for example, the perturbations due to the static and dynamical part (tides) of the J_{2n} coefficients; most of the Earth direct infrared-radiation driven nodal drag; part of the albedo radiation pressure nodal precession, etc.).[62,64,67,68] Several other periodical sources of uncertainty have a negligible average error over a period of approximately three years of the satellite nodal motion.

To give an idea of the kind of perturbations that have to be modeled in order to extract and measure the Lense-Thirring effect from the observed LAGEOS plus LAGEOS III nodal precessions, let us examine an interesting phenomenon: the anisotropic thermal radiation effect, or Yarkovsky effect, and a particular case of it, the "Rubincam effect," that is, the effect due to the coupling between the heating of the LAGEOS satellite by the Earth infrared radiation and the thermal inertia of the LAGEOS silica retroreflectors (see figs. 6.3, 6.6, and 6.7). Corresponding to these perturbations, there will be small modeling errors that have been evaluated in the final error budget. We first observe that the LAGEOS orbit has an observed, anomalous, long-period decrease in the semimajor axis[81] of about 1.2 mm/day, which corresponds to a partially "unknown" along-track acceleration of about -3.3×10^{-10} cm/s^2. However, the Rubincam effect, in addition to the standard neutral and charged particle drag, should explain the anomalous shrinkage of the orbit.

The electromagnetic radiation from the Sun and the radiation from Earth each instantaneously heat one hemisphere of LAGEOS. Because of the finite heat capacity and heat conductivity of the body, there is an anisotropic distribution

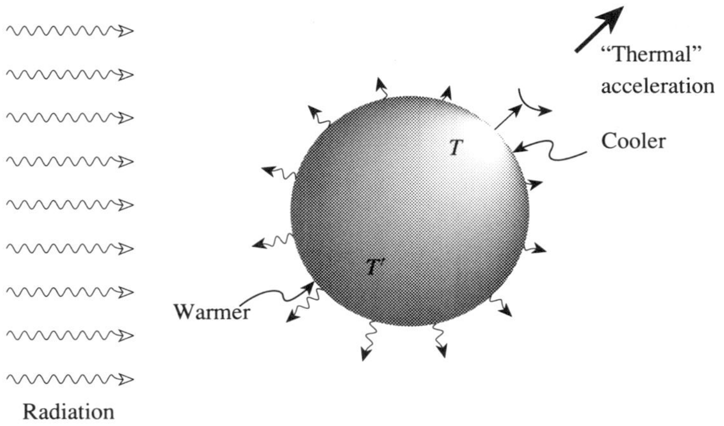

FIGURE 6.6. The Yarkovsky effect. The LAGEOS anisotropic thermal radiation $T' > T$ and the resulting "thermal thrust."[88]

of temperature on the satellite. Therefore, according to the Stefan-Boltzmann law, there is as anisotropic flux of energy ($\sim T^4$) and momentum from the surface of the satellite contributing to its acceleration.[82] However, if LAGEOS is spinning fast enough,[83] the anisotropy in the satellite temperature distribution should be mainly latitudinal (see fig. 6.6).[82,84]

The acceleration $a_{\Delta T}$ due to this "thermal thrust," acting along the satellite spin axis, is thus

$$a_{\Delta T} \sim \frac{4\varepsilon\pi r_L^2 \sigma T^3 \Delta T}{c\, m_L} \tag{6.8.1}$$

where $\varepsilon \cong 0.4$ is the emissivity coefficient of LAGEOS, $r_L = 30$ cm its radius, $m_L = 4.1 \times 10^5$ g its mass, $\sigma = 5.67 \times 10^{-8}$ W m^{-2} K^{-4} the Stefan's constant, T the average equilibrium temperature of LAGEOS, and ΔT the temperature gradient between the two hemispheres of LAGEOS. For example, for $T \cong 280$ K and $\Delta T = 5$ K,[85] we get an acceleration $a_{\Delta T}$ of the order of $a_{\Delta T} \sim 2 \times 10^{-9}$ cm/s^2, constant in direction along one LAGEOS orbit (the satellite spin axis is assumed to be substantially constant along one orbit).

We insert this value in the equation for the rate of change of the nodal longitude:

$$\frac{d\Omega}{dt} = \frac{1}{na \sin I}\left(1 - e^2\right)^{-\frac{1}{2}} fW \frac{r}{a} \sin(\nu + \omega) \tag{6.8.2}$$

where $n = \frac{2\pi}{P}$ is the orbital mean motion, $P = 3.758$ hr the orbital period, a =12,270 km the orbit semimajor axis, $I \cong 109^\circ\!.94$ the LAGEOS orbital inclination, $e = 4 \times 10^{-3}$ the orbital eccentricity, f the magnitude of the external force per unit satellite mass, W the direction cosine of the force f along the normal to the orbital plane, ν the true anomaly, and ω the argument of the pericenter (see § 6.1). We then integrate this expression over one orbital period $P = 3.758$ hr and neglecting, for the moment, Earth radiation and eclipses of the satellite by Earth, we have (in this calculation W is assumed to be constant along one orbit)

$$\dot{\Omega}^{\Delta T_{\odot}} \lesssim 2 \times 10^{-3} \dot{\Omega}^{\text{Lense–Thirring}}. \tag{6.8.3}$$

Therefore, over one orbit, the secular nodal precession due to the LAGEOS thermal anisotropy from Sun radiation is negligible when the satellite eclipses and the Earth radiation are not considered.

However, when the satellite orbit is partially in the shadow of Earth, the secular nodal precession (6.8.3) due to the satellite thermal anisotropy is no longer reduced, over one orbit, by a factor of the order of the orbital eccentricity, $\sim e = 4 \times 10^{-3}$.

Furthermore, Rubincam discovered that the infrared radiation from Earth plus the thermal inertia of the LAGEOS retroreflectors can cause a force on the satellite.[82,85,86] Infrared radiation from Earth is absorbed by the LAGEOS retroreflectors; therefore, due to their thermal inertia and to the rotation of the satellite there is a latitudinal temperature gradient on LAGEOS. The corresponding thermal radiation causes an along-track acceleration opposite to the satellite motion (see fig. 6.7). This anisotropic thermal radiation may cause an acceleration of LAGEOS of the order of 10^{-10} cm/s^2, variable in direction. This thermal drag *may* explain about 70% of the observed LAGEOS semimajor axis decrease of about 1.2 mm/day due to a peculiar along-track acceleration of about -3×10^{10} cm/s^2. The rest of the observed effect is probably due to neutral and charged particle drag (see fig. 6.7).[86,87]

However, some components of the nodal precession associated with the Rubincam effect are periodic with a nodal period and with half of a nodal period. Therefore these components give a negligible uncertainty when averaged over the period of the node. Furthermore, part of the secular nodal precession associated with the Rubincam effect is equal in magnitude and opposite in sign[88] for the two satellites, LAGEOS and LAGEOS III, with supplementary inclinations. Finally, these anisotropic thermal radiation perturbations can in part be modeled with measurements[154] of the LAGEOS spin axis orientation and spin rate. Therefore, after various computations, assuming the worst possible orientations of the LAGEOS and LAGEOS III spin axes,[154] one finds for these two effects:[67,152]

$$\delta\dot{\Omega}^{\Delta T} = \delta\dot{\Omega}^{\Delta T+}_{\text{eclipses}} + \delta\dot{\Omega}^{\Delta T,}_{\text{Earth radiation}} \lesssim 2\% \dot{\Omega}^{\text{Lense–Thirring}}. \tag{6.8.4}$$

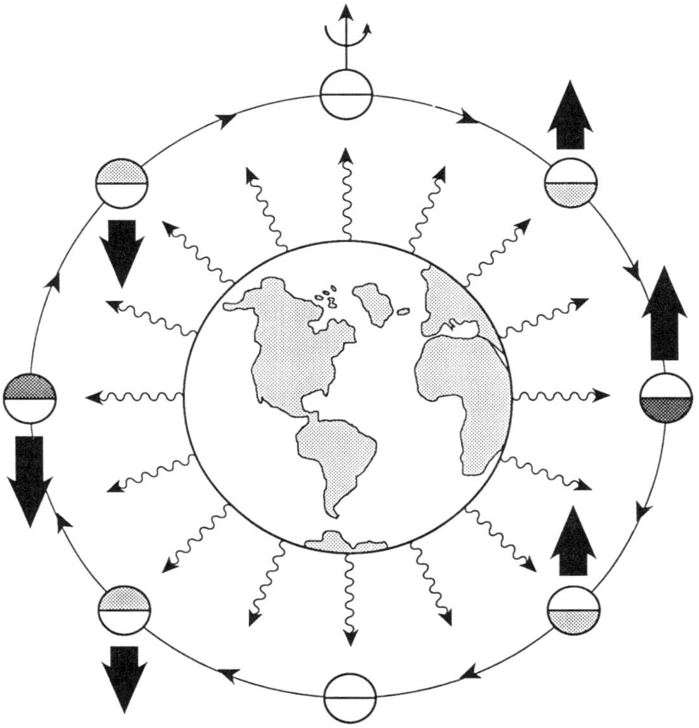

FIGURE 6.7. The Rubincam effect, that is, the coupled effect of the LAGEOS heating by the Earth infrared radiation and the thermal inertia of the satellite retroreflectors. The acceleration $a_{\Delta T,\mathrm{Earth}}$ is opposite to the satellite orbital motion (the satellite spin axis is, for simplicity, assumed to be lying on the orbital plane) (adapted from Rubincam 1987).[82]

Error Budget

(1) Taking the errors in the J_{2n} from the gravity field solutions GEM–L2 and GEM-T1, and assuming a special launching vehicle[67,89,90,153], a two-stage Delta II, one gets the statistical error in the measurement of $\dot{\Omega}^{\text{Lense–Thirring}}$ from orbital injection errors (however, see after formula 6.8.10):

$$\delta\dot{\Omega}\left(\begin{smallmatrix}\text{total injection}\\ \text{error in } \delta I \text{ and } \delta a\end{smallmatrix}\right) \lesssim 1\%\dot{\Omega}^{\text{Lense–Thirring}}. \tag{6.8.5}$$

(2) As already observed, among the gravitational and general relativistic perturbations, the only important source of uncertainty is in the modeling of solid and ocean Earth tides.

It is well known that the shape of Earth changes because of the inhomogeneous gravitational field generated by the Sun and by the Moon. This change in shape changes in turn the potential and the gravity field generated by Earth.[91–93] This variation of the Earth gravity field contributes to the nodal motion of a satellite. The phenomenon of tides can be mainly described as solid tidal deformations of Earth and oceanic tides. These tidal effects may be represented[94] as time variations of the spherical harmonic coefficients of the Earth gravitational field,[78] that is, the total tidal variation can be written as variation $\Delta\overline{C}_{lm} - i\Delta\overline{S}_{lm}$ of the coefficients $\overline{C}_{lm}$ and $\overline{S}_{lm}$, of degree l and order m, used to describe the Earth field. In the LAGEOS III experiment, the major source of uncertainty is associated with the parameters corresponding to tides with periods longer than the period of observation. From a list of the main Earth tides,[92] we learn that such long-period Earth deformations are due to the motion of the Moon node, taking place in $\cong$ 18.6 years. The main secular and long-period tides[67] affecting the LAGEOS node are a constant one, $l = 2$ and $m = 0$, and two periodic ones, with periods of 18.6 years and its half, 9.3 years, corresponding to a $l = 2$ and $m = 0$ Earth deformation. By far, the largest part of the uncertainty in the LAGEOS nodal tidal perturbations is due to this secular drift and to these two periodical effects, with periods of 18.6 and 9.3 years, corresponding to $l = 2$ and $m = 0$. However, the nodal perturbations induced by the Earth tidal deformations with $l =$ even and $m = 0$, and therefore the nodal perturbations due to the 18.6 and 9.3 year tides with $l = 2$ and $m = 0$, are equal in magnitude and opposite in sign for the two LAGEOS satellites; consequently they give a negligible uncertainty in the measurement of $\dot{\Omega}^{\text{Lense–Thirring}}$. Because of this cancellation, using two satellites with supplementary inclinations, and assuming a partial improvement in the tidal models and parameters, after some computations we find, over the period of the node:

$$\delta\dot{\Omega}^{\text{tides}} \lesssim 3\%\dot{\Omega}^{\text{Lense–Thirring}}. \qquad (6.8.6)$$

(3) The main nongravitational error sources are[65,67] the Earth albedo radiation pressure[95–97] and anisotropic thermal radiation from the satellite. In regard to the albedo, assuming that the Earth's surface reflects sunlight according to the Lambert's law, but considering also anisotropically reflected radiation and specular reflection, we find[67]

$$\delta\dot{\Omega}^{\text{albedo}} \lesssim 1\%\dot{\Omega}^{\text{Lense–Thirring}}. \qquad (6.8.7)$$

In regard to the effect of anisotropic thermal radiation, the Yarkovsky effect, including the satellite heating due to the Earth's infrared radiation, and the effect due to direct solar radiation heating and satellite eclipses by

Earth, assuming a modeling[154] of part of these effects, we found (6.8.4)[67]

$$\delta\dot{\Omega}^{\Delta T} \lesssim 2\%\dot{\Omega}^{\text{Lense–Thirring}}. \tag{6.8.8}$$

It is important to observe that the secular effect from neutral and charged particle drag due to the rotation of the exosphere has been shown to give an average nodal precession, for LAGEOS and LAGEOS III, negligible relative to the Lense-Thirring effect.[98]

(4) The main errors in the determination of the satellite orbital parameters, affecting the LAGEOS gravitomagnetic experiment, are the errors in determining the inclination relative to the instantaneous rotation axis of Earth, and the nodal longitude relative to an asymptotic inertial frame. Part of these errors is due to the uncertainties in determining the position of the satellites relative to the Earth's crust, that is, relative to the satellite-laser-ranging-stations frame, because of position uncertainties of the stations, observation biases, and other measurement uncertainties. Part of these errors is due to the uncertainties in the knowledge of the Earth orientation relative to an asymptotic inertial frame; that is, to the uncertainties in the X and Y coordinates of the pole and in the Earth rotational orientation, $UT1$. Fortunately, error of this kind is greatly reduced by using the VLBI determinations[99,100] of the Earth orientation parameters, $UT1$ and X, Y of the pole, having accuracies of less than 1 milliarcsec/yr.

Considering various sources of observation biases and measurement errors, including possible measurement errors in the LAGEOS position and errors in the determination of the LAGEOS inclination and nodal longitude, I and Ω, due to laser light atmospheric refraction mismodeling, we find[67,152]

$$\delta\dot{\Omega}\left[\delta\dot{\Omega}(\delta UT1),\ \delta I(\delta X,\delta Y),\ \begin{array}{c}\text{Earth crust}\\ \text{measurement}\\ \text{errors}\end{array},\ \begin{array}{c}\text{atmospheric}\\ \text{refraction}\\ \text{mismodeling}\end{array}\right] \lesssim 3\%\dot{\Omega}^{\text{Lense–Thirring}}. \tag{6.8.9}$$

In conclusion, from formulae (6.8.5), (6.8.6), (6.8.7), (6.8.8), and (6.8.9), we obtain an upper bound (in the sense that have been assumed the worst possible cases and configurations of the orbital parameters of Earth, Sun and LAGEOS, such as the LAGEOS spin orientation or its longitude of pericenter) for the statistical (1σ) **error in the measurement of the gravitomagnetic field,** using two supplementary-inclination LAGEOS-type satellites, over the period of the node of 1046 days[65–67] (see table 6.1):

$$\delta\dot{\Omega}_{\text{error}} \lesssim \sqrt{(1\%)^2 + (3\%)^2 + (1\%)^2 + (2\%)^2 + (3\%)^2}$$

$$\delta\dot{\Omega}_{\text{error}} \lesssim 5\%\dot{\Omega}^{\text{Lense–Thirring}} \tag{6.8.10}$$

TABLE 6.1 An upper bound to the statistical error (see § 6.8) in the measurement of the gravitomagnetic field with the LAGEOS III experiment.[67]

Source	*Upper Bound to Statistical Error*	
Orbital injection errors:		
δI and δa:	1%	$\longrightarrow 1\% \dot{\Omega}^{LT}$
δe:	negligible	
Gravitational Perturbations:		
Earth tides:	3 %	$\longrightarrow 3\% \dot{\Omega}^{LT}$
Others:	negligible	
Errors in determination of orbital parameters:		
$\delta\dot{\Omega}(\mathrm{SLR})$		
$\delta I(\mathrm{SLR})$		
$\delta I\left(\begin{smallmatrix}\text{atmospheric}\\ \text{refraction}\end{smallmatrix}\right)$		$\longrightarrow 3\% \dot{\Omega}^{LT}$
$\delta\dot{\Omega}(\delta UT1(\mathrm{VLBI}))$		
$\delta I(\delta X, \delta Y(\mathrm{VLBI}))$		
Nongravitational perturbations:		
Earth albedo radiation pressure:		$1\% \dot{\Omega}^{LT}$
Anisotropic thermal radiation, Earth infrared radiation, Sun radiation plus eclipses:		$\longrightarrow 2\% \dot{\Omega}^{LT}$
Others:	negligible	
Upper bound to total statistical error: less than 5% $\dot{\Omega}^{\text{Lense–Thirring}}$		

over $\approx$ 3 year data period and **improvable in the future**, by using longer periods, by improving the various perturbation models, and by using data from other laser ranged satellites, yet to be launched or recently launched, such as LAGEOS II (October 1992). For the 1993 error budget of the LAGEOS III gravitomagnetic measurement, see below.

The error analysis has been repeated and confirmed by comprehensive and extensive computer error analyses, simulations of the experiment, and "blind tests" performed in a joint study by **NASA**, **ASI** (Italian Space Agency), the US Air Force, University of Texas at Austin, and several other universities and research centers in the US and Italy,[101] under the supervision of a joint NASA-ASI committee (1989), and has also been confirmed and improved by several following analyses (1989–1993).

Present (1993) error analyses by the University of Texas at Austin,[152] and other studies,[153–156] show a *total statistical error* in the LAGEOS III experiment of *about 3%* $\dot{\Omega}^{\text{Lense–Thirring}}$ over a 3 yr period. The substantial reduction of the total statistical error is due to (a) improvements in the modeling of geopotential and Earth tides, also due to the data from the LAGEOS II satellite[152] launched in 1992; (b) improvements in the modeling of seasonal variations in low degree geopotential harmonics, also due to the LAGEOS II data;[152] (c) a McDonnell Douglas-NASA study (1990) on Delta II, two-stage, launch vehicles, together with the improvements in the geopotential determination, which imply a substantially negligible injection error in the LAGEOS III experiment;[153] (d) accurate measurements and theoretical studies of the spin axis orientation and spin rate of LAGEOS, and of its optical and thermal properties, in order to model and reduce the anisotropic thermal radiation uncertainties;[154] and (e) accurate computations of the Earth albedo effect on the LAGEOS satellites orbits[155] (using data from meteorological satellites) and especially on the LAGEOS nodal longitude and on the bisector of the nodal lines of two LAGEOS satellites with supplementary inclinations,[156] which showed a substantially negligible error in the LAGEOS III experiment due to Earth albedo.

Finally, we observe that the **3% total statistical error** budget in the LAGEOS III experiment is relative to a 3 yr period data analysis; however, since the lifetime of LAGEOS-type satellites is several orders of magnitude longer, the total error will decrease with longer periods of observations.

(For a detailed description of the LAGEOS III experiment see refs. 60–68, 90, 101, and 152–156, especially 62, 65, 67, 68, 90 and 101).

6.9 [OTHER EARTH "LABORATORY" EXPERIMENTS]

In section 6.4, formula (6.4.3), we have given the gravitomagnetic precession of a Foucault pendulum at the pole:

$$\dot{\Omega}^{\text{Lense–Thirring}}_{\text{pole pendulum}} \cong 220 \text{ milliarcsec/yr}. \tag{6.9.1}$$

Braginsky, Polnarev, and Thorne note that gravitomagnetism is "so weak in the solar system that it has never been detected. This is sad for astrophysics as well as fundamental physics, since some theories of quasars and galactic nuclei rely on the GM (gravitomagnetic) field of a supermassive black hole for energy storage, power generation, jet formation, and jet alignment."[57] Therefore, they propose a *Foucault pendulum* operated within a few kilometers of the South Pole. Braginsky, Polnarev, and Thorne analyze the problems presented by magnetic forces, frictional damping, Pippard precession, position-dependent forces, frequency anisotropy, seismic noise, atmospheric refraction, and distortion and tilt of the telescope. They propose a substantial program of laboratory development to clarify these problems and, if possible, bring them under control, with a view to a future pendulum experiment.

Another Earth-laboratory experiment, proposed by a group at the University of Trento,[59] would try to detect and measure the gravitomagnetic field by using **superconducting** and superfluid **gyroscopes**. Let us briefly describe the idea of the experiment.

On Earth, the angular velocity of a laboratory with respect to a local inertial frame $\dot{\boldsymbol{\Omega}}_{\text{local}}$ is given by

$$\dot{\boldsymbol{\Omega}}_{\text{local}} = \dot{\boldsymbol{\Omega}}_{\text{VLBI}} - \dot{\boldsymbol{\Omega}}^{\text{Thomas}} - \dot{\boldsymbol{\Omega}}^{\text{de Sitter}} - \dot{\boldsymbol{\Omega}}^{\text{Lense–Thirring}} \tag{6.9.2}$$

where $\dot{\boldsymbol{\Omega}}_{\text{VLBI}}$ is the angular velocity of the laboratory with respect to an asymptotic inertial frame, $\dot{\boldsymbol{\Omega}}^{\text{Thomas}}$, $\dot{\boldsymbol{\Omega}}^{\text{de Sitter}}$, and $\dot{\boldsymbol{\Omega}}^{\text{Lense–Thirring}}$ are, respectively, the contributions of the Thomas precession, of the de Sitter or geodetic precession (see § 3.4.3 and the next section), and of the Lense-Thirring drag. To measure $\dot{\boldsymbol{\Omega}}^{\text{Lense–Thirring}}$ one should measure $\dot{\boldsymbol{\Omega}}_{\text{local}}$ and then subtract from it the VLBI (Very Long Baseline Interferometry), independently measured value of $\dot{\boldsymbol{\Omega}}_{\text{VLBI}}$. The contributions due to the Thomas and de Sitter precessions can then be calculated and subtracted, since they originate from already tested parts (see § 3.4.3 and next section).

From expression (6.1.28), the Lense-Thirring term along the Earth spin axis is given by

$$-\dot{\Omega}^{\text{Lense–Thirring}} = \frac{2}{5}\left(\frac{M_\oplus}{R_\oplus}\dot{\Omega}_\oplus\right)(1 - 3\cos^2\theta) \tag{6.9.3}$$

where $M_\oplus$ and $R_\oplus$ are Earth mass and radius, and θ is the co-latitude of the laboratory. This correction amounts to $\sim 3 \times 10^{-14}$ rad/s at the highest reasonable latitudes achievable with a European laboratory. $\dot{\Omega}_{\text{VLBI}}$ can be measured to better than 2×10^{-16} rad/s with VLBI in a one year measurement time. However, global and local Earth crust movements have to be studied and modeled. $\dot{\boldsymbol{\Omega}}_{\text{local}}$ has to be measured by some rotation sensor with comparable accuracy.

The University of Trento *rotation sensor*, called the Gyromagnetic Electron Gyroscope (GEG),[59] is based on the idea of measuring the magnetization induced by rotation of a ferromagnetic material.

Magnetization by rotation[102] was demonstrated for ferromagnetic materials by Barnett.[103,104] For *superconductors* it was predicted by Becker, Heller, and Sauter[105] and London[106] and demonstrated in subsequent experiments:[107,108] a body that becomes magnetized in the presence of an applied field $\boldsymbol{B}^{\text{ext}}$, can also be magnetized by rotation, with angular velocity $\dot{\boldsymbol{\Omega}}$, with respect to a local inertial frame. The magnetization of the rotating magnetic body is the same that would be induced by an external magnetic field $B_i^{\text{ext}} = \gamma_{ij}\dot{\Omega}_j$. The magnetomechanical tensor γ_{ij} is related to the microscopic degrees of freedom associated with the magnetization. For isotropic ferromagnets one can be write $\gamma_{ij} = \left(\frac{2m}{ge}\right)\delta_{ij}$, thus $\boldsymbol{B}^{\text{ext}} = \left(\frac{2m}{ge}\right)\dot{\boldsymbol{\Omega}}$, where $g = 1$ for superconductors (magnetization of orbital origin) and $g \approx 2$ for ferromagnets (magnetization almost entirely of spin origin). The magnetization of a superconductor rotating with respect to a local inertial frame is the so-called *London moment*.[102,103] If the body is a closed superconducting *shield*, one of the consequences of this rotationally induced magnetization is to create in the interior of the shield (rotating with angular velocity $\boldsymbol{\delta}\dot{\boldsymbol{\Omega}}$) a true magnetic field $\boldsymbol{\delta B}_{sh}$ opposed to the polarizing field

$\delta\boldsymbol{B}^S_\Omega$. Due to the difference in the value of g, a ferromagnetic body, enclosed in such a shield and rigidly corotating with it, senses a polarizing field resulting from the field $\delta\boldsymbol{B}^F_\Omega = (\frac{m}{e})\delta\dot{\boldsymbol{\Omega}}$ (assuming $g \cong 2$ for ferromagnets) and from the true magnetic field $\delta\boldsymbol{B}_{sh} = -(\frac{2m}{e})\delta\dot{\boldsymbol{\Omega}}$. Thus it will magnetize as if in a total field:

$$\delta\boldsymbol{B}_{\text{tot}} = -\left(\frac{m}{e}\right)\delta\dot{\boldsymbol{\Omega}}. \tag{6.9.4}$$

A low-sensitivity version of the GEG has already been built.[59] The magnetization change of the ferromagnetic body, a long rod, due to the field of equation (6.9.4), is measured by means of a SQUID. The sensitivity of this rotation sensor is limited by thermal magnetic noise due to dissipation in the ferromagnetic body due to eddy currents and magnetic viscosity.

Another proposal is to measure the gravitomagnetic field of Earth using **ring laser gyroscopes.**[109–111] A ring laser gyroscope,[112–114] really a ring laser **rotation sensor**, consists of a laser light source inside a rotating ring cavity. It measures the *frequency* difference between two laser beams counterrotating in the ring cavity. In contrast, a "passive ring interferometer"[112,115] operates in a passive mode, that is, with the light source outside the resonator. It measures the *phase* difference between the two counterrotating beams in the rotating cavity (see appendix, § 6.12).

Assuming, for simplicity, that one uses a ring laser gyroscope oriented along the spin axis of Earth, as shown in figure 6.8, from formulae (6.9.2) and (6.12.6), one then gets a frequency difference between the two counterrotating laser

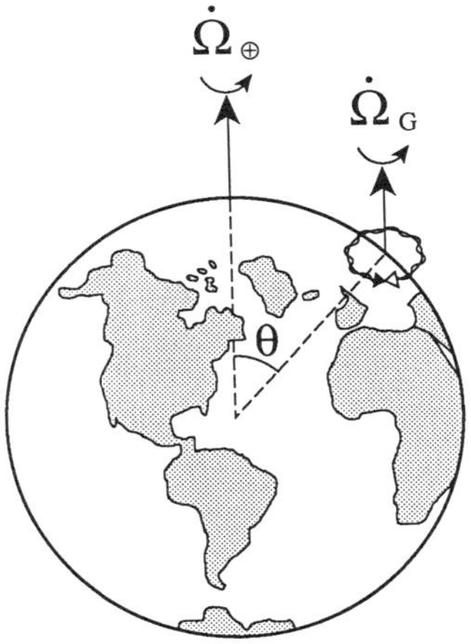

FIGURE 6.8. A ring laser rotation sensor on the Earth surface. θ = co-latitude, $\dot{\Omega}_\oplus$ = Earth angular velocity, $\dot{\Omega}_G$ = gyroscope angular velocity.

beams:[110]

$$\Delta\omega = \frac{4A}{\lambdabar L}\Big[\dot{\Omega}_G + \dot{\Omega}_\oplus - (1+\gamma)\frac{M_\oplus}{R_\oplus}\dot{\Omega}_\oplus \sin^2\theta + \mu\frac{J_\oplus}{R^3_\oplus}\Big(1 - 3\cos^2\theta\Big)\Big]. \tag{6.9.5}$$

Here μ is the **gravitomagnetic** post-Newtonian **parameter** introduced in section 3.4.3, equal to 1 in general relativity. Here, for simplicity, we have used the PPN metric containing the parameters γ and β only, and θ is the co-latitude of the gyroscope, $J_\oplus$ and $\dot{\Omega}_\oplus$ are Earth angular momentum and angular velocity, $\dot{\Omega}_G$ the ring angular velocity (zero for fixed apparatus), A the area bounded by the laser beams in the ring, L its optical length, and $\lambdabar \equiv \frac{\lambda}{2\pi} = \frac{1}{\omega}$ (where $c \equiv 1$) the reduced wavelength. The first two terms are the standard Sagnac effect[112] due to the ring gyroscope's angular velocity with respect to an asymptotic inertial frame, sum of $\dot{\Omega}_G$ and $\dot{\Omega}_\oplus$. The third term is minus the de Sitter or geodetic effect (including Thomas precession, see §§ 3.4.3 and 6.10) due to the velocity of the gyroscope $\dot{\Omega}_\oplus R_\oplus \sin\theta$ in the Earth gravity field $M_\oplus/R^2_\oplus$. The fourth term is minus the Lense-Thirring or frame-dragging effect (6.1.28), due to the Earth angular momentum $J_\oplus$ (see fig. 6.8).

Theoretically, using a proposed, special type of ring laser gyroscope it has been argued[109] that one may be able to reach a sensitivity of 10^{-10} of the Earth angular velocity $\dot{\Omega}_\oplus$, over an integration time of 10^3 s (present optical gyroscopes may reach a sensitivity of the order of $\sim 10^{-6}\dot{\Omega}_\oplus$; see appendix, § 6.12). The gravitomagnetic effect in expression (6.9.5) $\sim J_\oplus/R^3_\oplus \sim 10^{-10}\dot{\Omega}_\oplus$ may then be detectable with such a ring laser gyroscope. Practical problems are however the long-term stability of the laser frequency, misalignment effects due to thermal and mechanical stresses, and the backscattering from the mirrors in the ring cavity.[109]

We finally report a proposal to use **superconducting gravity gradiometers** in polar orbit around Earth to measure magnetic components of the Riemann tensor, with the accuracy needed to detect the gravitomagnetic field.[56,58,116] In 1980 a proposal was made to test the existence of the gravitomagnetic field by measuring magnetic components of the Riemann tensor using gravity gradient resonant detectors orbiting Earth.[56]

To cancel out the much larger classical contributions, the gradiometers must be in a special configuration. For example, to cancel out the large gravity gradients generated by the Earth mass, three gravity gradiometers with orthogonal axes could be used. In fact, since the Ricci tensor is in general relativity null in vacuum, we have $R_{00} \equiv \sum_i^{1-3} R^i{}_{0i0} = 0$. Classically this corresponds to measuring the Laplacian of the gravitational potential which is null in a vacuum, $\Delta U = 0$. A similar three-axes-gradiometer configuration[58] has also allowed experimenters to test further the validity of the inverse square law and to detect any deviation from it (see also § 3.2.1).

The gravitomagnetic effect to be measured has been calculated to be of about $1.7 \times 10^{-16}\,\mathrm{s}^{-2} \equiv 1.7 \times 10^{-7}$ E (1 Eötvös $\equiv 10^{-9}\,\mathrm{s}^{-2}$), or about 10^{-10} of the classical Earth gravity gradient. If one could remove, cancel, or model all the classical effects with enough accuracy, that is, in the case of the Earth gravity gradient with an accuracy of about 10^{-12} of the classical background, then a noise level of the superconducting gradiometers of less than $10^{-5}\,\mathrm{E\,Hz}^{-1/2}$, at the signal frequency of $\cong 3.4 \times 10^{-4}$ Hz, may allow a $\sim$ 1% measurement of the gravitomagnetic field over a one-year period.[58,116]

6.10 THE DE SITTER OR GEODETIC PRECESSION

Besides the gravitomagnetic drag, the other important general relativistic perturbation of an orbiting gyroscope, relative to an asymptotic inertial frame, is the **de Sitter or geodetic precession.**[117–119] As we have seen in section 3.4.3 this precession is due to the coupling between the velocity of a gyroscope orbiting a central body and the static part of the field (Schwarzschild metric) generated by the central mass. In the weak field, slow motion limit, the de Sitter effect for an orbiting gyroscope is given by expression (3.4.38)

$$\dot{\boldsymbol{\Omega}}^{\text{de Sitter}} = -\frac{3}{2}\boldsymbol{v} \times \boldsymbol{r}\frac{M}{r^3} \cong (19.2\ \text{milliarcsec/yr})\hat{\boldsymbol{n}}; \tag{6.10.1}$$

for an Earth gyroscope orbiting the Sun,

where $\boldsymbol{v}$ is the velocity of the orbiting gyroscope, $\boldsymbol{r}$ the distance from the central mass to the gyroscope, and M the mass of the central body as measured in the weak field region. For a gyroscope orbiting the Sun, this precession is about an axis perpendicular to the ecliptic plane, $\hat{\boldsymbol{n}}$. Interesting dragging effects should arise in a strong gravitational field for an observer orbiting a central mass.[159,160]

The de Sitter precession changes the nodal longitude of any satellite orbiting Earth, measured relative to an asymptotic inertial frame, by

$$19.2\ \text{milliarcsec/yr} \times \cos 23.5^\circ \cong 17.6\ \text{milliarcsec/yr} \tag{6.10.2}$$

where 23.5° is the obliquity of the ecliptic.

We observe that the de Sitter precession may be *formally* decomposed in two parts.[4,120] One part, contributing with factor 1/2, is due to the g_{00} component of the metric tensor. If one writes $g_{00} \cong -1 + 2U - 2\beta U^2$, this effect is due to the second term in g_{00}. The other part, contributing with factor 1, is due to the spatial curvature $g_{ij} \cong (1 + 2\gamma U)\delta_{ij}$ and may be parametrized by γ (equal to unity in general relativity) (however, see physical interpretation in § 3.4.3).[120–124]

By comparing Earth's rotational orientation as measured relative to the distant stars—quasars—with VLBI (Very Long Baseline Interferometry), and locally, relative to the Moon, with Lunar Laser Ranging, in 1987, Bertotti, Ciufolini, and Bender[125,126] obtained a $\sim$ 10% test of the de Sitter effect or geodetic precession (see also refs. 127, 128 and 161), for the Earth-Moon gyroscope orbiting the Sun (§ 3.4.3):

$$\dot{\boldsymbol{\Omega}}^{\text{de Sitter}} = \frac{3}{2}\boldsymbol{v}_{\oplus} \times \boldsymbol{\nabla} U_{\odot} = -\frac{3}{2}\boldsymbol{v}_{\oplus} \times \boldsymbol{r}_{\oplus}\frac{M_{\odot}}{r_{\oplus}^{3}} \tag{6.10.3}$$

(this test may also be considered as a test that the distant quasar frame, or VLBI frame, is not rotating relative to the local inertial frames, apart from local dragging effects, see § 4.8). Furthermore, in 1988 Shapiro et al. obtained a test of the de Sitter precession with about 2% accuracy, using Lunar Laser Ranging analyses of the Moon perigee rate relative to the planetary system (see also Dickey et al. 1988),[127,128] in 1991 Müller et al. and in 1993 Dickey et al. obtained a test of this effect with about 1% accuracy.[157,161]

From the conservation of the total angular momentum, one concludes that not only the spin but also the orbital angular momentum and therefore the orbit of a spinning particle must be affected by the coupling between the spin of the gyroscope and the background geometry.

For a *spinning particle*[129] (thus, one with negligible influence on the external gravitational field) of size very small compared with the typical radius of curvature of the external gravitational field (see § 2.5), the deviation from geodesic motion is described by the **Papapetrou equation**:[129]

$$\frac{D}{ds}\left(mu^{\alpha} + u_{\sigma}\frac{DJ^{\alpha\sigma}}{ds}\right) = -\frac{1}{2}R^{\alpha}{}_{\sigma\mu\nu}u^{\sigma}J^{\mu\nu} \tag{6.10.4}$$

coupled with the equation for the spin tensor $J^{\mu\nu}$

$$\frac{D}{ds}J^{\alpha\beta} - u^{\alpha}u_{\sigma}\frac{DJ^{\sigma\beta}}{ds} - u^{\beta}u_{\sigma}\frac{DJ^{\alpha\sigma}}{ds} = 0. \tag{6.10.5}$$

Here one chooses a point with coordinates $\overline{x}^{\alpha}$ inside the spinning particle (for example, for an extended solar system body, its center of mass) and $u^{\alpha} \equiv d\overline{x}^{\alpha}/ds$; $J^{\alpha\beta}$ is the angular momentum tensor of the spinning particle and is defined relative to $\overline{x}^{\alpha}$ as

$$J^{\alpha\beta} = \int (\delta x^{\alpha}T^{\beta 0} - \delta x^{\beta}T^{\alpha 0})\, d^{3}x \tag{6.10.6}$$

where $\delta x^{\alpha} \equiv x^{\alpha} - \overline{x}^{\alpha}$ and the integral is calculated over the volume of the particle at time $\overline{x}^{0}$ = constant. The intrinsic *spin* four-vector, S^{α}, of the spinning particle is then defined as $S^{\alpha} \equiv \frac{1}{2}\epsilon^{\alpha\beta\rho\sigma}u_{\beta}J_{\rho\sigma}$, where the $\overline{x}^{\alpha}$ are the coordinates of the center of mass.

In addition to equations (6.10.4) and (6.10.5) one may give the constraints $J^{i0} = 0$ (Corinaldesi-Papapetrou),[130] or $u_{\sigma}J^{\alpha\sigma} = 0$ (Pirani),[131] or $P_{\sigma}J^{\alpha\sigma}$, where $P^{\alpha} \equiv mu^{\alpha} + u_{\sigma} DJ^{\alpha\sigma}/ds$ (Tulczyjew and Møller).[132,133] For a discussion on these supplementary conditions we refer to O'Connel.[134]

In the next section we shall treat the interesting topic of the relations between the de Sitter precession and the Lense-Thirring effect and gravitomagnetism.

6.11 [GRAVITOMAGNETISM, DRAGGING OF INERTIAL FRAMES, STATIC GEOMETRY, AND LORENTZ INVARIANCE]

In this section, to analyze the meaning of **dragging of inertial frames** and related gravitational field, that we call here intrinsic gravitomagnetic field (§ 6.1), or simply **gravitomagnetic field**, we propose a characterization of **gravitomagnetism** independent from the frame and the coordinate system used, and based on spacetime curvature invariants.[135]

Let us first investigate the question described in some papers:[135–137] whether the existence of the dragging of the inertial frames and of the related gravitomagnetic field can be inferred as a consequence[136,137] of the existence of the standard gravitoelectric field (for example, the Schwarzschild solution) plus local Lorentz invariance, that is, as a consequence of local Lorentz invariance on a static background. In other words, can the Lense-Thirring effect (6.1.28) be inferred as a consequence of the de Sitter effect (6.10.1)?[136,137]

The first step in order to answer this question is to characterize dragging of inertial frames and related gravitomagnetism precisely, in a way independent from the frame and from the coordinates used. Intuitively intrinsic gravitomagnetism may be thought of as that phenomenon such that the spacetime geometry and curvature change due to currents of mass-energy relative to other matter.

Let us return to the analogy with electromagnetism,[138] but stressing again that apart from several formal analogies, general relativity, even the linearized theory, and electromagnetism are fundamentally different. Of course the main difference is the equivalence principle: locally (in spacetime) it is possible to eliminate (in the sense of making arbitrarily small) the effects of the gravitational field. This is true for gravity only.

In general relativity, the spacetime geometry $g_{\alpha\beta}$, where the various physical phenomena take place, is determined by the energy and by the energy-currents in the universe via the Einstein field equation (since the gravity field $g_{\alpha\beta}$ can carry energy and momentum, the gravitational energy contributes itself, in a loop, to the spacetime geometry $g_{\alpha\beta}$). However, in special relativistic electrodynamics, the spacetime geometry $\eta_{\alpha\beta}$, where the electromagnetic phenomena take place, is unaffected by these phenomena. As a consequence of the equivalence principle, we have the known phenomenon, explained in section 2.7, true for gravity only, that, locally, it does not have meaning to define the energy of the gravity field. Indeed, locally, in the freely falling frames, the gravity field can always be eliminated, in the sense of eliminating the first derivatives of the metric $g_{\alpha\beta}$ and having $\overset{(0)}{g}_{\alpha\beta} \rightarrow \eta_{\alpha\beta}$ at a point, and in the sense of locally eliminating any measurable effect of gravity including the gravitational energy. This clearly shows the known characteristic of general relativity (§ 2.9) that, to study the spacetime structure, one must not analyze a single metric component in a particular coordinate system but one has to analyze coordinate-independent geometrical quantities such as quantities built with the Riemann curvature tensor (see below), and to describe the energy-momentum of the gravity field it is useful to define a nonlocal energy-momentum pseudotensor for the gravity field (§ 2.7).

Let us now briefly recall a very well known feature of electromagnetism.[21] In electromagnetism, in the frame of a charge q at rest we only have a nonzero electric field $\boldsymbol{E}^{\mathrm{o}}$ but no magnetic field $\boldsymbol{B}^{\mathrm{o}}$. However, if we consider an observer moving with velocity $\boldsymbol{v}$ relative to the charge q, in this new frame we have a magnetic field $\boldsymbol{B}'$, according to the formula: $\boldsymbol{B}' = -\gamma(\boldsymbol{v})(\boldsymbol{\beta} \times \boldsymbol{E}^{\mathrm{o}})$.

Similarly, in general relativity, in the frame of a static mass M at rest we only have the nonzero metric components $g_{00} = -g_{rr}^{-1} = -\left(1 - \frac{2M}{r}\right)$, and $g_{\theta\theta} = g_{\phi\phi}/\sin^2\theta = r^2$, but we do not have the "magnetic" metric components g_{0i}. However, if we consider an observer moving with velocity $\boldsymbol{v}$ relative to the mass M, in his, or her, local frame we have a "magnetic" metric component g_{0i} according to the formula $g'_{0i} = \Lambda^{\alpha}_{0'}\Lambda^{\beta}_{i'}g_{\alpha\beta} \sim \frac{Mv}{r}$.

The question is whether the existence of the g_{0i} components in this local boosted frame, proportional to the orbital angular momentum $\sim Mvr$, as measured by the moving observer, can be considered as an indirect proof that the intrinsic angular momentum $\boldsymbol{J}$ of a mass distribution affects the spacetime geometry and changes the spacetime curvature.

The answer, we see, is no. Of course, for every spacetime solution we can always secure nonzero metric components g_{0i} via a mere coordinate transformation. To analyze the single metric components at a point is not sufficient to describe the spacetime structure and curvature. In section 2.9 we have seen that on the Schwarzschild event horizon one has a coordinate singularity (in Schwarzschild coordinates), but a well-behaving spacetime geometry and curvature (as it is clear by analyzing the curvature invariants and by using other coordinate systems). Furthermore, any metric, for example the flat Minkowski metric $\eta_{\alpha\beta}$, can be changed with a coordinate transformation into a metric with all the components complicated functions of the coordinates.

However, it is also well known[4] that in order to distinguish between a true curvature singularity and a mere coordinate singularity (§ 2.9), or between a flat Minkowskian spacetime and a curved manifold, one has to analyze the Riemann tensor, to see if it is different from zero or not and to see if its invariants are well behaved or if they diverge in some region. One might therefore think to test if the so-called "magnetic components" of the Riemann tensor, R_{i0jk}, are different from zero or not. However, again, out of the six "electric components" R_{i0j0}, via a local Lorentz transformation one can locally get magnetic components $R'_{i0jk} = \Lambda^{\alpha}_{i'}\Lambda^{\beta}_{0'}\Lambda^{\mu}_{j'}\Lambda^{\nu}_{k'}R_{\alpha\beta\mu\nu}$ different from zero.

Therefore, following the standard method[4] of characterizing curvature singularities, and following the usual method to classify a spacetime solution,[139] we conclude that the correct approach is to inspect the invariants of the spacetime. In a vacuum the Ricci curvature scalar, $R = R^{\alpha}{}_{\alpha}$, is identically equal to zero as a consequence of the Einstein field equation. Another scalar invariant is the *Kretschmann invariant* $R_{\alpha\beta\mu\nu}R^{\alpha\beta\mu\nu}$; however, in the case of a metric characterized by mass and angular momentum, such as the Kerr metric, the Kretschmann invariant is a function of $\frac{M}{r^3}$ and $\frac{J}{r^4}$, with the leading term $\sim (\frac{M}{r^3})^2$, therefore this invariant is different from zero in the presence of a mass M, whether or not there is any angular momentum.

At this point we turn again to a formal analogy between electromagnetism and general relativity.[138] In electromagnetism to characterize the electromagnetic field one can calculate the scalar invariant $-\frac{1}{2}F_{\alpha\beta}F^{\alpha\beta} = E^2 - B^2$, which is analogous to the Kretschmann invariant $R_{\alpha\beta\mu\nu}R^{\alpha\beta\mu\nu} \sim (\frac{M}{r^3})^2 + C(\frac{J}{r^4})^2$ (see formula 6.11.6). However, in electrodynamics one can also calculate the scalar pseudoinvariant $\frac{1}{4}F_{\alpha\beta}{}^*F^{\alpha\beta} = \boldsymbol{E}\cdot\boldsymbol{B}$, where "$*$" is the duality operation: ${}^*F^{\alpha\beta} = \frac{1}{2}\epsilon^{\alpha\beta\mu\nu}F_{\mu\nu}$. We recall that if we have a charge q only, in its rest frame we have an electric field only, and the invariant $F_{\alpha\beta}{}^*F^{\alpha\beta}$ is zero, therefore, even in inertial frames where both $\boldsymbol{B} \neq \boldsymbol{0}$ and $\boldsymbol{E} \neq \boldsymbol{0}$, this invariant is zero. However, if in the inertial rest frame we have a charge q and a magnetic dipole $\boldsymbol{m}$, in this frame we have in general $F_{\alpha\beta}{}^*F^{\alpha\beta} \neq 0$ and of course this invariant is different from zero in any other inertial frame.

Therefore, to characterize the spacetime geometry and curvature generated by mass-energy currents and by the intrinsic angular momentum J of a central body (in § 6.1 we have seen that in the weak field limit the angular momentum of a mass-energy current plays a role, in general relativity, analogous to the magnetic dipole moment generated by a loop of charge current in electromagnetism) we should look for an analogous spacetime invariant.

This invariant should therefore be built out of the dual of the Riemann tensor ${}^*R^{\alpha\beta\mu\nu} \equiv \frac{1}{2}\epsilon^{\alpha\beta\sigma\rho}R_{\sigma\rho}{}^{\mu\nu}$ "multiplied" by $R_{\alpha\beta\mu\nu}$. This pseudoinvariant is of the type[135] $\frac{1}{2}\epsilon^{\alpha\beta\sigma\rho}R_{\sigma\rho}{}^{\mu\nu}R_{\alpha\beta\mu\nu}$. Because of the formal analogy with electromagnetism, and because this pseudoinvariant, ${}^*\boldsymbol{R}\cdot\boldsymbol{R}$, is built using the Levi-Civita pseudotensor $\epsilon^{\alpha\beta\sigma\rho}$, it should change sign for time reflections: $t \to -t$, and therefore, it should be proportional to J. A list of all the possible spacetime invariants built out of the Riemann tensor and of its dual is given in reference 139.

In electromagnetism the electromagnetic field is described by the tensor field $\boldsymbol{F}$, and the spacetime geometry by the Minkowski tensor $\boldsymbol{\eta}$. Then, of course, ${}^*\boldsymbol{F}\cdot\boldsymbol{F}$ characterizes the electromagnetic field only, but not the spacetime geometry. However, in general relativity the gravitational field and the spacetime geometry are both described by the metric tensor $\boldsymbol{g}$. Then, the meaningful and useful invariant $^*\boldsymbol{R} \cdot \boldsymbol{R}$ characterizes both the gravitational field and the spacetime geometry.

We can get some information about this invariant, without performing lengthy calculations, by using the *Newman-Penrose null-tetrad formalism*[140] and by considering the coefficients built out of the tetrad components of the Weyl tensor defined by expression (2.5.18): Ψ_1, Ψ_2, Ψ_3, Ψ_4, and Ψ_5. In the case of the Kerr metric, characterized by mass M and angular momentum J, there exists a complex null tetrad $(\boldsymbol{k}, \boldsymbol{l}, \boldsymbol{m}, \overline{\boldsymbol{m}})$, of the Newman-Penrose formalism, such that, relative to it, the complex coefficients built with the Weyl tensor are (the spacetime is of Petrov type D)[141–143]

$$\Psi_2 = \frac{1}{2} C_{\alpha\beta\mu\nu} k^\alpha l^\beta (k^\mu l^\nu - m^\mu \overline{m}^\nu) = \frac{M}{(r - ia\cos\theta)^3} \tag{6.11.1}$$

and $\Psi_1 = \Psi_3 = \Psi_4 = \Psi_5 = 0$. Here $C_{\alpha\beta\mu\nu}$ is the Weyl tensor, and $a \equiv \frac{J}{M}$ is the angular momentum per unit mass, as measured in the weak field region.

Since the Ψ_i are linear in the Weyl tensor, that is the Riemann tensor in a vacuum, and since relative to the above tetrad and for the Kerr metric: $\Psi_1 = \Psi_3 = \Psi_4 = \Psi_5 = 0$ and only $\Psi_2 \neq 0$, by squaring Ψ_2 and by taking its real part and its imaginary part, we can have some information about the **invariants** $\boldsymbol{R} \cdot \boldsymbol{R}$ and $^*\boldsymbol{R} \cdot \boldsymbol{R}$. At the lowest order we have

$$\mathrm{Re}\left(\psi_2^2\right) \cong \frac{M^2}{r^6} - 21\frac{J^2}{r^8}\cos^2\theta, \qquad \text{and} \qquad \mathrm{Im}\left(\psi_2^2\right) \cong 6\frac{JM}{r^7}\cos\theta. \tag{6.11.2}$$

Indeed, we one can calculate $^*\boldsymbol{R} \cdot \boldsymbol{R}$ by using some computer algebra system, such as STENSOR or MACSYMA. The result is[144,145]

$$\frac{1}{2}\epsilon^{\alpha\beta\sigma\rho} R_{\sigma\rho}{}^{\mu\nu} R_{\alpha\beta\mu\nu} = 1536\, JM\cos\theta\left(r^5\rho^{-6} - r^3\rho^{-5} + \frac{3}{16}r\rho^{-4}\right) \tag{6.11.3}$$

where $\rho = (r^2 + (\frac{J}{M})^2\cos^2\theta)$, and in the weak field limit

$$^*\boldsymbol{R} \cdot \boldsymbol{R} \cong 288\frac{JM}{r^7}\cos\theta + \cdots, \tag{6.11.4}$$

and the *Kretschmann invariant*, $\boldsymbol{R} \cdot \boldsymbol{R}$, *for the Kerr metric* is[144,145]

$$\begin{aligned}\boldsymbol{R} \cdot \boldsymbol{R} = &-768\, J^2 r^4 \rho^{-6}\cos^2\theta + 384 J^2 r^2 \rho^{-5}\cos^2\theta \\ &- 48\, J^2\rho^{-4}\cos^2\theta + 768 M^2 r^6 \rho^{-6} \\ &- 1152\, M^2 r^4 \rho^{-5} + 432 M^2 r^2 \rho^{-4},\end{aligned} \tag{6.11.5}$$

and in the weak field limit

$$\boldsymbol{R} \cdot \boldsymbol{R} \cong 48 \left(\frac{M^2}{r^6} - 21 \frac{J^2}{r^8} \cos^2 \theta \right). \tag{6.11.6}$$

Since the external gravitational field of a stationary black hole is determined by its mass M, charge Q, and intrinsic angular momentum J, and indeed for the Kerr-Newman metric (2.6.36) the **invariant** $^*\boldsymbol{R} \cdot \boldsymbol{R}$ is still **proportional to** J,[145] the above result[135] is general in the case of a black hole and is valid for any quasi-stationary solution, asymptotically, in the weak field limit.

Furthermore, the above result derived in Einstein theory, is generally valid in *any* metric theory of gravity (with no prior geometry) *not* necessarily described at the post-Newtonian order by the PPN formalism[120] (see § 3.7); this can be seen in two ways. Let us first write, in a metric theory of gravity, the weak field, slow motion, expression of an asymptotically flat metric in the form: $g_{00} \cong -1+2U+$higher order terms, $g_{ik} \cong \delta_{ik}(1+2\gamma U)+$higher order terms, and $\boldsymbol{h} \equiv (g_{01}, g_{02}, g_{03})$. Then, one can easily calculate the pseudoinvariant $^*\boldsymbol{R} \cdot \boldsymbol{R}$ that at the lowest order is $^*\boldsymbol{R} \cdot \boldsymbol{R} \cong \boldsymbol{\nabla}^2[\boldsymbol{\nabla} U \cdot (\boldsymbol{\nabla} \times \boldsymbol{h})] +$ higher order terms. In the case of a static distribution of matter and of a corresponding static metric, with $\boldsymbol{h} = \boldsymbol{0}$, we then have $^*\boldsymbol{R} \cdot \boldsymbol{R} = 0$. However, for a stationary distribution of matter and for a corresponding stationary metric, with $\boldsymbol{h} \neq \boldsymbol{0}$, for example with $\boldsymbol{h} \sim \boldsymbol{J}$, we have: $^*\boldsymbol{R} \cdot \boldsymbol{R} \neq 0$. In the first case of a static distribution of matter and of a static metric, with a boost with velocity $\boldsymbol{v}$ we have: $\boldsymbol{h} \sim \boldsymbol{v} U$; however $^*\boldsymbol{R} \cdot \boldsymbol{R}$ is, of course, still zero: $^*\boldsymbol{R} \cdot \boldsymbol{R} \sim \boldsymbol{\nabla}^2[\boldsymbol{\nabla} U \cdot (\boldsymbol{\nabla} \times \boldsymbol{v}\gamma U)] = 0$. A second argument confirms the validity of the above result in any metric theory of gravity not necessarily described by the PPN formalism. For a generic source, in any metric theory of gravity (with no prior geometry), the full expression of the scalar $^*\boldsymbol{R} \cdot \boldsymbol{R}$ must be dependent from some of the intrinsic physical quantities characterizing the source, such as the total mass-energy of the source, its intrinsic angular momentum, its multipole mass moments, etc., that is, must be dependent from some integrals of the mass-energy density ε and of the mass-energy currents εu^i. In particular, since $^*\boldsymbol{R} \cdot \boldsymbol{R}$ must change sign for time reflections, its full expression, for a generic source, must be proportional to some odd functions of the intrinsic mass-energy currents εu^i (not eliminable with a change of origin or with a Lorentz transformation) characterizing the system, such as the intrinsic angular momentum of the source of expression (6.11.4).

Therefore, independently from the field equations of a particular metric theory, the pseudoinvariant $^*\boldsymbol{R} \cdot \boldsymbol{R}$ determines the existence and the presence of gravitomagnetism in that metric theory of gravity. Indeed, using this invariant $^*\boldsymbol{R} \cdot \boldsymbol{R} \sim \frac{JM}{r^7} \cos \theta$ we can determine whether or not there is a gravitomagnetic contribution to the spacetime curvature and geometry. We just need to calculate $^*\boldsymbol{R} \cdot \boldsymbol{R}$; if it is different from zero we have a contribution of the mass-energy currents to the curvature; if it is zero there is no gravitomagnetic contribution. It does not matter of local Lorentz transformations or other frame and coordinate transformations on a static background, either $^*\boldsymbol{R} \cdot \boldsymbol{R}$ is zero, as it is in the Schwarzschild case, or it is different from zero as in the Kerr case. A spacetime solution with $^*\boldsymbol{R} \cdot \boldsymbol{R} \neq 0$ is, of course, conceptually and qualitatively different from a spacetime solution with $^*\boldsymbol{R} \cdot \boldsymbol{R} = 0$, regardless of frame and coordinate transformations.

We can then characterize what may be called intrinsic gravitomagnetism,[135] a new feature of Einstein general relativity with respect to classical gravity theory. By taking a static background, in some metric gravity theory, and by assuming local Lorentz invariance, one can locally generate the so-called magnetic *components* of the curvature tensor. However, a fundamentally new concept that Einstein introduced in general relativity is that the spacetime geometry and the corresponding curvature invariants are affected and determined, not only by mass-energy, but also by mass-energy currents relative to other mass, that is, mass-energy currents not generable or eliminable by a Lorentz transformation (for example the intrinsic angular momentum of a body that cannot be generated or eliminated by a Lorentz transformation).

This is a new feature of general relativity that may be called intrinsic gravitomagnetism and that has never been experimentally directly measured.

It is also interesting and useful to define gravitomagnetism practically and operationally.[135]

Let us consider some observers inside a spaceship (with negligible mass in comparison with the external field). Can the observers determine by "*local*" measurements only, without looking outside, if they are in the field of a rotating central mass, or if they are just moving with velocity $\boldsymbol{v}$ on a static background (e.g., moving on a Schwarzschild background)? See figure 6.9.

The answer is yes; they can determine the existence of mass-energy currents and of rotating masses and distinguish them from mere local boosts and local Lorentz transformations on a static background.

How? They cannot use local gyroscopes and measure their orientations relative to some distant reference points. They do not know anything about the proper motion of the external reference points and anyhow they might not be able to see or to use any external reference point.

However, they know about the geodesic deviation equation (2.5.1) (valid in any metric theory in which test particles follow geodesic motion): $\frac{D^2(\delta x^\alpha)}{ds^2} = -R^\alpha{}_{\beta\gamma\delta} u^\beta \delta x^\gamma u^\delta$, and by using it and by measuring the relative accelerations between a number of test particles inside the spaceship (for the minimum number of test particles that would be needed, see § 2.5) they can determine all the components of the Riemann tensor. Then, they combine these components in such a way to construct the invariant $^*\boldsymbol{R} \cdot \boldsymbol{R} \propto \frac{JM}{r^7} \cos\theta$.

If $^*\boldsymbol{R} \cdot \boldsymbol{R}$ turns out to be zero, then, it does not matter whether or not they have measured magnetic components R_{i0jk} different from zero, there is not a

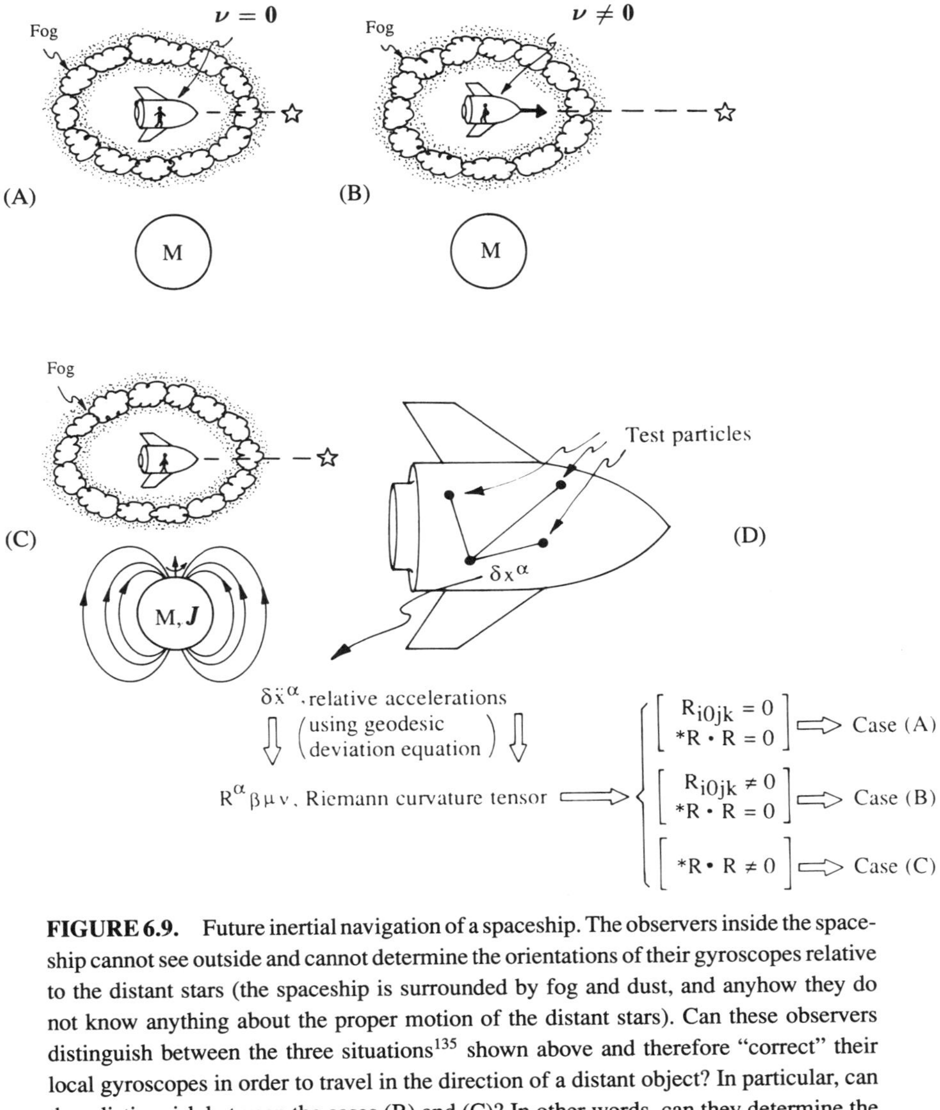

FIGURE 6.9. Future inertial navigation of a spaceship. The observers inside the spaceship cannot see outside and cannot determine the orientations of their gyroscopes relative to the distant stars (the spaceship is surrounded by fog and dust, and anyhow they do not know anything about the proper motion of the distant stars). Can these observers distinguish between the three situations[135] shown above and therefore "correct" their local gyroscopes in order to travel in the direction of a distant object? In particular, can they distinguish between the cases (B) and (C)? In other words, can they determine the presence of a central *rotating* object and the existence of an intrinsic gravitomagnetic field (case C) and distinguish it from a mere motion on a static background (case B)? Can they determine, without looking outside, if their gyroscopes have (A) no precession, (B) de Sitter precession, or (C) Lense-Thirring precession relative to an asymptotic inertial frame? Yes! See section 6.11 and above (D).

rotating central mass with intrinsic angular momentum different from zero. If $R_{i0jk} \neq 0$ but $^*\boldsymbol{R} \cdot \boldsymbol{R} = 0$, what they have measured is just the effect of a local Lorentz transformation on a static background. However, if they get $^*\boldsymbol{R} \cdot \boldsymbol{R} \neq 0$ there is a source with intrinsic angular momentum J or a mass-energy current relative to other local mass; that is, the spacetime curvature is affected by an intrinsic gravitomagnetic field not generable or eliminable with any Lorentz transformation or any other frame and coordinate transformation.

In summary, only by local measurements inside the spaceship and by using the geodesic deviation equation and the scalar invariants built out of the Riemann tensor, the observers are able to determine, without any observation of the external world, whether or not they are in the field of a *rotating* central mass, and they are able to distinguish this situation from a mere motion on a static background.

This example clearly shows that the de Sitter effect and the Lense-Thirring drag are two fundamentally different phenomena and, even assuming that the gravitational interaction must be described by a metric theory, the Lense-Thirring drag cannot be inferred as a consequence of the de Sitter effect. Counterexamples of metric theories of gravity that show this might be constructed. In other words, the standard gravitoelectric field plus local Lorentz invariance do not imply the existence of the intrinsic gravitomagnetic field,[135] unless one makes some additional theoretical hypotheses on the structure of a metric theory.

The method outlined above: (1) determine the Riemann curvature tensor components by measuring the accelerations between test particles and by using the geodesic deviation equation and (2) construct the scalar curvature invariants $\boldsymbol{R} \cdot \boldsymbol{R}$ and $^*\boldsymbol{R} \cdot \boldsymbol{R}$, can be used by a local observer, in the weak field region, to quasi-locally determine M, J, and r, for an isolated body, without performing any surface integral at infinity. The observers just need to repeat the process (1) plus (2) at different points in a neighborhood of a point P_1, by contemporarily measuring the distance, δl, between these points (or use a few gradiometers at known relative distances). Then, by a Taylor expansion: $^*\boldsymbol{R} \cdot \boldsymbol{R}|_{P_2} = {^*\boldsymbol{R}} \cdot \boldsymbol{R}|_{P_1} + ({^*\boldsymbol{R}} \cdot \boldsymbol{R})_{,\alpha}|_{P_1} \delta x^\alpha$, they can get: $\frac{({^*\boldsymbol{R}} \cdot \boldsymbol{R}|_{P_2} - {^*\boldsymbol{R}} \cdot \boldsymbol{R}|_{P_1})}{\delta l} \sim \frac{(JM)}{r^8}$, from this and from $^*\boldsymbol{R} \cdot \boldsymbol{R}|_{P_1} \sim \frac{(JM)}{r^7}$ and $\boldsymbol{R} \cdot \boldsymbol{R}|_{P_1} \sim \frac{M^2}{r^6}$ they can get, in the weak field region, r, M, and J with quasi-local measurements only (without any observation of the external world).

In summary, to precisely characterize and analyze the phenomenon of intrinsic gravitomagnetism and its meaning, one can use the frame- and coordinate-independent method described above, based only on spacetime curvature invariants. One can also apply this method to quasi-locally determine the quantities characterizing a one-body external metric.

6.12 APPENDIX: GYROSCOPES AND INERTIAL FRAMES

Local inertial frames have a fundamental role in Einstein geometrodynamics (see § 2.1). The spatial axes of a local inertial frame, along the world line of a freely falling observer, are mathematically defined using Fermi-Walker transport (eq. 3.4.25); that is, along his, or her, geodesic they are defined using parallel transport. These axes are physically realized with gyroscopes.

A brief description of the gyroscopes today available may then be useful. We have already described the most advanced gyroscopes which are developed to measure the very tiny effect due to the gravitomagnetic field of Earth: the "dragging of inertial frames," that is, the precession of the gyroscopes by the Earth angular momentum, which, in orbit, is of the order of a few tens of milliarcsec/yr.

There are two main types of gyroscopes commercially available: mechanical and optical. The optical gyroscopes (really optical rotation sensors) are usually built with optical fibers or with "ring lasers" (see below).

Mechanical Gyroscopes

A **mechanical gyroscope**[146] is essentially made of a wheel-like rotor, torque-free to a substantial level, whose spin determines the axis of a local, nonrotating, frame. Due to the very tiny general relativistic effects described in the previous sections, that is, the "dragging of inertial frames" and the geodetic precession, this spin direction may differ from a direction fixed in "inertial space" that may be defined by a telescope always pointing toward the same distant galaxy assumed to be fixed with respect to some asymptotic quasi-inertial frame (see § 4.8).

Mechanical gyroscopes are based on the principle of conservation of angular momentum of an isolated system, that is, a system with no external forces and torques. Since mechanical gyroscopes are small and massive and are rotating at high speed (about 6000 rpm in some commercial mechanical gyroscopes) they are, in practice, relatively stable to small torques. A mechanical gyroscope is usually mounted on three gimbals (three concentric rings) that give the physical support to the gyroscope and that give it the three needed degrees of freedom (see fig. 6.10). Therefore, if the bearings were perfectly frictionless, the spin axis of the gyroscope would be totally unaffected by the motion of the laboratory (fig. 6.10).

In 1851 Léon **Foucault** showed the rotation of Earth by using a **pendulum**. However, in order to perform an experiment that did not use the Earth gravitational attraction, in 1852 he showed again the rotation of Earth by using a rotating toy top. Since the spinning rotor maintains its direction fixed in "space" (apart from dragging effects) as Earth rotates but, however, a vector with general

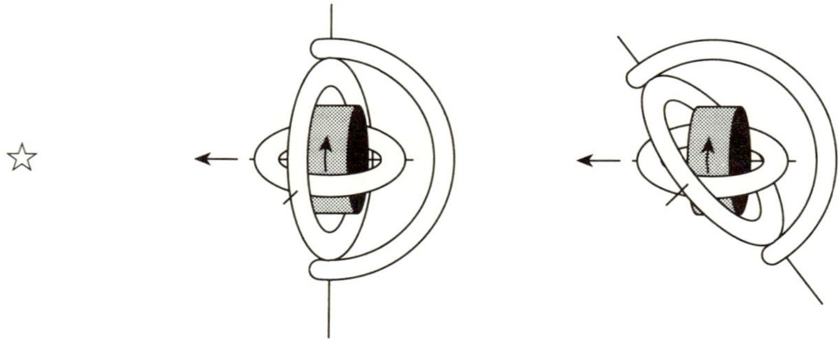

FIGURE 6.10. A spinning mechanical gyroscope mounted on three gimbals.

orientation, fixed with respect to the laboratory walls, describes a circle on the celestial sphere in 24 hours, a spinning rotor with general orientation describes a circle with respect to the laboratory walls in 24 hours (see fig. 6.11). The rotating toy top, previously known as "*rotascope*," was renamed "*gyroscope*" by Foucault.

In a moving laboratory, using three inertial sensors, that is, three gyroscopes to determine three "fixed" directions (apart from relativistic effects, see §§ 3.4.3 and 6.11) plus three accelerometers to measure linear accelerations and a clock

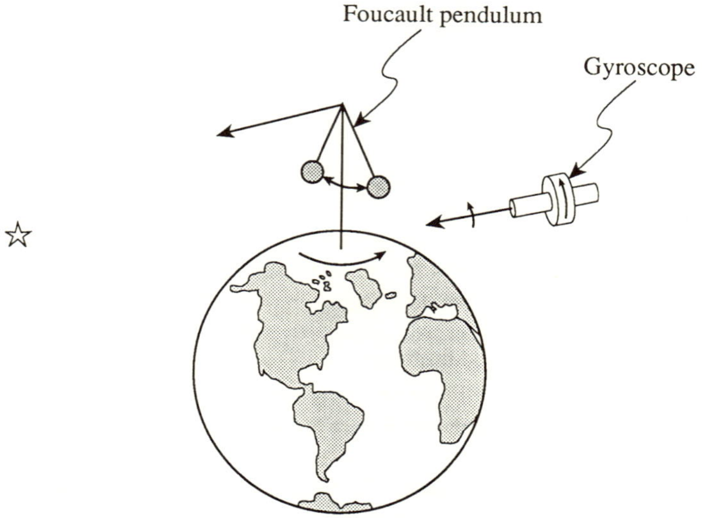

FIGURE 6.11. A spinning gyroscope and a Foucault pendulum (the normal to the plane of oscillation of the Foucault pendulum is a gyroscope) which define the orientation of the axes of local inertial frames.

(and possibly three gradiometers to correct for torques due to gravity gradients), one can determine the position of the moving laboratory with respect to the initial position. This can be done by a simple integration of the accelerations measured by the three accelerometers along the three fixed directions determined by the gyroscopes (fig. 6.12). Position can thus be determined solely by measurements internal to the laboratory. This process of internally measuring an object velocity, attitude, and position with respect to the initial position, by measurements within the object, a priori independently of external information, is called "*inertial navigation*" (fig. 6.12).[146] In practice, an on-board computer integrates the accelerations measured by the accelerometers along the fixed directions determined by the gyroscopes (held by gimbals), and therefore one is able to find velocity, attitude, and position of the object.

For an orbiting satellite is not in principle necessary to have a gyroscope for the building of a device to preserve "direction." Instead, one may use gradiometers (see § 6.11) or, apart from gravitational perturbations, a "conscience-guided" or drag-free satellite (fig. 6.13).

A satellite in orbit around Earth is subject to small accelerations due to solar radiation pressure, residual atmospheric drag, and other more complex combined effects (see § 6.8). These accelerations may typically range from about $10^{-6}g$ to values several orders of magnitude smaller than $10^{-6}g$ (§ 6.8), where g is the acceleration of gravity of Earth's surface. The acceleration was reduced to $5 \times 10^{-12}g$, averaged over a few days, for more than a year in orbit by use of a conscience or proof mass and the Disturbance Compensation System (DISCOS) mounted on a TRIAD US Navy satellite. The conscience, a gold-platinum sphere 2.2 cm in diameter, floats freely inside a spherical housing. Any nongravitational force results in an incremental velocity change. The floating proof mass continues in its original state of motion in an ideal friction-free

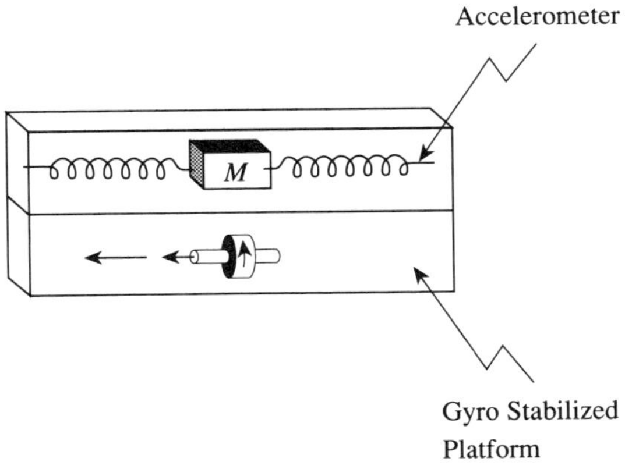

FIGURE 6.12. Inertial navigation.

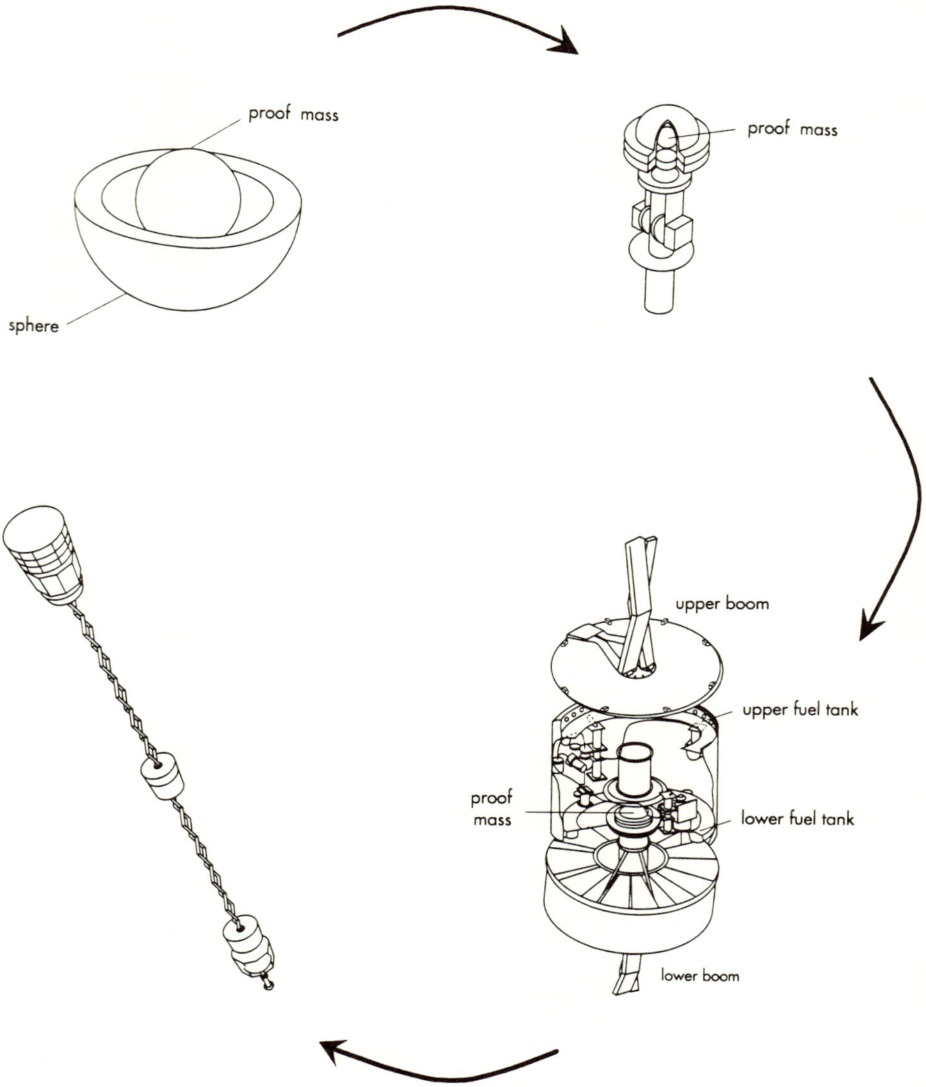

FIGURE 6.13. "Conscience-guided" satellite.

environment. Observing the proof mass through capacitor sensing devices, the satellite becomes aware that it is not keeping up with the motion demanded by the proof mass. An opposite vernier rocket fires long enough to bring the spaceship back into concord with its proof mass—its conscience. To reduce gravitational effects of the satellite itself on the proof mass, fuel for the vernier rockets is stored in donut-shaped tanks placed symmetrically above and below the proof mass; power supply and radio transmitter are each held at the end of a boom 2.7 m long on either side of the control unit (figure and caption adapted by permission from E. F. Taylor and J. A. Wheeler 1991).[148]

Optical Gyroscopes

If Christopher Columbus used the skills of an Arab navigator to meet his demand to sail "west, straight west; nothing to the north, nothing to the south," then it is not surprising that the dozens of captains who at this very moment guide transoceanic aircraft towards safe landings also insist on taking with them reliable navigational guides. For them, the magnetic compass is not acceptable. It is too susceptible to local magnetic fields and to Sun-driven magnetic storms. It is no wonder that the needs of air navigation have generated a powerful drive for a compact, light-weight gyroscopic compass of high accuracy. For years, a spinning mass provided the active element of every such gyroscopic compass. It locked on to a direction in space, a direction relative to the distant stars, not the magnetic pull of Earth. Today, optical gyros have displaced the mechanical gyro as the heart of the compass. A wave guide is bent into a circle. A beam-splitter takes light from a laser and sends it around the circle in two opposite directions. Where the beams reunite, interference between them gives rise to wave crests and troughs. If the wave-guide sits on a turning platform, the wave crests reveal the rotation of the platform or the airplane that carries it.

While mechanical gyroscopes are based on the principle of conservation of angular momentum, **optical gyroscopes** (really, optical *rotation sensors*) are essentially based on the principle of constancy of the speed of light equal to c in every inertial frame[112] (in this section we keep the symbol c for the speed of light). Therefore, in a rotating circuit, and relative to the observers moving with it, the round-trip travel time of light depends on the sense of propagation of light with respect to the circuit angular velocity relative to a local inertial frame, figure 6.14 (see § 3.2.2).

There are two main types of optical rotation sensors: **ring interferometers**,[112,115] which measure the relative phase shift between two counterrotating electromagnetic waves, and **ring lasers**, or *fiber resonators*,[112,113] which measure the frequency difference between two counterrotating laser beams. Picture 6.8 shows one of today's commercial gyros built on this principle.

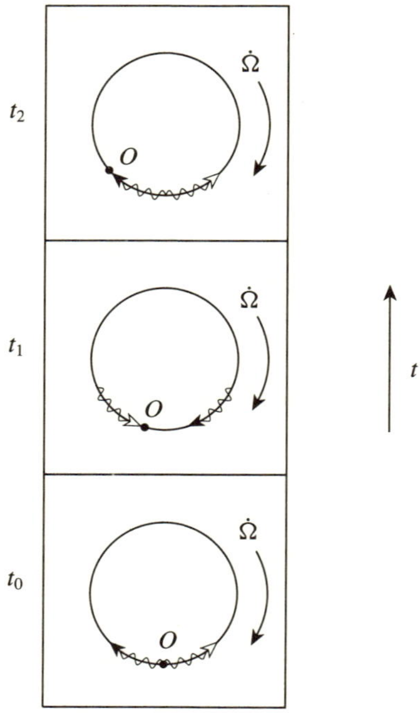

FIGURE 6.14. The Sagnac effect for two electromagnetic pulses propagating in opposite directions around a rotating wave guide or mirror array. t_0: emission at O of both signals; t_1: detection at O of the counterrotating light beam at time $T_{CR} = T - \frac{\dot{\Omega}RT}{c}$; t_2: detection at O of the corotating light beam at time $T_R = T + \frac{\dot{\Omega}RT}{c}$.

Both are based on the so-called *Sagnac effect*: the optical path difference and therefore the difference in round-trip travel time for light beams counterrotating in a rotating circuit is proportional to the angular velocity of the circuit.

Let us consider the simple circular circuit of figure 6.14, of area A and radius R, rotating with angular velocity $\dot{\Omega}$, where, for simplicity, we assume $R\dot{\Omega} \ll c$. When $\dot{\Omega} = 0$, the round-trip time for light is $T = \frac{2\pi R}{c}$. However, if the circuit rotates with angular velocity $\dot{\Omega}$, the fraction, L_R, of the circuit traveled by the light corotating with the rotating circuit, as measured in the circuit and corresponding to the time interval T, is

$$L_R = 2\pi R - \dot{\Omega}RT \tag{6.12.1}$$

while the fraction, L_{CR}, of the circuit traveled by the light counterrotating with respect to the rotating circuit, as measured in the circuit and corresponding to the time interval T, is

$$L_{CR} = 2\pi R + \dot{\Omega}RT. \tag{6.12.2}$$

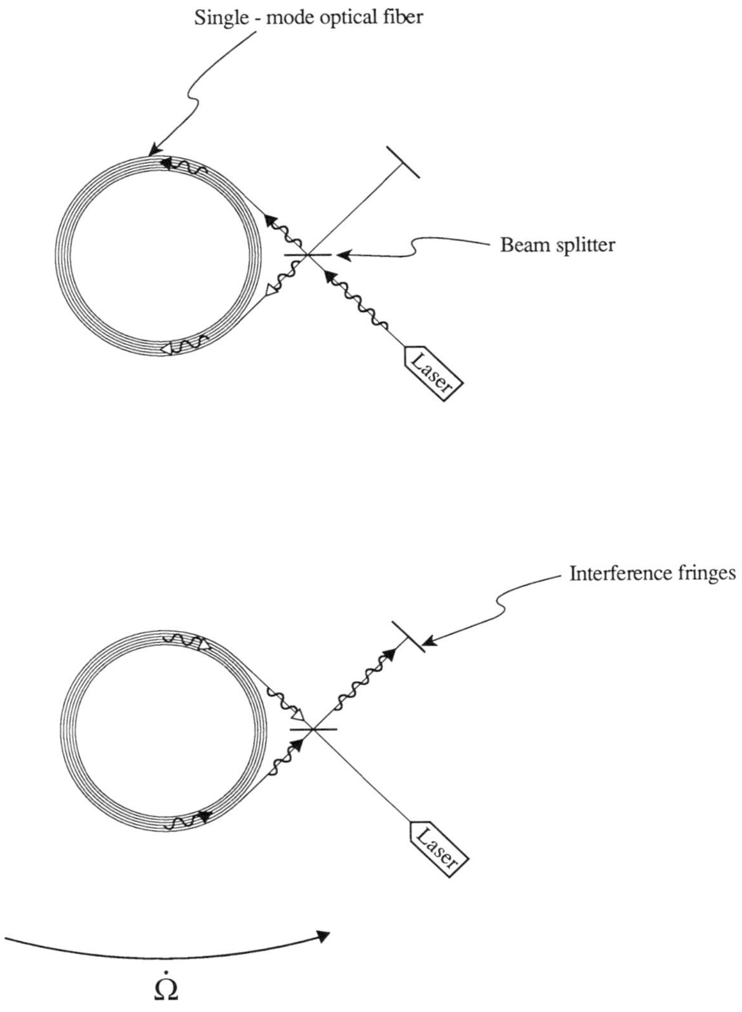

FIGURE 6.15. A fiber ring interferometer.

Therefore, the difference between the path lengths traveled in the circuit by the counterrotating light beam and by the corotating light beam in the interval T is

$$\Delta L = 2\dot{\Omega} RT \tag{6.12.3}$$

or

$$\Delta L = 4A \frac{\dot{\Omega}}{c}. \tag{6.12.4}$$

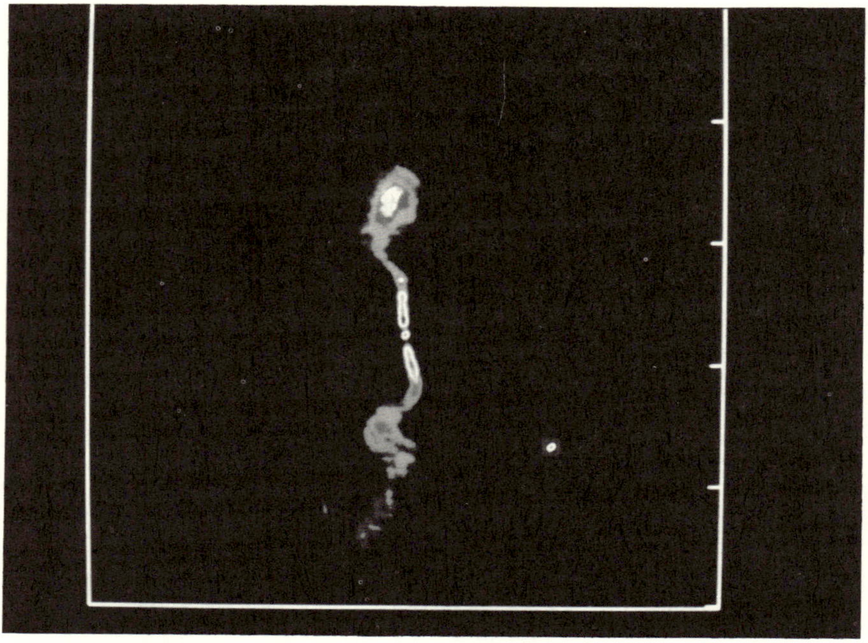

PICTURE 6.1. Radio Jets emerging from opposite sides of the nucleus of an elliptical galaxy, 3C 449. The distance to the galaxy is approximately 160 million lt-yr. The radio jets are about 60,000 lt-yr long (National Radio Astronomy Observatory, observers: R. A. Perley and A. G. Willis. The Astronomical Society of the Pacific).

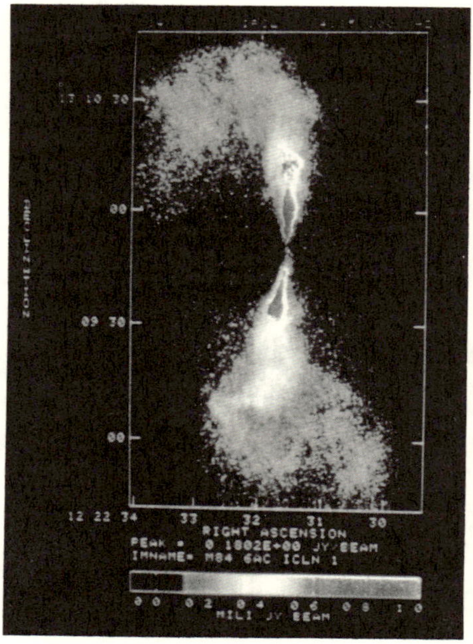

PICTURE 6.2. A double jet from the radio galaxy NGC 4374. The entire radio structure is contained within the visible galaxy M 84, approximately 30 million lt-yr away (National Radio Astronomy Observatory, observers: R. A. Laing and A. H. Bridle. The Astronomical Society of the Pacific).

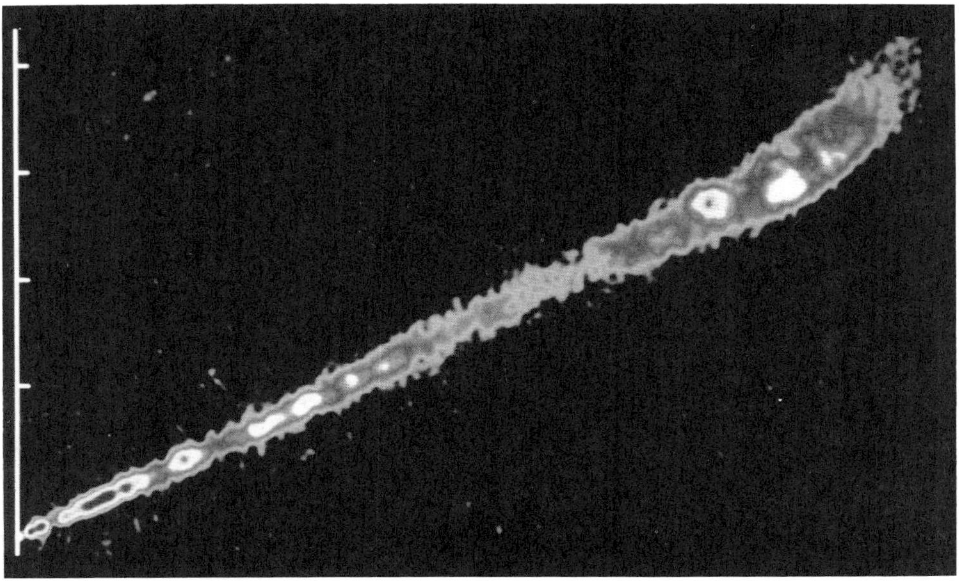

PICTURE 6.3. The long radio jet in the radio galaxy NGC 6251 at a distance of approximately 220 million lt-yr. The region imaged is over 300,000 lt-yr long. The flow direction is from lower left to upper right in this image. (National Radio Astronomy Observatory, observers: R. A. Perley, A. H. Bridle, and A. G. Willis. The Astronomical Society of the Pacific).

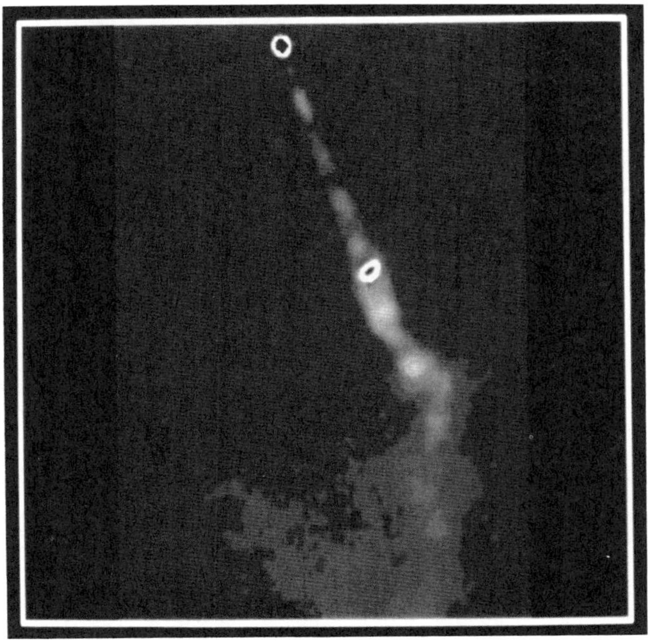

PICTURE 6.4. The one-sided, 6000 lt-yr long, radio jet in the elliptical galaxy M 87, approximately 50 million lt-yr away (National Radio Astronomy Observatory, observers: F. N. Owen, J. A. Biretta and P. E. Hardee. The Astronomical Society of the Pacific).

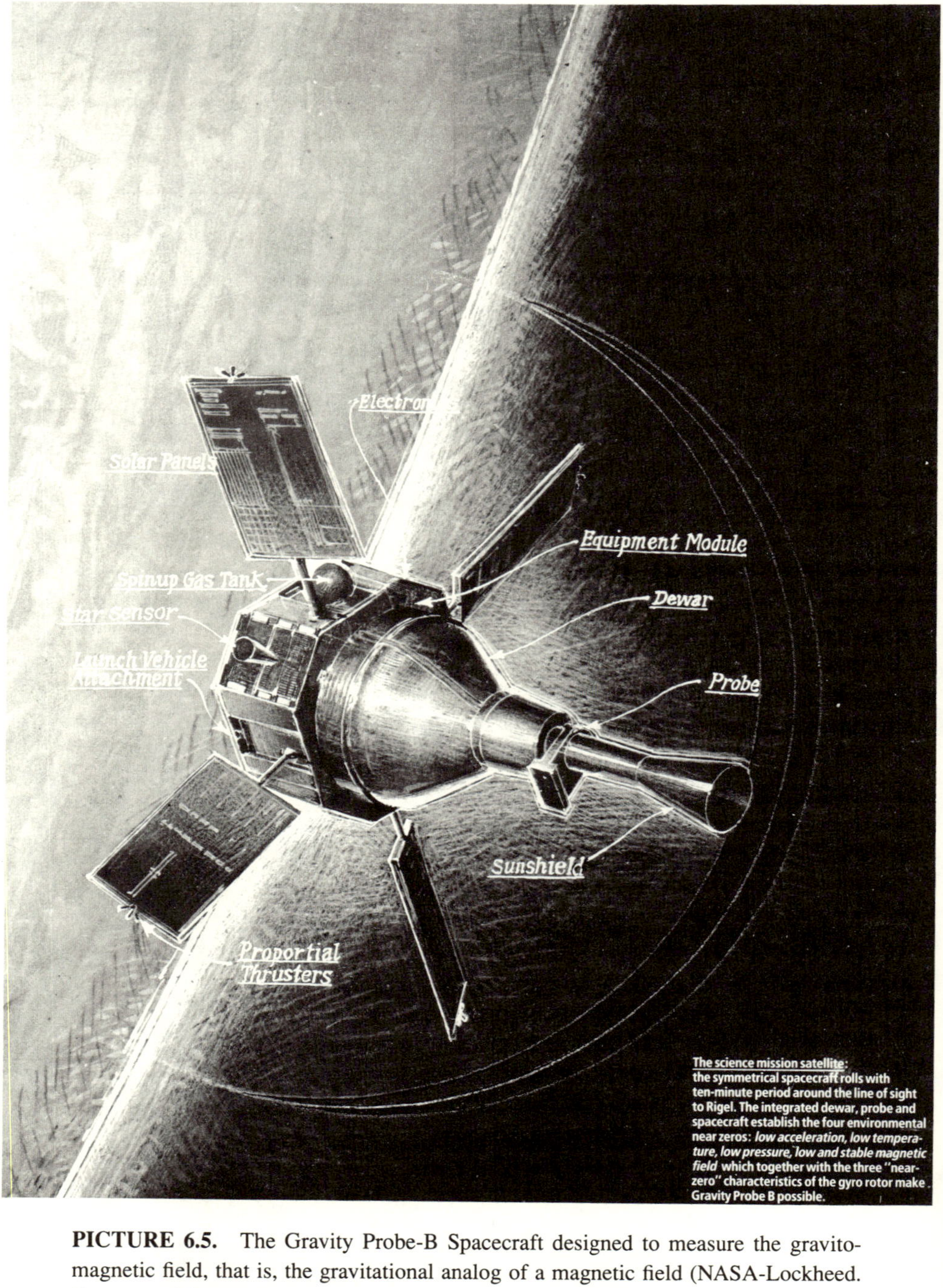

PICTURE 6.5. The Gravity Probe-B Spacecraft designed to measure the gravitomagnetic field, that is, the gravitational analog of a magnetic field (NASA-Lockheed. Courtesy of Francis Everitt).

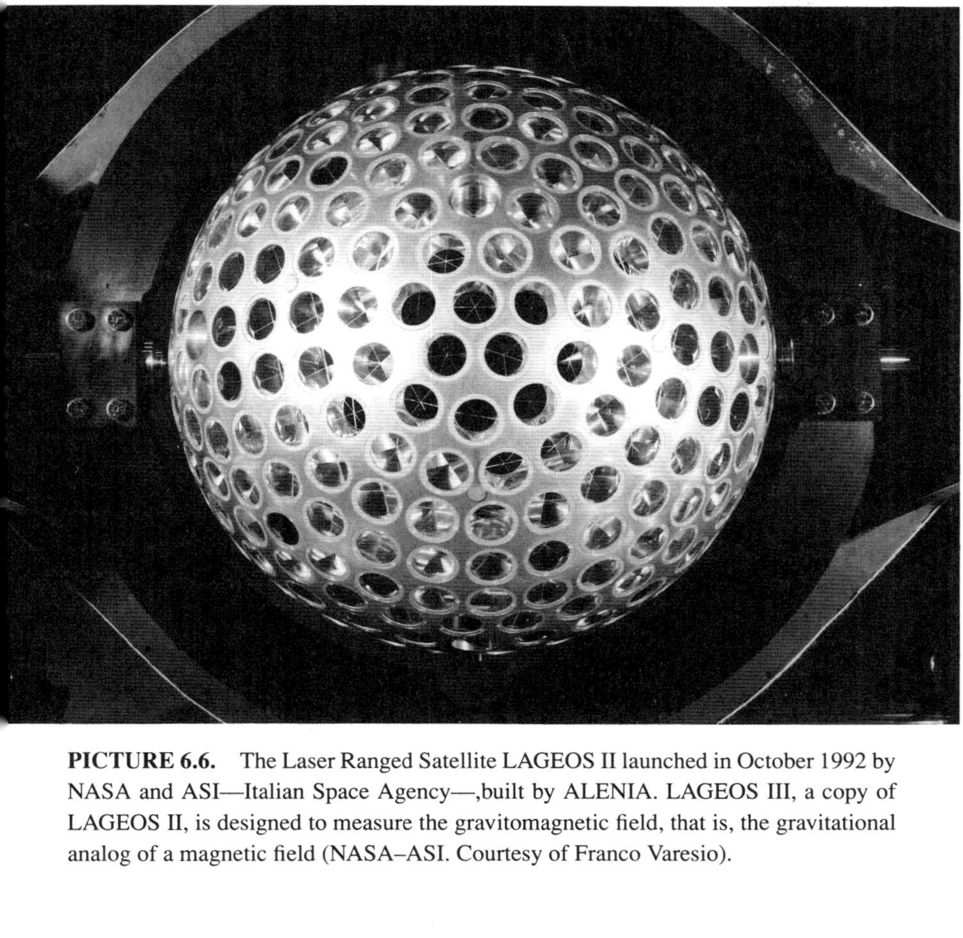

PICTURE 6.6. The Laser Ranged Satellite LAGEOS II launched in October 1992 by NASA and ASI—Italian Space Agency—,built by ALENIA. LAGEOS III, a copy of LAGEOS II, is designed to measure the gravitomagnetic field, that is, the gravitational analog of a magnetic field (NASA–ASI. Courtesy of Franco Varesio).

In other words, the corotating beam will return to the emission point after a time $T_R = T + \frac{\dot{\Omega}RT}{c}$, and the counterrotating beam will return to the emission point after a time $T_{CR} = T - \frac{\dot{\Omega}RT}{c}$. Therefore, the difference between the circulation times of the corotating beam and of the counterrotating beam is $\Delta T = \frac{2\dot{\Omega}RT}{c}$.

In modern optical ring interferometers the light propagates in optical fibers.[115] When the light leaves the fiber, because of the optical path difference between the corotating light beam and the counterrotating light beam, there is a phase difference $\Delta\phi$ between the two waves which manifests itself in a shift of the interference fringes observed by superposing the two counterpropagating beams on a screen (see fig 6.15); from expression (6.12.4) the phase shift between the

PICTURE 6.7. Artist's view of LAGEOS II laser ranging (ASI-NASA).

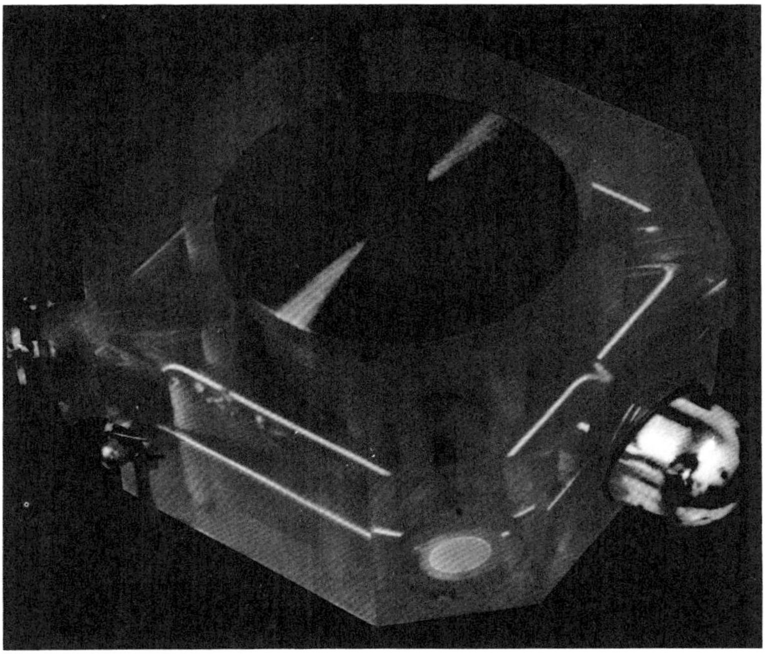

PICTURE 6.8. The Ring Laser Gyro LG-9028 of about 7 cm size (28 cm path-length), with a bias stability of better than 0.01°/hr, built by LITTON (courtesy of Tom Hutchings).

two waves is

$$\Delta\phi = \frac{\Delta L}{\lambda\!\!\!^{-}} = \frac{4A\dot{\Omega}}{\lambda\!\!\!^{-} c}. \qquad (6.12.5)$$

The first measurement of the fringe shift between counterrotating light beams was performed by Harress in 1911 and by Sagnac in 1913.[113]

In a ring laser gyroscope the number of wavelengths in the loop remains constant.[113] Therefore, from expression (6.12.4) one can derive the relative difference between the frequencies of the counterrotating light beam and of the rotating light beam: $\frac{\Delta\omega}{\omega} \cong \left|\frac{\Delta T}{T}\right|$, or

$$\frac{\Delta\omega}{\omega} = \frac{4A\dot{\Omega}}{cL}. \qquad (6.12.6)$$

In ring laser gyroscopes (RLG), or fiber resonators, one measures this relative frequency shift that is proportional to the ring angular velocity.

Optical rotation sensors may reach[110] an accuracy of about 10^{-10} rad/s, that is, about 10^{-5} degrees/hr, or about 10^{-6} of the Earth rotation rate.

REFERENCES CHAPTER 6

1. H. C. Oersted, *Experimenta circa effectum conflictus electrici in acum magneticum* (Copenhagen, 1820).
2. A. Einstein, *The Meaning of Relativity*, 5th ed. (Princeton University Press, Princeton, 1956).
3. S. Weinberg, *Gravitation and Cosmology: Principles and Applications of the General Theory of Relativity* (Wiley, New York, 1972).
4. C. W. Misner, K. S. Thorne, and J. A. Wheeler, *Gravitation* (Freeman, San Francisco, 1973), 2101.
5. S. W. Hawking and G. F. R. Ellis, *The Large Scale Structure of Space-Time* (Cambridge University Press, Cambridge, 1973).
6. L. D. Landau and E. M. Lifshitz, *The Classical Theory of Fields* (Pergamon Press, New York, 1975).
7. J. A. Wheeler, *Geometrodynamics, Topics of Modern Physics* (Academic, New York, 1962), vol. 1; expanded from *Rendiconti della Scuola Internazionale di Fisica Enrico Fermi XI Corso*, July 1959 (Zanichelli, Bologna, 1960).
8. K. S. Thorne, Multipole expansions of gravitational radiation, *Rev. Mod. Phys.* 52:299–339 (1980).
9. K. S. Thorne, R. H. Price, and D. A. MacDonald, eds., *Black Holes, the Membrane Paradigm* (Yale University Press, New Haven and London, 1986), 72.
10. J. A. Wheeler, View that the distribution of mass and energy determines the metric, in *Onzième Conseil de Physique Solvay: La structure et l'évolution de l'univers* (Éditions Stoops, Brussels, 1959), 97–141.
11. J. A. Wheeler, Geometrodynamics and the issue of the final state in *Relativity, Groups and Topology: 1963 Les Houches Lectures*, ed. C. DeWitt and B. DeWitt (Gordon and Breach, New York, 1964), 315–520.
12. J. A. Wheeler, Gravitation as geometry II, and Mach's principle as boundary condition for Einstein's equations, in *Gravitation and Relativity*, ed. H. Y. Chiu and W. F. Hoffman (Benjamin, New York, 1964), 65 and 303, respectively.
13. J. Isenberg and J. A. Wheeler, Inertia here is fixed by mass-energy there in every W model universe, in *Relativity, Quanta, and Cosmology in the Development of the Scientific Thought of Albert Einstein*, ed. M. Pantaleo and F. de Finis (Johnson Reprint Corp., New York, 1979), vol. 1, 267–93.
14. A. M. Ampère, Recueil d'observations èlectrodynamiques *Mém. de l'Acad.* 6:175 ff. (1825).
15. E. Whittaker, *A History of Theories of Aether and Electricity. I. The Classical Theories* (Nelson, London, 1910; revised ed., 1951), 86–87.

16. J. Lense and H. Thirring, Über den Einfluss der Eigenrotation der Zentralkörper auf die Bewegung der Planeten und Monde nach der Einsteinschen Gravitationstheorie, *Phys. Z.* 19:156–63 (1918).
17. J. Lense and H. Thirring, On the gravitational effects of rotating masses: The Thirring-Lense papers, trans. B. Mashhoon, F. W. Hehl, and D. S. Theiss, *Gen. Relativ. Gravit.* 16:711–50 (1984).
18. R. P. Kerr, Gravitational field of a spinning mass as an example of algebraically special metrics, *Phys. Rev. Lett.* 11:237–38 (1963).
19. E. T. Newman, E. Couch, K. Chinnapared, A. Exton, A. Prakash, and R. Torrence, Metric of a rotating charged mass, *J. Math. Phys.* 6:918–19 (1965).
20. R. H. Boyer and R. W. Lindquist, Maximal analytic extension of the Kerr metric, *J. Math. Phys.* 8:265–81 (1967).
21. J. D. Jackson, *Classical Electrodynamics* (Wiley, New York, 1975).
22. J. Weber, *General Relativity and Gravitational Waves* (Wiley-Interscience, New York, 1961).
23. K. S. Thorne, Gravitational radiation, in *300 Years of Gravitation*, ed. S. W. Hawking and W. Israel (Cambridge University Press, Cambridge, 1987), 330–458.
24. P. M. Fitzpatrick, *Principles of Celestial Mechanics* (Academic Press, New York, 1970).
25. W. de Sitter, On Einstein's theory of gravitation and its astronomical consequences, *Mon. Not. Roy. Astron. Soc.* 76:699–728 (1916).
26. H. Thirring, Über die Wirkung rotierender ferner Massen in der Einsteinschen Gravitationstheorie, *Z. Phys.* 19:33–39 (1918).
27. H. Thirring, Berichtigung zu meiner Arbeit: Über die Wirkung rotierender ferner Massen in der Einsteinschen Gravitationstheorie, *Z. Phys.* 22:29–30 (1921).
28. D. R. Brill and J. M. Cohen, Rotating masses and their effect on inertial frames, *Phys. Rev.* 143:1011–15 (1966).
29. H. Pfister and K. H. Braun, A mass shell with flat interior cannot rotate rigidly, *Class. Quantum Grav.* 3:335–45 (1986).
30. A. Lausberg, On the inertial effects induced by a shell of finite thickness, *Bull. Acad. Roy. Belgique* (*Cl. Sci.*) 57:125–53 (1971).
31. P. Teyssandier, On the precession of locally inertial systems in the neighbourhood of a rotating sphere, *Lett. Nuovo Cim.* 5 (2): 1038–43 (1972).
32. J. M. Cohen, W. J. Sarill, and C. V. Vishveshwara, An example of induced centrifugal force in general relativity, *Nature* 298:829 (1982).
33. L. Bass and F. A. E. Pirani, On the gravitational effects of distant rotating masses, *Phil. Mag.* 46:850–56 (1955).
34. M. Heller, Rotating bodies in general relativity, *Acta Cosm.* 3:97–107 (1975).
35. C. H. Brans, Mach's principle and the locally measured gravitational constant in general relativity, *Phys. Rev.* 125:388–96 (1962).
36. R. H. Dicke, *The Theoretical Significance of Experimental Relativity* (Gordon and Breach, New York, 1964).

37. J. M. Bardeen and J. A. Petterson, The Lense–Thirring effect and accretion disks around Kerr black holes, *Astrophys. J. Lett.* 195:L65–L67 (1975).

38. D. C. Wilkins, Bound geodesics in the Kerr metric, *Phys. Rev. D* 5:814–22 (1972).

39. C. W. Allen, Astrophysical Quantities, 3d ed. with corrections (Athlone, London, 1976).

40. I. Newton, *Philosophiae naturalis principia mathematica*, (London, Streater, 1687); trans. A. Motte (1729) and revised by F. Cajori as *Sir Isaac Newton's Mathematical Principles of Natural Philosophy and His System of the World* (University of California Press, Berkeley and Los Angeles, 1934; paperback, 1962).

41. S. Weinberg, *Gravitation and Cosmology: Principles and Applications of the General Theory of Relativity* (Wiley, New York, 1972), 17.

42. J. H. Oort, Observational Evidence Confirming Lindblad's Hypothesis of a Rotation of the Galactic System, *B.A.N.* 3 (120): 275–82 (1927).

43. W. Fricke and A. Kopff, 4te *Fundamental Katalog* (Astronomisches Rechen Institut, Heidelberg, 1963).

44. P. D. Hemenway, R. L. Duncombe, W. H. Jefferys, and P. J. Shelus, Using space telescope to tie the Hipparcos and extragalactic reference frames together, in *Proc. Colloq. on the European Astrometry Satellite Hipparcos, Scientific Aspects of the Input Catalogue Preparation*, Aussois, 3–7 June 1985, ESA SP-234 (1985).

45. V. L. Ginzburg, Artificial satellites and the theory of relativity, *Sci. Am.* 200:149–60 (1959).

46. H. Yilmaz, Proposed test of the nature of gravitational interaction, *Bull. Am. Phys. Soc.* 4:65 (1959).

47. H. Yilmaz, *Introduction to the Theory of Relativity and the Principles of Modern Physics* (Blaisdell, New York, 1965), 172.

48. G. E. Pugh, Proposal for a satellite test of the Coriolis prediction of general relativity, *WSEG Research Memorandum No.* 11 (The Pentagon, Washington, DC, 1959).

49. L. I. Schiff, Motion of a gyroscope according to Einstein's theory of gravitation, in *Proc. Nat. Acad. Sci.* 46:871–82 (1960); and L. I. Schiff, Possible new test of general relativity theory, *Phys. Rev. Lett.* 4:215–17 (1960).

50. C. W. F. Everitt, The gyroscope experiment. I. General description and analysis of gyroscope performance, in *Experimental Gravitation*, ed. B. Bertotti (Academic Press, New York, 1974), 331–60.

51. J. A. Lipa, W. M. Fairbank, and C. W. F. Everitt, The gyroscope experiment. II. Development of the London-moment gyroscope and of cryogenic technology for space, in *Experimental Gravitation*, ed. B. Bertotti (Academic Press, New York, 1974), 361–80; see also papers on the Stanford Relativity Gyroscope Experiment, in *Proc. SPIE*, vol. 619, Cryogenic Optical Systems and Instruments II, Los Angeles (Society of Photo-Optical Instr. Eng., 1986), 29–165; and J. P. Turneaure et al., The Gravity-Probe-B relativity gyroscope experiment: Approach to a flight mission, in *IV Marcel Grossmann Meeting*, ed. R. Ruffini (Elsevier, Rome, 1986), 411–64.

52. R. A. Van Patten and C. W. F. Everitt, Possible experiment with two counter-orbiting drag-free satellites to obtain a new test of Einstein's general theory of relativity and improved measurements in geodesy, *Phys. Rev. Lett.* 36:629–32 (1976).

53. R. A. Van Patten and C. W. F. Everitt, A possible experiment with counter-orbiting drag-free satellites to obtain a new test of Einstein's general theory of relativity and improved measurements in geodesy, *Celest. Mech.* 13:429–47 (1976).

54. V. B. Braginsky, C. M. Caves, and K. S. Thorne, Laboratory experiments to test relativistic gravity, *Phys. Rev. D* 15:2047–68 (1977).

55. M. O. Scully, Suggestion and analysis for a new optical test of general relativity, in *Laser Spectroscopy IV*, ed. H. Walther and K. W. Rothe (Springer-Verlag, Berlin, 1979), 21–30.

56. V. B. Braginsky and A. G. Polnarev, Relativistic spin quadrupole gravitational effect, *Pis'ma Zh. Eksp. Teor. Fiz.* (*USSR*), 31:444–47 (1980); translation in *JETP Lett.* (*USA*).

57. V. B. Braginsky, A. G. Polnarev, and K. S. Thorne, Foucault pendulum at the South Pole: Proposal for an experiment to detect the Earth's general relativistic gravitomagnetic field, *Phys. Rev. Lett.* 53:863–66 (1984).

58. H. J. Paik, B. Mashhoon, and C. M. Will, Detection of gravitomagnetic field using an orbiting superconducting gravity gradiometer, in *Int. Symposium on Experimental Gravitational Physics*, Guangzhou, China, August 1987, ed. P. F. Michelson (World Scientific, Singapore), 229–44.

59. S. Vitale, M. Bonaldi, P. Falferi, G. A. Prodi, and M. Cerdonio, Magnetization by rotation and gyromagnetic gyroscopes, *Phys. Rev. B*, 39:11993–12002 (1989).

60. I. Ciufolini, Measurement of general relativistic perturbations on high altitude laser ranged artificial satellites, *Bull. Am. Phys. Soc.* 6:1169 (1985).

61. I. Ciufolini, Mach principle, Lense-Thirring effect and gyrogravitation via two laser ranged satellites, *Bull. Am. Phys. Soc.* 6:1162 (1985).

62. I. Ciufolini, Measurement of Lense-Thirring drag on high-altitude laser ranged artificial satellites, *Phys. Rev. Lett.* 56:278–81 (1986).

63. I. Ciufolini, New relativistic measurements with laser ranged satellites, in *VI Int. Workshop on Laser Ranging Instrumentation*, September 1986, ed. J. Gaignebet and F. Baumont (GRGS-Cerga, France, 1987), 1–9.

64. I. Ciufolini, The LAGEOS Lense-Thirring precession and the LAGEOS non-gravitational nodal perturbations. I. *Celest. Mech.* 40:19–33 (1987).

65. I. Ciufolini, LAGEOS III and the gravitomagnetic field, in *NASA Workshop on Relativistic Gravitation Experiments in Space*, Annapolis, MD, June 1988 (NASA Conf. Publ. 3046), 126–31.

66. I. Ciufolini, Preliminary error budget of the LAGEOS gravitomagnetic experiment, *Bull. Am. Phys. Soc.* 33:1466 (1988) (Austin, TX, March 1988).

67. I. Ciufolini, A comprehensive introduction to the LAGEOS gravitomagnetic experiment: From the importance of the gravitomagnetic field in physics to preliminary error analysis and error budget, *Int. J. Mod. Phys. A* 4:3083–3145 (1989).

68. I. Ciufolini, The physics of the LAGEOS-III gravitomagnetic experiment: A summary and an introduction to the main scientific results of an Italian study, in *Proc. 1st William Fairbank Int. Meeting on Relativistic Gravitational Experiments in Space*, Rome, 10–14 September 1990 (World Scientific, Singapore, 1993), 170–77.

69. S. C. Cohen and D. E. Smith, LAGEOS scientific results: introduction, *J. Geophys. Res.* 90 (B11): 9217–20 (1985).

70. D. E. Smith and P. J. Dunn, Long term evolution of the LAGEOS orbit, *J. Geophys. Res. Lett.* 7:437–40 (1980).

71. C. F. Yoder, J. G. Williams, J. O. Dickey, B. E. Schutz, R. J. Eanes, and B. D. Tapley, Secular variation of Earth's gravitational harmonic J_2 coefficient from Lageos and nontidal acceleration of Earth rotation, *Nature* 303:757–62 (1983).

72. LAGEOS Scientific Results, ed. S. C. Cohen et al., *J. Geophys. Res.* 90:9215–9438 (1985).

73. D. P. Rubincam, General relativity and satellite orbits: The motion of a test particle in the Schwarzschild metric, *Celest. Mech.* 15:21–33 (1977).

74. N. Ashby and B. Bertotti, Relativistic perturbations of an Earth satellite, *Phys. Rev. Lett.* 52:485–88 (1984).

75. I. Ciufolini and R. A. Matzner, Non-Riemannian theories of gravity and lunar and satellite laser ranging, *Int. J. Mod. Phys. A* 7:843–52 (1992).

76. J. G. Williams, R. H. Dicke, P. L. Bender, C. O. Alley, W. E. Carter, D. G. Currie, D. H. Eckhardt, J. E. Faller, W. M. Kaula, J. D. Mulholland, H. H. Plotkin, S. K. Poultney, P. J. Shelus, E. C. Silverberg, W. S. Sinclair, M. A. Slade, and D. T. Wilkinson, New test of the equivalence principle from lunar laser ranging, *Phys. Rev. Lett.* 36:551–54 (1976).

77. I. I. Shapiro, C. C. Counselman, and R. W. King, Verification of the principle of equivalence for massive bodies, *Phys. Rev. Lett.* 36:555–58 (1976); with a correction, 1068 (1976).

78. W. M. Kaula, *Theory of Satellite Geodesy* (Blaisdell, Waltham, 1966).

79. F. J. Lerch, S. M. Klosko, G. B. Patel, and C. A. Wagner, A gravity model for crustal dynamics (GEM-L2), *J. Geophys. Res.* 90 (B11): 9301–11 (1985).

80. F. J. Lerch, S. M. Klosko, C. A. Wagner, and G. B. Patel, On the accuracy of recent Goddard gravity models, *J. Geophys. Res.* 90 (B11): 9312–34 (1985).

81. D. P. Rubincam, On the secular decrease in the semimajor axis of LAGEOS's orbit, *Celest. Mech.* 26:361–82 (1982).

82. D. P. Rubincam, LAGEOS orbit decay due to infrared radiation from Earth, *J. Geophys. Res.* 92 (B2): 1287–94 (1987).

83. B. Bertotti and L. Iess, The rotation of LAGEOS, *J. Geophys. Res.*, 96 (B2): 2431–40 (1991).

84. F. Barlier et al., Non-gravitational perturbations on the semimajor axis of LAGEOS, *Ann. Geophys.* 4 (A): 193–210 (1986).

85. D. P. Rubincam, Yarkovsky thermal drag on LAGEOS, *J. Geophys. Res.* 93 (B11): 13805–810 (1988).

86. D. P. Rubincam, Drag on the LAGEOS satellite, *J. Geophys. Res.* 95 (B11): 4881–86 (1990).

87. D. P. Rubincam, The LAGEOS along-track acceleration: a review, in *Proc. 1st William Fairbank Int. Meeting on Relativistic Gravitational Experiments in Space*, Rome, 10–14 September 1990.

88. P. Farinella et al., Effects of thermal thrust on the node and inclination of LAGEOS, *Astr. and Astrophys.* 234:546–54 (1990).

89. I. Ciufolini, D. Lucchesi, and F. Vespe, Status of the LAGEOS III gravitomagnetic experiment, in *Proc. 10th Italian Conf. on General Relativity and Gravitation*, Bardonecchia, Italy, September 1992.

90. I. Ciufolini, Gravitomagnetism and status of the LAGEOS III experiment, *Class. Quantum Grav.* 11:A73–A81 (1994). See also *Proc. Int. Symp. on Experimental Gravitation*, Nathiagali, 1993, ed. M. Karim and A. Qadir (Institute of Physics, Bristol & Philadelphia, 1994).

91. A. T. Doodson, Tide-Generating Potential, *Proc. Roy. Soc. London* A 100:305–29 (1921).

92. P. Melchior, *The Tides of the Planet Earth* (Pergamon, New York, 1980).

93. R. Eanes, B. E. Schutz, and B. Tapley, Earth and ocean tide effects on LAGEOS and Starlette, in *Proc. 9th Int. Symp. on Earth Tides*, ed. J. T. Kuo (E. Schweizerbart'sche Verlagsbuchhandlung, 1983), 239–49.

94. B. V. Sanchez, Rotational dynamics of mathematical models of the nonrigid Earth, *Appl. Mech. Res. Lab.*, Report 1066 (University of Texas at Austin, 1974).

95. P. Bender and C. C. Goad, The use of satellites for geodesy and geodynamics, vol. 2, in *Proc. 2nd Int. Symp. on the Use of Artificial Satellites for Geodesy and Geodynamics*, ed. G. Veis and E. Livieratos (National Technical University of Athens, 1979), 145.

96. D. P. Rubincam, P. Knocke, V. R. Taylor, and S. Blackwell, Earth anisotropic reflection and the orbit of LAGEOS, *J. Geophys. Res.* 92 (B11): 11662–68 (1987).

97. D. Lucchesi and P. Farinella, Optical properties of the Earth's surface and long term perturbations of LAGEOS' semimajor axis, *J. Geophys. Res.* 97 (B5): 7121–28 (1992).

98. I. Ciufolini et al., Effect of particle drag on the LAGEOS node and measurement of the gravitomagnetic field, *Nuovo Cimento B* 105:573–88 (1990).

99. W. H. Cannon, The classical analysis of the response of a long baseline radio interferometer, *Geophys. J. Roy. Astr. Soc.* 53:503–30 (1978).

100. D. S. Robertson and W. E. Carter, Earth orientation determinations from VLBI observations, in *Proc. Int. Conf. on Earth Rotation and the Terrestrial Reference Frame*, Columbus, OH, ed. I. Mueller (Department of Geodetic Science and Survey, Ohio State University, 1985), 296–306.

101. B. D. Tapley, I. Ciufolini et al., Measuring the Lense-Thirring precession using a second LAGEOS satellite, Results of a Joint NASA/ASI Study on the LAGEOS III Gravitomagnetic Experiment (30 September 1989).

102. L. D. Landau and E. M. Lifshitz, *Electrodynamics of Continuous Media* (Pergamon Press, Oxford, 1960).

103. S. J. Barnett, Magnetisation by rotation, *Phys. Rev.* 6:171–72 (1915).

104. S. J. Barnett, Magnetisation by rotation, *Phys. Rev.* 6:239–70 (1915).

105. R. Becker, G. Heller, and F. Sauter, Über die Stromverteilung in einer supraleitenden Kugel, *Z. Phys.* 85:772–87 (1933).

106. F. London, *Superfluids* (Wiley, New York, 1950), vol. 1.

107. A. F. Hildebrandt, Magnetic field of a rotating superconductor, *Phys. Rev. Lett.* 12:190–91 (1964).

108. A. F. Hildebrandt and M. M. Saffren, Superconducting transition of a rotating superconductor: The hollow cylinder, in *Proc. 9th Int. Conf. on Low Temperature Physics*, Columbus, OH, 1964, ed. J. G. Daunt, D. O. Edwards, F. J. Milford, and N. Yaqub (Plenum, New York, 1965), 459–65; see also C. A. King, J. B. Hendricks, and H. E. Rorschach, The magnetic field generated by a rotating superconductor, ibid., 466–70; and M. Bol and W. M. Fairbank, Measurement of the London moment, ibid., 471–74.

109. M. O. Scully, M. S. Zubairy, and M. P. Haugan, Proposed optical test of metric gravitation theories, *Phys. Rev. A* 24:2009–16 (1981).

110. W. Schleich and M. O. Scully, General relativity and modern optics, in *Les Houches 1982, New Trends in Atomic Physics*, ed. G. Grynberg and R. Stora (North-Holland, Amsterdam, 1984), 995–1124; see also ref. 111.

111. H. Dehnen, Proving of the general relativistic rotation effect by means of a ring laser, *Z. Naturforsch.* 22A:816–21 (1967).

112. E. J. Post, Sagnac effect, *Rev. of Mod. Phys.* 39:475–93 (1967).

113. C. V. Heer, History of the laser gyro, in *Proc. SPIE Vol. 487, Physics of Optical Ring Gyros*, Snowbird, UT, 1984, ed. S. F. Jacobs, M. Sargent III, M. O. Scully, J. Simpson, V. Sanders, and J. E. Killpatrick (Soc. of Photo-Optical Instr. Eng.) 2–12.

114. W. W. Chow, J. Gea-Banacloche, L. M. Pedrotti, V. E. Sanders, W. Schleich, and M. O. Scully, The ring laser gyro, *Rev. Mod. Phys.* 57:61–104 (1985).

115. S. Ezekiel, An overview of passive optical gyros, in *Proc. SPIE Vol. 478, Fiber Optic and Laser Sensors II*, Arlington, VA, May 1984 (Soc. of Photo-Optical Instr. Eng.), 2.

116. B. Mashhoon, H. J. Paik, and C. M. Will, Detection of the gravitomagnetic field using an orbiting superconducting gravity gradiometer: Theoretical principles, *Phys. Rev. D* 39:2825–38 (1989).

117. W. de Sitter, On Einstein's theory of gravitation and its astronomical consequences, *Mon. Not. Roy. Astron. Soc.* 77:155–84 (1916); see also ref. 118.

118. J. A. Schouten, Die direkte Analysis zur neueren Relativitätstheorie, *Verh. der Kon. Ak. van Wet. te Amsterdam*, Deel 12, No. 6 (1919).

119. A. D. Fokker, The geodesic precession; a consequence of Einstein's theory of gravitation, *Proc. K. Ned. Akad. Amsterdam* 23.5:729–38 (1921).

120. C. M. Will, Theory and experiment in gravitational physics, rev. ed. (Cambridge University Press, Cambridge, 1993).

121. C. M. Will and K. Nordtvedt, Jr., Conservation laws and preferred frames in relativistic gravity. I. Preferred-frame theories and an Extended PPN Formalism, *Astrophys. J.* 177:757–74 (1972).

122. K. Nordtvedt, Jr., and C. M. Will, Conservation laws and preferred frames in relativistic gravity. II. Experimental evidence to rule out preferred-frame theories of gravity, *Astrophys. J.* 177:775–92 (1972).

123. C. M. Will, The confrontation between general relativity and experiment: An update, *Phys. Reports* 113:345–422 (1984).

124. C. M. Will, Experimental gravitation from Newton's Principia to Einstein's General Relativity, in *300 Years of Gravitation*, ed. S. W. Hawking and W. Israel (Cambridge University Press, Cambridge, 1987), 80–127.

125. B. Bertotti, I. Ciufolini, and P. L. Bender, New test of general relativity: Measurement of de Sitter geodetic precession rate for lunar perigee, *Phys. Rev. Lett.* 58:1062–65 (1987).

126. B. Bertotti, I. Ciufolini, and P. L. Bender, Test of the de Sitter-geodetic precession, *Bull. Am. Phys. Soc.* 33:1466 (1988).

127. I. I. Shapiro, R. D. Reasenberg, J. F. Chandler, and R. W. Babcock, Measurement of the de Sitter precession of the moon: A relativistic three-body effect, *Phys. Rev. Lett.* 61:2643–46 (1988).

128. J. O. Dickey, X. X. Newhall, and J. G. Williams, Investigating relativity using lunar laser ranging: Geodetic precession and the Nordtvedt effect, *Adv. Space Res.* 9:75–78 (1989); see also *Relativistic Gravitation, Symp. 15 of the COSPAR 27th Plenary Meeting*, Espoo, Finland, 18–29 July 1988.

129. A. Papapetrou, Spinning Test-Particles in General Relativity. I. *Proc. Roy. Soc. London A* 209:248–58 (1951).

130. E. Corinaldesi and A. Papapetrou, Spinning test-particles in general relativity. II., *Proc. Roy. Soc. London A* 209:259–68 (1951).

131. F. A. E. Pirani, On the physical significance of the Riemann tensor, *Acta Phys. Polon.* 15:389–405 (1956).

132. C. Møller, *The Theory of Relativity* (Oxford University Press, London, 1972).

133. W. Tulczyjew: Equations of motion of rotating bodies in general relativity theory, *Acta Phys. Polon.* 18:37–55 (1959).

134. R. F. O'Connel, Spin rotation and C, P and T effects in the gravitational interaction and related experiments, in *Proc. Int. School of Physics "Enrico Fermi," Course 56, Experimental Gravitation*, Varenna, 1972, ed. B. Bertotti (Academic Press, New York, 1974), 496–514.

135. I. Ciufolini, On gravitomagnetism and dragging of inertial frames, submitted to *Phys. Rev. D* (1995).

136. K. Nordtvedt, The gravitomagnetic interaction and its relationship to other relativistic gravitational effects, NASA report, April 1991, to be published.

137. N. Ashby and B. Shahid-Saless, Geodetic precession or dragging of inertial frames? *Phys. Rev. D* 42:1118–22 (1990).

138. J. A. Wheeler, A few specializations of the generic local field in electromagnetism and gravitation, *Trans. NY Acad. Sci., Ser. II*, 38:219–43 (1977).

139. A. Z. Petrov, *Einstein Spaces* (Pergamon Press, Oxford, 1969).

140. E. T. Newman and R. Penrose, An Approach to Gravitational Radiation by a Method of Spin Coefficients, *J. Math. Phys.* 3:566–78 (1962).

141. S. Chandrasekhar, An introduction to the theory of the Kerr metric and its perturbations, in *General Relativity: An Einstein Centenary Survey*, ed. S. W. Hawking and W. Israel (Cambridge University Press, Cambridge, 1979), 370–453.

142. S. Chandrasekhar, *The Mathematical Theory of Black-Holes* (Clarendon Press, Oxford, 1983).

143. D. Kramer, H. Stephani, M. MacCallum, and E. Herlt, *Exact Solutions of Einstein's Field Equations* (VEB Deutscher Verlag der Wissenschaften, Berlin, 1980).

144. These scalar invariants were first calculated by Guido Cognola and Luciano Vanzo of the University of Trento, using MACSYMA.

145. These scalars have been then independently calculated, even for the Kerr-Newman metric, by Lars Hörnfeldt of the University of Stockholm using the powerful STENSOR.

146. N. Bowditch, *American Practical Navigator, an Epitome of Navigation*, vol. 1 (Defense Mapping Agency Hydrographic Center, 1977).

147. J. C. Ries, C. Huang, and M. M. Watkins, Effect of general relativity on a near-Earth satellite in the geocentric and barycentric reference frames, *Phys. Rev. Lett.* 61:903–6 (1988).

148. E. F. Taylor and J. A. Wheeler, *Spacetime Physics: Introduction to Special Relativity*, 2d ed. (Freeman and Co., New York, 1991), 277.

149. B. and I. Friedländer, *Absolute und relative Bewegung*? (Simion-Verlag, Berlin, 1896).

150. A. Föppl, Übereinen Kreiselversuch zur messung der Umdrehungsgeschwindigkeit der Erde, *Sitzb. Bayer. Akad. Wiss* 34:5–28 (1904); also in *Phys. Z.* 5:416–25 (1904). See also A. Föppl, Über absolute und relative Bewegung, *Sitzb. Bayer. Akad. Wiss* 34:383–95 (1904).

151. Translation of some paragraphs of the Friedländers' book in H. Pfister, Dragging effects near rotating bodies and in cosmological models, *Proc. Int. Conf. on Mach's Principle, From Newton's Bucket to Quantum Gravity*, Tübingen, Germany, July 1993, ed. J. Barbour and H. Pfister.

152. J. C. Ries, R. J. Eanes, B. D. Tapley, and M. M. Watkins, LAGEOS III mission analysis, in *Proc. LAGEOS II Science Working Team Meeting*, Matera, Italy, May 1993.

153. G. Ousley, LAGEOS III Delta launch vehicle support, paper presented at LAGEOS III meeting, ASI, Rome, Italy, 1991.

154. D. Currie and R. A. Matzner, private communications (1993).

155. C. F. Martin and D. P. Rubincam, Effects on the LAGEOS I along track acceleration due to Earth albedo as calculated from Earth radiation budget experiment (Erbe) data, in *Proc. LAGEOS II Science Working Team Meeting*, Matera, Italy, May 1993.

156. I. Ciufolini, Effect of Earth albedo in the LAGEOS III experiment, to be published (1994).

157. J. Müller, M. Schneider, M. Soffel, and H. Ruder, Testing Einstein's theory of gravity by analyzing Lunar Laser Ranging data, *Astrophys. J. Lett.* 382:L101–L103 (1991).

158. D. Wilkins, Is there a gravitational Thomas precession?, University of Nebraska, Omaha, report, unpublished (1990).

159. M. A. Abramowicz and J. P. Lasota, On traveling round without feeling it and uncurving curves, *A. J. Phys.* 54:936–39 (1986).

160. M. A. Abramowicz, B. Carter, and J. P. Lasota, Optical reference geometry for stationary and static dynamics, *Gen. Rel. Grav.* 20:1173–83 (1988).

7

Some Highlights of the Past and a Summary of Geometrodynamics and Inertia

In this chapter, after a brief historical review of concepts and ideas on inertia and gravitation, we summarize the interpretation of the origin of inertia, that is, local inertial frames, in general relativity described in the book.

7.1 SOME HIGHLIGHTS OF THE PAST

The idea is very old that happenings on Earth are governed by the configuration of the stars. How small or how great that influence is could first be assessed after **Newton** (1642–1727)[1] wrote that every particle in the universe acts on every other particle with a force of attraction proportional to the product of the two masses and inversely proportional to the square of the distance between them. The word inertia was first used in physics by **Kepler** (1571–1630). The concepts of inertial frame and law of inertia were introduced by **Galilei** (1564–1642).[2] "I wish I could tell this philosopher, in order to remove him from error, to take with him a very deep vase filled with water some time when he goes sailing, having prepared in advance a ball of wax or some other material which would descend very slowly to the bottom—so that in a minute it would scarcely sink a yard. Then, making the boat go as fast as he could, so that it might travel more than a hundred yards in a minute, he should gently immerse this ball in the water and let it descend freely, carefully observing its motion. And from the first, he would see it going straight toward that point on the bottom of the vase to which it would tend if the boat were standing still. To his eye and in relation to the vase its motion would appear perfectly straight and perpendicular. . . " (see also his famous passage on the dynamics in the cabin inside a uniformly moving ship).[3]

In the theory of Newton the inertial frame of reference, one particular frame of reference, is absolute space. The rate of change of position with respect to this absolute space, times mass, determined momentum, and the rate of change of momentum gave force. **Leibniz** (1646–1716)[4] argued against it in strong terms.

At rest or in uniform motion? Relative to what? Accelerated by a force? Accelerated relative to what? To answer by saying "relative to absolute space" was a workaday answer that provided a workaday way of doing physics. The special relativity of a later day made clear that one could never have hoped to find a mark that would give preference to one inertial frame of reference over another moving relative to it. However, Newton[1] gave two physical means to distinguish a rotating reference frame from a nonrotating reference frame: (1) "if a vessel, hung by a long cord, is so often turned about that the cord is strongly twisted, then filled with water, and held at rest together with the water; thereupon, by the sudden action of another force, it is whirled about the contrary way, and while the cord is untwisting itself, the vessel continues for some time in this motion; the surface of the water will at first be plain, as before the vessel began to move; but after that, the vessel, by gradually communicating its motion to the water, will make it begin sensibly to revolve, and recede by little and little from the middle, and ascend to the sides of the vessel, forming itself into a concave figure (as I have experienced), and the swifter the motion becomes, the higher will the water rise, till at last, performing its revolutions in the same times with the vessel, it becomes relatively at rest in it. This ascent of the water shows its endeavor to recede from the axis of its motion; and the true and absolute circular motion of the water, which is here directly contrary to the relative, becomes known, and may be measured by this endeavor." (2) "If two globes, kept at a given distance one from the other by means of a cord that connects them, were revolved about their common centre of gravity, we might, from the tension of the cord, discover the endeavor of the globes to recede from the axis of their motion, and from thence we might compute the quantity of their circular motions." And if one can believe in rotation with respect to absolute space can one not believe in velocity relative to absolute space? Regardless of these philosophical issues Newton's scheme worked, or appeared to work (see ref. 39).

However, Christian **Huygens** (1629–1695), Gottfried Leibniz,[4] and then George **Berkeley** (1685–1753)[5,6] objected strongly to Newton's scheme. Leibniz stressed that physics should, in our words, deal with relations between things, not relations to an unobservable absolute: "space and time are orders of things, and not things" and "I will here show how men come to form to themselves the notion of space. They consider that many things exist at once, and they observe in them a certain order of coexistence, according to which the relation of one thing to another is more or less simple. This order is their situation or distance. When it happens that one of those coexistent things changes its relation to a multitude of others which do not change their relations among themselves, and that another thing, newly come, acquires the same relation to the others as the former had, we then say it is come into the place of the former; and this change we call a motion. And that which comprehends all those places is called space. Which shows that in order to have an idea of place, and consequently of space, it is sufficient to consider these relations and the rules of their changes, without needing to fancy any absolute reality out of the things whose situation we consider" (Fifth Letter of Leibniz).

Berkeley wrote: "it does not appear to me that there can be any motion other than *relative*, so that to conceive motion there must be at least conceived two bodies, whereof the distance or position in regard to each other is varied. Hence, if there was only one body in being it could not possibly be moved. . . . [We] find all absolute motion we can frame an idea of to be at bottom no other than relative motion. . . . [A]bsolute motion, exclusive of all external relation, is incomprehensible. . . . As to what is said of the centrifugal force, that it does not at all belong to circular relative motion, I do not see how this follows from the experiment which is brought to prove it. For water in the vessel at that time wherein it is said to have the greatest relative circular motion, hath, I think, no motion at all. . . . For, to denominate a body *moved* it is requisite . . . that it changes its distance or situation with regard to some other body. . . . [It] follows that the philosophic consideration of motion does not imply the being of an *absolute Space*" . . . . "Let us imagine two globes, and that besides them nothing else material exists, then the motion in a circle of these two globes round their common center cannot be imagined. But suppose that the heaven of the fixed stars was suddenly created and we shall be in a position to imagine the motion of the globes by their relative position to the different parts of the heaven."

The concept of absolute space remained fuzzy, but the tool for distinguishing a frame locally rotating from one not rotating was sharpened in 1851. In that year Léon **Foucault** (1819–1868) demonstrated the diurnal motion of Earth by the rotation of the plane of oscillation of a freely suspended, long, heavy pendulum. It is not far conceptually from doing the experiment at Paris with clear skies to doing it at the North Pole on a stationary ice island in a fog. In the imagination one sees the plane of swing of the pendulum at the North Pole slowly turn, as one's clock ticks on, returning to its original direction in about 24 hours. What a strange result! Then the fog lifts. One comes to realize that the pendulum has been maintaining its plane of vibration fixed the whole time with respect to the faraway stars.

No wonder that Ernst **Mach** (1838–1916),[7,8] physicist, and philosopher and conscience of physics, translated rotation with respect to absolute space into the observable, verifiable rotation with respect to the stars and **masses** of the universe, and translated acceleration with respect to absolute space into acceleration with respect to the average frame of reference determined by the faraway stars and **masses** of the universe. In his *Science of Mechanics* (1883),[7] Mach wrote: "The principles of mechanics can, indeed, be so conceived, that even for relative rotations centrifugal forces arise. Newton's experiment with the rotating vessel of water simply informs us, that the relative rotation of the water with respect to the sides of the vessel produces no noticeable centrifugal forces, but that such forces are produced by its relative rotation with respect to the mass of the Earth and the other celestial bodies. No one is competent to say how the experiment would turn out if the sides of the vessel increased in thickness and mass till they were ultimately several leagues thick. The one experiment only lies before us, and our business is, to bring it into accord with the other facts known to us, and not with the arbitrary fictions of our imagination."

"I consider it possible that the law of inertia in its simple Newtonian form has only, for us human beings, a meaning which depends on space and time."

"When, accordingly, we say, that a body preserves unchanged its direction and velocity *in space*, our assertion is nothing more or less than an abbreviated reference to *the entire universe*."

"Instead of saying the direction and velocity of a mass μ in space remain constant, we may also employ the expression, the mean acceleration of the mass μ with respect to the masses $m, m', m'' \ldots$ at the distances $r, r', r'' \ldots$ is $= 0$, $d^2(\Sigma mr/\Sigma m)/dt^2 = 0$."

"For me only relative motions exist. . . . When a body rotates relatively to the fixed stars, centrifugal forces are produced; when it rotates relatively to some different body not relative to the fixed stars, no centrifugal forces are produced. I have no objection to calling the first rotation as long as it be remembered that nothing is meant except relative motion with respect to the fixed stars."

"On this account we must not underestimate even experimental ideas like those of Friedländer and Föppl, even if we do not yet see any immediate result from them." Mach also wrote (1872):[40] "By this it will be evident that, in its expression [the law of inertia], regard must be paid to the masses of the universe. . . . Now what share has every mass in the determination of direction and velocity in the law of inertia?" Mach's words were widely read.

How was one to translate Mach's program into concrete and well-defined mathematical physics? E. Kretschmann[9] had a proposal. He did not dream of anything so far-reaching as Einstein's general relativity. He presupposed from the start a flat spacetime. He used the "center of mass" of all the particles in the universe to define absolute rest. He further introduced a set of mutually orthogonal axes at this special point in such a way that relative to these axes the angular momentum of all the particles in the universe were zero. Some of the difficulties with Kretschmann's picture are obvious on a little inspection. First, electromagnetic mass-energy is not included. Second, gravitational interactions are taken to transfer momentum instantaneously from one mass to another. Third, demands of special relativity are not recognized. However Kretschmann's work had the great merit of making clear to the community that Mach's words could be more than complaints; they might be positive guides to a new framework for physics.

Einstein found Mach's words to be a guide to the theory of general relativity.[10,11,23] What does general relativity say about the mechanism through which faraway stars influence the local frame of reference in the here and the now?

Briefly put, it says (1) there is such an influence, (2) it is not some mysterious new natural phenomenon, and (3) instead it is a manifestation, in the subtle sense, of the very mechanism which transmits gravitation itself.

D. Sciama[12–14] and W. Davidson[15] have given us a model, linearized, and schematized, but illuminating, of how we can think of mass there governing inertia here. The model continues to think of spacetime as flat. However, it assumes that gravitational forces are, like electromagnetic forces, propagated with the speed of light. This feature of the propagation implies that in addition to the familiar static component of the force between particle and particle, inversely proportional to the square of the distance there will be a radiative component inversely proportional to the distance and linearly proportional to the acceleration of the source. A closer study of this component gives one reason to suspect that "mass there governs inertial frames here."

Why deal with suspicions? Why not go directly to general relativity and see what it says about the matter? There are two reasons for pursuing the question of inertia briefly within the framework of the older views of action at a distance. First, this approach enables one to see the quasi-Newtonian antecedents of "mass there governs inertia here." In consequence one sees Einstein's general relativity and what it has to say about the topic in terms of continuity with the past, not radical and inexplicable break. Second, the more ways one has to look at an important part of physics, the deeper the understanding one acquires of that topic. Instead of either the quasi-Newtonian approach or the general relativity treatment, may it not be more appropriate to look at the subject both ways and then compare and contrast the two approaches? This philosophy of science is a modest step in the direction of the ideal of theoretical physics enunciated by Richard Feynman (1918–1988) in his book *The Character of Physical Law*.[16] In brief, new views have a rich tissue of connection and correspondence with old views, a correspondence which helps one understand not only what in the new framework is new but what is old.

To picture the inertial frames as carried by the radiative component of the gravitational force exerted by the faraway matter it is helpful to have a quick look at the reason why there should be any radiative component at all.

One can spare himself detail and see the main point most quickly by asking the analogous question for electromagnetism. A distant charge exerts here not only a force proportional to the inverse square of its distance but also a force proportional to its acceleration and to the inverse first power of the distance. But why? For answer we turn to J. J. Thomson's (1856–1940)[17] simple explanation depicted in updated form in figure 7.1.

Assume that a charge slowly moving at uniform velocity to the left experiences in the course of a very short time Δt a force which causes it to reverse and move at an equal velocity to the right. The electric lines of force associated with a charge and uniform motion make the familiar Coulomb pattern (the very circumstance that they diverge outward as straight lines from the source

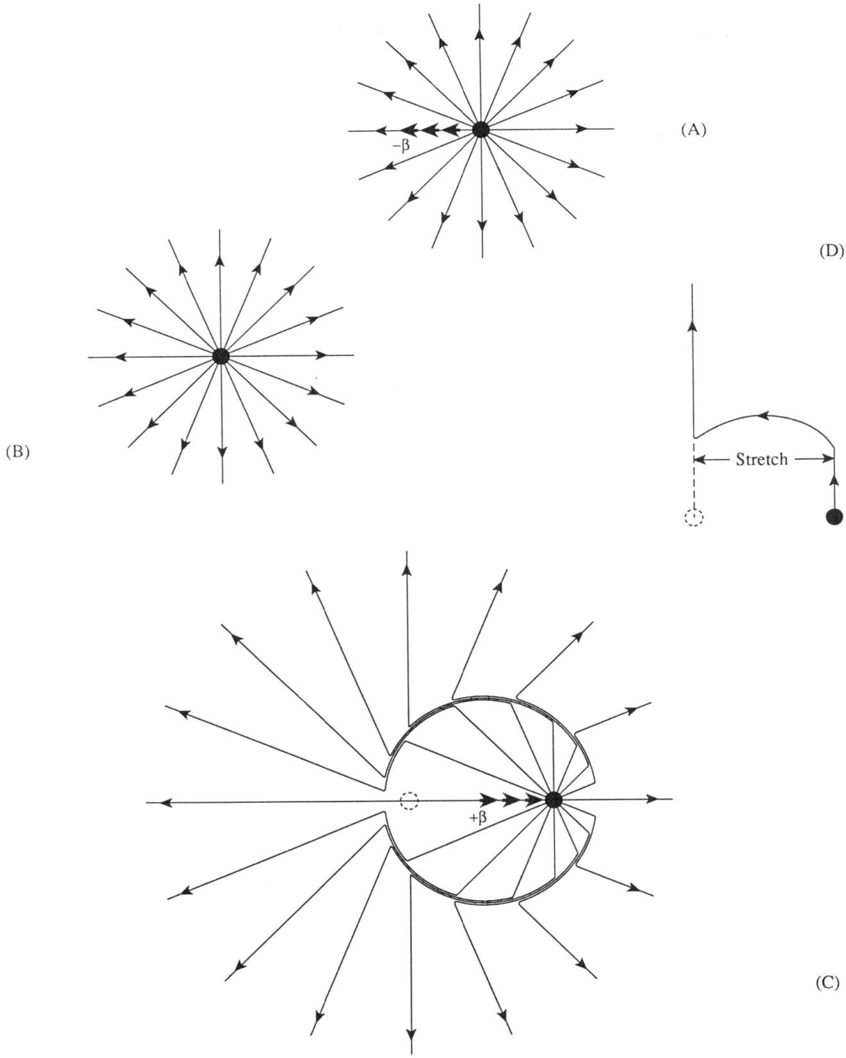

FIGURE 7.1. A pulse of electromagnetic radiation generated by an accelerated charge. (A) A charge is slowly moving at uniform velocity to the left: $v \ll c$. (B) It experiences, for a very short time Δt, a force which reverses its velocity. (C) After a time t, the perturbation of the lines of force has propagated to a distance ct (the path covered by the charge shown in the figure has been exaggerated with respect to the path covered by the spherical pulse of radiation traveling at the speed of light). (D) Electric line of force which "stretches" to arrive at the point where it would have been, had not the electric charge changed its motion.

provides the well-known explanation for the fact that a doubling of the distance from the charge makes a reduction in the strength of the field by a factor 4, if the velocity of the charge is very close to the speed of light, the lines of force are "pancaked" or denser in the plane at right angles to the direction of motion than elsewhere; but the inverse square law still applies). When the charge is accelerated, lines of force have to undergo a readjustment to connect the two patterns of lines of force; the disturbances travel with the speed of light; this means that the readjustment of the lines of force is confined to a spherical shell of thickness $c\Delta t$ located at a distance $r = ct$ from the point of acceleration, where t is the interval of time from acceleration. The line of force no longer travels radially outward over this distance. Instead, that part of the line of force is stretched out into an arc. The arc differs in length between one line of force and another and is greatest for the lines of force directed at right angles to the motion. In other words, a stationary test particle located at a distance r off the line of travel of the moving charge e experiences a field that rises gradually to the full strength e/r^2 and then falls off again as the charge turns around and moves back to its original position. Moreover this field is directed away from the charge. However, a little after the charge is turned around and started away, the test particle experiences an additional field which points in quite another direction, at right angles to the line connecting the place where the test particle is with the place where the charge was when it turned around. This "radiative component" of the field lasts for the time Δt and is:[18]

$$E = -ea \sin\theta / c^2 r \tag{7.1.1}$$

directed at a right angle to the line connecting the point of observation with the point where the acceleration took place; here $a = \Delta v/\Delta t$ is the acceleration, r is the radius of the spherical shell, and θ is the angle between the line of acceleration of the source and the line connecting the source and the test particle. This analysis presupposed that the source was in slow motion. This result is quite sufficient for our purpose. When the source is in rapid motion then the pattern of lines of force is "pancaked" so strongly that important corrections have to be made to this formula. It is not the purpose here to discuss electromagnetic radiation. We know that a radio receiver miles away from the broadcasting station gets an absolutely negligible drive from the $(1/r^2)$–proportional part of the force arising from the electrons that course up and down the antenna tower, and that the radiative component proportional to $1/r$ is the only one that needs to be considered. We have the same dominance of the radiative component in the energy of sunlight and in the X-rays that come from the sudden deceleration of electrons as they strike the tungsten target in an X-ray tube.

How many of these results from electromagnetism can we take over to gravitation? Why not everything except for units? In the cgs-Gaussian system of units the measure of charge is so defined that the force of repulsion between charges e_1 and e_2 is

$$F_{\text{electr}} = e_1 e_2 / r^2; \quad \text{whereas in Newtonian gravitation the force is} \qquad (7.1.2)$$
$$F_{\text{grav}} = -G m_1 m_2 / r^2.$$

A distinguished physicist even published in his very last years works, the main point of which is to claim that gravitation follows the pattern of electromagnetism. This thesis we cannot accept, and the community of physics, quite rightly, does not accept.

Gravitation[19,20] in important respects is utterly different from *electromagnetism*. Of course, the main difference is the equivalence principle, valid for gravity only among the four interactions of nature, that is, the equality between the inertial mass and the *one* gravitational charge. An electric field has very different effects, attractive or repulsive, on different kinds of particles, with positive or negative electric charge. At a point in a gravitational field objects of the most diverse variety and all kinds of particles fall at identical rates. In special relativistic electrodynamics, the spacetime where the dynamics of particles and fields take place is static and completely independent from the electromagnetic phenomena and the electromagnetic field. In general relativity, the spacetime that governs the dynamics of particles and fields is dynamical and dependent on the energy and momentum of particles and fields, and in fact itself represents the gravity field. Moreover the dependence of "force" on velocity is very different for electric forces and for gravitational "forces":

$$m \frac{du^\alpha}{ds} = eF^\alpha{}_\beta u^\beta \qquad \text{and} \qquad m \frac{du^\alpha}{ds} = -m\Gamma^\alpha_{\beta\mu} u^\beta u^\mu. \qquad (7.1.3)$$

The field equations of electromagnetism are linear differential equations for the electromagnetic four-potential A^α, whereas the field equations of general relativity are nonlinear differential equations for the spacetime metric $g_{\alpha\beta}$. Furthermore, an isolated source can radiate electric dipole radiation (as well as electric quadrupole radiation and radiation of higher polarity), with emitted power proportional to the square of the second time derivative of the electric dipole moment $\boldsymbol{d}_e \equiv \sum_i e_i \boldsymbol{x}^i$, that is, $L_{e,\text{dipole}} = \frac{2}{3}\ddot{\boldsymbol{d}}_e^2 = \frac{2}{3}(\sum_i e_i \boldsymbol{a}^i)^2$. However, an isolated source cannot radiate gravitational dipole radiation, only gravitational radiation of higher polarity. The reason is simple. The electric dipole moment can move around with respect to the center of mass; but the mass dipole moment, $\boldsymbol{d}_g \equiv \sum_i m_i \boldsymbol{x}^i$, is identical in location with the center of mass and due to the law of conservation of momentum cannot accelerate, or radiate $\ddot{\boldsymbol{d}}_g = \dot{\boldsymbol{p}} = \boldsymbol{0}$. Furthermore, there are various other differences (see also §§ 2.7, 6.1, and 6.11).

Important as these lessons of gravitation theory are in accounting for its great difference from electromagnetism, one still comes to some important, and correct, lessons about gravitation by treating it on the incorrect basis that they behave the same.

One lesson is this. It takes a force to accelerate a particle with respect to the other particles in the universe. Suppose the particle in question is being accelerated (by a force or impact) relative to all the other particles in space. These other particles are idealized as at rest with respect to a geometrical framework that is idealized as flat and Euclidean. Imagine now the same particles, but this time the one particle viewed as at rest and the others viewed as accelerated relative to it.[12–14] Just to stay at rest with respect to this tide of other particles undergoing acceleration will require force, if we are right in relying on the analogy between gravitation and electromagnetism. The forces on the maverick particle due to the various faraway particles are in quite different directions. However, if the distribution of the faraway particles is fairly uniform, the result of all these forces has no other natural direction in which to point except the direction of the tide itself. Therefore it is reasonable to add up on the components of the individual forces that point in this direction. Because the force originates from a faraway mass we consider only the radiative or "$1/r$" component of the force. In electromagnetism it is proportional to $\sin\theta$; but this factor should itself be multiplied with another factor of $\sin\theta$ to get the component of force in the direction of interest. We end up with the following result[12–14] for the force that seeks to carry the particle in question m in the direction of the common acceleration a of all the other particles m_i:

$$\text{Force} \sim \sum_i -m\left(\frac{-Gm_i a f(\theta_i)}{c^2 r_i}\right) = ma\sum_i \frac{Gm_i f(\theta_i)}{c^2 r_i} \tag{7.1.4}$$

where the quantity $f(\theta_i)$ is some dimensionless function of the angle θ_i between the line of acceleration of the source and the line connecting the source and the receptor. The angular dependence $\sin^2\theta_i$ would apply for electromagnetism;[18] however, for the radiative component of the gravitational pull we should not expect any difference in the order of magnitude of the final result from this term. The first minus sign comes from the same source as the minus sign in the expression for the radiative component of the electric force: the line of force being stretched backward as the charged particle was accelerated forward. The second minus sign is the one that accompanies the Newtonian constant of gravitation in the expression for the attractive force of gravitation and distinguishes it from the force of electricity. The sign of the force therefore is such as to drag the particle in question along with all the other particles in the "universe." In other words, we ourselves have to exert "externally" a force forward if we want to hold the particle in question "stationary" against the tide of particles accelerating backward.

We can translate this result back into a statement about what happens in the frame of reference at rest with the other particles. There also, a force is required to accelerate the particle relative to all the others. But that force is not new in the world. It is the familiar force required to impart to a particle of mass m an acceleration a in the familiar frame of reference in which the faraway matter of the universe is at rest, $F = ma$.

Here we have what on the surface are two entirely different calculations of the force. Yet they agree in making this force proportional to the mass and acceleration in question.

Therefore, Sciama suggests identifying the two results.[12–14] If this identification is to work out satisfactorily, then it is necessary that the coefficients of "ma" on the right-hand sides of the second law of dynamics and of equation (7.1.4) should be identical, that is, the right-hand side coefficient of ma in equation (7.1.4), called the "sum for inertia," should be one.

We have ended up with two ways of arriving at the expression, $F = ma$, required to accelerate a mass m. One takes this relation as primordial; as the starting point of Newtonian mechanics. It makes no reference to any matter anywhere else in the universe. It is content to accept the concept of "absolute frame of reference" as did Newton. The other approach takes as for granted the Newtonian expression, $-\frac{Gmm_i}{r_i^2}$, for the law of force between mass and mass.

However, we know that gravitation, like electromagnetism and every other primordial force, must propagate at the characteristic speed c. The elementary sum in equation (7.1.4) for the coefficient of inertia envisages a radiative interaction between particle and particle which propagates instantaneously. But how can stars at distance of 10^9 and 10^{10} lt-yr respond to the acceleration of a test particle here and now in such a way as to react back upon this test particle at this very moment? This difficulty alone should be sufficient cause for dropping the elementary formulation of equation (7.1.4) and to observe that one would, at least, need to use retarded potentials.

Therefore, what can we conclude from this analysis? First, nothing provable. Without using a sound theory of gravitation, we cannot expect a soundly founded theory of gravitational radiative reaction. Second, we have translated the Machian idea that "inertia here arises from mass-energy there" from a vague idea to a concrete suggestion with two components: first, the mechanism of coupling is not some new feature of nature but long-known gravitation and, second, it is the radiative or acceleration-proportional component of this interaction that counts. Finally, we recognize that we cannot go further in translating these intimations and suspicions into mathematically sound conclusions until we make use of a properly relativistic theory of gravitation at our disposition for the doing of it. Why "relativistic"? Because the $1/c^2$ factor in the expression for the radiative component of the force is clearly of relativistic origin, and the gravitational interaction must propagate at speed c, and in identifying the very different physics of the previous example, in identifying the second law of classical mechanics with a radiative force, one was tacitly using relativistic arguments in a context not justified by special relativity.

7.2 GEOMETRODYNAMICS AND INERTIA

Could the motive of the previous section be clearer for Einstein to go on from special relativity to general relativity? He had been driven to special relativity[21,22] by physical considerations having to do with the electromotive force generated in a coil moving relative to a bar magnet part way entering into it. The physics of his day gave very different treatment according as the coil was regarded as stationary and the magnet moving, or the magnet stationary and the coil moving. In special relativity he brought the two treatments together in one larger vision of spacetime.

A still deeper vision of *spacetime* was to result from Einstein's concern with this other physical problem, the **origin of inertia**.[11] Already before Einstein had the final formulation of *general relativity* but at a time when he was well on the track of that final formulation he wrote Ernst Mach[23] to felicitate Mach on the happiness he should feel that his longtime concern with, and writings on, "acceleration relative to the distant stars" had at last led Einstein to a theory that, treating local accelerated and local nonaccelerated frames on the same footings, enables one, he felt, to understand how the precession of a completely free gyroscope is affected by the movement of nearby and distant matter. It does not matter that Mach having rejected special relativity also rejected general relativity.

Today our understanding of the foundations of physics is deep enough to consider general relativity (chap. 2), as the correct classical theory of gravitation. No purported disagreements between its predictions and observations have ever stood the test of time. Indeed, we have seen the experimental triumphs of general relativity in chapter 3.

As we have explained in chapters 1 and 2, Einstein combined two great currents of thought, that, together with the powerful *equivalence principle* (§2.1), he brought together into the present day *geometrical description of gravitation and motion: geometrodynamics*.

One current went back to Bernard Riemann.[24] In modern language it goes like this: Matter gets its moving orders from spacetime: otherwise it would not know what kind of track to follow. Therefore spacetime geometry acts on matter. Therefore according to the "principle of action-reaction," matter must act back on *spacetime geometry*.

The second great current of thought goes back to Mach:[7,8] There is no meaning to "absolute space" and "acceleration relative to absolute space." What makes physical sense in general relativity is only *acceleration relative to the local inertial frames determined by the mass-energy in the universe*.

Einstein brought in new and decisive points but none more decisive than this: Gravitation[11] is not something foreign and physical acting through space; it is a manifestation of the geometry of spacetime. Further, what spacetime geometry acts on is not this, that, or the other esoteric and unnamed property of the moving matter, but its mass-energy; and what acts back on spacetime geometry is also mass-energy.

Einstein provided a principle that continues to guide progress: Physics is only then simple when it is analyzed locally. That is why the line of thought going back to Riemann is so central: *Mass-energy there curves space here.*

That is why it is harder to trace out how **mass energy there influences geometry here**, therefore motion here, therefore **inertia here.** This very circumstance, however, means that one gets a better view of the significance of *Einstein's geometric theory of gravitation* in the large through analyzing the *origin of inertia.*

No wonder that the study of the origin of inertia in the framework of Einstein's geometrodynamics leads to questions of global mathematical analysis and of global topology of space as well as physical issues concerned with the amount of matter in the universe and the beginning and end of time.

To all these issues, general relativity is the key that opens the door.

The analysis of Thirring[25–28] and Einstein[11] brings the argument of expression (7.1.4) into closer connection with the ideas of general relativity. On the one hand, in Einstein geometrodynamics the local inertial properties are expressed through the metric tensor $g_{\mu\nu}$. On the other hand, what is responsible for driving the metric $g_{\mu\nu}$ is not merely the density of matter, but the entire energy-momentum tensor $T_{\mu\nu}$. Thus Thirring and Einstein expressed the change of the metric in a Minkowskian background, owing to a change $\delta T_{\mu\nu}$, in the form (see §2.10 and formula 2.10.10):

$$h_{\mu\nu} = g_{\mu\nu} - \eta_{\mu\nu} \tag{7.2.1}$$

where $g_{\mu\nu}$ and $\eta_{\mu\nu}$ are the components of the new and old metric respectively. $h_{\mu\nu}$ is obtained from the equation (see formula 2.10.10)

$$h_{\mu\nu} - \frac{1}{2}\eta_{\mu\nu}h = 4\int \frac{\{\delta T_{\mu\nu}\}_{\text{ret}}\, d^3x}{r} \tag{7.2.2}$$

where $h \equiv \eta^{\mu\nu}h_{\mu\nu}$. This expression remains a good approximate solution of Einstein's field equation so long as the geometry does not differ substantially from the Minkowski background.

From equation (7.2.2) and the fact that in geometrodynamics the local inertial frames are determined by the metric, one is led to the following interpretation of the origin of inertia in general relativity: the *geometry of the spacetime* and therefore *local inertia* in the sense of *local inertial frames* at each event along the world line[29] of every test particle are *determined by the distribution and flow of energy* throughout all space.

However, there are some questions:

(1) The Einstein account of the Lense-Thirring effect[27,28] uses an approximate solution of the field equation (2.3.14). The departure $h_{\mu\nu}$ of the metric $g_{\mu\nu}$ from the metric of flat space $\eta_{\mu\nu}$ is taken to be small and is evaluated as a retarded potential. But Einstein's field equation is nonlinear and especially so in a compact space. Therefore it seems wrong in principle to try to express the solution $g_{\mu\nu}$ as a linear superposition of effects caused by the $T_{\mu\nu}$ throughout the compact space.[30,31]

(2) The quantity $1/r$ in the integrand of (7.2.2) is not a well defined quantity in an irregularly curved space. There more than one geodesic may connect an event on the world line of a given mass with the place where the metric is being evaluated.

(3) Will not the $(1/r)$–dependence of the supposed interaction make local inertia depend in a physically unreasonable way on the expansion and recontraction of the universe and the proximity of nearby masses?

(4) To evaluate the retarded potential of expression (7.2.2) caused by energy-momentum at points more and more remote, one is called upon to go to times farther and farther back in the past. If one does this calculation in a Friedmann model universe, one ultimately comes to a time when the calculated size of the system was zero, and the curvature was infinite, that is, a singularity. Under these conditions how can the integral (7.2.2) possibly have any well-defined meaning?

(5) How can it make sense to speak of distribution of mass-energy density and of flow of mass-energy density as determining the geometry? One cannot specify these quantities until one has been given the geometry! But what is there then to be determined?

To all these questions the *geometrodynamics initial-value formulation*,[32–34] described in chapter 5, is the key that opens the door (we have assumed here a closed—compact and without boundary—Cauchy surface; see related discussions in chaps. 4 and 5).

(1) Is not an integral formulation of the solution to Einstein's field equation wrong in principle? In the initial value formulation one is dealing no longer with a linearized approximation to Einstein's field equations, but with the accurate—and nonlinear—initial-value equations on a spacelike hypersurface (5.3.1) and (5.3.9).

(2) No natural replacement for the $1/r$ influence function? In place of the $1/r$ that comes from solving the elliptic equation of Poisson in a flat space one has the proper relativistic substitute for this influence function when one solves the equations (5.3.1) and (5.3.9) for the *gravitomagnetic vector potential* W_i and for the *conformal factor* ψ.

(3) $1/r$ "influence function"? The influence of a local conformal (i.e., up to a conformal factor $\psi(x, y, z)$) increment $\delta\overset{(c)}{\rho}$, $\delta\overset{(c)}{j}{}^i$ on the gravitomag-

netic vector potential W_i and on the conformal factor ψ is nonlocal. It shows up in the formalism neither as a retarded effect nor as an advanced effect. Instead, because the analysis deals with an everywhere *spacelike hypersurface*, the influence appears ostensibly as instantaneous. But superposed on this direct effect are radiation effects. They are specified by the conformal $\overset{(c)}{g}_{mn}$ and $\overset{(c)}{A}{}^{mn}_{\text{free}}$ (see §5.2). There is no conflict with causality. Effects traced off the initial hypersurface forward into the future or backward into the past by integrating the dynamical equations never propagate at a speed in excess of the speed of light. Similarly, in the initial-value formulation of electrodynamics[35] one could be troubled about the propagation of the effect of the charge ρ at one point on Σ to the quite different point of measurement of the electric field on the same spacelike hypersurface. Does not this kind of propagation contradict the most elementary ideas of causality? Ought not retarded potentials to be used? And if they are used, then what has happened to the program of finding initial-value data which are at the same time independent of each other—and freely specifiable—and yet adequate to determine the history completely? Actually there is no conflict with causality in specifying initial-value data over a spacelike hypersurface. An instantaneous action and a retarded action, as would follow from usual considerations of causality, differ by a field which can be—and is—described as a *radiation field* in the analysis. Everything is in order. One has a good and natural way to treat electrodynamics.[35] As soon as one has restricted the analysis to a spacelike hypersurface, there is no way to represent the effect of a charge at one point on the field at another point except by an instantaneous action (elliptic equation!) plus radiation terms. What will be called "radiation" then depends upon the choice of hypersurface Σ—but this dependence is no harder to understand than the dependence of the velocity of a particle upon the Lorentz frame of the observer who looks at it! It is not only natural physically but also desirable mathematically to be dealing with conditions on a spacelike hypersurface. The kind of equation to which one comes, as the equation $\Delta\Phi =$ (a given quantity), is elliptic in character, and well suited to establishing proofs of the *existence* and *uniqueness* of a solution. Of course it is required for this purpose (1) that suitable conditions be imposed on the fall off at large distances of the quantities to be specified on Σ—if the hypersurface Σ is asymptotically flat—or (2) that the three-manifold in question be compact and without boundary or (3) that some other appropriate boundary condition ("inner" boundaries and "boundary at spatial infinity") be given.

(4) Meaning of the integral (7.2.2) in a model universe with a big bang or a crunch? The singularities that develop in the three-geometry of a closed

model universe as its evolution is followed forward and backward in time by way of the field equation have no bearing on the initial-value problem itself. When the data on the Cauchy surface are once given, then in the vicinity of any test particle the local inertial frame is determined up to a Lorentz transformation. Moreover, it is determined in a way free of all inconsistency, because the analysis follows directly from Einstein's field equation and the geodesic postulate.

(5) Specification of geometry and other physical quantities in what logical order? On the initial-value[32–34] spacelike compact hypersurface Σ one specifies the conformal metric $\overset{(c)}{g}_{mn}(x, y, z)$ as a function of three coordinates x, y, z, and the conformal, traceless, transverse, symmetric, distortion tensor $\overset{(c)}{A}{}^{mn}_{\text{free}}(x, y, z)$, representing part of the "rate of change" of the conformal three-geometry $\overset{(c)}{g}_{mn}$. One also specifies the conformal energy density $\overset{(c)}{\rho}(x, y, z)$ and the conformal energy density current $\overset{(c)}{j}{}^{i}(x, y, z)$. Solving the initial-value equations (5.3.1) and (5.3.9) one then finds the conformal scale factor ψ, thus fully knowing on Σ the metric g_{mn}, its "rate of change" or the extrinsic curvature K^{mn}, the energy density ρ and the energy density current j^i. There is a reason for considering it natural to give g_{mn} and A^{mn}_{free} (up to a conformal factor ψ) on Σ. We know that "real mass" is not the only source of mass-energy and of gravitational effects. Radiation also has mass-energy. A cloud of electromagnetic radiation has surrounding it at large distances a Schwarzschild geometry. The mass associated with this geometry measures the energy of the cloud of radiation (examples: electromagnetic geon and the general theory of the asymptotic form of the metric associated with a cloud of radiation.)[36] The same is true for a cloud of gravitational radiation. Thus a gravitational geon[36–38] (see §2.10) has to produce the same effect at a distance as does mass of any other kind. But viewed close up, a gravitational geon, or a more general collection of gravitational radiation, consists of nothing at all except curved empty space. The conformal effective mass-energy associated with this radiation is naturally evaluated directly from $\overset{(c)}{g}_{mn}$ and $\overset{(c)}{A}{}^{mn}_{\text{free}}$. In this sense *the specification of the conformal three-geometry and of part of its "rate of change" is a necessary part of the task of giving the distribution of conformal energy and conformal energy current.*

We conclude by recalling our distinction (chaps. 4 and 5) among three possible interpretations, more or less strong, of the **origin of inertia in Einstein general relativity**, clarified by the formulation of the initial-value problem.

(1) The spacetime geometry and therefore the *local inertial frames* along the world line of every test particle are influenced and at least *in part deter-*

mined by the energy density and the energy density currents throughout the hypersurface Σ.

(2) The spacetime geometry is *completely determined by the energy density and the energy density currents* on Σ.

(3) The interpretation (2) is satisfied, and, *in addition*, some *cosmological requirements* are satisfied: the closure in time of the universe and the average global nonrotation of the mass-energy in the universe relative to local gyroscopes.

We have discussed these interpretations of inertia in geometrodynamics in chapters 4 and 5. We expect that the experimental verification of the weaker interpretation (1), always satisfied in general relativity, that is, the nonexistence of an absolute inertial frame and in particular the influence of mass-currents on (local) inertial frames, should eventually directly come from one of the experiments described in chapter 6 for the measurement of the "*dragging of inertial frames*." We further observe that according to the interpretation (2), one requires the absence of any part of the spacetime geometry that is unaffected by the mass-energy content in the universe, such as an asymptotically flat metric, $\boldsymbol{\eta}$, as a part of $\boldsymbol{g}$; one may then require the *space* Σ to be *compact* (see chaps. 4 and 5). The cosmological conditions of the stronger interpretation (3) are discussed in chapter 4.

In conclusion we may summarize: **mass-energy "tells" spacetime how to curve and spacetime "tells" mass-energy how to move.** From here the step is small in concept, if long in mathematics, to the influence of mass-energy there on inertia here: Mass-energy there (and here) curves "space" there (and here), and "space" there has to join on smoothly to "space" elsewhere and "space" elsewhere to "space" here. "Space" and mass-energy there and here determine spacetime. But spacetime here "tells" matter here, and gyroscopes here, how to move. Therefore: **mass-energy there rules inertia (local inertial frames) here**.

REFERENCES CHAPTER 7

1. Isaac Newton, *Philosophiae naturalis principia mathematica* (Streater, London, 1729); trans. A. Motte (1729) and revised by F. Cajori as *Sir Isaac Newton's Mathematical Principles of Natural Philosophy and His System of the World* (University of California Press, Berkeley and Los Angeles, 1934; paperback 1962).

2. Galileo Galilei, *Dialogo dei due massimi sistemi del mondo* (Landini, Florence, 1632); trans. S. Drake as *Galileo Galilei: Dialogue Concerning the Two Chief World Systems—Ptolemaic and Copernican* (University of California Press, Berkeley and Los Angeles, 1953).

3. E. F. Taylor and J. A. Wheeler, *Spacetime Physics* (Freeman, San Francisco, 1966), 176–78.
4. G. W. Leibniz, *Philosophical Papers and Letters*, ed. L. E. Loemker (Reidel, Dordrecht, 1969).
5. G. Berkeley, *The Works of George Berkeley*, ed. A. Campbell Fraser (Clarendon Press, Oxford, 1901).
6. G. Berkeley, *The Principles of Human Knowledge* (1710), 111–17; and, *De Motu* (1726) (A. Brown & Sons, London, 1937).
7. E. Mach, *Die Mechanik in Ihrer Entwicklung Historisch-Kritisch Dargestellt* (Brockhaus, Leipzig, 1912); trans. T. J. McCormack with an introduction by Karl Menger as *The Science of Mechanics* (Open Court, La Salle, Il, 1960); see also ref. 8.
8. E. Mach, *The Analysis of Sensations*, trans. C. M. Williams (Dover, New York, 1959).
9. E. Kretschmann, Über den physikalischen Sinn der Relativitätspostulate, A. Einsteins neue und seine ursprüngliche Relativitätstheorie, *Ann. Phys.* 53:575–614 (1917).
10. A. Einstein, Die Grundlage der allgemeinen Relativitätstheorie; *Ann. Phys.* 49:769–822 (1916); translation in *The Principle of Relativity* (Methuen, New York, 1923); reprint ed. (Dover, New York, 1952).
11. A. Einstein, *The Meaning of Relativity*, 5th ed. (Princeton University Press, Princeton, 1955).
12. D. W. Sciama, Inertia, *Sci. Am.* 196:99–109 (1957).
13. D. W. Sciama, On the origin of inertia, *Mon. Not. Roy. Astron. Soc.* 113:34–42 (1953).
14. D. W. Sciama, *The Unity of the Universe* (Doubleday and Co., New York, 1959).
15. W. Davidson, General relativity and Mach's principle, *Mon. Not. Roy. Astron. Soc.* 117:212–24 (1957).
16. R. P. Feynman, *The Character of Physical Law* (MIT Press, Cambridge, MA, 1965).
17. J. J. Thomson, *Electricity and Matter* (Archibald Constable, London, 1907).
18. J. D. Jackson, *Classical Electrodynamics* (Wiley, New York, 1962).
19. C. W. Misner, K. S. Thorne, and J. A. Wheeler, *Gravitation* (Freeman, San Francisco, 1973).
20. S. Weinberg, *Gravitation and Cosmology: Principles and Applications of the General Theory of Relativity* (Wiley, New York, 1972).
21. A. Einstein, Zur elektrodynamik bewegter Körper, *Ann. Phys.* 17:891–921 (1905); translation in ref. 10.
22. A. Einstein, Ist die Trägheit eines Körpers von seinem Energiehalt abhängig?, *Ann. Phys.* 18:639–41 (1905); translation in ref. 10.
23. A. Einstein, letter to E. Mach, 25 June 1913, Zurich; see, e.g., ref. 19, p. 544.
24. G. F. B. Riemann, Über die Hypothesen welche der Geometrie zu Grunde liegen, in *Gesammelte Mathematische Werke* (1866); 2d ed. reprint, ed. H. Weber (Dover,

New York, 1953); see also the translation by W. K. Clifford, *Nature* 8:14–17, 36–37 (1973).

25. H. Thirring, Über die Wirkung rotierender ferner Massen in der Einsteinschen Gravitationstheorie, *Phys. Z.* 19:33–39 (1918).
26. H. Thirring, Berichtigung zu meiner Arbeit: Über die Wirkung rotierender ferner Massen in der Einsteinschen Gravitationstheorie, *Phys. Z.* 22:29–30 (1921).
27. J. Lense and H. Thirring, Über den Einfluss der Eigenrotation der Zentralkörper auf die Bewegung der Planeten und Monde nach der Einsteinschen Gravitationstheorie, *Phys. Z.* 19:156–63 (1918).
28. J. Lense and H. Thirring, On the Gravitational Effects of Rotating Masses: The Thirring-Lense Papers, trans. B. Mashhoon, F. W. Hehl, and D. S. Theiss, *Gen. Relativ. Grav.* 16:711–50 (1984).
29. E. Fermi, Sopra i fenomeni che avvengono in vicinanza di una linea oraria, *Atti R. Accad. Lincei Rend. Cl. Sci. Fis. Mat. Nat.* 31: (I), 21–23, 51–52, 101–103 (1922).
30. D. W. Sciama, P. C. Waylen, and R. C. Gilman, Generally covariant integral formulation of Einstein's field equations, *Phys. Rev.* 187:1762–66 (1969).
31. D.J. Raine and M. Heller, *The Science of Space-time* (Pachart Publishing House, Tucson, AZ, 1981).
32. J. W. York, Jr., Kinematics and Dynamics of General Relativity, in *Sources of Gravitational Radiation: Proc. Battelle Seattle Workshop*, 24 July–4 August 1978, ed. L. L. Smarr (Cambridge University Press, Cambridge, 1979), 83–126.
33. Y. Choquet-Bruhat and J. W. York, The Cauchy Problem, in *General Relativity and Gravitation*, ed. A. Held (Plenum, New York, 1980), 99–172.
34. J. W. York, Jr., The Initial Value Problem and Dynamics, in *Gravitational Radiation*, Les Houches (1982), ed. N. Deruelle and T. Piran (North-Holland, Amsterdam, 1983), 175–201.
35. E. Fermi, Quantum theory of radiation, *Rev. Mod. Phys.* 4:87–132 (1932).
36. J. A. Wheeler, Geons, *Phys. Rev.* 97:511–36 (1955).
37. J. A. Wheeler, *Geometrodynamics* (Academic Press, New York, 1962).
38. See, e.g., R. K. Sachs, Gravitational Radiation, in *Relativity, Groups and Topology*, Les Houches, ed. C. DeWitt and B. DeWitt (Gordon and Breach, New York, 1963), 521–62.
39. J. B. Barbour, *Absolute or Relative Motion? The Discovery of Dynamics*, vol. 1 (Cambridge University Press, Cambridge, 1989).
40. E. Mach, History and Root of the Principle of the Conservation of Energy (1872), trans. (Open Court, La Salle, Il, 1911).

Mathematical Appendix

In this mathematical appendix we briefly review some basic definitions and concepts of differential geometry.

In the first column is given the name and the symbol of a mathematical object, in the second column its rigorous definition and main properties, and in the third column its intuitive meaning and some examples. At the end of the appendix is given an alphabetical list of the mathematical quantities defined.

This appendix may be used as a review or as a brief introduction to the mathematical objects needed in the book; however, a full introduction to these topics requires a separate text. For example, we suggest: Y. Choquet-Bruhat and C. De Witt-Morette with M. Dillard-Bleick, *Analysis, Manifold and Physics* (North-Holland, Amsterdam, 1982); S. W. Hawking and G. F. R. Ellis, *The Large Scale Structure of Space-time* (Cambridge Univ. Press, Cambridge, 1973); S. Kobayashi and K. Nomizu, *Foundations of Differential Geometry* (Wiley-Interscience, New York, 1963); M. Spivak, *A Comprehensive Introduction to Differential Geometry*, 2nd ed. (Publish or Perish, Berkeley, 1979); and B. Schutz, *Geometrical Methods of Mathematical Physics* (Cambridge Univ. Press, Cambridge, 1980).

MATHEMATICAL OBJECT AND SYMBOL	DEFINITION AND MAIN PROPERTIES	MEANING, EXAMPLES, AND APPLICATIONS IN THE BOOK
Topology $\mathcal{U}$ on a set S	A **topology** on a nonempty set S is a system $\mathcal{U}$ of subsets of S, **open sets**, such that: (1) S and Φ (empty set) are elements of $\mathcal{U}$. (2) The union of any number of elements of $\mathcal{U}$ is an element of $\mathcal{U}$. (3) The intersection of a finite number of elements of $\mathcal{U}$ is an element of $\mathcal{U}$.	Topology may be thought of as the study of sets of elements, based on the concepts of neighborhood and continuous mappings.
Topological space $(S, \mathcal{U})$	A **topological space** $(S, \mathcal{U})$ is a set S with a **topology** $\mathcal{U}$.	Examples: The ***n*-dimensional Euclidean space** $\Re^n$, with the topology defined by the open disks. An **open *n*-disk** of a point x^i of $\Re^n$ is defined as the set of all the points y^i of $\Re^n$ such that $\left(\sum_{i=1}^n (x^i - y^i)^2\right)^{1/2} < r$, where r is a positive real number.

MATHEMATICAL OBJECT AND SYMBOL	DEFINITION AND MAIN PROPERTIES	MEANING, EXAMPLES, AND APPLICATIONS IN THE BOOK
		The ***n*-dimensional sphere** S^n subset of $\Re^{n+1}$, defined as the set of all the points of $\Re^{n+1}$ such that $\sum_{i=1}^{n+1}(x^i)^2 = r^2$, where r is a positive real number, with the topology of the open neighborhoods of the points on the sphere S^n, defined for any point P of S^n as the sets of all the points of S^n with distance less than ε from P. Ordinary surfaces of $\Re^3$ such as: a standard two-sphere in $\Re^3$, a standard two-torus in $\Re^3$, an infinite cylinder, with topology defined by the open neighborhoods. See figs. 1.1 and 4.5–4.8. The non-Euclidean spacetime of general relativity. The discrete set: $S_1 = \{S, \Phi, \{a\}, \{b\}, \{a, b\}\}$ where $S = \{a, b, c\}$. Counterexample: The discrete set: $S_2 = \{S, \Phi, \{a\}, \{c\}, \{a, b\}\}$ is not a topology on S, since $\{a\}$ union $\{c\} = \{a, c\}$ is not an element of S_2.
Open set	Each element of the topolgy $\mathcal{U}$ is called an **open set**.	Examples: The **open interval** $(0, 1)$ of $\Re$. An open n-disk in $\Re^n$.
Neighborhood of a point P in S	A **neighborhood** of a point P is a set of S containing an open set containing P.	Examples: The open interval $(-1, 1)$ of $\Re$ and the **closed interval** $[-1, 1]$ of $\Re$ are neighborhoods of 0. An open, or closed, n-disk of a point x^i of $\Re^n$ is a neighborhood of x^i. A **closed *n*-disk** is defined as the set of all the points y^i of $\Re^n$ such that $\left(\sum_{i=1}^{n}(x^i - y^i)^2\right)^{1/2} \leq r$.
Closed set	A **closed set** C may be defined as a set of S such that its complement is an open set U in S: $C = \mathcal{C}(U)$ (the complement	Examples: The closed interval $[0, 1]$ of $\Re$. Any closed n-disk of a point x^i of $\Re$.

MATHEMATICAL OBJECT AND SYMBOL	DEFINITION AND MAIN PROPERTIES	MEANING, EXAMPLES, AND APPLICATIONS IN THE BOOK
	$\mathcal{C}(U)$ of a set U in S is the set of all the points of S not contained in U).	
Topological product: $S \times T$	The **topological product** of two topological spaces $(S, \mathcal{U})$ and $(T, \mathcal{V})$ is the topological space on the Cartesian product set $S \times T$ with **product topology** formed by the sets $U \times V$, where U and V are respectively elements of $\mathcal{U}$ and $\mathcal{V}$.	Examples: $\mathfrak{R}^2 = \mathfrak{R} \times \mathfrak{R}$: real plane. $\mathfrak{R}^n = \mathfrak{R} \times \cdots \times \mathfrak{R}$, n times. $S^1 \times S^1$: **two-torus**. $S^1 \times \mathfrak{R}$: **cylinder**, each with product topology.
Hausdorff, or **separated** topological **space** S	A **space** S is called **Hausdorff**, or **separated**, if for every two distinct points of S, there are two disjoint neighborhoods of the two points.	Example: The Euclidean space $\mathfrak{R}^n$.
Covering $\{U_\alpha\}$ of S (**Open covering**)	A (open) **covering** $\{U_\alpha\}$ of S is a collection of subsets (open subsets) of S such that each element in S is contained in some U_α.	The union of all the elements of $\{U_\alpha\}$ is the whole S, that is, $\{U_\alpha\}$ covers the whole space S.
Compact space	A **space** S is called **compact** if for every open covering of S there exists a subset of the covering that is also a covering of S (**subcovering**) and has a finite number of elements (**finite subcovering**).	Compact subsets of S are a generalization of closed and bounded subsets of the Euclidean space $\mathfrak{R}^n$. (In a space on which there is a properly defined distance, $d(x, y)$, between any two points x and y, a **set** is **bounded** if it is contained in some ball of finite radius. An open **ball of radius** r about x is the set of all the points y such that $d(x, y) < r$.) In $\mathfrak{R}^n$, compact is equivalent to closed and bounded. However, for $S \neq \mathfrak{R}^n$, closed and bounded sets are not necessarily compact (for example, in some infinite dimensional spaces). Examples of compact space: A space with a finite number of points. The closed interval $[0, 1]$ of $\mathfrak{R}$. Every closed and bounded subset of $\mathfrak{R}^n$. An n-sphere S^n. An ***n*-torus**: $S^1 \times S^1 \times \cdots \times S^1$, n times (where $\times$ is the topological product). A closed subset of a compact space is compact.

MATHEMATICAL OBJECT AND SYMBOL	DEFINITION AND MAIN PROPERTIES	MEANING, EXAMPLES, AND APPLICATIONS IN THE BOOK
		Counterexamples: The n-dimensional Euclidean space, $\Re^n$, is not compact. The real line, $\Re$, is not compact. The open interval (0, 1) is not compact. An infinite cylinder is not compact.
Locally compact	A space is **locally compact** if every point has a compact neighborhood.	$\Re^n$ is locally compact but not compact.
Paracompact	A Hausdorff space S is called **paracompact** if for every open covering $\{U_\alpha\}$ of S there exists an open covering $\{V_\beta\}$ such that each V_β is contained in some U_α ($\{V_\beta\}$ is a **refinement** of $\{U_\alpha\}$), and such that at each point of S there is a neighborhood that intersects only a finite number of elements of $\{V_\beta\}$ ($\{V_\beta\}$ is **locally finite**).	$\Re^n$ is paracompact but not compact.
Connected space	A topological **space** S is called **connected** if it is not the union of two disjoint, nonempty, open subsets of S.	Examples: The open interval (0, 1) of $\Re$. The closed interval [0, 1] of $\Re$. The real line $\Re$. The n-dimensional Euclidean space, $\Re^n$. An n-sphere S^n . $\Re^2 - \{0\}$. Counterexamples: The topological space union of two disjoint open disks of $\Re^2$ is not connected. $\Re - \{0\}$.
Locally connected space	A topological **space** S is called **locally connected** if every neighborhood of every point of S contains a connected neighborhood of the point.	
Continuous mapping	A **mapping** f from a topological space S to a topological space T is **continuous** at a point P of S if for every neighborhood V of $f(P)$ in T, there exists a neighborhood U of P in S such that $f(U)$ is contained in V. A mapping f is continuous on S if it is continuous at each point of S. A mapping f is continuous iff, for every open set A of T, $f^{-1}(A)$ is an open set of S.	The continuity of a mapping between two topological spaces is a generalization of the ordinary definition of continuity of a function from $\Re^n$ to $\Re$. Under a continuous mapping f, "neighboring" points of S are "carried" into "neighboring" points of T.
Arcwise connected space	A topological **space** is called **arcwise connected** if for every two points p and q of S there is a continuous arc (path)	

MATHEMATICAL OBJECT AND SYMBOL	DEFINITION AND MAIN PROPERTIES	MEANING, EXAMPLES, AND APPLICATIONS IN THE BOOK
	$c(s)$ of S, with $s \epsilon [0, 1]$, joining the two points, such that $c(0) = p$ and $c(1) = q$. A topological space arcwise connected is connected; the converse is not necessarily true.	
Homotopy $\simeq$	Given two **paths** $c(s)$ and $d(s)$, with $s \epsilon [0, 1]$, they are **homotopic**, $c \simeq d$, if there is a continuous function $f(s, t)$ with $t \epsilon [0, 1]$, such that $f(s, 0) = c(s)$ and $f(s, 1) = d(s)$. f is called **homotopy** between c and d.	Two curves are homotopic if it is possible to continuously "deform" one into the other. Counterexample: Consider a region of the real plane with a hole cut out. If two curves joined together form a closed path around the hole they are not homotopic; in fact, one cannot continuously "deform" one into the other.
Simply connected space	A **closed path** $c(s)$ (that is, a path with the same initial and final point $c(0) = c(1)$) is called **contractible to a point** if it is homotopic to the **constant path** $d(s) = p$. Then, a **simply connected space** is a connected space such that every closed path is continuously contractible to a point.	A closed path is contractible to a point if it can be continuously deformed to a point. Examples: A two-sphere is simply connected: every closed path on the two-sphere is contractible to a point. S^n, with $n \neq 1$, is simply connected. Counterexamples: The **circle** S^1 is not simply connected. An n-holed torus or a sphere with n handles (see fig. 4.8) are not simply connected, they are multiply connected. Indeed, a closed path around a hole cannot be continuously deformed to a point.
Homeomorphism between topological spaces	A **homeomorphism** is a mapping f from a topological space S into a topological space T that is a **bijection** (**one to one**, or **injective**, plus **onto**, or **surjective**) and **bicontinuous**, that is, continuous with inverse mapping, f^{-1}, from T to S, continuous.	The fundamental topological concept of homeomorphism between two topological spaces means that the two spaces are topologically "equivalent", that is, two homeomorphic topological spaces (such that there exists a homeomorphism between them) have the same topological properties, such as connectedness, compactness, Hausdorff

MATHEMATICAL OBJECT AND SYMBOL	DEFINITION AND MAIN PROPERTIES	MEANING, EXAMPLES, AND APPLICATIONS IN THE BOOK
		separation . . . (**topological invariant** properties). Under a homeomorphism f, "neighboring" points of S are "carried" into "neighboring" points of T and vice versa under f^{-1}. Examples: See figs. 4.6 and 4.8. Every compact two-sided surface of $\mathfrak{R}^3$ is homeomorphic to a sphere with a certain number of handles. For example, the ***n*-holed torus** is homeomorphic to a **sphere with *n*-handles**, fig. 4.8. A **flat two-torus**, $\{S^1 \times S^1\}$, subset of $\mathfrak{R}^4$, a **standard two-torus** subset of $\mathfrak{R}^3$, and a **two-sphere with one handle** are homeomorphic (even though they have different curvature); see figs 4.6 and 4.8.
***n*-manifold** M, or ***n*-dimensional topological manifold** M^n	A **topological *n*-manifold** M^n is a connected, Hausdorff, topological space such that every point of M^n has a neighborhood **homeomorphic** (**topologically equivalent**) to $\mathfrak{R}^n$ (or to an open set of $\mathfrak{R}^n$).	This definition of manifold is a generalization of the ordinary definitions of curve and surface in $\mathfrak{R}^3$, without use of an embedding in another space (such as $\mathfrak{R}^3$). An n-dimensional topological manifold is a topological space that is "locally" *topologically* equivalent to $\mathfrak{R}^n$ (the n-dimensional Euclidean space). Examples: Any open subset of $\mathfrak{R}^n$, any open subset of an n-manifold with the relative topology (the **relative topology** on a subset A of a topological space $(S, \mathcal{U})$ is given by all the intersections of the open sets U of S with A). A regular curve (open) in $\mathfrak{R}^3$. A regular surface (open) in $\mathfrak{R}^3$. An n-sphere S^n subset of $\mathfrak{R}^{n+1}$. $\mathfrak{R}^n$ itself. The product of an n-manifold with an m-manifold is an $(n + m)$-manifold. A

MATHEMATICAL OBJECT AND SYMBOL	DEFINITION AND MAIN PROPERTIES	MEANING, EXAMPLES, AND APPLICATIONS IN THE BOOK
		torus, $\{S^1 \times S^1\}$, subset of $\Re^4$ (compact). An ordinary two-torus subset of $\Re^3$ (compact). An infinite cylinder (noncompact). A **Möbius strip**: $[0, 2\pi] \times (-1, 1)$ with half a twist, (noncompact). A **Klein bottle** (compact). See fig. 2.3. Counterexamples: The **double cone**, $x^2 - y^2 - z^2 = 0$, subset of $\Re^3$ is not a manifold; in fact, the origin does not have a neighborhood homeomorphic to $\Re^2$. However, the **half cone**, $x^2 - y^2 - z^2 = 0$ with $x \geq 0$, is a C^0 manifold (see below). A **Möbius strip with boundary**: $[0, 2\pi] \times [-1, 1]$ with half a twist (compact), is a *manifold with boundary* (see below). The points on the boundary are not homeomorphic to $\Re^2$, but they are homeomorphic to $\frac{1}{2}\Re^2 \equiv \{(x, y) \epsilon \Re^2$, such that $x \geq 0\}$.
Chart (U, ϕ) of a manifold M, or **local coordinate system**: $(x^1 \cdots x^n)$	A **chart** (U, ϕ) is an open set U of the n-manifold M, together with a homeomorphism ϕ from U onto an open set of $\Re^n$. U is the domain of the chart. (U, ϕ) is also called a **local coordinate system**. The n real numbers $(x^1 \cdots x^n)$ of $\phi(P)$ in $\Re^n$ are the **coordinates** of P in the chart (U, ϕ).	
C^r **atlas** $\{(U_\alpha, \phi_\alpha)\}$ on a manifold M.	A C^r **atlas** on an n-manifold M^n is a set of charts $\{(U_\alpha, \phi_\alpha)\}$ covering the whole manifold, such that in the intersections between any two neighborhoods U_α and U_β, the maps $\phi_\beta(\phi_\alpha^{-1}(x^1 \cdots x^n))$ and $\phi_\alpha(\phi_\beta^{-1}(x^1 \cdots x^n))$, from $\Re^n$ to $\Re^n$, are C^r functions, that is, of class C^r.	An atlas on a manifold is a set of local coordinate systems covering the whole manifold, such that in the intersections between two coordinate systems the coordinates y^i of a point, in a system, are C^r functions of the coordinates x^i of the point, in another system; that is, the coordinate transformations $y^i(x^1 \cdots x^n)$ and $x^i(y^1 \cdots y^n)$

MATHEMATICAL OBJECT AND SYMBOL	DEFINITION AND MAIN PROPERTIES	MEANING, EXAMPLES, AND APPLICATIONS IN THE BOOK
		are **functions of class** C^r, that is, continuous with r continuous partial derivatives (a C^0 **function** is just a continuous function).
C^r **differentiable manifold** and **smooth manifold**	A C^r **differentiable manifold** is a topological manifold with a C^r atlas. **Smooth manifold** or C^∞ **differentiable manifold** (or just differentiable manifold) is *usually* called a manifold with a C^∞ atlas.	The definition of differentiable manifold generalizes the definition of differentiable curve, with tangent vector at each point, and differentiable surface, with tangent plane at each point, in $\Re^3$. However, the definition of differentiable manifold does not use any embedding in another space (such as $\Re^3$). A smooth manifold is a topological manifold covered with local coordinate systems, such that the coordinate transformations $y^i(x^1 \cdots x^n)$ and $x^i(y^1 \cdots y^n)$ between any two systems are C^∞ functions. Examples: A C^0 differentiable manifold is a topological manifold (by definition). The real line $\Re$; the real plane $\Re^2$; the n-dimensional Euclidean space $\Re^n$ (they can just be covered with the one chart $(\Re^k, Id)$, where Id is the **identity mapping**: $\boldsymbol{x} \to \boldsymbol{x}$); a circle or one-dimensional sphere, S^1; a two-sphere, S^2; an n-sphere, S^n (they can be covered with a minimum of two charts); a torus; the **group of the linear bijective mappings of** $\Re^n$ **onto** $\Re^n$, $\mathrm{GL}(n, \Re)$ (each covered with a suitable atlas) are differentiable manifolds.
Function on a manifold and C^r **differentiable function** on a C^r differentiable manifold	A **function** f on a manifold M^m, with real values, is a mapping from M^m into $\Re$. Given a chart (U, ϕ), f is C^r **differentiable** at a point P of a C^r manifold M^m if the function $f(\phi^{-1}(x^1 \cdots x^m))$, from $\Re^m$ into $\Re$, is a C^r differentiable function (this definition is independent of the chosen chart).	A function on M^m with real values is an assignment of one real number to each point of M^m.

MATHEMATICAL OBJECT AND SYMBOL	DEFINITION AND MAIN PROPERTIES	MEANING, EXAMPLES, AND APPLICATIONS IN THE BOOK
Differentiable mapping between two manifolds	Given two (C^r) differentiable manifolds M^m and N^n, a (C^r) **differentiable mapping**, f, from M^m into N^n is a mapping such that at each point P of M^m the function: $\psi(f(\phi^{-1}(x^1 \cdots x^m)))$ from $\Re^m$ into $\Re^n$ is a (C^r) differentiable function (where ϕ is a homeomorphism between an open neighborhood of P in M^m and an open set of $\Re^m$, and ψ is a homeomorphism between an open neighborhood of $f(P)$ in N^n and an open set of $\Re^n$.	For each point P of M^m, the coordinates, $\{y^i\}$, of the point $f(P)$ in N^n are (C^r) differentiable functions of the coordinates, $\{x^k\}$, of P in M^m, that is, $y^i(x^1 \cdots x^m)$ are (C^r) differentiable functions.
$C^r(C^\infty)$ **diffeomorphism** between two differentiable manifolds	A $C^r(C^\infty)$ **diffeomorphism** is a bijection f between two differentiable manifolds such that both f and the inverse f^{-1} are $C^r(C^\infty)$ differentiable mappings.	A diffeomorphism is a relation of equivalence between differentiable manifolds and is a generalization of the concept of homeomorphism (topological equivalence) between topological manifolds. Two homeomorphic differentiable manifolds may not be diffeomorphic (see below). The coordinates, $\{y^i\}$, of a point $f(P)$ in the manifold N are $C^r(C^\infty)$ differentiable functions of the coordinates, $\{x^k\}$, of the point P in the manifold M, and vice versa, the coordinates of a point $f^{-1}(Q)$ in M are $C^r(C^\infty)$ differentiable functions of the coordinates of the point Q in N: $$y^i(x^1 \cdots x^n)$$ and $$x^k(y^1 \cdots y^n)$$ are $C^r(C^\infty)$ functions.
Diffeomorphic manifolds	Two differentiable **manifolds** are called **diffeomorphic** if there exists a diffeomorphism between them.	For $n \neq 4$, every differentiable manifold homeomorphic to $\Re^n$ is also diffeomorphic to $\Re^n$ with the **standard differentiable structure** (standard atlas). However, there are **exotic differentiable manifolds**: $\Re^4_{\text{FAKE}}$, homeomorphic but not diffeomorphic to $\Re^4$ with the standard differentiable structure. It is amazing that in the case of $\Re^n$ these **exotic manifolds** exist in

MATHEMATICAL OBJECT AND SYMBOL	DEFINITION AND MAIN PROPERTIES	MEANING, EXAMPLES, AND APPLICATIONS IN THE BOOK
		four dimensions only: for $\Re^n$ with $n \neq 4$ any differentiable structure on $\Re^n$ is equivalent to the standard one on $\Re^n$. However, in general there are other nonunique differentiable structures, for example on S^7 and S^{31}.
Embedding, immersion and **submanifold**	A differentiable mapping f between two manifolds M^m and N^n, where $m \leq n$, is called **embedding** if f restricted to sufficiently small neighborhoods has an inverse which is also a differentiable mapping (**immersion**, that is, for every point P of M^m there is a neighborhood U of P such that f^{-1} restricted to $f(U)$ is a differentiable mapping) and f is a homeomorphism onto its image with the relative topology. The image $f(M)$ is called ***m*-dimensional** embedded **submanifold** of N or **submanifold** of N.	
Manifold M **with boundary** and **boundary of a manifold**, ∂M	An n-dimensional **manifold** M **with boundary** is a Hausdorff topological space such that every point has a neighborhood homeomorphic (topologically equivalent) to an open set in half $\Re^n$, that is, to the topological subspace H^n of $\Re^n$ of all the points $(x^1, x^2, \cdots, x^n)$ of $\Re^n$ such that $x^n \geq 0$. The **boundary** ∂M of this manifold M is the $(n-1)$-dimensional manifold (without boundary) with image on the subspace of H^n corresponding to the points $x^n = 0$.	Examples: A **solid sphere** of $\Re^n$, that is, the set of all points of $\Re^n$ such that $(\sum_{i=1}^n (x^i)^2)^{1/2} \leq r$, the boundary is the $(n-1)$-sphere S^{n-1}; that is, the set of all the points of $\Re^n$: $(\sum_{i=1}^n (x^i)^2)^{1/2} = r$. A solid torus in $\Re^3$; the boundary is the surface of the torus. A Möbius strip with boundary, $[0, 2\pi] \times [-1, 1]$ with half a twist (compact); its boundary is homeomorphic to a circle.
Closed manifold	A compact manifold without boundary is called a **closed manifold**.	
Orientable and **oriented manifold**	An **orientable manifold** is a manifold that can be covered by an atlas, that is, by a family of local coordinate systems $(x^1, \cdots, x^n), \cdots, (y^1, \cdots, y^n)$, such that at each point in the intersections between any two systems, the **Jacobian** determinant, that is, the determinant $\left\| \frac{\partial x^i}{\partial y^k} \right\| \equiv \det \left(\frac{\partial x^i}{\partial y^k} \right)$ of the derivatives of the coordinates, is	Examples: The standard n-dimensional Euclidean space $\Re^n$, an n-sphere S^n. Counterexamples: The Möbius strip is nonorientable. The **Klein bottle**, or twisted torus, is nonorientable. See fig. 2.3.

MATHEMATICAL OBJECT AND SYMBOL	DEFINITION AND MAIN PROPERTIES	MEANING, EXAMPLES, AND APPLICATIONS IN THE BOOK
	positive. A **manifold** with such an atlas is called **oriented**.	
Induced orientation on the boundary ∂M of an oriented manifold M	Given an oriented atlas covering M, for each chart $(U_\alpha, \varphi_\alpha)$ with nonzero intersection with the boundary ∂M (corresponding to the points $x^n = 0$), the coordinates $(x^1, x^2, \cdots, x^{n-1})$, in the intersections of each U_α with ∂M, define a naturally **induced orientation** on ∂M.	
Tangent vector to a differentiable manifold	*Coordinate-dependent definition*: A **tangent vector** $\boldsymbol{v}$ at a point P of a differentiable manifold is a mathematical object that, in a coordinate system, is represented by a set of n numbers v^i at P, components of $\boldsymbol{v}$, that, under a coordinate transformation $x'^k = x'^k(x^i)$, change according to the transformation law: $$v'^k = \left(\frac{\partial x'^k}{\partial x^i}\right)_P v^i.$$ *Definition independent of coordinates*: A **tangent vector** at a point P is a mapping $\boldsymbol{v}_P$ that to each differentiable function defined in a neighborhood of P assigns one real number, and which is linear and satisfies the **Leibniz rule**. That is: $\boldsymbol{v}_P(af + bg) = a\boldsymbol{v}_P(f) + b\boldsymbol{v}_P(g)$, linearity; and $\boldsymbol{v}_P(f \cdot g) = \boldsymbol{v}_P(f)g(P) + f(P)\boldsymbol{v}_P(g)$, Leibniz rule; where a, b are real numbers and f, g are differentiable functions. *Equivalent definition independent of coordinates*:	
	Given a differentiable curve $c(t)$, that is, a differentiable mapping from an interval of the real numbers into M, and given a function f on M differentiable at P, the **tangent vector** to the curve at $P = c(t_P)$ is defined by $$\boldsymbol{v}^c_P(f) = \left(\frac{df(c(t))}{dt}\right)_{t_P}$$ and one may write in a local coordinate system $(x^1, \cdots, x^n)$:	This definition of tangent vector is a coordinate independent generalization to differentiable manifolds M of the standard definition of directional derivative along a vector v^i, and of tangent vector to a curve $x^i(t)$ in $\Re^n$, with components $\frac{dx^i}{dt}$. It reduces to this standard definition in the case of $M = \Re^n$. We recall that the **directional derivative** of a function f along a vector v^i_P, or along a curve $x^i(t)$ of $\Re^n$, at each point P (corresponding to t_P), is

MATHEMATICAL OBJECT AND SYMBOL	DEFINITION AND MAIN PROPERTIES	MEANING, EXAMPLES, AND APPLICATIONS IN THE BOOK

$$\left(\frac{df(c(t))}{dt}\right)_{t_P} = \sum_i \left(\frac{dx^i(c(t))}{dt}\right)_{t_P} \left(\frac{\partial f(x)}{\partial x^i}\right)_P ,$$

generalization of the ordinary definition of tangent vector to a curve in $\Re^n$.

These definitions of tangent vector are equivalent.

respectively:

$$\boldsymbol{v}_P(f) = \sum_i v^i_P \left(\frac{\partial f}{\partial x^i}\right)_P$$

and

$$\left(\frac{df(x^i(t))}{dt}\right)_{t_P} = \sum_i \left(\frac{dx^i}{dt}\right)_{t_P} \left(\frac{\partial f}{\partial x^i}\right)_P .$$

Tangent vector space to the manifold at P:

$T_P(M)$ or T_P

The vector space of all the tangent vectors to M at P, with the operations of addition and scalar multiplication defined by

$$(a\boldsymbol{v}_P + b\boldsymbol{w}_P)(f) = a\boldsymbol{v}_P(f) + b\boldsymbol{w}_P(f)$$

is called **tangent vector space** at P.

Coordinate vector $\left(\frac{\partial}{\partial x^i}\right)_P$ at P

Given a chart (U, ϕ) in a neighborhood of a point P, with coordinates $(x^1, \cdots, x^n)$, a **coordinate vector** $\left(\frac{\partial}{\partial x^i}\right)_P$ at P is a linear mapping that to each differentiable function f, defined in a neighborhood of P, assigns the real number:

$$\left(\frac{\partial f}{\partial x^i}\right)_P \equiv \left(\frac{\partial(f(\phi^{-1}(x^1 \cdots x^n)))}{\partial x^i}\right)_{(x^1_P, \ldots, x^n_P)} .$$

Natural or **coordinate basis** of T_P, or **holonomic frame**: $\left\{\frac{\partial}{\partial x^i}\right\}_P$

The vectors $\left\{\frac{\partial}{\partial x^i}\right\}_P$ form a basis for the tangent vector space, called **coordinate** or **natural basis**, or **holonomic frame**: $\left\{\frac{\partial}{\partial x^i}\right\}_P$.

Components v^i **of a tangent vector** $\boldsymbol{v}$ at P

Given a chart, the coordinate vectors $\left\{\frac{\partial}{\partial x^i}\right\}_P$ are a basis for the tangent vector space T_P, therefore one can write every vector $\boldsymbol{v}$ at P:

$$\boldsymbol{v} = v^i \left(\frac{\partial}{\partial x^i}\right)_P \quad \left(\text{with summation over } i\right);$$

v^i are called the **components** of $\boldsymbol{v}$ at P with respect to the local coordinate system $(x^1, \cdots, x^n)$, or components **with respect to the natural basis** $\left\{\frac{\partial}{\partial x^i}\right\}_P$.

One can also write a vector $\boldsymbol{v}$ at P

MATHEMATICAL OBJECT AND SYMBOL	DEFINITION AND MAIN PROPERTIES	MEANING, EXAMPLES, AND APPLICATIONS IN THE BOOK

with respect to a *general basis* $\{\boldsymbol{e}_i\}_P$ of the tangent vector space T_P:

$$\boldsymbol{v} = v^a\, \boldsymbol{e}_a.$$

v^a are the **components of $\boldsymbol{v}$ with respect to the general basis $\{\boldsymbol{e}_a\}_P$.**

Given a local coordinate system, the coordinate vectors, $\frac{\partial}{\partial x^i}$, due to their definition, transform according to:

$$\left(\frac{\partial}{\partial \boldsymbol{x}'^k}\right)_P = \left(\frac{\partial x^i}{\partial x'^k}\right)_P \left(\frac{\partial}{\partial \boldsymbol{x}^i}\right)_P .$$

Therefore the components of a vector transform at a point according to

$$v'^k = \left(\frac{\partial x'^k}{\partial x^i}\right)_P v^i .$$

This is just the coordinate-dependent definition of **contravariant vectors.**

Given a general basis $\{\boldsymbol{e}_a\}$ of T_P, under a change of basis, $\{\boldsymbol{e}'_a\}$,

$$\boldsymbol{e}'_b = L^a{}_{b'} \boldsymbol{e}_a,$$

the components v_a of a vector transform according to

$$v'^b = L^{b'}{}_a v^a$$

where $L^{b'}{}_a$ is the inverse of $L^a{}_{b'}$

$$L^a{}_{s'} L^{s'}{}_b \equiv \delta^a{}_b$$

$(\delta^a{}_b$ Kronecker symbol$)$.

Vector field $\boldsymbol{v}$ on M and **differentiable vector field**

A **vector field $\boldsymbol{v}$** on M is a mapping that to each point P of M assigns one tangent vector $\boldsymbol{v}_P$ of T_P. The **vector field $\boldsymbol{v}$** is **differentiable** if the mapping $\boldsymbol{v}(P)$ is differentiable.

Commutator of two vector fields: $[\boldsymbol{v}, \boldsymbol{w}]$

The **commutator of two vector fields $\boldsymbol{v}$** and $\boldsymbol{w}$ is defined as the vector field $[\boldsymbol{v}, \boldsymbol{w}]$ such that:

$$[\boldsymbol{v}, \boldsymbol{w}](f) = \boldsymbol{v}(\boldsymbol{w}(f)) - \boldsymbol{w}(\boldsymbol{v}(f)).$$

Structure coefficients, or **commutation coefficients**: C^d_{ab}

Given a general basis $\{\boldsymbol{e}_a\}$ of T_P, the **structure coefficients**, C^d_{ab}, are the **commutation coefficients** of the basis vectors $\{\boldsymbol{e}_a\}$:

$$[\boldsymbol{e}_a, \boldsymbol{e}_b] \equiv C^d_{ab}\, \boldsymbol{e}_d$$

where

MATHEMATICAL OBJECT AND SYMBOL	DEFINITION AND MAIN PROPERTIES	MEANING, EXAMPLES, AND APPLICATIONS IN THE BOOK
	$$C^d_{ab} = -C^d_{ba}.$$	
1-form $\boldsymbol{\theta}_P$ **(differential 1-form)** at P, or **covariant vector** at P	*Coordinate dependent definition*: A **1-form $\boldsymbol{\theta}_P$** at P is a mathematical object that in a coordinate system is represented by a set of n numbers θ_i at P that, under a coordinate transformation $x'^k = x'^k(x^i)$, change according to $$\theta'_k = \left(\frac{\partial x^i}{\partial x'^k}\right)_P \theta_i.$$ *Definition independent of coordinates*: A **1-form $\boldsymbol{\theta}_P$** at P is a linear function $\boldsymbol{\theta}_P$ that to each vector $\boldsymbol{v}_P$ at P assigns one real number: $$\boldsymbol{\theta}_P(\boldsymbol{v}_P)\text{: real number.}$$	
Cotangent vector space: T^*_P	The set of all the 1-forms at P is an n-dimensional vector space, T^*_P, dual to the tangent vector space T_P, called **cotangent vector space.** Given a basis $\{\boldsymbol{e}_a\}$ of the tangent vector space T_P, the n 1-forms $\boldsymbol{\eta}^a$ defined by $$\boldsymbol{\eta}^a(\boldsymbol{e}_b) = \delta^a{}_b$$ form a basis (**dual basis** to $\{\boldsymbol{e}_a\}$) for the **dual space T^*_P.** Given a chart, the 1-forms of the basis **dual** to the **coordinate basis** $\{\frac{\boldsymbol{\partial}}{\boldsymbol{\partial x^i}}\}$, written $\{\boldsymbol{dx}^i\}$, (see below) are defined by $$\boldsymbol{dx}^i\left(\frac{\boldsymbol{\partial}}{\boldsymbol{\partial x^k}}\right) = \delta^i{}_k.$$ $\{\boldsymbol{dx}^i\}$ is a basis for T^*_P. One can then write a 1-form, or covariant vector, using the basis 1-forms, $$\boldsymbol{\theta} = \theta_a \boldsymbol{\eta}^a;$$ θ_a are the components of the 1-form, or **covariant vector**, $\boldsymbol{\theta}$. In a local coordinate system: $$\boldsymbol{\theta} = \theta_i\, \boldsymbol{dx}^i.$$ Given a local coordinate system, the 1-forms $\boldsymbol{dx}^i$ transform according to $$dx'^k = \frac{\partial x'^k}{\partial x^i}\, dx^i.$$	

MATHEMATICAL OBJECT AND SYMBOL	DEFINITION AND MAIN PROPERTIES	MEANING, EXAMPLES, AND APPLICATIONS IN THE BOOK
	Therefore the components of a 1-form, with respect to the natural basis, transform according to $$\theta'_k = \frac{\partial x^i}{\partial x'^k}\,\theta_i .$$ This is just the coordinate dependent definition of a **covariant vector**. Under a change of **general basis**, $$\boldsymbol{\eta}'^b = L^{b'}{}_a \boldsymbol{\eta}^a$$ the components of a 1-form transform according to $$\theta'_b = L^a{}_{b'}\,\theta_a$$ $$(L^a{}_{s'}\,L^{s'}{}_b \equiv \delta^a{}_b).$$	
Differential df of a differentiable function f	The **differential df** of a differentiable function f is a 1-form that to each vector $\boldsymbol{v}$ assigns the real number $$\boldsymbol{d}f(\boldsymbol{v}) = \boldsymbol{v}(f);$$ in particular the **differentials of the coordinate functions** x^i are the 1-forms $\boldsymbol{d}x^i$: $$\boldsymbol{d}x^i\left(\frac{\boldsymbol{\partial}}{\boldsymbol{\partial x}^k}\right) = \frac{\partial x^i}{\partial x^k} = \delta^i{}_k .$$ Given a local coordinate system, one has then $$\boldsymbol{d}f = \frac{\partial f}{\partial x^i}\,\boldsymbol{d}x^i .$$	
Tensor	*Coordinate dependent definition*: A ***p*-covariant and *q*-contravariant tensor $\boldsymbol{T}$**, or tensor of type $(q\ p)$, at P, is a mathematical object that, in a coordinate system, is represented by a set of $(n)^{p+q}$ functions $T_{i_1\cdots i_p}{}^{k_1\cdots k_q}$, the components of $\boldsymbol{T}$, that, under a coordinate transformation, change according to the transformation law: $$T'_{i_1\cdots i_p}{}^{k_1\cdots k_q} = \frac{\partial x^{m_1}}{\partial x'^{i_1}}\cdots\frac{\partial x'^{k_q}}{\partial x^{n_q}}\,T_{m_1\cdots m_p}{}^{n_1\cdots n_q}$$ under a transformation of a general basis: $\boldsymbol{e}'_b = L^a{}_{b'}\boldsymbol{e}_a$, $$T'_{b_1\cdots}{}^{\cdots a_q} = L^{m_1}{}_{b'_1}\cdots L^{a'_q}{}_{n_q}\,T_{m_1\cdots}{}^{\cdots n_q}.$$	Tensors are a generalization of contravariant vectors and covariant vectors. Examples: A (1 0) tensor is a vector. A (0 1) tensor is a 1-form. A (0 0) tensor is a scalar function. An example of (3 1) tensor is the Riemann curvature tensor used all through the book (see below).

MATHEMATICAL OBJECT AND SYMBOL	DEFINITION AND MAIN PROPERTIES	MEANING, EXAMPLES, AND APPLICATIONS IN THE BOOK
	Definition independent of coordinates: A **tensor** of type $(q\ p)$, ***p*-covariant and *q*-contravariant**, at a point P, is a multilinear function that to each ordered set of p vectors and q forms assigns a real number.	
Antisymmetrization of tensors: $T_{[a_1 \cdots a_n]}$	The operation of **antisymmetrization** of a tensor, $T_{a_1 \cdots a_n}$, written with the indices of the tensor within square brackets, is defined by $$T_{[a_1 \cdots a_n]} = \frac{1}{n!} \sum_{\substack{\text{all} \\ \text{permutations}, \pi}} \epsilon_\pi \, T_{a_1 \cdots a_n}$$ where the sum is extended to all the permutations of $a_1 \cdots a_n$, with plus sign for even permutations, $\epsilon_{\pi \text{ even}} \equiv +1$, and minus sign for odd permutations, $\epsilon_{\pi \text{ odd}} \equiv -1$.	Example: $S_{[ik]} = \frac{1}{2}(S_{ik} - S_{ki})$.
Symmetrization of a tensor: $T_{(a_1 \cdots a_n)}$	**Symmetrization of a tensor**, $T_{a_1 \cdots a_n}$, written with the indices of the tensor in parentheses, is defined by $$T_{(a_1 \cdots a_n)} = \frac{1}{n!} \sum_{\substack{\text{all} \\ \text{permutations}, \pi}} T_{a_1 \cdots a_n}.$$	Example $S_{(ik)} = \frac{1}{2}(S_{ik} + S_{ki})$.
***p*-forms** and ***q*-polyvectors**	***p*-forms**, $\boldsymbol{\theta}$, are completely antisymmetric p-covariant tensors, in components: $\theta_{a_1 \cdots bc \cdots a_p} = -\theta_{a_1 \cdots cb \cdots a_p}$, for exchange of any pair of nearby indices; p is the degree of the form. One may also define a p-form as $$\theta_{a_1 \cdots a_n} = \theta_{[a_1 \cdots a_n]}.$$ Similarly, ***q*-polyvectors**, or **multivectors**, are completely antisymmetric q-contravariant tensors.	An example of **2-form** is the electromagnetic tensor: $F_{\alpha\beta} = F_{[\alpha\beta]}$; see section 2.8. For other examples of p-forms and q-polyvectors see also section 2.8.
Scalar invariant	A **scalar invariant** field ϕ, or simply **scalar field** on a manifold M, with real values, is a function from M into $\Re$ such that at each point of M its value is independent of the chart chosen, that is, at each point its value is invariant under coordinate transformations.	Examples: Given a vector $\boldsymbol{v}$ and a 1-form $\boldsymbol{\theta}$ the interior product of $\boldsymbol{v}$ and $\boldsymbol{\theta}$, $\sum_i v^i \theta_i \equiv v^i \theta_i$, is a scalar invariant. Given the Riemann tensor, $R^i{}_{klm}$ (see below), $R \equiv R^{ik}{}_{ik}$ is a scalar invariant called the Ricci curvature scalar. Other examples of scalars, or pseudoscalars, are given in section 6.11.

MATHEMATICAL OBJECT AND SYMBOL	DEFINITION AND MAIN PROPERTIES	MEANING, EXAMPLES, AND APPLICATIONS IN THE BOOK
		Counterexamples: Each component of a vector, or of a tensor, is a function but *not* a scalar invariant: $v^i(\boldsymbol{x})$, $\theta_i(\boldsymbol{x})$, $R^i{}_{klm}(\boldsymbol{x})$.
Tensor field (**field of forms**)	A **tensor field** (**field of forms**) on a manifold M is a mapping that to each point P of M assigns one tensor $\boldsymbol{T}(P)$ (or a form $\boldsymbol{\theta}(P)$). A **1-form** field $\boldsymbol{\theta}$ is called **differentiable** if all the components θ_i are differentiable.	
Inner product (**scalar product**) on a vector space over $\Re$	The **inner product** on a vector space V over $\Re$ is a bilinear function that to any two vectors $\boldsymbol{v}$ and $\boldsymbol{w}$ of V assigns one real number, $\boldsymbol{g}(\boldsymbol{v}, \boldsymbol{w})$, such that $\boldsymbol{g}(\boldsymbol{v}, \boldsymbol{w})$ is **symmetric**: $\boldsymbol{g}(\boldsymbol{v}, \boldsymbol{w}) = \boldsymbol{g}(\boldsymbol{w}, \boldsymbol{v})$ and **nondegenerate**: $\boldsymbol{g}(\boldsymbol{v}, \boldsymbol{w}) = \boldsymbol{0}$, for every $\boldsymbol{v} \neq \boldsymbol{0}$, if and only if $\boldsymbol{w} = \boldsymbol{0}$.	
Tensor product, or **outer product**: $\boldsymbol{T} \otimes \boldsymbol{S}$	The **tensor product** of a $(q\ p)$ tensor $\boldsymbol{T}$ with a $(n\ m)$ tensor $\boldsymbol{S}$, written as $\boldsymbol{T} \otimes \boldsymbol{S}$, is a $(q+n\ ,\ p+m)$ tensor, with components: $$T^{a_1 \cdots a_q}{}_{b_1 \cdots b_p} S^{a_{q+1} \cdots a_{q+n}}{}_{b_{p+1} \cdots b_{p+m}}.$$	
Contraction and **contracted tensor product**	Given a $(q\ p)$ tensor $\boldsymbol{T}$, the **contraction** of the tensor on two indices, one contravariant and the other covariant, is a $(q-1,\ p-1)$ tensor defined by $$\sum_{i=1}^{n} T^{a \cdot\cdot i \cdot\cdot}{}_{b \cdot\cdot i \cdot\cdot} \equiv T^{a \cdot\cdot i \cdot\cdot}{}_{b \cdot\cdot i \cdot\cdot}$$ where, whenever an index is repeated up and down, summation over the index is understood. The **contracted product** of two tensors $\boldsymbol{S}$ and $\boldsymbol{T}$ is a tensor formed by tensor product of $\boldsymbol{S}$ and $\boldsymbol{T}$ followed by contraction on two of their indices.	
Interior product of a form $\boldsymbol{\theta}$ and a vector $\boldsymbol{v}$: $\boldsymbol{v} \lrcorner \boldsymbol{\theta}$	The **interior product**, $\boldsymbol{v} \lrcorner \boldsymbol{\theta}$, of a form $\boldsymbol{\theta}$ and a vector $\boldsymbol{v}$, is the contracted multiplication of $\boldsymbol{v}$ and $\boldsymbol{\theta}$. In components: $$(\boldsymbol{v} \lrcorner \boldsymbol{\theta})_{j \cdots m} = \sum_i v^i\, \theta_{ij \cdots m} \equiv v^i\, \theta_{ij \cdots m}.$$	

MATHEMATICAL OBJECT AND SYMBOL	DEFINITION AND MAIN PROPERTIES	MEANING, EXAMPLES, AND APPLICATIONS IN THE BOOK
Exterior product, or **wedge product** between forms: $\boldsymbol{\theta} \wedge \boldsymbol{\omega}$	The **exterior product**, or **wedge product**, is an operation that from a p-form $\boldsymbol{\theta}$ and a q-form $\boldsymbol{\omega}$ gives a $(p+q)$-form defined in components by $$(\boldsymbol{\theta} \wedge \boldsymbol{\omega})_{a_1 \cdots a_{p+q}} = \frac{(p+q)!}{p!\,q!}\,\theta_{[a_1 \cdots a_p}\omega_{a_{p+1} \cdots a_{p+q}]}$$ where $[a_1 \cdots a_{p+q}]$ means antisymmetrization. The exterior product satisfies the properties: • $(\boldsymbol{\theta}_1 \wedge \boldsymbol{\theta}_2) \wedge \boldsymbol{\theta}_3 = \boldsymbol{\theta}_1 \wedge (\boldsymbol{\theta}_2 \wedge \boldsymbol{\theta}_3)$ • $(\boldsymbol{\theta}_1 + \boldsymbol{\theta}_2) \wedge \boldsymbol{\omega} = \boldsymbol{\theta}_1 \wedge \boldsymbol{\omega} + \boldsymbol{\theta}_2 \wedge \boldsymbol{\omega}$ • $\boldsymbol{\theta} \wedge (\boldsymbol{\omega}_1 + \boldsymbol{\omega}_2) = \boldsymbol{\theta} \wedge \boldsymbol{\omega}_1 + \boldsymbol{\theta} \wedge \boldsymbol{\omega}_2$ • $f(\boldsymbol{\theta} \wedge \boldsymbol{\omega}) = f\boldsymbol{\theta} \wedge \boldsymbol{\omega} = \boldsymbol{\theta} \wedge f\boldsymbol{\omega}$ • $\boldsymbol{\theta} \wedge \boldsymbol{\omega} = (-1)^{pq}\boldsymbol{\omega} \wedge \boldsymbol{\theta}$.	
Exterior derivative: $\boldsymbol{d\theta}$	The **exterior derivative**, $\boldsymbol{d\theta}$ (of a p-form, $\boldsymbol{\theta}$) is an operator that maps p-forms into $(p+1)$-forms, defined in components by $$d\theta_{a_1 \cdots a_{p+1}} = (p+1)\,\frac{\partial}{\partial x^{[a_1}}\,\theta_{a_2 \cdots a_{p+1}]} = \frac{1}{p!} \sum_{\text{all permutations},\pi} \epsilon_\pi \frac{\partial}{\partial x^{a_1}}\theta_{a_2 \cdots a_{p+1}}$$ where $\epsilon_\pi \equiv +1$ for even permutations, and $\epsilon_\pi \equiv -1$ for odd permutations. The exterior derivative of the exterior product satisfies the property: $$\boldsymbol{d}(\boldsymbol{\theta} \wedge \boldsymbol{\omega}) = \boldsymbol{d\theta} \wedge \boldsymbol{\omega} + (-1)^p \boldsymbol{\theta} \wedge \boldsymbol{d\omega}$$ where p is the degree of $\boldsymbol{\theta}$. *Definition independent of coordinates*: The operator $\boldsymbol{d}$ on p-forms, $\boldsymbol{\theta}$, into $(p+1)$-forms, is defined by • $\boldsymbol{d}(k\boldsymbol{\theta} + c\boldsymbol{\omega}) = k\boldsymbol{d\theta} + c\boldsymbol{d\omega}$, where k and c are constants; • $\boldsymbol{d}(\boldsymbol{\theta} \wedge \boldsymbol{\omega}) = \boldsymbol{d\theta} \wedge \boldsymbol{\omega} + (-1)^p \boldsymbol{\theta} \wedge \boldsymbol{d\omega}$, where p is the degree of $\boldsymbol{\theta}$; • $\boldsymbol{dd} = \boldsymbol{0}$; • Given a 0-form f, $\boldsymbol{df}$ is the ordinary differential of f.	The exterior derivative is a type of derivative on a manifold that when applied to an antisymmetric p-covariant tensor gives an antisymmetric $(p+1)$-covariant tensor, and may be thought of as an extension to arbitrary p-forms of the differential, $\boldsymbol{d}$, of a function, f, that, when applied to a **0-form**, or function f, gives the 1-form: $\boldsymbol{df} = \frac{\partial f}{\partial x^i}\,\boldsymbol{dx}^i$. This type of derivative on a manifold is defined only on forms. Example: In terms of the **electromagnetic four-potential 1-form** $\boldsymbol{A}$, the **electromagnetic field 2-form** $\boldsymbol{F}$ is given by $$\boldsymbol{F} = \boldsymbol{dA}$$ or $$F_{\alpha\beta} = A_{\beta,\alpha} - A_{\alpha,\beta}.$$ See section 2.8.
Lie derivative along a vector field $\boldsymbol{v}$: $\mathcal{L}_{\boldsymbol{v}}$	Given a differentiable vector field $\boldsymbol{v}$, in a neighborhood of a point, P, the **Lie derivative** (of a tensor field, $\boldsymbol{T}$), at	The Lie derivative is a type of derivative on a manifold that is defined at a point by

MATHEMATICAL OBJECT AND SYMBOL	DEFINITION AND MAIN PROPERTIES	MEANING, EXAMPLES, AND APPLICATIONS IN THE BOOK
	the point P along the vector $\boldsymbol{v}$, is an operator that maps $(m\ n)$ tensors into $(m\ n)$ tensors defined in components by $$(\mathcal{L}_{\boldsymbol{v}}\boldsymbol{T})^{a\cdots}{}_{b\cdots} = T^{a\cdots}{}_{b\cdots,s}v^s - T^{s\cdots}{}_{b\cdots}v^a{}_{,s} + T^{a\cdots}{}_{s\cdots}v^s{}_{,b} + \cdots .$$ The Lie derivative of a vector field $\boldsymbol{w}$ along a vector field $\boldsymbol{v}$ is just the **commutator** of $\boldsymbol{v}$ and $\boldsymbol{w}$. In a general basis $\{\boldsymbol{e}_a\}$, in components: $$(\mathcal{L}_{\boldsymbol{v}}\,\boldsymbol{w})^b = [\boldsymbol{v}, \boldsymbol{w}]^b \equiv v^a\boldsymbol{e}_a(w^b) - w^a\boldsymbol{e}_a(v^b) + v^a w^d C^b_{ad}$$ and in a natural basis $\{\frac{\partial}{\partial x^i}\}$: $$(\mathcal{L}_{\boldsymbol{v}}\,\boldsymbol{w})^i = v^k w^i{}_{,k} - w^k v^i{}_{,k}.$$	using a vector field $\boldsymbol{v}$, defined in a neighborhood of the point. It may be thought of as an extension to arbitrary tensors of the **directional derivative** of a function along $\boldsymbol{v}$: $$\mathcal{L}_{\boldsymbol{v}} f = \frac{\partial f}{\partial x^i} v^i .$$ A vector field $\boldsymbol{v}$, defined in an open neighborhood, determines a congruence of curves, in the open neighborhood, with tangent vector at each point equal to the vector field $\boldsymbol{v}$ at that point. The Lie derivative of a tensor field, $\boldsymbol{T}(\boldsymbol{x})$, along the vector field $\boldsymbol{v}$, measures at each point, with coordinates x^i, the deviation of the tensor field $\boldsymbol{T}(\boldsymbol{x})$ from its formal invariance along the curves: $x'^i = x^i + \varepsilon v^i$ where $\varepsilon \ll 1$, and where **formal invariance** of $\boldsymbol{T}(\boldsymbol{x})$ along the curves x'^i is defined by $$\boldsymbol{T}(y^i \equiv x^i) \equiv \boldsymbol{T}'(y^i \equiv x'^i).$$ See sections 2.6 and 4.2.
Lie dragging	A **tensor** such that: $$\mathcal{L}_{\boldsymbol{v}}\boldsymbol{T} = \boldsymbol{0}$$ is called **Lie dragged** along the congruence of curves determined by the field $\boldsymbol{v}$.	Lie dragging of a tensor along a curve means formal invariance of the tensor $\boldsymbol{T}(\boldsymbol{x})$ along the curve.
Isometry on a manifold	An **isometry** is a transformation under which the metric tensor $\boldsymbol{g}$ (see below) is formally invariant. An **infinitesimal isometry** is a transformation $x'^i = x^i + \varepsilon\xi^i$, where $\varepsilon \ll 1$, such that $$\mathcal{L}_{\boldsymbol{\xi}}\boldsymbol{g} = \boldsymbol{0}.$$	An **infinitesimal isometry** is an infinitesimal transformation $x'^i = x^i + \varepsilon\xi^i$ under which the metric $\boldsymbol{g}$ is formally invariant, that is, $$\boldsymbol{g}(y^i \equiv x^i) = \boldsymbol{g}'(y^i \equiv x'^i).$$ See sections 2.6 and 4.2.
Killing vector and **Killing equation**	A vector field $\boldsymbol{\xi}$ such that $\mathcal{L}_{\boldsymbol{\xi}}\boldsymbol{g} = \boldsymbol{0}$ is called **Killing vector**. One can write the formal invariance of the metric, $\boldsymbol{g}$, along $\boldsymbol{\xi}$, in components, as $$\xi_{(i;k)} = 0,$$ called **Killing equation**.	See sections 2.6 and 4.2.

MATHEMATICAL OBJECT AND SYMBOL	DEFINITION AND MAIN PROPERTIES	MEANING, EXAMPLES, AND APPLICATIONS IN THE BOOK
Lie group	A **Lie group** G is a group that is also a differentiable manifold, such that the mapping from $G \times G$ into G: $(x, y) \rightarrow xy^{-1}$, product of x with the inverse y^{-1} of y, is a C^∞ differentiable mapping (x and y are elements of G).	$\mathfrak{R}^n$ with the standard operation of sum is a Lie group. The group of isometries of a Riemannian manifold (with a finite number of connected components) is a Lie group. The **group of all nonsingular real $n \times n$ matrices:** $GL(n, \mathfrak{R})$, the **general linear group**, is a Lie group. The **orthogonal group** $O(n)$, of all $O \in GL(n, \mathfrak{R})$ such that $O \cdot O^T = I$.
Lie algebra	A **Lie algebra** is a finite dimensional vector space with a bilinear operation [,] such that: $$[\boldsymbol{v}, \boldsymbol{w}] = -[\boldsymbol{w}, \boldsymbol{v}],$$ $$[[\boldsymbol{v}, \boldsymbol{w}], \boldsymbol{u}] + [[\boldsymbol{w}, \boldsymbol{u}], \boldsymbol{v}] + [[\boldsymbol{u}, \boldsymbol{v}], \boldsymbol{w}] = \boldsymbol{0} \text{ (Jacobi identity).}$$	The set of all the Killing vector fields on a Riemannian manifold with the operation $[\boldsymbol{v}, \boldsymbol{w}] \equiv \mathcal{L}_{\boldsymbol{v}}\boldsymbol{w}$ form a Lie algebra. See section 4.2.
Connection, ∇, and **connection coefficients**: Γ^i_{jk}	A **connection** (**classical**) on an n-dimensional manifold M is an assignment, at each point of M, for each coordinate system, of n^3 numbers, Γ^i_{jk}, called **connection coefficients**, called in a coordinate basis **Christoffel symbols**, such that the connection coefficients in different coordinate systems transform into each other according to $$\Gamma'^l_{mn} = \frac{\partial x'^l}{\partial x^i}\frac{\partial x^j}{\partial x'^m}\frac{\partial x^k}{\partial x'^n}\Gamma^i_{jk} + \frac{\partial x'^l}{\partial x^s}\frac{\partial^2 x^s}{\partial x'^m \partial x'^n}.$$ *Definition independent of coordinates*: A linear **connection** (**Koszul**) on a smooth manifold M is a mapping $\boldsymbol{\nabla}$ that for every two C^∞ vector fields, $\boldsymbol{v}$ and $\boldsymbol{w}$, gives a C^∞ vector field, $\boldsymbol{\nabla}_{\boldsymbol{v}}\boldsymbol{w}$, such that: • $\boldsymbol{\nabla}_{\boldsymbol{v}_1+\boldsymbol{v}_2}\boldsymbol{w} = \boldsymbol{\nabla}_{\boldsymbol{v}_1}\boldsymbol{w} + \boldsymbol{\nabla}_{\boldsymbol{v}_2}\boldsymbol{w}$ • $\boldsymbol{\nabla}_{\boldsymbol{v}}(\boldsymbol{w}_1 + \boldsymbol{w}_2) = \boldsymbol{\nabla}_{\boldsymbol{v}}\boldsymbol{w}_1 + \boldsymbol{\nabla}_{\boldsymbol{v}}\boldsymbol{w}_2$ • $\boldsymbol{\nabla}_{f\boldsymbol{v}}\boldsymbol{w} = f \cdot \boldsymbol{\nabla}_{\boldsymbol{v}}\boldsymbol{w}$ • $\boldsymbol{\nabla}_{\boldsymbol{v}}(f\boldsymbol{w}) = f \cdot \boldsymbol{\nabla}_{\boldsymbol{v}}\boldsymbol{w} + \boldsymbol{v}(f) \cdot \boldsymbol{w}$ where f is a differentiable function.	By introducing a classical connection, or affine connection (see parallel transport below) on a manifold M, one is able to (parallel) transport and to compare vectors and tensors at different points on M, even on non-Euclidean manifolds. Therefore, one is able to define a type of derivative, the covariant or absolute derivative, with tensorial character, and that reduces to the ordinary partial derivative when the connection coefficients are zero.

MATHEMATICAL OBJECT AND SYMBOL	DEFINITION AND MAIN PROPERTIES	MEANING, EXAMPLES, AND APPLICATIONS IN THE BOOK
	In a local coordinate system, by defining $$\nabla_{\frac{\partial}{\partial x^i}} \frac{\partial}{\partial \boldsymbol{x}^j} = \Gamma^k_{ij} \frac{\partial}{\partial \boldsymbol{x}^k},$$ one has the components Γ^k_{ij} of a classical connection. A third definition of **connection (Cartan)**, using moving frames, is given in section 2.8.	
Torsion tensor: T^i_{jk}	Given a connection, Γ^i_{jk}, on a manifold M, the difference $$T^i_{jk} = \Gamma^i_{jk} - \Gamma^i_{kj}$$ is called **torsion tensor**. The tensorial character of T^i_{jk} immediately follows from the transformation law of the connection coefficients, Γ^i_{jk}. A connection is said to be symmetric if the torsion tensor is zero.	See section 2.8.
Covariant derivative: $(\ )_{;k}$ or $\boldsymbol{\nabla}$	The **covariant derivative**, or **absolute derivative** (written with a semicolon followed by a letter "; k"), is a derivative operator on a differentiable manifold that maps $(m\ n)$ tensors into $(m, n+1)$ tensors, and is defined, in components, as $$T^{i\cdots}{}_{j\cdots;k} \equiv T^{i\cdots}{}_{j\cdots,k} + \Gamma^i_{sk} T^{s\cdots}{}_{j\cdots} - \Gamma^s_{jk} T^{i\cdots}{}_{s\cdots}.$$ Indeed, it immediately follows from the definition of connection coefficients that the covariant derivative of a $(m\ n)$ tensor is a $(m, n+1)$ tensor. *Definition independent of coordinates*: The **covariant derivative** of a differentiable $(q\ p)$ tensor field $\boldsymbol{T}$ is a $(q, p+1)$ tensor field $\boldsymbol{\nabla T}$, such that: (1) $\boldsymbol{\nabla} f = \boldsymbol{d}f$, for any differentiable function f (2) $\boldsymbol{\nabla}(\boldsymbol{S}+\boldsymbol{T}) = \boldsymbol{\nabla S} + \boldsymbol{\nabla T}$: linearity of $\boldsymbol{\nabla}$ (3) $\boldsymbol{\nabla}(\boldsymbol{S}\otimes\boldsymbol{T}) = \boldsymbol{\nabla S}\otimes\boldsymbol{T} + \boldsymbol{S}\otimes\boldsymbol{\nabla T}$: **Leibniz rule**, where $\boldsymbol{S}$ is any differentiable tensor field or differentiable function (4) $\boldsymbol{\nabla}$ commutes with the operation of contracted product of tensors,	The covariant derivative of a tensor on a differentiable manifold M is a type of derivative with tensorial character that generalizes to an arbitrary smooth manifold M the ordinary partial derivative in Euclidean space, $\Re^n$; in fact, if $\Gamma^i_{jk} = 0$ it reduces to the standard partial derivative. The covariant derivative is a type of derivative on a smooth manifold that has tensorial character and is defined on any tensor, not only on forms as the exterior derivative, and does not need an additional vector field $\boldsymbol{v}$ as does the Lie derivative.

MATHEMATICAL OBJECT AND SYMBOL	DEFINITION AND MAIN PROPERTIES	MEANING, EXAMPLES, AND APPLICATIONS IN THE BOOK
	where the covariant derivative of a vector field $\boldsymbol{w}$ is the (1 1) tensor field $\nabla \boldsymbol{w}$ that contracted with a vector field $\boldsymbol{v}$ is equal to the vector field $\nabla_{\boldsymbol{v}} \boldsymbol{w}$.	
Directional covariant derivative: $(\,)_{;k} v^k$, or $\nabla_{\boldsymbol{v}}$, or $\frac{D}{ds}$	Given a vector field $\boldsymbol{v}$, the **covariant derivative in the direction of** $\boldsymbol{v}$, $\nabla_{\boldsymbol{v}}$, is a derivative operator that maps $(m\ n)$ differentiable tensor fields into $(m\ n)$ tensor fields defined, in components, as $$(\nabla_{\boldsymbol{v}} \boldsymbol{T})^{i \ldots}{}_{k \ldots} = T^{i \ldots}{}_{k \ldots ; s} v^s .$$	
Torsion tensor: $\boldsymbol{T}$	*Definition independent of coordinates*: The **torsion tensor field** $\boldsymbol{T}$ is a mapping that at every point P of M and for every two vectors $\boldsymbol{v}$ and $\boldsymbol{w}$ of T_P gives the vector of T_P: $$\boldsymbol{T}(\boldsymbol{v}, \boldsymbol{w}) \equiv \nabla_{\boldsymbol{v}} \boldsymbol{w} - \nabla_{\boldsymbol{w}} \boldsymbol{v} - [\boldsymbol{v}, \boldsymbol{w}],$$ that is, $\boldsymbol{T}$ is a (1, 2) tensor field such that $$\boldsymbol{T}(\boldsymbol{v}, \boldsymbol{w}) = -\boldsymbol{T}(\boldsymbol{w}, \boldsymbol{v}).$$	
Parallel transport of a vector: $\nabla_{\boldsymbol{u}} \boldsymbol{v} = \boldsymbol{0}$	By definition, a **vector** field $\boldsymbol{v}$, defined in a neighborhood of a curve (however, $\nabla_{\boldsymbol{u}} \boldsymbol{v}$ depends only on the value of the field $\boldsymbol{v}$ along the curve) with tangent vector $\boldsymbol{u}$, is **parallel transported** along the curve if, at each point on the curve, the covariant derivative of $\boldsymbol{v}$ in the direction of $\boldsymbol{u}$ is zero: $$v^i{}_{;k} u^k = 0$$ or $$\frac{D\boldsymbol{v}}{dt} \equiv \nabla_{\boldsymbol{u}} \boldsymbol{v} = \boldsymbol{0}.$$ A **geodesic** (parametrized with an affine parameter) is a curve, with tangent vector $\boldsymbol{u}$, such that: $$\nabla_{\boldsymbol{u}} \boldsymbol{u} = \boldsymbol{0} \text{ or } u^i{}_{;k} u^k = 0$$ (see below). Since parallel transport depends, in general, on the path followed, by parallel transporting a vector around a loop back to the original point one gets a vector, in general, different from the original one. The **Riemann curvature tensor** $R^i{}_{jkl}$ (see below) describes the change of a	On a differentiable manifold M, parallel transport allows one to compare vectors and tensors at different points on M, in a way independent of the coordinate systems used. However, it depends, in general, on the path followed. It extends to curved manifolds (see below) the ordinary parallel transport of a vector on a Euclidean manifold: $\frac{d\boldsymbol{v}}{dt} = \boldsymbol{0}$ (in a coordinate system with constant g_{ik}); in fact, if $\Gamma^i_{jk} = 0$ it reduces to the standard parallel transport in Euclidean space.

MATHEMATICAL OBJECT AND SYMBOL	DEFINITION AND MAIN PROPERTIES	MEANING, EXAMPLES, AND APPLICATIONS IN THE BOOK
	vector parallel transported around an infinitesimal loop.	
Metric tensor on a smooth manifold: $\boldsymbol{g}$ or g_{ij}	A **metric tensor** $\boldsymbol{g}$ on a smooth manifold M is a continuous, symmetric, 2-covariant tensor field on M. In components, at each point P of M: $$g_{ij}(P) = g_{ji}(P).$$ The metric tensor is called **nondegenerate** if $\det(g_{ik}(P)) \neq 0$ (see below). The contravariant components g^{ij} of the metric $\boldsymbol{g}$ are defined by $g^{is} g_{sj} = \delta^i{}_j$ where $\delta^i{}_j$ is the Kronecker symbol. The indices of a tensor $T^{i\cdots}{}_{j\ldots}$ are "raised" and "lowered" by using g^{ij} and g_{ij}: $$T^{ij} = g^{js} T^i{}_s,$$ $$T_{ij} = g_{is} T^s{}_j,$$ where it is implicit summation on the repeated index s (contraction).	For any two vector fields $\boldsymbol{v}$ and $\boldsymbol{w}$ on M, the metric tensor $\boldsymbol{g}$ assigns, at each point, the real number $\boldsymbol{g}(\boldsymbol{v}, \boldsymbol{w})$, in components: $$g_{ik} v^i w^k \text{ (real number)}.$$ Therefore, at each point of M, the metric $\boldsymbol{g}$ defines a **scalar** or **inner product** for the vectors of the tangent space T_P: $$\boldsymbol{v} \cdot \boldsymbol{w} \equiv g_{ik} v^i w^k.$$ Two **vectors** $\boldsymbol{v}$ and $\boldsymbol{w}$ are called **orthogonal** if $$g_{ik} v^i w^k = 0.$$ For any vector $\boldsymbol{v}$, the metric $\boldsymbol{g}$ assigns the real number $$\|\boldsymbol{v}\|^2 = g_{ik} v^i v^k;$$ $\|\boldsymbol{v}\|$ is called the **norm of $\boldsymbol{v}$**.
Proper Riemannian manifold and **pseudo-Riemannian manifold**	A **Riemannian manifold** is a smooth manifold M with a continuous 2-covariant tensor field $\boldsymbol{g}$, the **metric** tensor, such that $\boldsymbol{g}$ is **symmetric** and **nondegenerate**, $$g_{ik} = g_{ki}$$ and $$\det(g_{ik}) \neq 0,$$ or, at each point, $$\boldsymbol{g}(\boldsymbol{v}, \boldsymbol{w}) = \boldsymbol{g}(\boldsymbol{w}, \boldsymbol{v}),$$ and $\boldsymbol{g}(\boldsymbol{v}, \boldsymbol{w}) = 0$ for every $\boldsymbol{v}$, if and only if $\boldsymbol{w} = \boldsymbol{0}$. M is called **proper Riemannian manifold** if, at each point, for every nonzero vector $\boldsymbol{v}$, $\|\boldsymbol{v}\|^2$ is always positive: $$g_{ik} v^i v^k > 0.$$ M is called **pseudo-Riemannian manifold** if, at each point, for every nonzero vector $\boldsymbol{v}$, $g_{ik} v^i v^k$ can be positive, negative, or null (the manifold has **indefinite metric**). In the spacetime,	The spacetime of general relativity is a pseudo-Riemannian manifold (the metric $\boldsymbol{g}$ is assumed to be at least C^2). Examples of pseudo-Riemannian solutions are given all through the book; in particular see sections 2.6 and 4.2.

MATHEMATICAL OBJECT AND SYMBOL	DEFINITION AND MAIN PROPERTIES	MEANING, EXAMPLES, AND APPLICATIONS IN THE BOOK
	such a **vector** is then, respectively, called **spacelike**, **timelike**, or **null** (using the convention $(-+++)$). By diagonalizing g at a point P of M, using a suitable basis for T_P, the difference between the number of positive and negative diagonal elements of g is called **signature of** g. Spacetime has signature $+2$ (or -2 using the other convention $(+---)$).	
Induced metric h_{ij} and **Extrinsic curvature tensor** K_{ij}, or **first** and **second fundamental "forms"**, of a hypersurface Σ	Given a smooth orientable manifold M^n with metric g and an orientable, timelike or spacelike, hypersurface Σ of M^n (that is, an orientable, timelike, or spacelike $(n-1)$-submanifold of M^n), with a unit vector field n normal to Σ (that is, $g_{ij}n^i n^j = \pm 1$, and for every tangent vector v to Σ: $g_{ij}n^i v^j = 0$), one may define the **first fundamental "form"** h on Σ, or **induced metric** on Σ, in components, by $$h_{ij} = g_{ij} - \epsilon\, n_i n_j,$$ where $\epsilon \equiv n \cdot n = \pm 1$, and $n_i = g_{ij}n^j$. The **second fundamental "form"** of the hypersurface Σ, or **extrinsic curvature tensor** K, is then defined, in components, by $$K_{ij} = -\frac{1}{2}\mathcal{L}_n h_{ij} = -\frac{1}{2} h^r{}_i\, h^s{}_j \mathcal{L}_n g_{rs} = -\frac{1}{2} h^r{}_i\, h^s{}_j\, (n_{r;s} + n_{s;r}),$$ where n is any vector field defined in an open neighborhood of Σ that on Σ is equal to its unit normal vector field.	See sections 2.3, 4.5, and 5.2 and chap. 5.
Lorentzian manifold	A **Lorentzian manifold** is a four-dimensional pseudo-Riemannian manifold with signature $+2$ (or -2, depending on the convention used: $(-+++)$ or $(+---)$).	The spacetime of general relativity is a Lorentzian manifold, in fact, at an event, using a suitable basis for T_P: $$\overset{(i)}{g}_{\alpha\beta} = \eta_{\alpha\beta} \equiv \mathrm{diag}(-1, +1, +1, +1),$$ with signature $+2$.

MATHEMATICAL OBJECT AND SYMBOL | **DEFINITION AND MAIN PROPERTIES** | **MEANING, EXAMPLES, AND APPLICATIONS IN THE BOOK**

Riemannian connection (or **Levi-Civita connection**) and **Christoffel symbols**, $\{ {}^{\,i}_{k\,l} \}$

On a Riemannian manifold there exists a unique linear **connection, symmetric**, that is, with zero torsion, and **metric compatible**, that is, with zero covariant derivative of the metric tensor g_{ik}; it is called the **Riemannian** (or **Levi-Civita**) **connection**.

In a general basis, the Riemannian connection Γ^i_{kl} is

$$\Gamma^i_{kl} = \frac{1}{2} g^{is} \big(g_{sk,l} + g_{sl,k} - g_{kl,s} + C_{skl} + C_{slk} - C_{kls} \big),$$

where $C_{lsk} = g_{kj}\, C^j_{ls}$ are the commutation coefficients.

In a coordinate basis the connection coefficients Γ^i_{kl} are equal to the symmetric **Christoffel symbols**, $\{ {}^{\,i}_{k\,l} \}$:

$$\left\{ \begin{matrix} i \\ k\,l \end{matrix} \right\} = \left\{ \begin{matrix} i \\ l\,k \end{matrix} \right\} \equiv \frac{1}{2} g^{is} \left(g_{sk,l} + g_{sl,k} - g_{kl,s} \right),$$

symmetric with respect to k and l.

In an orthonormal basis, where $g_{ik} = \pm\delta_{ik}$:

$$\Gamma^i_{lk} = -\Gamma^l_{ik},$$

antisymmetric with respect to i and k.

Riemann curvature tensor: $R^i{}_{kmn}$

Given a connection Γ^i_{kl} on a smooth manifold M, the **Riemann curvature tensor** $R^i{}_{kmn}$ is defined in components as

$$R^i{}_{kmn} = \Gamma^i_{kn,m} - \Gamma^i_{km,n} + \Gamma^s_{kn}\Gamma^i_{sm} - \Gamma^s_{km}\Gamma^i_{sn} - C^s_{mn}\Gamma^i_{ks}.$$

In a coordinate basis, where $C^i_{kl} = 0$, the Riemann curvature tensor can be symbolically written:

$$R^i{}_{kmn} = \begin{vmatrix} \dfrac{\partial}{\partial x^m} & \dfrac{\partial}{\partial x^n} \\ \Gamma^i_{km} & \Gamma^i_{kn} \end{vmatrix} + \begin{vmatrix} \Gamma^i_{sm} & \Gamma^i_{sn} \\ \Gamma^s_{km} & \Gamma^s_{kn} \end{vmatrix}.$$

On a Riemannian n-manifold, the change $\boldsymbol{\delta v}$ of a vector $\boldsymbol{v}$ parallel transported around an infinitesimal

On a smooth manifold M by parallelly transporting a vector from a point P to another point one gets a vector that, in general, depends from the path followed. Therefore, on a smooth manifold M by parallel transporting a vector from a point P back to the same point P following a closed loop, one does not, in general, get the same original vector. One can easily see this effect on two-dimensional curved surfaces, for example, on the two-sphere of fig. 2.1. On two-dimensional surfaces, the change of a vector parallel transported back to the original point P, around an infinitesimal closed loop, depends

MATHEMATICAL OBJECT AND SYMBOL	DEFINITION AND MAIN PROPERTIES	MEANING, EXAMPLES, AND APPLICATIONS IN THE BOOK

closed quadrilateral, determined by two infinitesimal displacements δx^i and $\widetilde{\delta x}^i$ (infinitesimal "quadrilateral" which is closed apart from higher order infinitesimals in δx), is given by the Riemann curvature tensor:

$$\delta v^i = -R^i{}_{kmn} v^k \delta x^m \widetilde{\delta x}^n.$$

A **manifold** is **locally flat** in a neighborhood U if and only if the Riemann curvature tensor is zero in U:

$$R^i{}_{kmn} = 0.$$

Since, unlike ordinary partial derivatives, covariant derivatives, in general, do not commute, on a smooth manifold one may define the Riemann curvature tensor as the **commutator of the covariant derivatives**:

$$u^i{}_{;kl} - u^i{}_{;lk} = R^i{}_{slk} u^s - T^s_{kl} u^i{}_{;s}.$$

One may also derive this relation from the coordinate independent definition of Riemann tensor.

Definition independent of coordinates:

The **Riemann curvature tensor** is a (1 3) tensor field that to each set of three vector fields $\boldsymbol{u}$, $\boldsymbol{v}$, and $\boldsymbol{w}$ and one 1-form (field) $\boldsymbol{\theta}$ assigns the function defined by

$$\boldsymbol{R}(\boldsymbol{u}, \boldsymbol{v}, \boldsymbol{w}, \boldsymbol{\theta}) \equiv \boldsymbol{\theta}\left[(\boldsymbol{\nabla}_{\boldsymbol{u}} \boldsymbol{\nabla}_{\boldsymbol{v}} - \boldsymbol{\nabla}_{\boldsymbol{v}} \boldsymbol{\nabla}_{\boldsymbol{u}} - \boldsymbol{\nabla}_{[\boldsymbol{u},\boldsymbol{v}]})\boldsymbol{w}\right],$$

where the expression between square parentheses is a vector, and, at each point, the 1-form $\boldsymbol{\theta}$ maps this vector into a real number.

Sectional curvature:

Given two linearly independent vectors $\boldsymbol{v}$ and $\boldsymbol{w}$, at a point P of a Riemannian manifold M^n, the scalar invariant

$$K_\pi = \frac{R_{ikmn} v^i w^k v^m w^n}{(g_{im} g_{kn} - g_{in} g_{km}) v^i w^k v^m w^n}$$

is called sectional curvature of M at P relative to $\boldsymbol{v}$ and $\boldsymbol{w}$, and is the Gaussian curvature of the two-surface of M at P generated by all the geodesics through

on the Gaussian curvature of the surface at P and on the path followed (the **Gaussian curvature** is given by the product of the two extreme curvatures at P, that is, the maximum and minimum curvatures of all the curves generated by intersection of the surface with the normal planes at P).

To extend the definition of curvature of two-manifolds to manifolds of higher dimensionality one introduces the Riemann curvature tensor. Applications of the curvature tensor are all through the book.

MATHEMATICAL OBJECT AND SYMBOL	DEFINITION AND MAIN PROPERTIES	MEANING, EXAMPLES, AND APPLICATIONS IN THE BOOK

P with tangent vector at P equal to any linear combination (with real coefficients) of $\boldsymbol{v}$ and $\boldsymbol{w}$. For a space with constant curvature, that is, $K_\pi(P)$ independent of P and π (tangent two-plane at P determined by $\boldsymbol{v}$ and $\boldsymbol{w}$), we have

$$R_{ikmn} = K(g_{im}g_{kn} - g_{in}g_{km})$$

(see section 4.2).

Geodesic deviation equation:

In general relativity the Riemann tensor represents the spacetime curvature (in the spacetime, it has, in general, 20 independent components) and one can determine it by measuring the relative covariant accelerations between test particles and by using the geodesic deviation equation. The **geodesic deviation equation** relates the spacetime geometry with observable quantities:

$$\frac{D^2(\delta x^\alpha)}{ds^2} = -R^\alpha{}_{\beta\mu\nu}u^\beta \delta x^\mu u^\nu.$$

This equation generalizes to an n-dimensional manifold the classical **Jacobi equation** for the distance between two geodesics on a two-dimensional surface:

$$\frac{d^2y}{d\sigma^2} + Ky = 0,$$

where y is the distance between the two geodesics, σ is the arc of the base geodesic, and $K(\sigma)$ is the **Gaussian curvature** of the surface (see section 2.5).

Properties of the Riemann tensor:

The Riemann tensor has the following properties:

$$R_{ikmn} = R_{[ik][mn]}$$

$$R_{i[kmn]} = 0$$

$$(R_{ikmn} = R_{mnik} \text{ and } R_{[ikmn]} = 0),$$

that is, the Riemann tensor is antisymmetric on the first pair of indices, antisymmetric on the last pair of indices, and it satisfies the **first Bianchi identity**, that is, the sum of the cyclic permutations of its last three indices is zero. It follows that it is symmetric under exchange of the first pair of indices with the last pair of indices and that its completely antisymmetric part vanishes. Therefore, on a Riemannian manifold, the total number of independent components of the Riemann tensor is

$$\frac{n^2(n^2-1)}{12}.$$

MATHEMATICAL OBJECT AND SYMBOL	DEFINITION AND MAIN PROPERTIES	MEANING, EXAMPLES, AND APPLICATIONS IN THE BOOK
	Furthermore, it satisfies the **second Bianchi identity** (see § 2.8) $$R^{i}{}_{k[mn;l]} = 0.$$ One can write the Riemann tensor R_{ikmn} as a function of the Ricci tensor R_{ik} and the Weyl tensor C_{ikmn} (see below).	
Ricci tensor: R_{ik}	In components, the **Ricci tensor** R_{ik} is defined by contraction on the first and third indices of the Riemann tensor: $$R_{ik} \equiv R^{s}{}_{isk}.$$ The Ricci tensor is symmetric: $$R_{ik} = R_{ki},$$ and in the spacetime has, in general, 10 independent components.	In general relativity, one can write the **vacuum Einstein field equation** by equating the Ricci tensor to zero: $$R_{\alpha\beta} = 0.$$ Examples of solutions with zero Ricci tensor are given in section 2.6.
Ricci or **curvature scalar**: R	In components, the **Ricci curvature scalar** R is defined by contraction of the indices of the Ricci tensor: $$R \equiv R^{r}{}_{r} = R^{rs}{}_{rs}.$$	
Weyl tensor: C_{ikmn}	In components, the **Weyl tensor** is defined by $$C_{ikmn} = R_{ikmn} + g_{i[n}R_{m]k} + g_{k[m}R_{n]i} + \frac{1}{3}Rg_{i[m}g_{n]k}.$$ In the spacetime, the Weyl tensor has, in general, 10 independent components. For $n = 1$, $n = 2$, and $n = 3$ the Weyl tensor is zero.	The Weyl tensor is also called **conformal tensor**. Indeed, the components of the Weyl tensor are invariant: $C'_{ikmn}(\mathbf{g}') = C_{ikmn}(\mathbf{g})$, for **conformal transformations**: $$g'_{ik} = \phi^2 g_{ik},$$ where ϕ is a never null differentiable function on M. Therefore, the Weyl tensor is zero *if* (and, for $n \neq 3$, *only if*) the **manifold** is **conformally flat**, that is, if there is a conformal transformation such that, on M, $$\tilde{\eta}_{ik} = \phi^2 g_{ik},$$ where $\tilde{\eta}_{ik}$ is a diagonal matrix with elements equal to ± 1.
Einstein tensor: G_{ik}	In components, the **Einstein tensor** is defined by $$G_{ik} \equiv R_{ik} - \frac{1}{2}Rg_{ik}.$$	In general relativity the **Einstein field equation** is defined by the Einstein tensor $G_{\alpha\beta}$ equal to the energy-momentum tensor of matter and fields $T_{\alpha\beta}$: $$G_{\alpha\beta} = 8\pi T_{\alpha\beta}.$$

MATHEMATICAL OBJECT AND SYMBOL	DEFINITION AND MAIN PROPERTIES	MEANING, EXAMPLES, AND APPLICATIONS IN THE BOOK

Examples of solutions with zero Einstein tensor $G_{\alpha\beta} = 0$ (and zero Ricci tensor) are given in section 2.6. Examples of solutions with $G_{\alpha\beta} \neq 0$ are given in chapter 4.

Geodesic

On a Riemannian manifold one can define a **geodesic** as the **extremal curve** $x^i(t)$—or **critical point**—for the integral of the squared interval E_a^b between two events $a^i = x^i(t_a)$ and $b^i = x^i(t_b)$:

$$E_a^b(x(t)) \equiv \frac{1}{2}\int_{t_a}^{t_b} g_{ik}(x(t))\frac{dx^i}{dt}\frac{dx^k}{dt}\,dt,$$

that is, $\delta E_a^b(x^i(t)) = 0$, for any first order variation $\delta x^i(t)$ of the C^∞ curve $x^i(t)$ such that $\delta x^i(t_a) = \delta x^i(t_b) = 0$ (keeping the end points fixed). Taking the variation of E_a^b, one then gets

$$\frac{d^2x^i}{dt^2} + \left\{ {i \atop kl} \right\}\frac{dx^k}{dt}\frac{dx^l}{dt} = 0.$$

The geodesic equation keeps this form for every transformation of the parameter t of the type $s = ct + d$, where $c \neq 0$ and d are two constants; when the geodesic equation has this form, t is called **affine parameter**. In particular $s(p) = L_a^p(x) = \int_a^p \sqrt{\pm g_{ik}(x(p'))\frac{dx^i}{dp'}\frac{dx^k}{dp'}}\,dp'$ is the **arc-length** (+ sign for spacelike geodesics and − sign for timelike geodesics), where p is a parameter along the curve. When $s = p +$ constant, the geodesic is said to be parametrized by arc-length, then $g_{ik}\frac{dx^i}{ds}\frac{dx^k}{ds} = \pm 1$. One sometimes writes the "norm" of an infinitesimal displacement dx^i as

$$ds^2 = g_{ik}dx^i dx^k.$$

On a proper Riemannian manifold there is a variational principle that gives the geodesic equation parametrized with any parameter. This variational principle defines a geodesic as the **extremal curve**

The geodesic is an extension to curved manifolds of the concept of straight line of Euclidean space, $\Re^n$. One finds:

(1) A geodesic is an extremal curve for the length between two points (see below). In particular, on a proper Riemannian manifold any point P has a neighborhood U such that for any point Q of U there is a unique geodesic in U joining the two points P and Q; this geodesic is the shortest curve between the two points.

(2) A geodesic is a curve with tangent vector transported parallel to itself all along the curve.

(3) In general relativity, the geodesic equation is the equation of motion of test particles and photons.

See section 2.4.

MATHEMATICAL OBJECT AND SYMBOL	DEFINITION AND MAIN PROPERTIES	MEANING, EXAMPLES, AND APPLICATIONS IN THE BOOK
	for the length $L^a_b(x(p))$ (see § 2.4): $L^b_a(x(p)) = \int_{p_a}^{p_b} \sqrt{g_{ik}(x(p)) \frac{dx^i}{dp} \frac{dx^k}{dp}}\, dp.$ On a proper Riemannian manifold, a curve $x^i(p)$ is an extremal curve for E^b_a if and only if is an extremal curve for the length L^b_a and is parametrized proportionally to arc-length: $\frac{d^2s}{dp^2} = 0$. It is possible to reparametrize a curve that is extremal for the length, with $\frac{dx^i}{dp} \neq 0$ everywhere, to give an extremal curve for E^b_a (see § 2.4).	
Exponential mapping: $\exp_P(\boldsymbol{v})$	At any point P of a Riemannian manifold M, for any vector $\boldsymbol{v}$ of the tangent vector space $T_P(M)$, there exists a unique geodesic $\boldsymbol{x}(t)$ (or $\boldsymbol{x}_v(t)$) with $t \in (-\varepsilon, \varepsilon)$, which satisfies $\boldsymbol{x}(0) = \boldsymbol{P}$ and $\frac{d\boldsymbol{x}}{dt}(0) = \boldsymbol{v}$. The **exponential mapping** of $\boldsymbol{v}$ is then defined: $\exp_P(\boldsymbol{v}) \equiv \boldsymbol{x}_v(1)$ (*if* $\boldsymbol{x}_v(1) \equiv \boldsymbol{x}(1)$ is defined for $t = 1$). Therefore: $\boldsymbol{x}_v(t) = \boldsymbol{x}_{tv}(1) = \exp_P(t\boldsymbol{v})$ and $\exp_P(s\boldsymbol{v}) \exp_P(t\boldsymbol{v}) = \exp_P((s+t)\boldsymbol{v}).$ Given a vector $\boldsymbol{v}$, if $\exp_P \boldsymbol{v}$ is defined for every t the geodesic is called **complete geodesic**; if $\exp_P \boldsymbol{v}$ is defined for every $\boldsymbol{v} \in T_P(M)$ the **manifold** is called **geodesically complete at $\boldsymbol{P}$**. If $\exp_P \boldsymbol{v}$ is defined for every P of M and for every $\boldsymbol{v} \in T_P(M)$ the **manifold** is called **geodesically complete** (every geodesic can be extended to any value of its affine parameter).	The exponential mapping is a mapping from the tangent vector space $T_P(M)$, at a point P, into the manifold M, that to a vector $\boldsymbol{v} \in T_P(M)$ assigns the point Q of M at $t = 1$ distance from P along the geodesic through P with tangent vector $\boldsymbol{v}$.
Geodesically complete manifold	A **manifold** is called **geodesically complete** if one can extend every geodesic to any arbitrarily large value of its affine parameter, that is, one can extend it to an infinite geodesic.	Counterexamples: See the discussion of singularities in section 2.9.

MATHEMATICAL OBJECT AND SYMBOL	DEFINITION AND MAIN PROPERTIES	MEANING, EXAMPLES, AND APPLICATIONS IN THE BOOK
Fermi-Walker transport	Given a timelike curve $x^\alpha(s)$ with tangent vector (four-velocity) u^α, a **vector** v^α is said to be **Fermi-Walker transported** along the curve if $$\frac{Dv^\alpha}{ds} \equiv v^\alpha{}_{;\beta}u^\beta = u^\alpha(a_\beta v^\beta)$$ where $a^\beta \equiv \frac{Du^\beta}{ds} = u^\beta{}_{;\alpha}u^\alpha$ is the four-acceleration of the curve.	In general relativity, the Fermi-Walker transport of the spin (angular momentum vector) of a test gyroscope determines the orientation of the spin of the test gyroscope all along the world line of an accelerated observer carrying the test gyroscope (see § 3.4.3 and chap. 6). If the four-acceleration is zero, the Fermi-Walker transport is just the parallel transport along the corresponding geodesic.
Integration on a smooth manifold: Partition of unity	Given a manifold M, a **partition of unity** $\{f_\alpha\}$ is a collection of C^∞ functions f_α on M, such that (1) $0 \leq f_\alpha(P) \leq 1$, at each point P of M and for each f_α. (2) At each point P of M there are only a finite number of functions: $f_\alpha(P) \neq 0$. (3) $\sum_\alpha f_\alpha(P) = 1$ at each point P of M. A **partition of unity** $\{f_\alpha\}$ is called **subordinate to an open covering** $\{U_\alpha\}$ if, for each f_α, the smallest closed set outside which f_α is zero, called the **support** of f_α, is contained in some U_α. On a paracompact manifold, given a locally finite open covering, it is possible to prove the existence of a partition of unity subordinate to the covering, such that the support of each f_α is compact.	A partition of unity is used to define integration on smooth manifolds, to sum the results of the integrations on each chart. Another theory (equivalent) of integration of forms using *chains* is briefly described in § 2.8.
Integration of a form on a chart of a manifold	Given a paracompact, smooth, orientable, and oriented manifold M^n and given an n-form $\boldsymbol{\theta}$ with *compact support* (that is, the smallest closet set outside which $\boldsymbol{\theta}$ is zero, the support of $\boldsymbol{\theta}$, is compact) contained in the domain U of a chart (U, ϕ), the **integral of $\boldsymbol{\theta}$**, with components $\theta_{i_1 \cdots i_n}$, is defined $$\int_U \boldsymbol{\theta} \equiv \int_{\phi(U)} \theta_{1\cdots n}(\phi^{-1}(x^1, \cdots, x^n))\, dx^1 \cdots dx^n$$ where $\phi(U)$ is an open set of $\Re^n$ and the integral on $\phi(U)$ is a standard multiple integral on $\Re^n$. This definition does not depend on the local coordinate system	This definition extends to a neighborhood of a smooth manifold the ordinary definition of multiple integral of a function (0-form) and of a p-form on a neighborhood of the Euclidean space $\Re^n$.

MATHEMATICAL OBJECT AND SYMBOL	DEFINITION AND MAIN PROPERTIES	MEANING, EXAMPLES, AND APPLICATIONS IN THE BOOK
	chosen (assuming the same orientation of the coordinate systems, that is, positive Jacobian determinant).	
Integration of a form on a manifold	Given a paracompact, smooth, orientable manifold M with a locally finite oriented atlas $\{U_\alpha, \phi_\alpha\}$ and a partition of unity $\{f_\alpha\}$ subordinate to $\{U_\alpha\}$, the **integral of** $\boldsymbol{\theta}$, with compact support (for example, when M is a compact manifold), **on** $\boldsymbol{M}$ is defined by $$\int_M \theta \equiv \sum_\alpha \int_{U_\alpha} f_\alpha \theta = \sum_\alpha \int_{\phi_\alpha(U_\alpha)} f_\alpha \theta_{1\cdots n} (\phi_\alpha^{-1}(x_1, \cdots, x_n)) dx^1 \cdots dx^n$$ where the integrals on the open sets $\phi_\alpha(U_\alpha)$ of $\Re^n$ are just ordinary multiple integrals on $\Re^n$. This definition does not depend on the partition of unity and on the atlas chosen.	This definition extends to smooth manifolds the ordinary definition of multiple integral of a function (0-form) and of a p-form on the Euclidean space $\Re^n$.
Stokes theorem	Given a paracompact, smooth, oriented manifold M^n with boundary ∂M^{n-1} with the induced orientation, and given an $(n-1)$-form $\boldsymbol{\theta}$, on M^n, with compact support, one has $$\int_M d\theta = \int_{\partial M} \theta,$$ **Stokes theorem** (see § 2.8).	Particular cases of the Stokes theorem are the well-known **divergence theorem** (**Ostrogradzky-Green formula** or **Gauss theorem**): $$\int_V \nabla \cdot \boldsymbol{W} d^3V = \int_{\partial V = S} \boldsymbol{W} \cdot \boldsymbol{n}\, d^2S,$$ and the **Riemann-Ampère-Stokes formula**: $$\int_S (\nabla \times \boldsymbol{W}) \cdot \boldsymbol{n}\, d^2S = \int_{\partial S = l} \boldsymbol{W} \cdot d^1\boldsymbol{l},$$ where d^3V, d^2S, and $d^1\boldsymbol{l}$ are the standard volume, surface, and line (with positive orientation) elements, and $\boldsymbol{n}$ is the outward (with positive orientation) unit normal to the surface S.
Bundle: (E, B, π)	A **bundle** (E, B, π) is a triple of two topological spaces, E and B, and a continuous surjective mapping or	

MATHEMATICAL OBJECT AND SYMBOL	DEFINITION AND MAIN PROPERTIES	MEANING, EXAMPLES, AND APPLICATIONS IN THE BOOK
	projection π from E onto B. E is called the **total space**, B the **base space**.	
Fibre Bundle: (E, B, π, G)	A **fibre bundle** (E, B, π, G) is a bundle with a space F called **typical fibre**, a **topological group** G (a group with a topology and with continuous group mappings: ab and a^{-1}) of homeomorphisms of F onto itself called **structural group**, and an open covering $\{U_\alpha\}$ of the base space B, such that • The **bundle** is locally homeomorphic to a product bundle, that is, it is **locally trivial**: for every U_α, $\pi^{-1}(U_\alpha)$ is homeomorphic to the topological product $U_\alpha \times F$. The homeomorphism, φ_α, of $\pi^{-1}(U_\alpha)$ onto $U_\alpha \times F$ is of the form $\varphi_\alpha(e) = (\pi(e), h_\alpha^P(e))$, where e is an element of $E_P = \pi^{-1}(P)$, $P = \pi(e)$ is an element of B, and h_α^P is a homeomorphism from the **fibre** $\pi^{-1}(P)$ onto F. • Given an element P of B contained in two open sets U_α and U_β, the homeomorphism $h_\alpha^P((h_\beta^P)^{-1})$ of F onto F is an element of the group G, where h_α^P and h_β^P are homeomorphisms from $\pi^{-1}(P)$ onto F.	The concept of fibre bundle is a generalization of the concept of topological product. Locally, a fibre bundle is a topological product of two topological spaces, however, globally is not necessarily a topological product. Examples: The **product bundle** $(S^1 \times I, S^1, \pi)$, the total space is a cylinder, $S^1 \times I$, topological product of a circle S^1 with a real open interval I. The cylinder is a **globally trivial fibre bundle**. A Möbius strip is locally the topological product of a proper subset of a circle S^1 with a real open interval I. However, globally is not a topological product. The Möbius strip is a **nontrivial fibre bundle** (only locally is a product bundle), with fibre I.
	• The **transition functions**, that is, the mappings that for each element P of the intersection of U_α and U_β assign the element $h_\alpha^P((h_\beta^P)^{-1})$ of the group G are continuous.	The structural group G with its elements $h_\alpha^P((h_\beta^P)^{-1})$ gives information about the structure of the fibre bundle. For example, if G has only one element the bundle is trivializable.
Vector bundle	A **vector bundle** is a fibre bundle such that the typical fibre F is a vector space and the group G is the linear group.	Example: The Möbius strip is a vector bundle over S^1.
Tangent bundle: $(T(M), M, \pi, \mathrm{GL}(n, \Re))$	The **tangent bundle** $(T(M), M, \pi, \mathrm{GL}(n, \Re))$ is a fibre bundle such that the base space M is an n-dimensional differentiable manifold; the total space $T(M)$ is the space of all the pairs $(P, \mathbf{v}_P)$, where P is a point of M and $\mathbf{v}_P$ is a vector of the tangent space $T_P(M)$ at P; the projection π is $\pi(P, \mathbf{v}_P) = P$; the fibre at P is $T_P(M)$; the typical fibre F is $\Re^n$; and the structural group is the group of the linear bijective mappings of $\Re^n$ onto $\Re^n$:	Using the language of fibre bundles, a C^∞ **pseudo-Riemannian metric** $\boldsymbol{g}$ on a differentiable manifold M^4 for the tangent bundle $(T(M^4), M^4, \pi, \mathrm{GL}(4, \Re))$ is a C^∞ mapping $\boldsymbol{g}$ (that is, $\boldsymbol{g}$ is C^∞ on M) which to each $P \in M^4$ assigns an inner product $\boldsymbol{g}_P$ on the fibre $\pi^{-1}(P) = T_P(M)$.

MATHEMATICAL OBJECT AND SYMBOL	DEFINITION AND MAIN PROPERTIES	MEANING, EXAMPLES, AND APPLICATIONS IN THE BOOK
	$\mathrm{GL}(n, \mathfrak{R})$, with matrix representation given by the set of all the $n \times n$ nonsingular matrices with real elements.	
Bundle of linear frames: $(L(M), M, \pi, \mathrm{GL}(n, \mathfrak{R}))$ and **bundle of orthonormal frames**: $(O(M), M, \pi, \mathrm{O}(n))$	The **bundle of linear frames** $(L(M), M, \pi, \mathrm{GL}(n, \mathfrak{R}))$ is a fibre bundle such that the total space $L(M)$ is the space of all the pairs $(P, \{\boldsymbol{e}_a\})$, where P is a point of the n-dimensional differentiable manifold M and $\{\boldsymbol{e}_a\}$ is an **ordered basis (linear frame)** for the tangent vector space $T_P(M)$, that is, an ordered set of n (nonzero) linearly independent vectors $\{\boldsymbol{e}_a\}$ of $T_P(M)$; and the projection π is $\pi(P, \{\boldsymbol{e}_a\}) = P$. If there is a Riemannian metric $\boldsymbol{g}$ on M, the **bundle of orthonormal frames** $(O(M), M, \pi, \mathrm{O}(n))$ is a fibre bundle with total space $O(M)$, of all the pairs $(P, \{\boldsymbol{n}_a\})$, where $\{\boldsymbol{n}_a\}$ is an orthonormal basis with respect to $\boldsymbol{g}$ for $T_P(M)$ and with $\mathrm{O}(n)$ the **n-dimensional orthogonal group**, subgroup of $\mathrm{GL}(n, \mathfrak{R})$.	

Symbols and Notations

The general relativity sign convention and notation of this book follow the text *Gravitation* by Misner, Thorne, and Wheeler. In particular:

Greek indices are four-dimensional and run from 0 to 3.

Latin indices run from 1 to 3.

$\eta_{\alpha\beta} = \text{diag}(-1, +1, +1, +1)$.

Riemann tensor $R^{\alpha}{}_{\beta\gamma\delta}$ and Einstein tensor $G_{\alpha\beta}$ are "positive" (see §§ 2.2 and 2.3, and mathematical appendix).

Geometrized units are used, mass and intervals of time are measured in units of length, that is, intervals of time Δt are multiplied by c: $c\,\Delta t$, and masses M by $\frac{G}{c^2}$: $\frac{GM}{c^2}$. Gravitational constant and speed of light are equal to unity, $G = c = 1$.

In this book the symbol $\partial^{\alpha}_{\beta'}$ means partial derivative of x^{α} with respect to x'^{β}: $\partial^{\alpha}_{\beta'}{}^{\gamma'\cdot\cdot}_{\delta\cdot\cdot} \equiv \frac{\partial x^{\alpha}}{\partial x'^{\beta}} \frac{\partial x^{\gamma'}}{\partial x^{\delta}} \cdot \cdot$.

$\dot{\Omega}$ means angular frequency or rate of change of the nodal longitude, whereas Ω is the nodal longitude or an angle in radians.

SYMBOL	NAME	SECTION
$\boldsymbol{H}$	gravitomagnetic field	1.1, and 6.1
$\boldsymbol{J}$	angular momentum	1.1, 6.1, and 2.6
$R^{\alpha}{}_{\beta\gamma\delta}$	Riemann curvature tensor	2.2, 2.6, 2.8, 4.2, 6.11, and mathematical appendix
$g_{\alpha\beta}$ and $g^{\alpha\beta}$	metric tensor	2.2, and mathematical appendix
$\Gamma^{\alpha}{}_{\beta\gamma} \equiv \Gamma^{\alpha}_{\beta\gamma}$	connection coefficients	2.2, 2.8, and mathematical appendix
$\partial^{\alpha}_{\beta'} \equiv \frac{\partial x^{\alpha}}{\partial x'^{\beta}}$	partial derivative of x^{α} with respect to x'^{β}	2.2, and mathematical appendix
$\partial^{\alpha\cdots\gamma'}_{\beta'\cdots\delta} \equiv \frac{\partial x^{\alpha}}{\partial x^{\beta'}} \cdots \frac{\partial x^{\gamma'}}{\partial x^{\delta}}$	partial derivatives	2.2, and mathematical appendix
$(\;)_{;\alpha}$ or ∇_{α}	covariant derivative with respect to x^{α}	2.2, and mathematical appendix
$F^{\alpha\beta}$	electromagnetic field tensor	2.3, 2.8, 6.11, and mathematical appendix
A^{α}	electromagnetic vector potential	2.3, 2.8, and mathematical appendix

SYMBOL	NAME	SECTION
L	Lagrangian	2.3
$\mathcal{L}$	Lagrangian density	2.3
$R_{\alpha\beta}$	Ricci tensor	2.3, and 2.6, 4.2, and mathematical appendix
R	Ricci curvature scalar	2.3, 4.2, and mathematical appendix
∂ and ∂M	boundary operator and boundary of manifold M	2.8, and mathematical appendix
$\boldsymbol{n}$	unit vector field normal to a hypersurface	2.3, and mathematical appendix
$K_{\alpha\beta}$, or $\boldsymbol{K}$	second fundamental form or extrinsic curvature	2.3, 5.2, and mathematical appendix
$\pounds_{\boldsymbol{v}}$	Lie derivative with respect to $\boldsymbol{v}$	2.3, 4.1, 4.4, and mathematical appendix
$T_{\alpha\beta}$	energy-momentum tensor	2.3, 2.6, and 4.2
$\chi = \frac{8\pi G}{c^4}$	factor converting energy-momentum density to spacetime curvature	2.3
$\{ {}^{\alpha}_{\beta\gamma} \}$	Christoffel symbols	2.3, 2.8, and mathematical appendix
u^α	four-velocity	2.3, and 4.2
ε	energy density	2.3, and 4.2
ε_c	cosmological critical density	4.2
p	isotropic pressure	2.3, and 4.2
q^α	energy flux vector, or heat flow vector	2.3
$\pi^{\alpha\beta}$	anisotropic pressure tensor or viscous stress tensor	2.3
η	coefficient of shear viscosity	2.3, and 6.3
ζ	coefficient of bulk viscosity	2.3
$\eta^{\alpha\beta}$	Minkowski tensor	2.6, and 2.8
$\delta^\alpha{}_\beta$	Kronecker delta	2.2, and 2.8
g	determinant of $g_{\alpha\beta}$	2.3
$(\ )_{,\alpha}$ or ∂_α	partial derivative with respect to x^α	2.2
$G^{\alpha\beta}$	Einstein tensor	2.3, 2.8, and mathematical appendix
ξ^α or $\boldsymbol{\xi}$	Killing vector	2.3, 4.2, 4.4, and mathematical appendix
s	arc-length	2.4, and mathematical appendix

SYMBOL	NAME	SECTION
τ	proper time	2.4
$\frac{D}{ds}$, or ∇, or $(\,)_{;\beta}u^{\beta}$	covariant derivative along u^{β}	2.4, and mathematical appendix
$C_{\alpha\beta\gamma\delta}$	Weyl tensor	2.5, and mathematical appendix
M^n	n-dimensional manifold	mathematical appendix
M^4	four-dimensional manifold	2.2
ds^2	square of spacetime interval	2.2, and 2.6
M	mass of central body	2.6, and 2.7
$\overset{(c)}{M}$	conformal density of gravitational-wave effective energy	5.3
Q	charge of a central body, charge	2.6, and 2.7
Q	quality factor of a resonant antenna	3.6.1
$\overset{(c)}{Q}$	conformal density of mass-energy	5.2, and 5.3
e	charge of a particle	2.5
e	orbital eccentricity	6.1, and 6.7
P^{α}	four-momentum	2.7
$J^{\alpha\beta}$	angular momentum tensor	2.7, and 6.10
$t^{\alpha\beta}$	pseudotensor for the gravitational field	2.7
$E \equiv P^0$	energy	2.7
$T^{\alpha\beta}_{\text{eff}}$	effective energy-momentum pseudotensor	2.7
T_{eff}	effective temperature of a resonant detector	3.6.1
j^k	momentum density vector, or mass-energy current-density vector	2.2, 2.7, and 5.2
j^{α}	electric charge current-density four-vector	2.8
$\mathfrak{R}^n$	n-dimensional Euclidean space	2.8
$\mathfrak{R}^3$	ordinary three-dimensional Euclidean space	2.8
∂	boundary	2.8, and mathematical appendix
$T_{[\alpha\cdots\beta]}$	antisymmetrization of a tensor $T_{\alpha\cdots\beta}$	2.8, and mathematical appendix
$\boldsymbol{\theta} \wedge \boldsymbol{\omega}$	exterior product or wedge product of $\boldsymbol{\theta}$ with $\boldsymbol{\omega}$	2.8, and mathematical appendix

SYMBOL	NAME	SECTION
$\boldsymbol{d}$	exterior derivative	2.8, and mathematical appendix
$\epsilon^{\alpha\beta\gamma\delta}$ and $\epsilon_{\alpha\beta\gamma\delta}$	Levi-Civita pseudotensor	2.8, and 6.11
$[\alpha\beta\gamma\delta]$	alternating symbol	2.8
$\delta^{\alpha\beta\gamma\lambda}{}_{\mu\nu\rho\sigma}$	alternating symbol	2.3, and 2.8
$\delta^{\alpha\beta\gamma}{}_{\mu\nu\rho}$	alternating symbol	2.8
$\delta^{\alpha\beta}{}_{\mu\nu}$	alternating symbol	2.8
$d^2S^{\alpha\beta}$	two-dimensional surface element	2.8
$d^3\Sigma^{\alpha\beta\gamma}$	three-dimensional hypersurface element	2.8
$d^4\Omega^{\alpha\beta\gamma\delta}$	four-dimensional volume element	2.8
$^*\boldsymbol{\theta}$	dual of a form $\boldsymbol{\theta}$	2.8
$^*\boldsymbol{v}$	dual of a polyvector $\boldsymbol{v}$	2.8
$d^3\Sigma_\alpha$	dual of three-dimensional hypersurface element	2.3, and 2.8
$d^4\Omega$	dual of four-dimensional volume element	2.8
d^4x	dual of four-dimensional coordinate volume element	2.3, and 2.8
$d^3\Sigma_0$ or d^3V	dual of three-dimensional coordinate element of spacelike hypersurface	2.8
$\boldsymbol{\omega}^\alpha{}_\beta$	connection 1-forms	2.8
$\boldsymbol{\Omega}^\alpha{}_\beta$	curvature 2-forms	2.8
$\boldsymbol{D}$	exterior covariant derivative	2.8
$T^\alpha_{[\beta\gamma]}$, and $\boldsymbol{\Theta}^\alpha$	torsion tensor, and torsion 2-forms	2.8, and mathematical appendix
$^\star\boldsymbol{T}$	left star operator on $\boldsymbol{T}$	2.8
$\boldsymbol{T}^\star$	right star operator on $\boldsymbol{T}$	2.8
$R_{\alpha\beta\gamma\delta}\ R^{\alpha\beta\gamma\delta}$	Kretschmann invariant	2.9, 6.11
Λ	cosmological constant	2.9, 4.2, and 4.6
λ and $\lambda\!\!\!^{-} \equiv \frac{\lambda}{2\pi}$	wavelength and reduced wavelength	3.6.2
λ	range of hypothetical fifth force	3.2.1
$h_{\alpha\beta}$	perturbation of metric tensor	2.10
$h_{\alpha\beta}$	space-projection tensor	4.2, and 4.5
$\Box$	d'Alambertian operator	2.10, and 3.7
T	trace of energy-momentum tensor	2.10

SYMBOL	NAME	SECTION
T	trace of extrinsic curvature tensor	4.3, and 5.2
$\boldsymbol{h}^{\mathrm{TT}}$	tensor $\boldsymbol{h}$ in the transverse-traceless gauge	2.10
A_+, $A_\times$	amplitudes of components of polarized gravitational waves	2.10
$\text{\it{I}}\!\!\!-_{ij}$	reduced quadrupole moment	2.10
G	Newtonian gravitational constant	2.3, 3.2.1, and 3.2.3
U	classical gravitational potential	2.3, 3.2.2, and 3.7
α	coupling constant of hypothetical fifth force	3.2.1
GPS	Global Positioning System	3.2.2
DSN	Deep Space Network	3.2.2
ϕ	scalar field of Jordan-Brans-Dicke theory	3.2.3
ω	Brans-Dicke theory parameter	3.2.3
$\dot{\omega}$	rate of pericenter shift	3.5.1
$\dot{\tilde{\omega}}$	rate of change of longitude of pericenter (periastron)	6.1
$\dot{\Omega}$	rate of change of angle Ω (angular frequency)	3.2.2, 6.9, and 6.12
$\dot{\Omega}$	rate of change of nodal longitude	6.1, and 6.7
Ω	nodal longitude	6.1
Ω	angle	3.2.2
Ω	gravitational binding energy	3.2.5
$\Omega_\circ$	density parameter for universe-wide average mass-energy density relative to critical density	4.2
η	Nordtvedt effect parameter	3.2.5
β	PPN parameter	3.4.1, and 3.7
γ	PPN parameter	3.4.1, and 3.7
VLBA	Very Long Baseline Array	3.4.1
VLBI	Very Long Baseline Interferometry	3.4.1
LLR	Lunar Laser Ranging	3.2.5, and 3.4.1
S^α	spin vector of a particle	3.4.3, and 6.10
SLR	Satellite Laser Ranging	3.4.1, and 6.7
μ	gravitomagnetic field parameter	3.4.3, and 6.9
$\boldsymbol{\lambda}_{(\mu)}$	orthonormal tetrad or vierbein	3.4.3
$\Lambda^\alpha{}_\beta$	matrix of post-Galilean transformations	3.4.3

SYMBOL	NAME	SECTION
$\oplus$	Earth	3.2.5
$\odot$	Sun	3.2.5
☽	Moon	3.2.5
☿	Mercury	3.5.1
J_2	Quadrupole moment coefficient	3.5.1, and 6.7, see also astronomical constants
$\sum_0$	"resonance integral of the absorption cross section" of a gravitational-wave detector	3.6.1
$\sum$	spacelike hypersurface	2.7, and mathematical appendix
U_{ik}, V_i, χ, W_i, $\phi_1, \phi_2, \phi_3, \phi_3,$ ϕ_W, A, B	PPN potentials	3.7
$\alpha_1, \alpha_2, \alpha_3$, $\zeta_1, \zeta_2, \zeta_3, \zeta_4, \xi$	PPN parameters	3.7
$R(t)$	expansion factor in Friedmann cosmological models	4.2
K_Σ	sectional curvature	4.2
k	curvature index ($= -1, 0,$ or $+1$)	4.2
ω^α	rotation or vorticity vector	4.2, and 4.5
$\sigma^{\alpha\beta}$	shear tensor	4.2, 4.5, and 2.3
$\omega^{\alpha\beta}$	rotation or vorticity tensor	4.5
Θ	expansion scalar	4.5, and 2.3
$H_\circ$	Hubble constant at present time	4.2, see also astronomical constants
$q_\circ$	deceleration parameter at present time	4.2
r_{PH}	particle horizon coordinate	4.2
r_{EH}	event horizon coordinate	4.2
GUT	Grand Unified Theories	1.2, and 4.2
S^n	n-sphere	4.3, and mathematical appendix
T^n	n-torus	4.3, and mathematical appendix

SYMBOL	NAME	SECTION
$\times$	topological product	4.3, and mathematical appendix
$\otimes$	tensor product	mathematical appendix
$\mathrm{Tr}\,\boldsymbol{K}$ or T	Trace of extrinsic curvature tensor	4.3, and 5.2
${}^{(3)}R$	curvature scalar of three-dimensional hypersurface	4.2
#	connected sum of manifolds	4.3
S^n/P_i	n-sphere quotient the subgroup P_i	4.3
$\mathfrak{R}^4_{\mathrm{FAKE}}$	exotic differentiable four-manifold	4.3, and mathematical appendix
$k(S^n)$	connected sum of k copies of S^n	4.3
C^c_{ab}	structure constants	4.4, and mathematical appendix
J	Jacobian	4.4
P_t	time-projection tensor	4.5
P_Σ	space-projection tensor	4.5
$\psi(\boldsymbol{x})$	conformal scalar factor	5.2
$\overset{(c)}{\boldsymbol{g}}_{mn}$	conformal three-metric	5.2
$\overset{(c)}{G}_{mn}$	A tensor G_{mn} up to a conformal factor	5.2
${}^{(3)}\boldsymbol{g}$	three-metric of spacelike hypersurface	5.2
W^i	gravitomagnetic three-vector potential	5.2, and 5.3
$A_{\mathrm{total}\,ij}$ or A_{ij}	distortion tensor	5.2
$A_{\mathrm{free}\,ij}$	covariant divergence-free part of distortion tensor	5.2
$\overset{(c)}{\Delta}{}^*$	conformal vector Laplacian	5.3
$\overset{(c)}{\Delta}$	conformal Laplacian	5.3
α	lapse function	5.4
β_i	shift function	5.4
$\boldsymbol{m}$	magnetic moment	6.1
m	mass of a particle	2.5
$\boldsymbol{h}$ or h_{0i}	weak field, gravitomagnetic potential	6.1
$\dot{\Omega}^{\mathrm{Lense–Thirring}}$	Lense-Thirring precession	3.4.3, and 6.1
$\dot{\Omega}^{\mathrm{de\ Sitter}}$	de Sitter or geodetic precession	3.4.3, and 6.10

SYMBOL	NAME	SECTION
$\dot{L}_0$	rate of change of mean orbital longitude	6.1
a	semimajor axis	6.1, and 6.7
P	orbital period	6.1, and 6.7
I	orbital inclination	6.1, and 6.7
J_{2n}	coefficients measuring multipole mass moments	6.7
UT1	rotational orientation of Earth (rotation angle of Earth relative to faraway stars)	6.8
γ_{ij}	magnetomechanical tensor	6.9
$\boldsymbol{F} \cdot \boldsymbol{F}$, or $F_{\alpha\beta}\, F^{\alpha\beta}$	electromagnetism invariant	6.11
$\boldsymbol{F} \cdot^{*} \boldsymbol{F}$, or $F_{\alpha\beta}{}^{*}F^{\alpha\beta}$	electromagnetism pseudoinvariant	6.11
$\boldsymbol{R} \cdot \boldsymbol{R}$, or $R^{\alpha\beta\gamma\delta}\, R_{\alpha\beta\gamma\delta}$	general relativity invariant	6.11
${}^{*}\boldsymbol{R} \cdot \boldsymbol{R}$, or ${}^{*}R^{\alpha\beta\gamma\delta}\, R_{\alpha\beta\gamma\delta}$	general relativity pseudoinvariant	6.11
$\psi_1, \psi_2, \psi_3, \psi_4, \psi_5$	coefficients of Newman-Penrose formalism	6.11
$\boldsymbol{d}$	electric dipole moment	7.1

Author Index

Abramowicz, 383
Adamowicz, 178
Adams, 174
Adelberger, 169, 172, 183
Albrecht, 261
Alcock, 267
Allen, 179, 266
Alley, 79, 173, 175, 378
Alpher, 258
Altshuler, 313
Amaldi, 181
Ames, 311
Ampère, 374
Ander, 171, 172
Andersen, 175
Anderson, A., 181, 182
Anderson, J. L., 79
Anderson, J., 169, 174, 175, 177, 180, 181, 183
Armstrong, 182
Arnowitt, 82, 311
Aronson, 171
Arp, 263
Ash, 179
Ashby, 174, 378, 381
Ashtekar, 86
Aubourg, 267

Babcock, 178, 265, 381
Bahcall, 268
Baierlein, 309
Bailyn, 86
Barbour, 401
Bardeen, 376
Barlier, 378
Barnebey, 309
Barnes, 263
Barnett, 379
Barrow, 261, 262, 309, 310
Bartlett, 171, 172, 176
Bass, 375
Batchelor, 262
Baugher, 169
Becker, 379
Behr, 262
Bekenstein, 266
Belasco, 311
Belinski, 264, 313, 314
Beltran-lopez, 175
Bender, 79, 174, 175, 178, 181, 182, 184, 265, 378, 379, 381
Bennet, C., 258, 267
Bennet, W., 172
Berkeley, 400
Berthias, 170
Bertotti, 79, 178, 184, 265, 378, 381
Beruff, 172
Bessel, 170
Bethe, 258
Bianchi, 80, 179, 262, 312
Birkhoff, 81
Bizzeti, 171
Bizzeti-Soma, 171
Bjerre, 175
Blackwell, 379
Blair, 181
Blamont, 173
Blau, 260
Blomberg, 169
Bol, 380
Bollinger, 175
Boltzmann, 312
Bonaldi, 377
Bond, 267
Bondi, 85, 265, 307, 314
Bos, 179
Bowditch, 382
Bower, 312
Boyer, 81, 375
Boynton, 171
Braginsky, 84, 85, 169, 182, 377

Brans, 173, 266, 309, 375
Brault, 169
Braun, 375
Breidenthal, 169
Brenkle, 169, 177
Brill, 261, 310, 311, 375
Brillet, 184
Broadhurst, 26
Brown, 180
Buchhorn, 266
Burbidge, 257
Burstein, 256, 257

Cacciani, 180
Cain, 169, 177
Campbell, 169, 179
Cannon, 176, 379
Canuto, 174
Carfora, 308
Carswell, 177
Cartan, 12, 83, 178, 311
Carter, B., 383
Carter, W., 169, 177, 378, 379
Carusotto, 183
Casares, 85
Cattaneo, 262
Cavallo, 310
Caves, 377
Cerdonio, 377
Chandler, 178, 265, 381
Chandrasekhar, 178, 263, 382
Chapman, 171
Charles, 85
Chave, 172
Chincarini, 256
Chinnapared, 81, 375
Choquet-Bruhat, 80, 82, 265, 307, 310, 401
Chow, 380
Christensen-Dalsgaard, 180
Christodoulou, 85, 311
Christoffel, 80
Chupp, 175
Ciufolini, 12, 81, 172, 175, 178, 180, 183, 184, 265, 377, 378, 379, 381, 382
Clarke, 80
Clayton, 179
Cocconi, 174
Cognola, 382
Cohen, J., 12, 375
Cohen, S., 378
Coleman, 314
Collins, 265, V. 68
Colombo, 180
Condon, 265
Contopoulos, 178
Corinaldesi, 381
Couch, 81, 375
Counselman, 169, 378
Cowie, 267
Cowsik, 172
Crittenden, 267
Crosby, 171
Cruz, 172
Cuddihy, 177
Curott, 174
Currie, 79, 175, 378, 382

Dalsgaard, 180
Damour, 81
Danzmann, 184
Darmois, 310
Davidson, C., 176
Davidson, J., 265
Davidson, K., 267
Davidson, W., 400
Davies, 182, 256, 257
Davis, 267
De Felice, 80
de Lapparent, 256
de Sitter, 11, 176, 179, 260, 380, 375
Decher, 169
Dehnen, 264, 308, 380
Demianski, 81
Deser, 82, 311
DeWitt, B., 84
DeWitt-Morette, C., 80
Dicke, 79, 169, 173, 174, 175, 179, 180, 257, 265, 307, 309, 375, 378
Dickey, 12, 170, 175, 184, 266, 381
Dickinson, 256
Dillard-Bleick, 80
Dillinger, 169
Dirac, 173, 174
Donaldson, 261
Doodson, 379
Doroshkevich, 258, 313
Dressler, 256, 257
Drever, 175, 184
Duncombe, 376
Dunn, 11, 378
Duvall, 180
Dyce, 179

Dymnikova, 260
Dyson, 176, 311
Dziembowski, 180

Eanes, 12, 378, 379, 382
Eardley, 310
Eby, 169
Echeverria, 85, 263
Eckhardt, 79, 171, 172, 175, 378
Eddington, 83, 176, 182, 259
Efstathiou, 266, 267
Ehlers, 85, 176, 265
Einasto, 256
Einstein, 10, 11, 78, 81, 84, 85, 172, 179, 263, 264, 307, 308, 309, 374, 400
Eisenhart, 259
Ekers, 177
Ekstrom, 171
Ellis, G., 79, 176, 259, 261, 262, 263, 309, 314, 374
Ellis, R., 266
Eötvös, 170
Esposito, 177
Estabrook, 182, 262
Evans, 310
Everitt, 11, 171, 376, 377
Exton, 81, 375
Ezekiel, 380

Faber, 256, 257
Fabian, 257
Fackler, 171
Fairbank, 11, 376, 380
Falferi, 377
Faller, 79, 169, 171, 175, 172, 181, 182, 378
Farinella, 379
Farrel, 169
Fazzini, 171
Fekete, 170
Felske, 171
Fermi, 177, 262, 309, 314
Feynman, 85, 311, 313, 400
Finkelstein, 83
Fischbach, 171, 172
Fitch, 172
Fitzpatrick, 375
Flynn, 268
Fock, 79, 264
Fokker, 176, 380
Fomalont, 177
Föppl, 12, 382
Ford, H., 86, 267
Ford, V., 266
Fordyce, 181
Fortson, 175
Forward, 181, 184
Francaviglia, 311
Franklin, 184
Frautschi, 311
Freed, 261
Freedman, M., 261
Freedman, W., 267
Frenk, 266
Fricke, 376
Friedländer, B. and I., 12, 382
Friedman, J., 85, 263
Friedman, L., 180
Friedmann, A., 257
Frolov, 85, 263
Fujii, 171
Fuller, 84

Galilei, 79, 170, 399
Galloway, 262
Gamow, 258, 259, 309
Gauss, 80
Gea-Banacloche, 380
Geller, 256, 266, 263
Geroch, 84
Gertsenshtein, 181, 182
Giacconi, 85, 259
Giazotto, 181
Gillespie, 257
Gilman, 401
Gilmore, 267
Ginzburg, 376
Giovanelli, 256
Gliner, 260
Goad, 379
Göckeler, 82
Gödel, 262, 263
Goenner, 307
Gold, 265
Golden, 257
Goldenberg, 179
Goldman, 174
Goldschmidt, 312
Goldstein, 169, 177
Gompf, 261
Goode, 179
Gorenstein, 85
Gott, 256

Gough, 179, 180
Gould, 184, 266, 268
Green, 12
Greenstein, 173
Gregory, P., 265
Gregory, S., 256
Grishchuk, 79, 84, 85, 313
Gundlach, 169, 172
Gurevich, 260
Gursky, 85, 257
Guth, 260, 310

Hafele, 173
Hall, J., 175, 181
Hamilton, A., 266
Hamilton, W., 183
Hancock, 267
Hanson, 171
Hardy, 260
Harms, 267
Harris, 171
Harrison, 83, 172
Hartle, 308, 311, 312
Harvey, 180
Haugan, 380
Hawking, 79, 84, 84, 176, 263, 265, 308, 311, 314, 374
Haynes, 256
Healy, 179
Heckel, 169, 172, 175, 183
Heer, 380
Hehl, 83, 178, 314
Helgason, 81, 259
Heller, G., 379
Heller, M., 307, 375, 401
Hellings, 170, 174, 180, 181, 184
Hemenway, 376
Hendricks, 380
Herlt, 79, 259, 382
Herman, 258
Hibbs, 85, 311
Hilbert, 80, 179, 311
Hildebrandt, 172, 380
Hill, 179
Hils, 181, 182
Hoare, 175
Hodge, 259
Hoffman, T., 169
Hoffmann, B., 81
Hojman, 178
Holz, 267
Hönl, V. 44, 264, 308
Hörnfeldt, 382
Hough, 183, 184
Hoyle, 266
Hu, 267
Huang, 267, 382
Hubble, 259
Huchra, 256, 266
Hughes, 175
Hulse, 177, 312
Huygens, 170

Iess, 378
Infeld, 81
Ingalls, 179
Isaacson, 84, 309
Isaila, 172
Isenberg, 11, 263, 306, 308, 310, 374
Itano, 175
Ivanenko, 82, 312

Jackson, 80, 312, 375, 400
Jacobs, 175
Jaffe, 173
Jantzen, 262
Jefferys, 376
Jekeli, 171, 172
Jenkins, 173
Jôeveer, 256
Jordan, 173
Jurgens, 169

Kaiser, 257, 266
Kammeraad, 171
Kasameyer, 171
Kaufman, 169
Kaula, 79, 175, 378
Keating, 173
Keesey, 174, 177
Keiser, 171, 172
Kellermann, 266
Kellogg, 85
Kerlick, 178
Kerr, 81, 375
Khalatnikov, 264, 313, 314
Kheyfets, 83
Killing, 81, 259, 312
Kim, 85, 266
King, C., 380
King, H., 264
King, R., 169, 378

Kirby, 266
Kirhakos, 268
Kirshner, 256
Klinkhammer, 85, 263
Klosko, 378
Knocke, 379
Kobayashi, 80, 176
Koester, 170
Komarek, 169, 177
Koo, 266
Kopff, 376
Korté, 314
Kovács, 84
Kramer, 79, 259, 382
Kretschmann, 312, 400
Kreuzer, 176
Krisher, 169, 175, 180, 181, 183
Krishnan, 172
Krotkov, 169
Kruskal, 83
Kuchar, 309, 311
Kuhn, 180
Kuroda, 172

Laing, 177
Lamoreaux, 175
Landau, 79, 173, 263, 374, 379
Lasota, 383
Lau, 169, 174, 177, 180
Lausberg, 375
Lawrence, 266
Lazarewicz, 171
Lee, 175
Leibacher, 180
Leibniz, 400
Leighton, 179
Lemaître, 258
Lenard, 176
Lense, 11, 81, 178, 264, 265, 314, 375, 401
Lerch, 378
Levi-Civita, 80, 81, 179
Levine, 169
Libbrecht, 180
Lichnerowicz, 310
Lifshitz, 79, 173, 262, 264, 313, 314, 374, 379
Lin, 262
Linde, 261, 310, 311, 312
Lindner, 172
Lindquist, 81, 310, 375
Lipa, 11, 376
Lobačevskij, 80
Lobo, 184
Logan, 175
London, 380
Lopresto, 183
Lorentz, 78
Lorenzini, 171
Loveman, 175
Lowe, 182
Lucchesi, 379
Lukash, 261
Lutes, 175
Lynden-Bell, 256, 257, 313

MacCallum, 79, 259, 262, 382
MacDonald, D., 11, 264, 374
MacDonald, G., 175
Mach, 11, 174, 264, 307, 400, 401
Machalski, 257
Macneil, 169, 177
Maitra, 264
Mäkinen, 172
Maleki, 175, 181
Man, 184
Markov, 313
Marsden, 262, 307, 309
Martin, 177, 382
Marzke, 176
Mashhoon, 314, 377, 380
Mat'tsev, 313
Mather, 265, 267
Mathewson, 266
Mattison, 169
Matzner, 180, 182, 183, 259, 261, 262, 265, 313, 314, 378, 382
McClintock, 86
McCrea, 83, 85
McDow, 179
McHugh, 169, 172
McMurry, 172
Melchior, 379
Melott, 256
Michael, 177
Miller, L., 181, 184
Miller, W., 82, 267
Millett, 171
Minkowski, 78
Mio, 172
Misner, 11, 12, 79, 82, 168, 261, 264, 308, 311, 374, 400
Moffat, 178, 179
Møller, 79, 381

Moody, 183
Moore, 171
Morabito, 183
Moreno, 181
Morris, 84, 85, 263
Morrow, 180
Moss, 181, 184
Mugge, 171
Muhleman, 177
Mulholland, 79, 175, 378
Müller, 172, 183, 266, 383
Murnaghan, 311
Murray, 257

Napier, 184
Narlikar, 309
Naylor, 85
Nester, 178
Newhall, 169, 170, 174, 175, 266, 381
Newman, 81, 375, 382
Newton, 79, 170, 264, 376, 399
Ni, 179
Nicolaci Da Costa, 267
Niebauer, 169, 172
Nomizu, 80, 176
Nordstrøm, 81
Nordtvedt, 79, 172, 173, 175, 176, 182, 380, 381
Novikov, 85, 86, 258, 261, 263
Noyes, 179
Nystrom, 169

O'Connel, 381
O'Dell, 259
O'Hanian, 311
O'Murchadha, 307, 312
Oda, 85
Oemler, 256
Oersted, 374
Oke, 173
Oleson, 179
Oort, 256, 376
Oppenheimer, 83
Ori, 267
Oteiza, 175
Ousley, 382
Ozsváth, 263, 264

Paczyński, 267
Pagels, 258
Paik, 183, 377, 380
Palatini, 80
Palmer, 172
Panov, 169
Papapetrou, 381
Parker, 172
Partridge, 257
Patel, 378
Patton, 310
Patz, 179
Pecker, 263
Pedrotti, 380
Peebles, 256, 257, 258, 263
Pekár, 170
Penrose, 82, 83, 84, 382
Penzias, 257
Perego, 171
Petrov, 79, 381
Pettengil, 179
Petterson, 376
Pfister, 375, 382
Pierce, A., 183
Pierce, M., 267
Pirani, 176, 181, 375, 381
Pizzella, 181
Plotkin, 79, 175, 378
Polnarev, 84, 85, 377
Popper, 173
Post, 380
Potter, 170
Poulsen, 175
Poultney, 79, 175, 378
Pound, 173
Prakash, 81, 375
Prestage, 175
Price, 11, 264, 374
Primas, 175
Prodi, 377
Pugh, 11, 376
Pustovoit, 181

Qadir, 310

Raab, 172, 175
Raine, 307, 313, 401
Randall, 266
Rauber, 312
Raychaudhuri, 259, 262
Reasenberg, 169, 174, 177, 178, 265, 381
Rebka, 173
Rees, 256
Regge, 312

Reichley, 177
Reinhardt, 307
Reisenberger, 183
Reissner, 81
Remillard, 86
Renner, 170
Ricci Curbastro, 80
Richardson, 175
Richter, 182, 183
Riemann, 11, 80, 400
Ries, 382
Riis, 175
Rindler, 79, 179, 260
Roberts, 263
Robertson, D., 11, 169, 177, 379
Robertson, H., 182, 258
Robinson, 175
Roddier, 173
Rogers, 169, 172
Roll, 169, 257
Romaides, 171, 172
Rood, 256
Rorschach, 380
Rösch, 172
Rosen, 84, 179
Rosenbaum, 178
Rovelli, 86
Rowan-Robinson, 259, 266
Rubincam, 378, 379, 382
Ruder, 183, 266, 383
Ruffini, 83
Ryan, 178, 259, 262, 264, 313

Saar, 256
Sachs, 401
Saffren, 380
Sahu, 267
Salpeter, 174
Sanchez, 379
Sandage, 260, 266
Sanders, 380
Sandford, 184
Sands, 171
Sarill, 375
Saunders, 266
Sauter, 379
Schechter, 256
Schiff, 11, 182, 265, 376
Schild, 80, 170, 176
Schleich, 380
Schneider, 183, 266, 383
Schoen, 82, 261
Schouten, 380
Schrader, 183
Schreiber, 261
Schreier, 85
Schücker, 82
Schücking, 263, 264
Schur, 259
Schutz, B. E., 12, 378, 379
Schutz, B., 80, 184, 259
Schwartz, 257
Schwarz, 12
Schwarzschild, 81, 176
Sciama, 267, 307, 309, 400, 401
Scully, 377, 380
Seielstad, 177
Seifert, 309
Seldner, 257
Shahid-Saless, 381
Shandarin, 256, 258
Shapiro, I., 169, 174, 177, 178, 179, 265, 378, 381
Shapiro, P., 256
Shapiro, S., 309
Sharp, 309
Shectman, 256
Shelus, 376, 378
Shepley, 178, 259, 262, 264, 313
Shipman, 173
Silk, 257, 258
Silverberg, 79, 175, 378
Simon, 179
Sinclair, 174, 378
Slade, 169, 378
Smarr, 310
Smith, D., 11, 378
Smith, G., 169, 183
Smith, H., 263
Smoot, 258, 267
Snider, J., 169, 173
Snyder, 83
Soffel, 172, 183, 266, 383
Sokolov, 82, 312
Soldner, 176
Solheim, 263
Sonnabend, 181
Southerns, 170
Spanier, 82
Speake, 172
Spencer, 312
Spiess, 172

Spivak, 80, 176, 259, 314
Sramek, 177
Stacey, 171
Standish, 169, 170, 174
Stark, 267
Stebbins, 179, 181, 182
Steinhardt, 261, 267
Stephani, 79, 259, 382
Stevenson, 172
Stoeger, 259
Stubbs, 169, 172
Su, 169, 183
Sudarsky, 171
Sussman, 308
Swanson, 169, 172, 183
Synge, 79, 178, 262
Szafer, 171
Szalay, 258, 266
Szekeres, 83
Szumilo, 171

Taccetti, 171
Tago, 256
Takeno, 81
Talmadge, 170, 171, 172
Tananbaum, 85
Tandon, 172
Tapley, 12, 378, 379, 382
Tarenghi, 256
Taub, 313
Taylor, A., 183
Taylor, E., 382, 400
Taylor, J., 177, 183, 312
Taylor, V., 379
Teitelboim, 82
Terlevich, 256, 257
Teuber, 169
Teukolsky, 309
Tew, 171, 172
Teyssandier, 375
Theiss, 314
Thieberger, 171
Thirring, 11, 81, 82, 178, 264, 265, 314, 375, 401
Thomas, 171, 183
Thompson, A., 175
Thompson, L., 256
Thomson, 400
Thorne, 11, 12, 79, 83 , 84, 85, 168, 180, 261, 263, 264, 268, 308, 312, 374, 375, 377, 400
Thornton, 177
Threlfall, 309
Tifft, 256
Tipler, 261, 262, 309, 310, 311
Tolman, 259
Tomonaga, 309
Toomre, 180
Torrence, 81, 375
Touboul, 184
Trautman, 178
Truehaft, 183
Tuck, 171
Tulczyjew, 381
Tully, 256
Turneaure, 376

Uhlenbeck, 261
Unnikrishnan, 172
Unruh, 309

Van Buren, 176
Van Patten, 376, 377
Vanzo, 382
Vespe, 379
Vessot, 169, 173
Vigier, 263
Vilenkin, 261, 312
Vincent, 174, 179, 181
Vishveshwara, 375
Vitale, 377
Vogel, 183
Vogt, 184
Von Der Heyde, 178

Wagner, 378
Wagoner, 180
Wagshul, 175
Wahlquist, 182, 262
Wahr, 174
Wakano, 83, 267
Wald, 262
Walker, 258
Walsh, 177
Warren, 262, 313
Watanabe, 172
Watkins, 382
Watt, 169
Waylen, 401
Webbink, 182
Weber, 84, 180, 375
Webster, 257
Wegner, 256, 257

Weiler, 177
Weinberg, 79, 174, 258, 374, 376, 400
Weiss, 184
Weyl, 78, 307
Weymann, 177
Wheeler, 11, 12, 79, 82, 83, 84, 168, 176, 261, 263, 264, 267, 306, 307, 308, 309, 310, 311, 312, 313, 374, 381, 382, 400, 401
Whittaker, 374
Wilkins, 376, 383
Wilkinson, 79, 175, 257, 378
Will, 79, 168, 169, 172, 175, 182, 183, 184, 309, 377, 380, 381
Williams, 12, 79, 170, 174, 175, 176, 266, 378, 381
Wills, 169
Wilson, 257
Wineland, 175
Wiseman, 184
Witten, 12, 82
Wolfe, 257
Woolgar, 179
Worden, 171
Wright, E., 258
Wright, J., 263

Yano, 81, 258
Yau, 82, 261
Yilmaz, 376
Yoder, 12, 174, 378
York, 80, 82, 265, 307, 311, 312, 401
Yudin, 313
Yurtsever, 84, 85, 263

Zanoni, 179
Zeeman, 170
Zel'dovich, 84, 86, 256, 258, 260, 263
Zubairy, 380
Zumberge, 172
Zürn, 172
Zygielbaum, 169, 177

Subject Index of Mathematical Appendix

Absolute, or covariant, derivative, 423
Affine connection, 422
Affine parameter, 431
Algebra, Lie, 422
Antisymmetrization, 418
Arc-length, 431
Arcwise connected space, 406–407

Ball, 405
Base space of a bundle, 435
Basis
  coordinate, 414
  dual, 416
  general, 415
  natural, 414
  ordered, 436
Bianchi
  first identity, 429
  second identity, 430
Bottle, Klein, 409, 412
Boundary
  manifold with boundary, 412
  Möbius strip with boundary, 409, 412
  of a manifold, 412
Bundle, 434
  fibre, 435
  globally trivial fibre bundle, 435
  locally trivial fibre bunde, 435
  nontrivial fibre bundle, 435
  of linear frames, 436
  of orthonormal frames, 436
  product, 435
  tangent, 435
  trivializable, 435
  vector, 435

C^r and C^∞ diffeomorphism, 411
C^r and C^∞ differentiable manifold, 410
C^r differentiable function, 410
C^0 function, 410
C^0 manifold, 410
Chain (section 2.8), 50
Chart, 409
Christoffel symbols, 427
Circle, 407
Closed
  interval, 404
  manifold, 412
  n-disk, 404
  path, 407
  set , 404–405
Coefficients
  commutation, 415–416
  connection, 422
  structure, 415–416
Commutation coefficients, 415–416
Commutator
  of covariant derivatives, 428
  of vector fields, 415
Compact
  locally, 406
  space, 405–406
  support, 433
Compatible with metric, connection, 427
Complete
  geodesic, 432
  geodesically, manifold, 432
  manifold geodesically complete at a point, 432
Components of tangent vector, 414–415
Cone
  double, 409
  half, 409
Conformal
  tensor, 430
  transformation, 430
Conformally flat manifold, 430
Connected space, 406
  arcwise, 406–407
  locally, 406

Connected space (*cont.*)
multiply, 407
simply, 407
Connection, 422–423
affine, 422
(Cartan), 423
(Koszul), 422–423
Levi-Civita, 427
linear, 422
metric compatible, 427
Riemannian, 427
symmetric, 427
Connection coefficients, 422
Continuous mapping, 406
Constant path, 407
Contracted tensor product, 419
Contractible closed path, 407
Contraction, 419
Contravariant
tensor, 417–418
vector, 415
Coordinate basis, 414
dual, 416
Coordinate system, local, 409
Coordinate vector, 414
Coordinates of a point, 409
Cotangent vector space, 416
Covariant derivative, 423–424
directional, 424commutator of covariant derivatives, 428
Covariant
tensor, 417–418
vector, 416, 417
Covering, 405
locally finite, 406
Critical point for squared spacetime interval, 431
Curvature
extrinsic curvature tensor, 426
Gaussian, 428
Ricci scalar, 430
Riemann tensor, 427–430
sectional, 428–429
Curve
extremal for squared spacetime interval, 431
extremal for length, 431–432
Cylinder, 405

Derivative
covariant, or absolute, 423–424
directional, 413–414, 421
exterior, 420
Lie, 420–421
Deviation, equation of geodesic deviation, 429
Diffeomorphic manifold, 411–412
Diffeomorphism, C^r and C^∞, 411
Differentiable
function, C^r, 410
manifold, C^r and C^∞, 410
mapping, 411
1-form, 416, 417, 419
structure of $\mathfrak{R}^n$, exotic, 411–412
structure of $\mathfrak{R}^n$, standard, 411–412
vector field, 415
Differential of function, 417
of coordinate function, 417
Directional
covariant derivative, 424
derivative, 413–414, 421
Disk, n-disk, 403, 404,
Divergence theorem, 434
Double cone, 409
Dragging, Lie, 421
Dual
basis, 416
coordinate basis, 416
space, 416

Einstein field equation, 430–431
vacuum, 430–431
Einstein tensor, 430
Electromagnetic
field 2-form, 420
potential 1-form, 420
Embedded submanifold, 412
Embedding, 412
Equation, Killing, 421–422
Euclidean space, n-dimensional, 403
Exotic
differentiable structure of $\mathfrak{R}^n$, 411–412
manifolds, 411–412
Exponential mapping, 432
Exterior
derivative, 420
product, 420
Extremal curve
for squared spacetime interval, 431
for length, 431–432
Extrinsic curvature tensor, 426

Fermi-Walker transport, 433
Fibre, 435

typical, 435
Fibre bundle, 435
globally trivial, 435
locally trivial, 435
nontrivial, 435
Field,
differentiable vector field, 415
of forms, 419
scalar invariant, 418
tensor, 419
vector, 415
Field equation
Einstein, 430–431
vacuum Einstein, 430–431
Finite, locally, covering, 406
Finite subcovering, 405
First Bianchi identity, 429
First fundamental form, 426
Flat
conformally, manifold, 430
locally flat manifold, 428
two-torus, 408
Form, 416, 417, 418
differentiable, 419
Formal invariance, 421
Forms
field of, 419
integration of, 433–434
Frame
holonomic, 414
linear, 436
Frames
bundle of linear frames, 436
bundle of orthonormal frames, 436
Function, 410
C^0, 410
differentiable, of class C^r, 410
Fundamental form
first, 426
second, 426

Gauss theorem, 434
Gaussian curvature, 428, 428
General basis, 415
General linear group, 422
Geodesic, 424, 431–432
complete, 432
Geodesic deviation equation, 429
Geodesically complete manifold, 432
at a point, 432
$GL(n, \mathfrak{R})$, 410, 422, 435–436
Globally trivial fibre bundle, 435
Group
Lie, 422
general linear, 422
of linear bijective mappings, 410, 422, 435–436
of nonsingular real matrices, 422, 436
orthogonal, 422, 436
structural, 435
topological, 435

Half cone, 409
Handle, 407, 408
Hausdorff space, 405
Holonomic frame, 414
Homeomorphic, 407, 408
Homeomorphism, 407–408
Homotopic curves, 407
Homotopy, 407

Identity mapping, 410
Immersion, 412
Indefinite metric, 425
Induced
metric, 426
orientation, 413
Infinitesimal isometry, 421
Injective mapping, 407
Inner product, 419, 425
Integration
of a form on a chart, 433
on smooth manifold, 434 (for integration theory using chains, see section 2.8, pages 49–55)
Interior product, 419
Invariance, formal, 421
Invariant
scalar, field, 418
topological, properties, 407–408
Isometry, 421
infinitesimal, 421

Jacobi equation, 429
Jacobian, 412

Killing
equation, 421–422
vector, 421
Klein bottle, 409, 412

Leibniz rule, 413, 423

Length, 431–432
  arc length, 431
Levi-Civita connection, 427
Lie
  algebra, 422
  derivative, 420–421
  dragging, 421
  group, 422
Linear connection, 422
Linear frame, 436
Linear frames, bundle of, 436
Linear group, general, 422
Local coordinate system, 409
Locally compact, 406
Locally connected space, 406
Locally finite covering, 406
Locally flat manifold, 428
Locally trivial fibre bundle, 435
Lorentzian manifold, 426

Manifold
  conformally flat, 430
  diffeomorphic manifolds, 411
  differentiable, 410
  exotic, 411–412
  geodesically complete, 432
  geodesically complete at a point, 432
  locally flat, 428
  Lorentzian, 426
  n-manifold, 408–409
  orientable, 412–413
  oriented, 412–413
  proper Riemannian, 425
  pseudo-Riemannian, 425
  Riemannian, 425
  smooth, 410
  with boundary, 412
Mapping
  continuous, 406
  differentiable, 411
  exponential, 432
  group of linear bijective mappings, 410, 422, 435–436
  identity, 410
  injective, 407
  one to one, 407
  onto, 407
  surjective, 407
Matrices, group of nonsingular real, 422, 436
Metric
  indefinite, 425
  induced, 426
  nondegenerate metric tensor, 425
  pseudo-Riemannian, 435
  signature of, 426
  symmetric, 425
  tensor, 425
Metric compatible connection, 427
Möbius strip, 409
  with boundary, 409
Multiply connected space, 407

Natural basis, 414
n-dimensional embedded submanifold, 412
n-dimensional orthogonal group, 436
Neighborhood, 404
Nondegenerate
  inner product, 419
  metric tensor, 425
Nontrivial fibre bundle, 435
Norm of vector, 425
Null vector, 426

O(n), 436
1-form, 416, 417
  differentiable, 419
  electromagnetic potential, 420
One to one mapping, 407
Onto, mapping, 407
Open
  covering, 405
  interval, 404
  n-disk, 403
  set, 404
Ordered basis, 436
Orientable manifold, 412–413
Orientation induced, 413
Oriented manifold, 412–413
Orthogonal
  group, 422, 436
  vectors, 425
Orthonormal frames, bundle of, 436
Ostrogradzky-Green formula, 434

Paracompact, 406
Parallel transport, 424
Parameter, affine, 431
Partition of unity, 433
  subordinate to a covering, 433
Path, 406–407
  constant, 407

p-covariant and q-contravariant tensor, 417–418
p-forms, 418
Product
  bundle, 435
  contracted tensor product, 419
  exterior, 420
  inner, 419, 425
  interior, 419
  scalar, 419, 425
  tensor, 419
  topological, 405
  topology, 405
  wedge, 420
Projection of bundle, 434–435
Proper Riemannian manifold, 425
Pseudo-Riemannian
  manifold, 425
  metric, 435

q-contravariant and p-covariant tensor, 417–418
$(q\ p)$ tensor, 417–418
q-polyvectors, 418

Refinement, 406
Relative topology, 408
Ricci
  curvature scalar, 430
  tensor, 430
Riemann curvature tensor, 427–430
Riemann-Ampère-Stokes formula, 434
Riemannian
  connection, 427
  manifold, 425

S^7 and S^{31}, 7-sphere and 31-sphere, 412
Scalar
  invariant, 418
  product, 425
  Ricci curvature scalar, 430
Second Bianchi identity, 430
Second fundamental form, 426
Sectional curvature, 428–429
Separated space, 405
Signature of a metric, 426
Simply connected space, 407
Smooth manifold, 410
Solid sphere, 412
Space
  base space of a bundle, 435
  cotangent vector space, 416
  dual space, 416
  tangent vector space, 414
  total space of a bundle, 435
Spacelike vector, 426
Sphere
  n-sphere, 404
  two-sphere, 407, 408
  with n-handles, 408
Standard differentiable structure of $\Re^n$, 411–412
Standard two-torus, 408
Stokes theorem, 434
Structural group, 435
Structure coefficients, 415–416
Subcovering, 405
Submanifold, 412
  embedded, 412
Subordinate to a covering, partition of unity, 433
Support, compact, 433
Surjective mapping, 407
Symbols, Christoffel, 422, 427
Symmetric
  connection, 427
  inner product, 419
  metric, 425
Symmetrization, 418

Tangent
  bundle, 435
  vector, 413–414
  vector, components of, 414–415
  vector space, 414
Tensor, 417–418
  extrinsic curvature tensor, 426
  field, 419
  metric, 425
  p-covariant and q-contravariant, 417–418
  product, 419
  product, contracted, 419
  Riemann curvature tensor, 427–430
  torsion, 423, 424
Timelike vector, 426
Topological
  group, 435
  invariant properties, 407–408
  manifold, 408–409
  product, 405
  space, 403–404
Topologically equivalent, 407–408

Topology, 403
  product, 405
  relative, 408
Torsion tensor, 423, 424
  field, 423, 424
Torus
  n-holed, 408
  n-torus, 405
  two-torus, 405, 408
Total space of a bundle, 435
Transformation, conformal, 430
Transition functions, 435
Transport
  Fermi-Walker, 433
  parallel, 424
Trivial
  globally, fibre bundle, 435
  locally, fibre bundle, 435
2-form, 418
  electromagnetic field 2-form, 420
two-sphere with one handle, 408
two-torus, 405
  flat, 408
  standard, 408
Typical fibre, 435

Vacuum Einstein field equation, 430–431
Vector
  bundle, 435
  components of tangent vector, 414–415
  contravariant, 415
  coordinate, 414
  covariant, 416, 417
  field, 415
  field, differentiable, 415
  Killing, 421
  norm of, 425
  null, 426
  orthogonal vectors, 425
  space, cotangent, 416
  space, tangent, 414
  spacelike, 426
  tangent, 413–414
  timelike, 426

Wedge product, 420
Weyl tensor, 430

0-form, 420
(0 0) tensor, 417

Subject Index

"Absolute frame of reference," 393
Absolute geometric elements, no absolute geometric elements in general relativity, 300
"Absolute inertial frame," 249, 250
"Absolute space," 249, 250, 384–387, 394
  in Galilei-Newton mechanics, 249, 250
Accelerometers, 362, 363
Accretion disk, 327–328
Action at a distance potentials of electrodynamics, 297
Action in geometrodynamics, 21–22, 280–281
Active and passive gravitational mass, tests of equivalence of, 115–116
Active galactic nuclei, 327, 347. See also jets
Active gravitational mass and pressure, 240
ADM energy in general relativity, 48
Advance of pericenter, 141–147
Affine parameter of geodesic equation, 30. See also mathematical appendix
"Age of universe" in Friedmann cosmological models, 212
Albedo, Earth, 339–340, 344, 347
Ampère, 316
Ampère's law, 317
Analytic extension of Schwarzschild solution, 67
Angular momentum
  and curvature invariants, 355–360
  and gravitomagnetic field, 319–324, 320
  and mass of a body, quasilocal measurement of, 360
  and spacetime geometry, 354–360
  charge and mass, 357
  contribution of angular momentum to orientation of local inertial frames, 249–252, 299, 321, 323. See also dragging of inertial frames
  geometry far from an object with, 42, 299, 316
  in general relativity, 41, 42, 47–48, 319–324, 352, 354–360
  in special relativity and conservation laws, 43–44
  metric with mass and angular momentum (Kerr), 42
  metric with mass, angular momentum and charge (Kerr-Newman), 41
  of a central object and jets in astrophysics, 327–328
  of Earth, 328, 330, 336, 350, dragging due to, 330, 335–338
  of matter and gravitational waves in Bianchi IX cosmological models, 248–249
  of a spinning particle, 352
Angular momentum four-vector of a spinning particle or gyroscope, 352
  Fermi-Walker transport of, 132
  rate of change of, 132–133
Anholonomic basis, 58. See also mathematical appendix
Anisotropic thermal radiation, 340–343, 344–347
Anisotropy of cosmic background radiation, measurements of, 186, 189–192, 255. See also COBE
Anomaly
  eccentric, 322
  true, 322
Antisymmetrization of a tensor, 50–51, 53. See also mathematical appendix
Apollo 11, 114
Arc-length, 30. See also mathematical appendix
Argument of latitude, 321
Argument of pericenter, 322
ASI, Italian Space Agency, 346
Astrometry, interferometric, space mission, 122
Astronomical frame, 134

Astrophysical processes and initial-value formulation, 283
Astrophysics
and dynamics of geometry, 294
and gravitomagnetic field, 315, 327–328
Asymptotic quasi-Minkowskian frame, 134–136, 299, 329, 345
Asymptotically flat metric, nondynamical, versus compact space, 271–274, 300–303, 304
Asymptotically flat space and inertia, 271–274
Atmospheric refraction, 345
Austin, University of Texas at, 346
"Axions" and cosmology, 193, 213

Baby-universes, 219
Bar. See resonant detector
Bardeen and Petterson effect, 327
Barnett, 348
Barrow and Tipler, 228
Basis
coordinate, or holonomic, 21, 58
noncoordinate, or anholonomic, 58
Becker, Heller, and Sauter, 348
Bekenstein-Hawking temperature, 69
Belinsky, Khalatnikov, and Lifshitz, 297, 298
Bender, 158
Berkeley, 296, 385
against absolute space, 385
β, post-Newtonian parameter, 119, 140, 146, 350
Bianchi, 234
Bianchi VIII cosmological models, 234, 242, 243
Bianchi I, VII_0, VII_h and IX model universes, experimental limits on rotation of, 254–255
Bianchi IX cosmological models, 231, 234, 244–249, 254–255, 297
general, 246, 247
nonrotating, 246, 247
nontumbling, 246, 247
rotating, 244–249
Bianchi identity, 27–28, 139–140. See also mathematical appendix
first, 57–58
second, 27–28, 57–58, 201
second, consequence of boundary of the boundary principle, 58–59
second contracted, 28, 57
second contracted from variational principle, 28
Bianchi spatially homogeneous models, 231–234, 234
Big bang, 192, 212, 217, 298
Big crunch, 298
Bimetric, Rosen theory, 147
Binary pulsar PSR 1913+16, 44, 77, 88, 128, 146–147, 295
Birkhoff theorem, 40
Blackbody, cosmic microwave radiation, 186, 189–192
Black hole, 61–69, 62
"a black hole has no hair," 65
and singularities, 65–67, 69–71
candidates, 62
evaporation, 67–69
features of, 64–65
mass, angular momentum, and charge of, 64
supermassive, in galaxy M87, 62, 204
supermassive, rotating, 327, 347. See also jets
with angular momentum, 65
Blueshift of electromagnetic waves in contracting Friedmann models, 203
Boltzmann, 296
Bolyai, 2, 19
Boötes the Shepherd, 186
"Bottom-up" cosmological scenario, 193
Boundary
of a manifold, 49. See also mathematical appendix
of a two-cube, 51
Boundary conditions
and general relativity, 300, 303. See also initial value formulation
inertia and matter in the universe, Einstein words on, 326
in general relativity, Einstein view of, 271–273
Boundary of the boundary
is zero, 49
of a four-simplex, 52
of a three-cube, 51
Boundary of the boundary principle, 49–61
and electromagnetism, 56–57
and geometrodynamics, 10, 49–61
and rotation of a vector induced by curvature, 61
physical consequences, 55–61
Boundary, quantities to be fixed at the boundary in variational principle for field equation, 24, 279–282

Boyer-Lindquist coordinates, 41–42
Braginsky and Panov, Moscow, experiment of weak equivalence principle, 93, 95
Braginsky, Polnarev, and Thorne, 347
Brans-Dicke parameter ω
and Mercury perihelion advance, 146
limits on, 111, 146, 219
Brans-Dicke theory, 109–111, 113, 146, 219, 275. See also Jordan-Brans-Dicke theory
limits on the ω parameter, 111, 146, 219
Bubbles, cosmological, production in old inflationary models, 219
Bubbliness, 185
Bundle complete manifold, or b-complete, 70
Burst of gravitational radiation, 153, 154

"Candles," standard, 205, 207
Cartan, 10, 280
Cartan, first, structure equation, 57
Cartan Repère Mobile, 57
Cartan, second, structure equation, 57
Cartan torsion tensor, 137
Cartan unit tensor, 59
CASSINI, 163
Cauchy surface, 5, 70, 230, 231, 250, 253, 277, 300, 301, 304, 306
compact, 230, 396–398
Cauchy problem, 252, 306. See also initial-value formulation
Cauchy problem in general relativity, existence, uniqueness and stability of solution, 306
Causality
and closed timelike curves, 243–244
and Gödel model universe of "first type," 243–244
and initial-value formulation in geometrodynamics and electrodynamics, 396–397
and initial-value problem, 274–275
condition of no causality violation, 304
in Bianchi IX cosmological models, 247
satisfied in Gödel rotating model universes of "second type," 244, 247
Cavendish balance, 115
Center for Astrophysics (CfA), 146
Centrifugal-type and Coriolis-type forces inside rotating shell, 323–324
Cepheid variables and estimation of Hubble constant, 207
Chandrasekhar and Contopoulos post-Galilean transformation, 131
Chandrasekhar limit for mass of a white dwarf, 63
Chaotic inflationary models, 219
Charge
angular momentum and mass, 357
in electrodynamics, 42
metric with mass and charge (Reissner-Nordstrøm), 41
metric with mass, charge, and angular momentum (Kerr-Newman), 41
Charge, electromagnetic, current density, 21, 56
Child-universes, 219
Choquet-Bruhat, 277–292
Christoffel, 19
Christoffel symbols, 21, 22–23, 25, 58. See also mathematical appendix
Chronology condition, 71
"Chronology Protection Conjecture," 243
Ciufolini, 334
Classical mechanics, inertia, and absolute space, 249, 250
Closed causal path (causality violation), condition of absence of, 304
Closed form, 55
Closed in time model universes
and inertia, 250, 252–254
conjectures of sufficient conditions for, 231
must have spatial topology of three-sphere, or two-sphere times the circle. . . , 226–230
sufficient conditions and spatial topology for, 230–231
"Closed," spatially, Friedmann cosmological models "open" in time, 225–226
Closed "space," 226, 290
Closed timelike curves
absence of, in Gödel rotating model universes of "second type," 244, 247
and causality, 243–244
Closed trapped surface, 70
"Closure in space," 224
versus "closure in time" of cosmological models, 220–231
"Closure in time," 224, 399
and spatial curvature in Friedmann models, 209–214
spatial topology and spatial compactness, 277, 278

"Closure in time (*cont.*)
versus "closure in space" of cosmological models, 220–231
COBE. See Cosmic Background Explorer
Coefficient of bulk viscosity, 26
Coefficient of shear viscosity, 26
Cold dark matter, 193
Collapse, gravitational, as seen by observers falling in and by faraway observers, 63–64
Collapsed star, 62–63, 71
Coma Cluster, 186
Commutator of two vector fields, 232
Comoving observer, 130–132, 235–238
Compact Cauchy surface, 230, 285, 396–398
Compact space, 270, 273–275, 396–399
Einstein ideas on, 271–273
Gödel universe of "first type," spatially noncompact, 243–244
hypothesis of, 273–275, 278
Compact spatial topology, 273–275, 278
and closure in time, 220–231, 277, 278
of cosmological models and inertia, 231, 250, 252–254
Compact, spatially, Friedmann cosmological models expanding for ever, 224–226
Compact, spatially, model universes, 220–231, 273–274, 300, 303, 304
and inertia, 250, 252–254
Compass of inertia, 248. See also gyroscopes
Composition-dependent new force, hypotheses of, and constraints on, 93–97, 98
Composition-independent deviations from inverse square law, 93–97, 98
Composition-independent new force, hypotheses of, and constraints on, 93–97, 98
Conditions
chronology condition, 71
dominance of energy condition, 49, 230, 286
generic condition, 71
null convergence condition, 70
positive pressure condition, 230, 231
strong energy condition, 71
timelike convergence condition, 71
weak energy condition, 71
Confinability, condition of
in electrostatics, 290
in general relativity, 290–291
Confinable source, 289–291, 290
Conformal
current of mass-energy, 283–284, 286, 289–291, 293, 305
"degrees of freedom" of conformal three-metric, 288
density of gravitational-wave energy, 292
density of mass-energy, 283–284, 285, 292, 293, 305
distortion tensor, 284–285, 286–287, 291, 293, 304–305, square of, 292, 305
extrinsic curvature, 24, 226, 227, 276, 286, 293. See also mathematical appendix
initial-value equation for conformal scale factor, 292
Killing vector, 290
quantities, 270
rate of change of conformal three-geometry, 283
scale factor, 270, 284, 288, 292, 306
three-geometry (three-metric), 279, 283–284, 285, 287, 289, 293, 304
Connected sum, 278
of manifolds, 227, 230
Connecting vector between fluid particles, 236–238
rate of change of, 237–238
Connection. See also mathematical appendix
metric compatible, 58
non-Riemannian, 136–137
Riemannian, 58
Connection coefficients, 21. See also mathematical appendix
difference between two sets of connection coefficients, 23
from Palatini variational principle, 25
Connection 1-forms, 57
Conscience-guided satellite, 363–365. See also drag free satellite
Conservation laws, 42–49
for charge, 42
local, in metric theories, and cosmology, 208
Conserved quantities in general relativity
angular momentum, 47–48
energy, 47–49
four momentum, 47–48
Constant, cosmological, 208–209. See also cosmological constant
Constants of motion
for geodesic motion, 117
for photon in spherically symmetric static spacetime, 117–118
"Constants," physical
hypotheses on variations in time and space of, 109, 335

hypothesis of time changes of, 277
Continuity equation, 29
Contorsion tensor, 137
Contravariant component of tensor, 20. See also mathematical appendix
Contravariant tensor, 20. See also mathematical appendix
Convergence condition, timelike, 230
Coordinate basis, 20. See also mathematical appendix
Coordinate time
in a static spacetime, 100, 102
in a stationary spacetime, 100, 101
Corinaldesi-Papapetrou, 352
Coriolis-type forces, 248
Coriolis-type and centrifugal-type forces inside rotating shell, 323–324
Corner reflectors on the Moon, 114
Cosmic Background Explorer, COBE, 186, 189–192, 217
Cosmic microwave background radiation, 186, 189–192, 217
anisotropy, measurements of, 186, 189–192. See also Cosmic Background Explorer
Cosmic strings, 219
Cosmological
density of mass-energy, 250, 254
dynamics of spatially homogeneous and isotropic models, 207–214
"fitting problem," 199
gravitational waves, 248–249
models, rotating, Bianchi IX, 244–249
principle, 185
term, 241. See also cosmological constant
Cosmological constant, 208–209, 218, 244, 245, 275
Einstein ideas on, 275
null, 248, 278
Cosmology
and dynamics of geometry, 207–214, 294
and inertia, 249–255
and "local conservation law" in metric theories, 208
and metric theories, 208
Coulomb gauge in electromagnetism, 317
Covariant component of a tensor, 20. See also mathematical appendix
Covariant derivative, 20. See also mathematical appendix
conformal, 287
of metric tensor (with Riemannian connection) is zero, 22, 26–27
Covariant tensor, 20. See also mathematical appendix
Creation, continuous, of matter, 242. See also "steady-state" model universe
Creation of particles at horizon of black hole, 68
Critical density of mass-energy, 211, 250, 254
Critical mass of a neutron star, 63, 71
Critical point for integral of squared spacetime interval (geodesic), 29. See also mathematical appendix
"Cross section," classical resonance integral of absorption cross section of a resonant detector, 152
Currents
conformal, of mass-energy, 283–284, 286, 289–291, 293, 305
in electromagnetism and magnetic field, 317–318
of mass-energy, 293, 358
of mass-energy and gravitomagnetism, 355–360
of mass-energy and inertia, 398–399
Curvature. See also mathematical appendix
conformal Ricci scalar of, 292
curvature index, energy density and deceleration parameter in Friedmann models, 213, 214
curvature invariants and singularities, 65
detector, 35
extrinsic curvature, 226, 227, 276, 286, 293. See also extrinsic curvature
intrinsic curvature, 226
invariants, 22, 355–360
negative spatial curvature Friedmann models and homogeneity hypersurfaces, 222
of space and pericenter advance, 144
of space, Gauss idea of, 2
of space, Riemann idea of, 4
of space, tests of, 116–117, 121, 127, 128, 138–139, 140, 352
of spacetime, tests of, 116–117
of spacetime versus curvature of space, 4
of two-surfaces, 3, 221
perturbation, 74
positive spatial curvature Friedmann models and homogeneity hypersurfaces, 220–231
Riemann curvature tensor, 21

Curvature (*cont.*)
scalar, 22, 280
singularities, curvature, parallelly propagated, 69
singularities, curvature, scalar polynomial, 69
2-forms, 57
zero spatial curvature Friedmann models and homogeneity hypersurfaces, 222
Curvature invariants
and angular momentum, 355–360
and gravitomagnetic field, 353–360
and mass, 355–360
and mass currents, 355–360
Cygnus X-1, 62
Cylinder, 221
Cylindrically symmetric solution, 244

d'Alambertian, 72
on a curved manifold, 168
Damping force in geodesic deviation equation with force term, 148
Damping time of resonant detector, 151
Dark matter, 192–193, 213–214. See also missing mass
hot, 192
cold, 193
Darmois, 277, 280
Data, initial-value, 293. See also initial value problem
Davidson, 388
Deceleration parameter, 207, 208, 211–214, 231, 250, 254
curvature index, and energy-density in Friedmann models, 213, 214
determination of, 212–213
Decomposition of a symmetric two-covariant tensor, 286
Decomposition of a tensor using projection tensors, 235
Deep pencil-beam survey, 186
Deflection of electromagnetic waves
by a mass, 117–122, formula of, 120, interpretation in general relativity, 120
by angular momentum of Sun, 122
by mass of Jupiter, 122
experimental results, 121–122
higher order, 122
tests of, with Very Long Baseline Interferometry, 121–122
Deformation of fluid element, 48, 238
"Degrees of freedom"
of conformal three-metric, 288
of electromagnetic field, 287–288
Dehnen and Hönl, 248
"Delay-Doppler-mapping," 127
Delay in propagation time of electromagnetic waves by a mass, 122–128
formula of, 126
interpretation of, 126, 128
tests of, 127–128
Density, critical, in Friedmann models, 211, 250, 254
Density of gravitational-wave energy, conformal, 292
Density of mass-energy, 293
conformal, 292, 293, 305
Density parameter, 212, 217
de Sitter, 7, 145, 331
de Sitter effect, or geodesic (geodetic) precession, 121, 128–136, 251, 255, 330, 338, 339, 348, 350, 351–353, 361
and Lense-Thirring effect, 353–360
formula of, 133
interpretation of, 133–134, 136, 351
measurements of, 134–136
on orbiting gyroscopes, 332–333
de Sitter universe, 218
Determinant of a two-tensor, 54
Deviation of spinning particle from geodesic motion, 352
Deviations from inverse square law, hypotheses and null results, 93–97, 98
Dialogues Concerning Two New Sciences, 13. See also Galilei
Dicke, 254
Differential form, 50. See also mathematical appendix
Differential geometry, 19–21, 403–436. See also mathematical appendix
Dipole gravitational radiation in alternative gravity theories, 147
limit on, 147
Dipole, magnetic, moment in electromagnetism, 318
Dipole radiation in electromagnetism, 391
absent in general relativity, 147, 391
Dirac cosmology, 109–110, 113
Discorsi e Dimostrazioni Matematiche Intorno a Due Nuove Scienze, 91. See also Galilei

DISCOS (DIsturbance COmpensation System), 363. See also drag free satellite
Displacement between two masses connected by spring and gravitational waves, 148–151, 149
Disquisitiones Generales Circa Superficies Curvas, 19. See also Gauss
Distance to astronomical objects, methods to determine, 204–205, 207
Distance, velocity versus distance relation, 186, 203
Distance versus redshift relation in Friedmann models, 203–207. See also Hubble law
Distortion tensor, 241, 270, 282, 286–287, 296, 304–305
  and energy currents, 398
  conformal, 283–284, 287, 291, 293, 304–305, "square" of, 292
  total, 293
Distribution and motion of mass-energy in the universe, determining geometry and local inertial frames, 250, 252–253, 300–303, 399
Distribution, redshift, of galaxies, 186
Divergence
  covariant, of vectors and tensors, 22, 27
  free, tensor, 286–287
  theorem, four-dimensional, 23, 24, 45. See also mathematical appendix
  theorem, or Ostrogradzky-Green formula, 48, 55. See also mathematical appendix
Dominance of energy condition, 49, 230, 286
Doppler tracking of spacecraft and gravitational wave detection, 162–163
Double image quasar Q0957+561, 122
Drag-free satellite, 333
Dragging
  of Foucault pendulum, 330
  of gyroscopes by Earth angular momentum, Gravity Probe B, GP-B, 8–9, 332–333
  of gyroscopes by Earth angular momentum, LAGEOS, 9, 334–339
  of gyroscopes by Earth mass rotation, 6–7
  of inertial frames, 249–255, 250–251, 303, 315–324, 321, 323, 353–360, 361, 399
  of inertial frames and dragging of nodal lines, 322, 338. See also Lense-Thirring effect
  of inertial frames and gravitomagnetism, 315–324, 330–331, 353–360
  of inertial frames and gyroscopes, 250–255, 321, 323
  of inertial frames and initial-value problem, summary of, 303–306
  of inertial frames and rotating sphere of fluid, 8
  of inertial frames by mass-energy currents, 248–249, 306
  of inertial frames due to gravitational waves, in cosmology, 248–249
  of inertial frames, gravitomagnetism and initial-value problem, 302
  of inertial frames, proposed measurements of, 148, 330–351
Drag, particle, on Earth satellite, 340, 342, 345
Drever, 112. See also Hughes-Drever
DSN, Deep Space Network, 103, 128
Dual basis of a frame, 57. See also mathematical appendix
Dual of Riemann tensor, 355
Dust like matter, 208, 286
Dust particles, equation of motion (geodesic), 29
Dynamical components of Einstein field equation and equations of motion, 306
Dynamical equations
  consequence of field equation and Bianchi identity in general relativity, 28, 44, 139–140
  of electromagnetism, from boundary of the boundary principle, 56–57
  of general relativity, consequence of boundary of the boundary principle, 59–61, 60
Dynamical field coordinate in geometrodynamics, 288
Dynamical field momentum in geometrodynamics, 288
Dynamical geometry in general relativity and inertia, 300–303
Dynamics
  of matter-energy in Friedmann models, 207–214
  of spatially homogeneous and isotropic cosmological models in general relativity, 207–214
  of three-geometry, 277, 293–295

Earth
  albedo, 339–340, 344, 347
  angular momentum, 328, 330, 336, 350
  angular velocity, 134–136, 349–350
  as gravitational-wave detector, 154

Earth (*cont.*)
comoving local frame, 134–136
crust, 345
crustal movements, plate motions, polar motion and rotation, 335
mass, 328, 330
mines and borehole measurements of G, 94–95
orbiting gyroscopes, 332–333
pole, coordinates of, 345
quadrupole moment, 335, uncertainty of, 335
rotation, 329–330, 361–362
rotation axis, 345
rotational orientation, UT1, 134–136, 345, 352, with VLBI, 121, 345, parameters, 121, 135
tides, solid and ocean, 339, 343–344
Earth-LAGEOS-Moon measurements, 96–97, 98
Earth-Moon
gyroscope, 352
local frame, 134–136, 352
system and tests of the strong equivalence principle, 113–115
Earth-Moon-Sun system, 110, 113, 114
Eccentric anomaly, 322
Eccentricity, orbital, 322
Eddington and Dyson expeditions, 120
Eddington-Finkelstein coordinates, 65
Effective energy of gravitational waves, 76, 78, 295–296, 297, 398
Effective temperature of resonant detector, 153
Einstein, 4, 14, 139, 209, 269, 300, 387, 394, 395
and Wheeler ideas on compact space, and Gödel universe of "first type," 244
and Wheeler ideas on inertia and boundary conditions of field equation, 253
and Wheeler ideas on origin of inertia and spatial compactness of universe, 224, 231, 253
equivalence principle, 14
equivalence principle, tests of, 90–115
field equation, 21–27, 24, 218, 241, 289
field equation and cosmology, 208–209
field equation and initial-value formulation, 284
field equation and intrinsic and extrinsic curvatures, 226
field equation, geometrical content and meaning as dual of moment of rotation equal to energy-momentum, 60–61
field equation, in vacuum, 39
field equation, nonlinearity of, 140, 292
field equation on spacelike hypersurface, 227
field equation, tests of, 88, 139–140
field equation with cosmological term, 208
general relativity and boundary of the boundary principle, 49–61
general theory of relativity, or geometrodynamics, 13–78
geometrodynamics and very strong equivalence principle, 13–18
ideas on inertia, 4, 271–273, 324–326
ideas on universe spatially compact, 224, 271–273
letter to Mach, 4, 394
static universe, 241
Einstein-Cartan theory, 137
Einstein-Cross, or gravitational lens G2237+0305, 123
Einstein-de Sitter cosmological model, 210
Einstein-Rosen bridge, 66
Electric components of Riemann tensor, 355
Electrodynamics
divergence-free vector field, 287–288
quantum, 274
Electromagnetic field
and boundary of the boundary principle, 56–57
and Reissner-Nordstrøm solution, 41
"degrees of freedom" of, 287–288
tensor, 21, 355
Electromagnetic radiation
and lines of force of accelerated charge, 388–390, 389
generated by accelerated charge, Thomson's explanation, 388–390, 389
Electromagnetic waves
deflection by a mass, formula of, 120
interpretation in general relativity, 120
delay in propagation time by a mass, 122–128
Electromagnetism
and boundary of the boundary principle, 56–57
and general relativity, analogies and differences, 44–45, 56–61, 287–289,

290–291, 297, 316–323, 324, 353–356, 390–391
invariants of, 355
Electrostatic field, 291
Electrostatics, condition of confinability, 290
and initial-value problem, 396–397
Elliptic initial-value differential equations, 284, 298
Ellis, 199, 274
Ellis, "Fitting problem" in cosmology, 199
Energy
and momentum of gravitational field, 44–48
detectable with resonant bar, minimum, 153
effective, of gravitational waves, 76, 78, 295–297
energy current, and distortion tensor, 398
energy flow and energy density (conformal) on a compact spacelike hypersurface, 270
energy flux of weak, monochromatic, plane gravitational wave, 76, 152
in general relativity, 47–49
in special relativity and conservation laws, 42–44
of gravitational field and equivalence principle, 45
Positive Energy Theorem, 48–49
total, in general relativity, ADM formula for, 48
Energy condition
dominance of energy, 230, 286
strong, 230, 231
weak, 228
Energy density
conformal, 285, 292
cosmological, 212–214
curvature index, and deceleration parameter in Friedmann models, 213, 214
function of scale factor in Friedmann cosmological models, 208
Energy, "Mass-Energy Rules Inertia," 303
in general relativity, and origin of inertia, 270
various interpretations of, 249–255
Energy-momentum
pseudotensor for gravitational field, 45–48
pseudotensor of weak, monochromatic, plane gravitational wave, 76, 152
tensor of electromagnetic field, 25–26
tensor of fluid, 25–26, 208
tensor of fluid, its relation with expansion, shear, and vorticity of the fluid, 239–240
tensor of matter and fields, 24–26
tensor of perfect fluid, 26, 208
tensor, post-Newtonian parametrized, 167
Eötvös, 90
Eötvös experiments, 93
reanalysis, 93–97, 98
Equation, initial-value
for conformal scale factor, 292
for gravitomagnetic vector potential, 289
Equation of motion of orbiting test particle, 141–144, 320
Equations of motion, 27–31
and Bianchi identity, 28
of test particles and photons (geodesic), 30–31
of test particles (geodesic), tests of, 88, 139–140
Equivalence principle, 1, 13–18, 97, 100, 139, 249, 353, 394
and energy of gravitational field, 44–45
and local Minkowskian structure of spacetime, 138
and Lorentz invariance, 111–112
and pericenter advance, 144
and spacetime location invariance, 109–111
medium strong form, 14–18
tests of, 87, 90–115, medium strong equivalence principle, 90–112, very strong equivalence principle, 90, 112–115, weak equivalence principle, 90–97
very strong form, 14–18
weak form, 13–14
Eridani B, 103
Error budget of LAGEOS III experiment, 343–347 table of, 346
ESA, European Space Agency, 122
Euclid, 1, 2
Euclid's 5th postulate, 19
Euclidean flat three-manifold in Friedmann cosmology, 222
Euler equations, 29
Evaporation of black hole, 67–69
Event horizon, 65–67, 214, 216, 354
Evolution equations, 293
of spacetime geometry, matter, and fields, 306
Evolution of fields, motion of matter, and initial-value problem, 294
Exact form, 55

Exact solutions of Einstein field equation, 36–42
Existence and uniqueness
  of initial-value equations, 291, 292, 397
Exotic differentiable structure, and manifolds, in four dimensions, 229
Exotic particles and cosmology, 193, 213
Expanding
  for ever, Friedmann cosmological models, 210–211
  for ever, spatially compact model universes, example of, 224–226, 230
  spatially homogeneous, and rotating model universes, 244–249
Expanding and recollapsing
  Friedmann cosmological models, 210
  rotating Bianchi IX models, 244–249
Expansion
  of universe, 186–192, 203–207, 212
  expansion scalar, 26, 235–236, 238, 246
  in Bianchi IX cosmological models, 247
  null in Gödel model of "first type," 241, 242, 244
  stage of maximum expansion, 285
Experiments
  of Einstein geometrodynamics, chapter 3, 87–184, and chapter 6, 315–383
  proposed to measure gravitational waves, 147, 148–163
  proposed to measure gravitomagnetic field, 330–351, 331, on Earth, 331, 347–350, orbital, 331, 332–347, 350–351
Exponential expansion of scale factor in inflationary models, 218
"Extended" inflation, 219
Extension, analytic, of Schwarzschild solution, 67
Exterior
  covariant derivative, 58, 60
  derivative, 52–53. See also mathematical appendix
  derivative of an exact form, 55
  product, or wedge product, 51–52. See also mathematical appendix
Extrinsic curvature, 24, 226, 227, 276–277, 286, 293, 306. See also mathematical appendix
  and Einstein equation, 226
  constant trace of, 276–277, 289, 293, 304, 305
"Extrinsic geometry," 294
Fabry-Perot interferometer, 154
Fairbank, 332
"False vacuum" in inflationary models, 218
Fermi, 274
Fermi, local, frame, 134, 135, 237–238, 241–243, 244, 249
Fermi normal coordinates, 16
Fermi-Walker transport, 128, 134, 137, 149, 237, 238, 242, 300, 302, 361. See also mathematical appendix
  of angular momentum vector of spinning particle, or gyroscope, 132
Ferromagnet, 348, 349
Feynman, 388
Feynman sum over all histories, 279
Fiber optical interferometer, 365–367, 371, 373
Fiber resonator, 365–367, 371, 373–374. See also ring laser
Field coordinate, geometrodynamic, 293
Field equation
  in general relativity, 21–26, 115
  in general relativity and variational principle, 21–25, 280
  linearized, 72
  linearized, in the Lorentz gauge, 72–73
  of general relativity, nonlinearity of, 140
  weak-field, classical limit of, 24–25
Field equations in general relativity and other metric theories, tests of, 139–140
"Fifth" force coupled to baryon number, hypotheses and null results, 93–97, 94, 98
Filamentariness, 185
Filaments, cosmological, 185, 192
"Fitting problem" in cosmology, 199
Five-point curvature detector, 35
Fixed stars, 329
Flat
  asymptotically flat, nondynamical geometry, 252–253
  asymptotically flat space, 274
  Euclidean three-manifold in Friedmann cosmology, 222
  spatially flat, Friedmann cosmological models, 209–210
Flatness problem, 217, 218
Flow. See currents
  mass flow, 291
  mass-energy flow, conformal, 286, 289–291, 305
Fluid element, deformation of, 238

Fokker, 134
Foliation
  and Schwarzschild geometry, 277
  of Bianchi IX models, 245
  of spacetime by spacelike hypersurfaces, 276–277
Force on moving charge, in electromagnetism, 318
"Force" on test particle, in geometrodynamics, 320–321, 323
Form, second fundamental form, 24, 276. See also extrinsic curvature, and mathematical appendix
Formal invariance of tensor field, 36, 193–194
Formation of galaxies, 217
Foucault, 361, 386
Foucault pendulum, 328–330, 361, 362, 386
  at Pole, 347
  dragging of, 330
  Einstein on, 270
Four-momentum
  in general relativity, 47–49
  in special relativity and conservation laws, 42–43
Four-simplex and its boundary, 52
Frame dragging, 300, 350. See also dragging of inertial frames
  of Foucault pendulum, 330
"Frame dragging" in Einstein letter to Mach, 7
Frame
  astronomical, 134–136
  asymptotic, 134–136
  Earth-Moon, 134–136
  Fermi, local, 134, 135, 237–238, 241–243, 244, 249
  local, comoving with Earth, 134–136
  local inertial, 13–18, 134
  of distant quasars, 352
  planetary, 134
Free-fall, to test weak equivalence principle, 91, 93, 95
Free part of conformal distortion tensor, 287, 291
Freely falling frame and observer, 13–18, 361
Frequency shift of counterrotating electromagnetic waves in rotating circuit, 365–367, 371, 373–374
Friedmann, 203
Friedmann equation, 209
Friedmann geometry, 297
Friedmann metric, or
  Friedmann-Robertson-Walker metric, 192, 198, 201–202, 240
  derivation from spatial isotropy about one point plus spatial homogeneity, 195–198
  expression in different coordinate systems, 198, 201–202
Friedmann model universes, 185, 186, 192, 193–214, 209–211, 240, 243, 246, 247, 254, 277
  are nonrotating, 202
  "best fit" Friedmann model, 199
  cosmological fluid in, 202
  dynamics of matter-energy in, 207–214
  expanding and recollapsing, or "closed" in time, 210
  expanding for ever, or "open" in time, 210–211
  matter-dominated, 209–213
  nonrecollapsing and inertia, 253–254
  propagation of photons in, 203–207
  radiation-dominated, 214
  Ricci curvature scalar of, 201
  spatially compact but expanding for ever, that is, spatially "closed" but "open" in time, 224–226, 230
  spatially flat, 201, 209–210, 222–226
  time of maximum expansion of recollapsing models, 277
  with negative spatial curvature, 201, 210–211, 222, 223
  with positive spatial curvature, 201, 210, 220–231, 247
  Friedmann prediction of dynamical universe, 203
Friedmann-Robertson-Walker metric, 198. See also Friedmann metric
Fundamental form, second, 276, 286. See also mathematical appendix

Galaxy formation, 217
Galaxy M87, and supermassive black hole, 62, 204
Galaxy M100, and estimation of Hubble constant, 207
Galilei, 1, 13, 90, 91, 384
Galilei equivalence principle, 13
  tests of, 91–97
Galilei experiments of uniqueness of free-fall, 91

Galilei-tower experiment (weak equivalence principle test), 93, 95
GALILEO mission, 163
and gravitational redshift, 103
γ post-Newtonian parameter, 119, 121, 127–128, 138, 146, 350, 351
Gamow, 190, 275
Gauge transformation, 72
Gauge, standard, post-Newtonian, 140, 166
Gauss, 2
Gauss, *General Investigations of Curved Surfaces* , 19
Gauss intrinsic description of surface, 19
Gauss-theorem (divergence theorem), 55. See also mathematical appendix
Gauss theorema egregium, 19
Gaussian curvature, 3, 21, 32, 200, 221. See also mathematical appendix
Gaussian normal coordinates, 227, 245
"Generalized Cartesian coordinates," 37
"Generalized polar coordinates," 37
General relativity and Newtonian gravity theory, 249, 250, 358
General relativity, or geometrodynamics, chap. 2, 13–86, chapters 1–7
and electromagnetism, analogies and differences, 44–45, 56–61, 287–289, 290–291, 297, 316–323, 324, 353–356, 390–391
and inertia, summary of, 249–252, 394–399
and inertia, table, 250–251
inertia, and dragging of inertial frames, 249–252, 250–251
Generic condition, 71
GEO-600, 156
GEOS-1, 103
Geodesic completeness, 69–70. See also mathematical appendix
Geodesic completeness of Gödel model universe of "first type," 243
Geodesic, curve with tangent vector parallel to itself, 31
Geodesic deviation equation, 16, 31–36, 358
generalized to particles with arbitrary four-velocity, 35
with force term, 148
Geodesic equation, 29–31, 320. See also mathematical appendix
and Lorentz force equation, 391
as extremal line for integral of squared interval, 29–30
as extremal line for length, 30
equation of motion of test particles and photons, 30–31
of a photon, 117
of particles of dust, 29
Geodesic motion, spinning particle deviation from, 352
Geodesically complete manifold, 70. See also mathematical appendix
Geodesics. See also geodesic equation and mathematical appendix
incomplete, 70
on curved surface, 2, 3
Geodetic (geodesic) precession, or de Sitter precession, 121, 128–136, 351–352, 361. See also de Sitter effect
formula of, 133
interpretation of, 133–134, 136, 351
Geometrical structure of general relativity, 19–21
tests of, 87–88, 116–128, 136–139
Geometrodynamics, or Einstein's theory of general relativity, 13–78
and inertia, summary of, 249–252, 300–303, 394–399
condition of confinability, 289–291
field coordinate, 293
inertia, and initial-value problem, 269–275, 294–295, 295–296, 300–306
nonlinearity of, 292
Geometry
dynamical, 1
dynamical in general relativity and inertia, 252–253, 300–303
dynamics of three-geometry, 282
Geon, 44, 77–78
electromagnetic, 398
gravitational, 295, 398
stability and instability of, 63
Geophysical measurements of G, 94–97
Global hyperbolicity, 278
Globally hyperbolic
model universe, 231
spacetime, 277
Global rotation, 243
Gödel, 241
Gödel metric, 241
Gödel rotating model universes, 240–244, 304
of "first type" (Bianchi VIII), 240–243, and causality, 243–244, and global rotation, 243, properties of, 242–243

of "second type" (Bianchi IX), 244 (see Bianchi IX), and causality satisfied in, 244, 247
GPA, Gravity Probe A, 103
GP-B, 250, 252, 331, 332–333, 370. See also Gravity Probe B
GPS, Global Positioning System, 103
Gradiometers, 15, 91, 97, 363
superconducting, 350–351
to measure gravitomagnetic field, 350–351
with three orthogonal axes, 350–351
Grand Unified Theories, GUTs, 10
and cosmology, 217
Graviphotons, 96
Graviscalars, 96
Gravitation
and electromagnetism, 44–45, 56–61, 287–289, 290–291, 297, 316–323, 324, 353–356, 390–391
and other interactions of nature, 9–10
and space curvature, ideas of Riemann on, 19
Gravitational binding energy
and inertial to gravitational mass ratio, 14, 113
and the very strong equivalence principle, 14, 113
Gravitational constant
and Brans-Dicke theory, 109–111
cosmology and hypothesis of decrease of, 254
hypotheses on space variations of, 111
hypotheses on time variations of, 109–110
limits on anisotropy of, 111, 335
limits on variations in time and space of, 110–111, 335. See also deviations from inverse square law
Gravitational geon, 77–78, 295. See also geon
Gravitational lensing, 122
and cosmic strings, 219
G2237+0305, or Einstein Cross, 123
in galaxy cluster AC114, 124
Gravitational mass, 109
inertial mass, and tests of the strong equivalence principle, 113
Gravitational perturbations on Earth satellite, 339, 343–344
Gravitational radiation from binary pulsar and decrease of its orbital period, 146–147. See also quadrupole gravitational radiation
Gravitational redshift, 97, 99–108
Gravitational time dilation of clocks, 97, 99–108
Gravitational waves, 71–78, 147–163, 297, 315
and cosmological singularities, 298
and displacement between two masses connected by spring, 148–151, 149
and dragging of inertial frames in cosmology, 248–249
and dynamics of geometry, 294–296
and electromagnetic waves, 317
and proper distance between test particles, 74–75
cosmological, 248–249
detectors, 147–163, 295
effective energy of, 76, 78, 295–296, 297, 398
energy-momentum pseudotensor of plane gravitational wave, 76
influencing effect on "inertia," 296
plane, 73–75
proposed measurement of, 88, 147, 148–163, with Doppler tracking in space, 160–163, with laser interferometer in space (LISA, or LAGOS), 158–160, with laser interferometer near Earth (LINE or SAGITTARIUS), 160–162, with laser interferometers on Earth (LIGO, VIRGO, GEO-600, TENKO, . . .), 154–157, with microwave interferometer in space (MIGO), 157–158, with resonant detectors, 147, 148–154
quadrupole formula for emission of, 76–77
with "position-coded" memory, 76
with "velocity-coded" memory, 76
"Gravitinos" and cosmology, 193, 213
Gravitoelectric field, Lorentz invariance and gravitomagnetism, 353–360
Gravitomagnetic contribution to pericenter advance, 144–145
Gravitomagnetic dragging of Foucault pendulum, 330
Gravitomagnetic effects, 315–324. See also dragging of inertial frames and Lense-Thirring effect
on gradiometers, 350–351
on laser ring gyroscopes, 349–350
on orbiting gyroscopes, 320–321, 322–323, 331, 332–333. See also Gravity Probe-B
on satellites, 321–322, 331, 334–339. See also LAGEOS III

Gravitomagnetic effects (*cont.*)
on superconducting gyroscopes, 347–349
Gravitomagnetic field, 8, 270, 288, 293, 315–374, 315–324, 326, 353–360
and counterpropagating pulses of radiation, 106, 349–350
and dragging of inertial frames, proposed tests of, 88, 330–351
and jets, 327–328
and magnetic field, 8, 315–324
and metric theories of gravity, 355
and synchronization, 105–107
far from a rotating sphere, 320
in astrophysics, 327–328
in physics, 330–331
inside a rotating shell, 322–324
of a spinning black hole, 65
proposed measurements of, 148, 330–351, 331, with laser ranged satellites, 331–332, 334–347 (see LAGEOS III), with orbiting gyroscopes, 331, 332–333 (see Gravity Probe-B), with orbiting superconducting gradiometer, 350–351, with ring laser gyroscopes, 349–350, with superconducting rotation sensors, 347–349
proposed measurement by general relativistic "Sagnac effect," 107
Gravitomagnetic parameter μ, 129–130, 350
Gravitomagnetic potential, 134, 270, 284–285, 289–291, 316, 319–320, 330, 396–397
Gravitomagnetism, 7–9, 315–374, 315–324, 353–360
absence of, in Galilei-Newton mechanics, 316
and curvature invariant defining it, 353–360
and dragging of inertial frames, 8, 315–324, 330–331, 353–360
and inertia, 250, 252
and inertia in general relativity, 250, 252, 324–326
and magnetism, 7–8, 315–324, 353–356
and metric theories of gravity, 355
characterization of, 353–360
dragging of inertial frames, and initial-value problem, 302
gravitoelectric field and Lorentz invariance, 353–360
invariant characterization of, 353–360, 356–358
of Earth, 330, 332–333, 336, 361
Gravitons, 96, 295
Gravity, poetry on, 9
Gravity Probe B, or GP-B, 65, 315, 331, 332–333, 370
"Great Attractor," 186
Great Wall, 186
Green function and initial-value problem, 275
GUTs, Grand Unified field Theories, 10, 217
Gyrogravitation, 7. See also gravitomagnetism
Gyroscope
Earth-Moon orbiting "gyroscope," 352
orbital, to measure gravitomagnetic field, 338
Gyroscopes, 241, 244, 248, 338, 352, 361–374
and cosmology and inertia, 249–255
and orbital angular momenta, dragging of, 338
and rotating mass-energy in the universe, 248–249, 254, 399. See also dragging of inertial frames
around spinning black hole, 65
at the North Pole, 6, 329–330
dragging of, by Earth mass rotation, 6–7, 330, 332–333, 336
effect on, due to gravitomagnetic field, 320–321, 322–323
frame dragging of, 300, 332–333. See also dragging of inertial frames
Gravity Probe B, 331, 332–333
in Bianchi IX rotating model universes, 248–249
influenced by moving matter in general relativity, 248–249, 306, 358, 394. See also dragging of inertial frames
in Gödel model universes, 242, 244
mechanical, 361–365
optical rotation sensors, 365–367, 371, 373–374
orbiting Earth, 331, 332–333
ring laser, 349–350, 361, 365–367, 371, 373–374, and gravitomagnetic effect, 105–108, 349–350
superconducting, 332–333, 347–349
transport of, 132–136
Gyro stabilized platform, 363

Handles, 228, 229
Harmonic coefficients, even zonal, 335
Harmonics of Earth, 336, 339, 344
Harress, 373
Hawking, 70, 243

Hawking evaporation process, 67–69
Hawking-Penrose singularity theorem, 71
Hierarchy of clusters, 193
Hilbert, 139, 280
Hilbert variational principle, 21–25, 280
Hipparcos, 122
Holonomic basis, 21. See also mathematical appendix
Homogeneity
  and isotropy, spatial, of a Riemannian manifold, 192, 193–202, 195, 201
  of universe on very large scale, 186, 187, 188
  plus isotropy about one point, 195, 196–199, 199, 223
Homogeneous
  and isotropic, spatially, model universes, 185–214
  hypersurface, 196–197, 220, 222
  metric, 195
  spacetime-homogeneous model universes, 240, 242
  spatially, cosmological models, 185, 193–214, 231–234, 240–249. See also Bianchi models
  spatially, expanding and rotating model universes, 244–249. See also Bianchi IX models
Horizon, 214–217
  event, 214, 216, particle, 214–216, Schwarzschild, 65–67, 354
Horizon problem, 217, 219
Hot dark matter, 192
Hoyle-Narlikar theory, 275
Hubble, 203, 209
  constant, 186, 203, 203–207, 211, 285
  expansion, law, 186, 203, 206, 209, deviations from, 186
  Space Telescope, 207, and observations of supermassive black hole, 62, 204, and estimation of Hubble constant, 207
  value of Hubble constant, 207
Hubble time and Friedmann model universes, 212, 240
Hughes-Drever, 90
Hughes-Drever experiment, 112
Hulse and Taylor, 146
Huygens, 1, 91, 385
Hydra-Centaurus supercluster, 186
Hydrodynamic equation, 327
Hydrogen-maser clocks, 103
Hyperbolic paraboloid, 221
Hyperboloid, three-hyperboloid in Friedmann cosmology, 222, 223
"Hyperextended" inflation, 219
Hypersurface
  Cauchy surface, or initial-value hypersurface, 230, 231, 285, 289
  element, three-dimensional, 54
  maximal hypersurface, 226, 226–228, 230

Icarus lander, 145
Inclination of LAGEOS, 335–337, 339, 341, 345
Inclination, orbital, 322
Incomplete geodesics, 70. See also mathematical appendix
Induced metric, 228. See also mathematical appendix
Inertia (local inertial frames)
  and absolute space, 249, 250
  and "closed" in time model universes, 250, 252–254
  and compact space, 250, 252–254
  and cosmology, 249–255
  and dragging of inertial frames, 249–252, 250–251
  and Friedmann model universes, 5, 250, 253–254
  and gravitomagnetism in general relativity, 324–326
  and gyroscopes, 249–255. See also dragging of inertial frames
  and gyroscope orientation determined by average flux of energy, 251, 254
  and mass-energy currents, 398–399
  and matter in the universe, and boundary conditions, Einstein ideas on, 326
  and noncompact space, 252–253
  and rotation in Bianchi IX cosmological models, 248–249
  and rotation of universe, 251, 254–255
  Einstein ideas on, 4, 271–273, 324–326
  in Einstein general relativity, 249–255, 296–297, a summary, 394–399
  inertia-determining power, 329–330
  inertia here influenced by mass and mass current there, 291, 399
  inertia here influenced by mass-energy there, 4, 5, 9, 274, 298–299, 393

Inertia (local inertial frames) (*cont.*)
"inertia here influenced, or determined by mass-energy there," summary of, 300–303
inertia of a body influenced by nearby masses and null Hughes-Drever experiment, 112
in Galilei-Newton mechanics, 249, 250
Kepler introduction of word "inertia" in physics, 384
"law of inertia," 384
local compass of inertia, 244. See also gyroscopes
local inertial frames, 13–18, 134, 249–255
local inertial frames influenced by masses, 298–299
origin of, and initial-value problem in Einstein geometrodynamics, 252, 269–306, 294–295
origin of, and initial-value problem, summary, 300–306
origin of, in Bianchi IX rotating models, 248–249
origin of, in general relativity and cosmology, various interpretations of, 249–255
Sciama sum for, 392–393
Inertia-influencing effect of gravitational wave, 248–249, 296
Inertia-influencing effect of mass in motion, 315. See also dragging of inertial frames
Inertial frames, 249–255, 250, 251, 298–300, 328, 384
asymptotic, quasi-inertial frames, 134–136
dragging by Earth angular momentum, 330, 332–339
local, 13–18, 134, 296, 303, 352, 361, 365
local, and distribution and flow of mass-energy, 395
local, and spacetime metric, 395
local, determined by energy and energy currents, 249–255, 270, 300–306, 398–399
local, influenced by mass-energy currents, 249–255, 300–306, 398–399. See also dragging of inertial frames
local, initial-value problem and cosmology, 253
Inertial mass, 14, 109, 113
and gravitational mass, and tests of the strong equivalence principle, 113
Inertial navigation, 363
and gravitomagnetism, 358–360, 359
future, of a spaceship, 358–360, 359
Inertial sensor, 349, 361, 362, 365. See also gyroscope
Inertial to gravitational mass ratio, 14, 113
and gravitational binding energy, 14, 113. See also inertial mass
Inextendible. See also mathematical appendix
curve, 70
manifold, 69–70
Inflation, 218–219, "extended," 219, "hyperextended," 219
Inflationary cosmological models, 217–219
chaotic, 219
new, 219
old, 218
Infrared radiation from Earth and artificial satellites, 342–343
Inhomogeneities, "local," of universe, 185–186
Initial conditions in Einstein geometrodynamics, 148, 269–306
Initial singularity in cosmology, 212. See also "big bang"
Initial-value problem, 252, 269–306
Initial-value data, 293
Initial-value data
plus dynamic equation, 298
and physically observable quantities, 306
Initial-value Einstein field equations, 289, 292, 304–305. See also initial-value equations
Initial-value electric field, 291
Initial-value equation
and effective density of gravitational waves, 296
scalar, for conformal scale factor, 292, 294, 305
vector, 289, 294, 305
Initial-value equations, 279, 284, 289, 292, 294, 297, 305, 396
and inertia, 298
solution of, 289–292
Initial-value formulation, 269, 269–306. See also initial-value problem
and causality, 274–275, 396–397
and singularities, 397
in electrodynamics, 274–275
in geometrodynamics, 269–306, 303, 396–399
in geometrodynamics and origin of inertia, 269–306, 271–275, 294–295

in geometrodynamics versus integral over past light "cone" formulation, 297–298
versus retarded potentials, 396–397
Initial-value hypersurface, 285, 289, 294. See also Cauchy surface
Initial-value problem, 5, 269–306. See also initial-value formulation
and astrophysics, 276
and causality, 274–275, 396–397
and dragging of inertial frames, summary of, 303–306
and origin of inertia, summary of, 300–306
gravitomagnetism and dragging of inertial frames, 302
in geometrodynamics, 269–306, 303, 396–399
in geometrodynamics and origin of inertia, 269–306, 271–275, 294–295
Integration of a form. See also mathematical appendix
on boundary of a k-chain, 54–55
on boundary of a manifold, 55
Interferometer, ring, 349, 365–367, 371, 373
Interferometric astrometry space mission, 122
Interplanetary Doppler Tracking, 147, 162–163
range of operation, 163
sensitivity, 163
sources of noise, 162–163
Intrinsic curvature, 226
and Einstein equation, 226
Intrinsic description of surface by Gauss, 2, 19
"Intrinsic geometry," 294
Invariance, formal, of metric tensor (isometry), 36–37, 193–195, 194, 232
Invariants
curvature, 355–360
electromagnetic, 355
Inverse square law
constraints on composition-dependent deviation from, 98
constraints on composition-independent deviation from, 98
deviations from, hypotheses and null results, 93–97, 98
tests of, 93–97, 98, 335, 350–351
Isometry, 36–37, 194, 232 infinitesimal, 232
Isotropic and homogeneous, spatially, cosmological models, 185–214
Isotropic coordinates, 122–123, 129
Isotropic metric, 195
Isotropy
about every point, 199, 201
about one point plus homogeneity, 195, 196–199, 199, 223
and homogeneity, spatial, of a Riemannian manifold, 192, 193–202
group, 195
of a manifold, 195, 199, 200–201
of cosmic blackbody radiation: anisotropies detected by COBE, 186, 189–192
of cosmological background radiation and limits on rotation of universe, 251, 254–255
large scale, of universe, 186–192
Italian Space Agency, ASI, 346
Ivanenko process, 44, 295

Jacobi
equation, 31–32
identity, 195, 232, 233. See also mathematical appendix
Jacobian, 49, 233
Jets, 327–328, 347, 368–369
alignment, 327–328
and gravitomagnetic field, 327–328
constant direction of, 327–328
Jordan-Brans-Dicke theory, 14, 109, 113, 114, 139, 146, 219. See also Brans-Dicke theory
JPL-Caltech, 146
Jupiter
fifth moon of, 331
deflection of radio waves by mass of, 122

Kantowski-Sachs models, 231
Kepler, introduction of word "inertia" in physics, 384
Kerr
solution, 42, 320, 356, 358
Kerr-Newman
solution, 41, 253, 316, 357
Killing, 290
Killing equation, 37, 194, 202, 241
Killing vector, 37, 194–198, 202, 232–233. See also mathematical appendix
algebra of Killing vectors, 195, 232
conformal, 290, 305
five Killing vectors of Gödel rotating solution, 242
maximum number on a n-manifold, 195, 199
representing spatial homogeneity, 196–197

Killing vector (*cont.*)
representing spatial spherical symmetry, 37, 196
representing spatial isotropy, 195–196
timelike, 40, 241, coordinate system adapted to a timelike Killing vector, 40, and static metric, 40, and stationary metric, 40
Klein bottle, or twisted torus, 50
Knots, cosmological, 185
Kretschmann, 296, 387
Kretschmann curvature invariant, 355–360
for Schwarzschild metric, 69
Kronecker tensor, 20. See also mathematical appendix
Kruskal-Szekeres coordinates, 65, 67
Kuchar, 281

LAGEOS (Laser Geodynamics Satellite), 9, 65, 88, 96–97, 98, 115, 146, 250, 252, 315, 334–347, 371, 372
perigee advance, 146
LAGEOS and LAGEOS III, 334–347, 337
and tests of gravitational theories, 96–97, 98, 335, 336–339
determination of orbital parameters, 345
inclination, 335–338, 339, 345, 347
Lense-Thirring precession, 336
nodal longitude, 345
nodal precession, 335–338, 337
orbital eccentricity, 335, 337
parameters, 335, 337, 340, 341
perigee, 335
perturbations, 339–347
retroreflectors, 340
semimajor axis decay, 342
spin, 341–343
LAGEOS-Earth-Moon measurements, 96–97, 98
LAGEOS satellites with supplementary inclinations to measure gravitomagnetism, 9, 331, 334–347. See also LAGEOS III
LAGEOS II, 346
picture of, 371
LAGEOS III experiment to measure gravitomagnetic field, 9, 331, 334–347, 337. See also LAGEOS and LAGEOS III
as enormous gyroscope, 338
error budget, 343–347, table of, 346
error sources of, 339–347
improved error budget, 347
LAGOS, or LISA, 158–160. See also Laser Gravitational-wave Observatory in Space
Lagrange equations for test particle, 29
Lagrangian density for field equation in geometrodynamics, 22
Lagrangian for field equation in geometrodynamics, 22
Landau-Lifshitz energy-momentum pseudotensor for gravitational field, 47
Laplacian, 350
Laplacian equation for gravitomagnetic vector, 289
Laplacian operators on curved manifold, 289, 292
Lapse function, 294
LAser Gravitational-wave Observatory in Space (LAGOS, or LISA), 147, 158–160
astrophysical events observable with, 160
range of operation, 158, 160
scheme of, 159
sensitivity, 160
sources of noise, 158, 160
Laser Interferometer Gravitational-wave Observatory (LIGO), 147, 154–157
laser interferometers in the world, 156
range of operation, 156–157
scheme of Michelson laser interferometer, 155
sensitivity, 156–157
sources of noise, 156
variation of proper length of arms due to gravitational waves, 155
Laser Interferometer Near Earth (LINE) or SAGITTARIUS, 147, 160–162
astrophysical events observable with, 162
scheme of, 161
sensitivity, 162
sources of noise, 162
range of operation, 162
Laser ranged satellites, 335. See also LAGEOS and Starlette
Laser, ring, gyroscope (rotation sensor), 349–350, 365–367, 371, 373–374
and gravitomagnetic effect, 349–350
Latitude, argument of, 321
"Law without law," Wheeler's view of physics, 49
Leibniz, 296, 384
against absolute space, 384–386
Lemaître, 193
Lense and Thirring, 7, 322, 326, 331

Lense-Thirring effect and de Sitter effect, 353–360
Lense-Thirring effect, or gravitomagnetic precession, 7, 133–134, 250–251, 252, 316, 321–323, 322, 331, 333, 336–338, 345, 346, 347, 350, 353–360, 396. See also gravitomagnetic effect and gravitomagnetism
on Foucault pendulum, 330
on longitude of nodes, 322, 326, 331, 332, 335–338, 337, 338, 346
on longitude of pericenter, 322, 331
on mean orbital longitude, 322
on orbiting gyroscopes, 332–333
on orbit of test particle, 321–323
Lensing, gravitational, 122
and cosmic strings, 219
G2237+0305, or Einstein cross, 123
in galaxy cluster AC114, 124
Lenz vector, 322. See also Runge-Lenz vector
Levi-Civita, 19, 31
Levi-Civita pseudotensor, 53, 238, 355
Lichnerowicz, 277, 280, 284, 292
Lie algebra, 195, 232. See also mathematical appendix
Lie derivative, 24, 193–194, 202, 231, 236. See also mathematical appendix
of metric tensor, 36, 194, 232
Lie group, 195, 232. See also mathematical appendix
Lienard and Wiechert, 297
potentials of electrodynamics, 297
Light deflection by a mass, 117–122
Light pulses counterpropagating in rotating circuit, 104–108, 366–374
LIGO, 154–157. See also Laser Interferometer Gravitational-wave Observatory
LINE, 160–162. See also Laser Interferometer Near Earth
Lines of force of accelerated charge and electromagnetic radiation, 388–390, 389
LISA, Laser Interferometer Space Antenna, 158–160. See also LAGOS
Lobačevskij, 2, 19
Local Earth-Moon frame, 352
Local inertial frames, 13–18, 134, 296, 303, 352, 361, 365
initial-value problem, and cosmology, 253
Local Minkowskian structure of spacetime and equivalence principle, 116–117, 138
Local, quasilocal system, 298–299
"Location invariance," 90
London magnetic moment, 333, 348. See also Becker, Heller and Sauter
Long Baseline Interferometry, 121
Longitude, nodal, 321, 322, 336, 341
Lorentz boost, 130–131, 354, 355, 358
Lorentz equation, 34, 318
and geodesic equation, 391
Lorentz gauge
in electromagnetism, 73, 317
in geometrodynamics, 72, 319
Lorentzian character of spacetime, tests of, 116–117, 138
Lorentzian manifold (spacetime), 20, 138. See also mathematical appendix
in general relativity and other metric theories, 88, 139
Lorentz invariance
and equivalence principle, 111–112
gravitoelectric field and gravitomagnetism, 353–360
local, 90
Lorentz transformations, 130–131, 358, 360
Luminosity distance, 205
function of cosmological redshift, 206, 207
versus redshift, observations of, 186, 207, 212
Luminosity of a galaxy, decreased due to expansion of universe, 205
formula of, 205
Lunar Laser Ranging, LLR, 113–115, 121, 134–136, 255, 352
and test of weak equivalence principle, 93
and test of the equivalence of active and passive gravitational mass, 116
and test of very strong equivalence principle, 14, 113–115

Mach, 1, 4, 296, 324, 326, 386, 394
against absolute space, 386–387
Mach principle, 1, 269–270
a general relativistic interpretation of, 249, 303, 324, 331
a general relativistic interpretation of, and Gödel model universe of "first type," 242
and Hughes-Drever experiment, 111–112
Brans-Dicke theory, and change in time of gravitational constant, 109
Einstein ideas on, 4, 271–273, 324
MACHOs, Massive Compact Halo Objects, 214

MACSYMA, 356
Magnetic components of Riemann tensor, 350, 354–355, 358–360
Magnetic moment
  in electromagnetism, 317–318, 320
Magnetic dipole vector potential, 317
Magnetic field
  and gravitomagnetic field, 8, 315–324, 330
  and spatial topology of three-sphere with wormholes, 274
Magnetic moment, London, 333. See also Becker, Heller and Sauter
Magnetic monopoles
  absence in classical electrodynamics, 317
  hypothetical cosmological production of, 217
Magnetism and gravitomagnetism compared, 8, 315–324, 353–356
Magnetization
  of ferromagnet by rotation, 348–349
  of superconductor by rotation, 348–349
Magnetogravitation, 315–374. See also gravitomagnetism
Magnetomechanical tensor, 348
Manifold. See also mathematical appendix
  bundle complete or b-complete, 70
  geodesically complete, 70
  with boundary, 49
Map of galaxies, 187
Map of radio sources, 188
Mariner 6 and 7, 127
Mariner 9, 110, 127
Mariner 10, 110
Mars, 127
Mass
  and curvature invariants, 355–360
  flow, 291. See also mass-energy currents
  metric generated by a mass (Schwarzschild), 40–41
  missing mass problem, cosmological, 192–193, 212–213, galactic, 192–193, 213–214
  of Earth, 328, 330
  "rules inertia here," 5, 294, 296–297, 388. See also dragging of inertial frames and inertia
Mass, active and passive gravitational mass, and nonconservation of momentum and energy, 115
Mass-energy currents, 293, 358
  and curvature invariants, 355–360
  and gravitomagnetism, 355–360
  and inertia, 398–399
  conformal, 283–284, 286, 289–291, 293, 305
Mass-energy density, 293
  conformal mass-energy distribution, 283–284, 285, 305
  cosmological, 211–214
  curvature index, and deceleration parameter in Friedmann models, 213, 214
Mass-energy "there"
  curves spacetime "here," 394–395, 399
  influences "inertia here," 249–255, 300–303, 394–395, 398
  determines "inertia here," 249–255, 300–303, 399
Massive Compact Halo Objects, MACHOs, 214
Massive neutrinos and cosmology, 192, 213
Mass without mass, 78. See also geon
"Matter dominated," spatially homogeneous and isotropic, Friedmann cosmological models, 209–213
Matzner, 335
Matzner, Shepley, and Warren, 185, 245, 247, 248
Matzner-Shepley-Warren, Bianchi IX cosmological models, 245–248
Maximal hypersurface, 226, 226–228, 230
Maximally extended Schwarzschild metric, 66
Maximally symmetric manifold, 195, 199
Maximum expansion of Friedmann model universe, 210, 277. See also "closed" Friedmann model
Maximum number of Killing vectors and of conformal Killing vectors, 195, 199, 291
Maxwell equations
  and magnetism, 317
  sourceless, from boundary of the boundary principle, 56–57
  with source, 21, 42, 56–57
Mechanical gyroscopes, 361–365
Medium strong equivalence principle, 13–18
  tests of, 90–112
M87, elliptical galaxy, 62, 204, 369. See also black hole
Mercury, 110, 127
  orbiter, 145
  perihelion advance, and gravitomagnetic field, 144–145, 331, and quadrupole moment of Sun, 144–145, interpretation

of, 144, measurements of, 144–145, per century, 144, per revolution, 144
proposal for a Mercury orbiting transponder, 111
Metric, 20, 36–42, 193–202. See also mathematical appendix
conformal, 283–284, 285
post-Newtonian parametrized, 166
tensor, 20
Metric-compatible connection, 58. See also mathematical appendix
Metric theories of gravity, 88, 139, 163–168
and cosmology, 208
and their post-Newtonian limit, 163–168
counterexamples of metric theories not described by the Parametrized-Post-Newtonian, PPN, formalism, 167–168
Michelson laser-interferometer, 154, 155. See also laser interferometer
Michelson millimeter wave Interferometer Gravitational-wave Observatory (MIGO), 147, 157–158
range of operation, 157
scheme of, 157
sensitivity, 157–158
sources of noise, 158
MIGO, 157–158. See also Michelson millimeter wave Interferometer Gravitational-wave Observatory
Millisecond pulsar 1937+21 and gravitational redshift, 103, 128. See also binary pulsar
Milky Way, 329
Minimum number of test particles to determine spacetime curvature, 34–36
Minimum number of test particles to measure electromagnetic field, 34
Minkowski metric, 15, 117, 138, 329. See also asymptotically flat metric
Missing mass, 192–192, 212–214
Missing mass problem
cosmological, 192, 212–213
galactic, 192, 213–214
Mixmaster cosmological models, 297
Möbius strip, 49, 50. See also mathematical appendix
Moffat theory, 137–138, 144–147
its contribution to pericenter advance, 144–145
strong limits on, 146–147
Møller-Tulczyjew, 353
Moment of rotation, dual of moment of rotation of a vector induced by Riemann curvature, 60, 61
Momentum density, conformal, 305
M 100, spiral galaxy, and estimation of Hubble constant, 207
Monopole problem, 217, 218
Monopoles, magnetic
absence in classical electrodynamics, 317
hypothetical cosmological production of, 217
Moon-Earth gyroscope, 134–136, 352
Moon-Earth local frame, 134–136, 352
and tests of strong equivalence principle, 113–115
Moon-Earth-Sun system, 110, 113, 114
Moon-LAGEOS-Earth measurements, 96–97, 98
Moon perigee, 135–136, 352
and de Sitter effect, 135–136, 251, 255, 352
rate of, 146
Moon's Sea of Tranquility, 114
Moving frame, 57
μ, gravitomagnetic parameter, 350

n-chains, 50
n-cubes, 50
n-holed torus, 227–229
NASA, National Aeronautics and Space Administration, 103, 122, 123, 189–191, 204, 346–347, 370–372
Navier-Stokes equation, 327
Negative spatial curvature, Friedmann cosmological models with, 210–211, 213, 220, 222–223
Neptune, 328
Neutrinos, 286
massive, and cosmology, 192, 213
New inflationary models, 219
Newman-Penrose null-tetrad formalism, 356
Newton, 1, 91, 249, 328, 384, 385, 393
Newton bucket, or vessel, 328, 385
critique by Mach, 386–387
Einstein ideas on, 269–270
Ni theory, 139
Nodal longitude, 321–322, 336, 341
equation for rate of change of, 336, 341
Nodal precession of Earth satellite
due to anisotropic thermal radiation, 340–343, 344–347
due to tides, 339, 343–344

Nodal precession of Earth satellite (*cont.*)
due to albedo, 339–340, 344, 347
LAGEOS total, 336–337
LAGEOS classical, 335–338, 336, 337
uncertainties in LAGEOS classical nodal precession, 336
Lense-Thirring drag, 322, 326, 331, 332, 335–338, 337, 338, 346
Nodes, longitude of, 321–322. See also nodal longitude
Noncompact, spatially, model universes and inertia, 252–253, 300–303, 304
Nondynamical, asymptotically flat metric versus compact space, 252–253, 300–303, 304
Non-Euclidean geometries, 2, 19. See also curvature and mathematical appendix
Nongravitational perturbations of Earth satellite, 339–347
Nonlinearity of Einstein field equation and general relativity, 140, 292
and pericenter advance, 138, 144
Nonorientable manifolds, 49–50. See also mathematical appendix
Nonpolar, supplementary inclination satellites to measure gravitomagnetic field, 332. See also LAGEOS III
Non-Riemannian connection, 136–137. See also mathematical appendix
Nonrotating Bianchi IX cosmological models, 246
Nonsymmetric gravitational theory, 137–138
Nordtvedt, 113
Nordtvedt and Will, 163
Nordtvedt effect, 14, 90
test of, 114–115, 335
Nova Muscae, black hole candidate, 62
Novikov, 243
Null convergence condition, 70
Null strut calculus, 294
Null tetrad formalism, 356
Null vector, 20

Oblateness of Sun, optical, and Sun quadrupole moment, 145
Obliquity of ecliptic, 351
Observer, comoving, 130–132, 235–238
Oersted, 315
ω parameter of Brans-Dicke theory, 109, 111, 146, 219
On the Hypotheses which Lie at the Foundations of Geometry, 3–4, 19. See also Riemann
Open in time, model universes, and cosmology, 220–231, 226, 253–254
Operators, Laplacian, on curved manifold, 289, 292
Oppenheimer-Snyder treatment of gravitational collapse, 63
Optical fiber interferometer, 365–367, 371, 373
Optical gyroscopes, or optical rotation sensors, 349–350, 361, 365–367, 371, 373–374
Optical rotation sensors, 349–350, 361, 365–367, 371, 373–374
Orbital
eccentricity, 322
inclination, 321, 336, 337, 339, 340, 345
parameters, 321–322, 336, 337, 339, 340, 345
period, 322
Orbital angular momentum dragged by gravitomagnetic field, 321. See also Lense-Thirring effect
Orbital plane dragged by gravitomagnetic field, 321. See also Lense-Thirring effect
Orbiting gravitational-wave detectors. See MIGO, LAGOS, LISA, and SAGITTARIUS or LINE
Orientable manifold, 49. See also mathematical appendix
Origin of inertia, 249–255, 252, 269–306, 294–295, 300–306
in general relativity and cosmology, various interpretations of, 249–255
Orthonormal tetrad, local, 130, 237
Ostrogradzky-Green formula (divergence theorem), 55. See also mathematical appendix
Ozsváth and Schücking, 243, 244, 248

Pail, Newton, 328, 385. See also Newton bucket
Palatini variational principle, 25
"Pancake" theory in cosmology, 192
Papapetrou equation, 18, 352
Parallel lines postulate, 1
Parallelly propagated curvature singularities, 69
Parallel transport, 31, 302. See also mathematical appendix

of a vector on a sphere, 32
Parameter β, 119, 140, 146, 150
Parameter γ, 119, 121, 127–128, 138, 146, 350, 351
Parameters characterizing the evolution of Friedmann models, 206–207
Parameters, post-Newtonian, PPN, 167. See also β and γ
Particle creation at horizon of a black hole, 68
Particle horizon, 214–216
Particle horizon distance, 217
Particle, spinning, equation of, 352
Passive and active gravitational mass, tests of equivalence of, 115–116
Passive ring interferometer, 349
Peebles, 193
Pendulum experiments to test weak equivalence principle, 91, 92
Pendulum, Foucault, 328–330, 361, 362, 386
dragging of, 330
Penrose, 48, 70
Penrose process, 67
Penrose singularity theorem, 70. See also Hawking-Penrose singularity theorem
Penzias and Wilson, 190
Periastron advance, 141–147, 322. See also pericenter, perigee and perihelion advance
Pericenter, argument of, 322
Pericenter, or Periastron advance, 141–147
due to gravitomagnetic field, 144, 322, 331
interpretation of, 144
post-Newtonian formula of, 143–144
Perigee, measurement of general relativistic shift of
of LAGEOS, 146, 335
of Moon, 146, 352
Perihelion advance, 144–146
of Mercury due to gravitomagnetic field, 144, 331
Period, orbital, 322
Perturbations, nodal, of Earth satellite
gravitational, 339, 343–344, 346
nongravitational, 339–347
Petrov, type-D, spacetime, 356
Phase shift of counterpropagating electromagnetic waves in a rotating circuit, 365–367, 371, 373. See also light pulses counterpropagating
Philosophiae Naturalis Principia Mathematica, 399. See also Newton
Phobos, 255
"Photinos" and cosmology, 193, 213
Photon. See also electromagnetic waves
equation of motion of photon in a spherically symmetric static metric, 118
post-Newtonian parametrized equation of motion of photon in a spherically symmetric, static metric, 119–120
propagating in Friedmann model universe, 203–206
Pirani, 352
Planck distribution of cosmic background radiation, 191
Plane gravitational wave, 73–75, 150, 152, 155, 159
Planetary frame, 134, 352
POINTS, Precision Optical INTerferometry in Space, 122
Poisson equation, 25
in electromagnetism, 317
Polarization of a plane gravitational wave, 74–75
Polar satellites to measure gravitomagnetic field, 331–332, 336–337. See also Yilmaz
Pole, coordinates of Earth pole, 345
Polyvector, or multivector, 50
dual to a form, 54
"Poor man's prescription" to determine a local inertial frame, 298–300
"Position-coded" memory, gravitational waves with, 76
Positive energy theorem in general relativity, 48–49
Positive pressure condition, 230, 231
Positive spatial curvature, Friedmann cosmological models with, 210, 213, 220, 222–223
Post-Galilean transformations, 131, 165–166
Post-Newtonian. See also PPN
equation of motion of photon in a spherically symmetric static metric, 119–120
metric with gravitomagnetic term and gravitomagnetic parameter μ, 129–130
parametrized, energy-momentum tensor, 167
parametrized (PPN) formalism, 163–168. See also PPN formalism
parametrized, (PPN) metric, 166
parametrized, spherically symmetric, static metric, 119

Post-Newtonian (*cont.*)
parametrized, spherically symmetric, static metric in isotropic coordinates, 122–123, 129
potentials, 165–166
Post-Newtonian terms, new type of, 168
Potential, gravitomagnetic vector, 270, 284–285, 289–291, 330, 396–397. See also gravitomagnetic potential
Pound-Rebka-Snider test of gravitational red-shift, 103
Poynting-Robertson effect, 340
PPN, Post-Newtonian Parametrized formalism, 163–168. See also post-Newtonian
conceptual limits of PPN formalism, 167–168
derivation of, 163–167
energy-momentum tensor of, 167
hypotheses of, 164–165
metric of, 166
metric theories not described by PPN formalism, 167–168
parameters of, 167. See also β and γ
potentials of, 165–166
Pressure, positive pressure condition, 230, 231
Pressure, "regeneration of pressure," 63, 240
Principle, cosmological, 185
"Principle of austerity," Wheeler's view of physics, 49
Prior geometry, 252–253, 300
Projection tensor
time-projection tensor, 234–235
space-projection tensor, 234–235
Propagation of electromagnetic waves in a rotating circuit, 106–108, 366–367, 371, 373–374
Proper distance between test particles and gravitational waves, 74–75, 148–151, 155, 157–163
Proper reference frame, local, 149–150
Proper time
and coordinate time, 100–102
in a static spacetime, 100–102
interval, between two events, 104
Pseudotensor, energy and momentum, of gravitational field, 45–47
Pseudotensor, Levi-Civita, 53, 238, 355
Pugh, 332
Pulsar, millisecond, 1937+21, 103, 128
and test of gravitational redshift
and test of time delay of radio signals
Pulsar. See binary pulsar

Quadrupole
anisotropy of cosmic microwave background radiation, 189–192, 255. See also COBE
formula for emission of gravitational radiation, 76–77
gravitational radiation and binary pulsar, 77, 146–147
Quadrupole moment of Earth, 336
uncertainty of, 336
Quadrupole moment of Sun
and perihelion advance, 144–146
measurements of, 145–146
Quadrupole moment, reduced, 76–77
Quality factor of a resonant detector, 150
Quantities to be fixed on a Cauchy surface in the initial-value problem, 293, 304–305
Quantum cosmology, 283
Quantum electrodynamics, 274
Quantum vacuum fluctuations at horizon of a black hole, 68
Quartz sphere, 333
Quasars, 121, 134–136, 251, 255, 327, 347, 352
Quasars, angular separation measured with VLBI, 121
Quasars, frame of, 134–136, 352
Quasi-Cartesian coordinate systems, 164, 165
Quasilocal definition of energy-momentum and angular momentum, 48
Quasilocal measurement of angular momentum and mass of a body, 360
Quasi-Lorentzian frames, 164–165
Q0957 + 561, double image quasar, 122

Radar delay
active reflection with Mariner 6, 7, and 9, and Mars Vikings, 127–128
passive reflection on Mercury or Venus, 127
Radiation, dipole
absent in general relativity, 147, 391
in electromagnetism, 391
"Radiation-dominated" Friedmann models, 214
Radiation, electromagnetic, generated by accelerated charge, Thomson's explanation of, 388–390, 389

Radiation field and initial-value formulation, 397
Radiation, gravitational, and its mass-energy, 76, 78, 295–296, 297, 398. See also gravitational waves
Radiation pressure, 339–346
Radiation pulses counterpropagating in a rotating circuit and in a gravitomagnetic field, 104–108, 106, 366–374
Radiation, thermal
 from black hole, 68, 69
 from a satellite, 340–343
Radiative component of electromagnetic field, 388–390, 389
Radio sources map, 188
Radio waves, 390
 deflection by a mass, 117–122
Rate of change of conformal three-geometry, 283
Rate of change of nodal longitude, equation for, 321–322, 322, 336, 341
Rate of change of travel time of photons by Sun, 126–128
 derivation in isotropic coordinates, 122–126
 derivation in standard coordinates, 126–127
Raychaudhuri equation, 234, 239–240, 241
"Recollapse topology," 278
Recollapsing model universes (closed in time), 210, 250, 252–254
 and inertia, 250, 252–254
 and spatially compact models, 220–231
 conjectures of sufficient conditions for, 230–231
 necessary spatial topology for, 226–230
 spatial topology of, 185
 sufficient conditions and spatial topology for, 230–231
Redshift, cosmological
 and Gödel model universe of "first type," 242, 244
 function of luminosity distance, 203–206. See also Hubble law
 in expanding Friedmann models, 203–206
 observations of, 186, 203, 207
 parameter of a galaxy, 203
Redshift distribution of galaxies, 186
Redshift, gravitational, 97, 99–108
 astrophysical test of, 103
 derivation in metric theories of gravity, 99–102
 from conservation of energy and equivalence principle, 97, 99
 general formula for, 108
 tests of, 103
Redshift survey of southern sky, 186
Redshift versus distance relation, 186, 203–206. See also Hubble law
Reduced quadrupole moment, 76–77
Reference frame, rotating, 106–108, 365–367, 371, 373, 385
"Regeneration of pressure," 63, 240
Regge calculus, 294
Reissner-Nordstrøm solution, 41
Relativistic "interpretation of Mach principle," 249, 303, 324, 331
Resonant detectors, 147, 148–154
 change in length due to gravitational waves, 148–153
 effective temperature of, 153
 energy of, 151–152
 fundamental mode of, 150, 151
 minimum energy variation detectable with, 153–154
 orbiting Earth, 350
 quality factor of, 150
 resonant detectors in the world, 154
 scheme of, 154
 sensitivity of, 153, 154
 sources of noise of, 153
 transducer, 153, 154
Resonator, fiber, 365–367, 371, 373–374. See also ring laser
Restoring force in geodesic deviation equation with force term, 148
Retarded potentials, 297, 396–397
 and compact space, 396–397
 and Friedmann model universe, 396
 versus initial-value formulation, 297, 396–397
Retroreflectors, 114, 340
Ricci Curbastro, 19
Ricci Curbastro and Levi-Civita, *Methods of Absolute Differential Calculus and Their Applications*, 19
Ricci curvature scalar, 22, 280, 355. See also mathematical appendix
 conformal, 292
 in Friedmann cosmological models, 201, 220
 of homogeneous and isotropic hypersurface, 201, 220

Ricci tensor, 22, 23, 24, 27, 39, 72, 350. See also mathematical appendix
its relation with expansion, shear and vorticity of a fluid, 239. See also Raychaudhuri equation
linearized, 72
Riemann, 1, 3–4, 19, 300, 394
Riemann-Ampère-Stokes formula, 55. See also mathematical appendix
Riemann conception of geometry, 273
Riemann curvature tensor, 3, 15–16, 21, 31, 34–36, 58, 69, 150, 220, 354–355, 358–360. See also curvature and mathematical appendix
and change of a vector parallel transported around an infinitesimal loop, 31
and curvature singularities, 69
determined by geodesic deviation equation, 34–36
dual of, 355
electric components of, 355
of weak plane gravitational wave, 74, 150
invariants built with, 355–357
magnetic components of, 350, 354–355, 358–360
of homogeneous and isotropic hypersurfaces, 200, 220
Riemannian connection, 58. See also non-Riemannian connection, and mathematical appendix
Riemannian manifold, 19, 116–117. See also mathematical appendix
Riemann, *On the Hypotheses Which Lie at the Foundations of Geometry*, 3–4, 19
Rigel, 333
Ring interferometer, 349, 365–367, 371, 373
Ring laser gyroscope (rotation sensor), RLG, 349–350, 361, 365–367, 371, 373–374
Roll, Krotkow and Dicke, Princeton, experiment on weak equivalence principle, 93, 95
Rosen theory, 139, 147
"Rotascope," 362. See also gyroscope
Rotating circuit, 365–367, 371, 373–374
and counterpropagating pulses of radiation, 106–108, 366–367, 371, 373–374
and propagation of light, 106–108
Rotating cosmological models, 240–249, 254–255
and origin of inertia, 244, 248–249, 251, 252, 254–255
Bianchi IX, 244–249
Gödel, "first type" (Bianchi VIII), 240–243
spatially homogeneous and expanding, 244–249
Rotating mass-energy in the universe and gyroscopes, 248–249, 254, 399
Rotating object, metric generated by, 41, 42, 299, 316, 320, 323, 356–358
Rotating reference frame, 106–108, 365–367, 371, 373, 385
Rotating shell, 322–324, 326. See also shell
Rotating sphere of fluid and dragging of inertial frames, 8
Rotation, 234–240, 238, 241–249
axis of Earth, 345
global, and Gödel model universe of "first type," 243
in Bianchi IX cosmological models, 245–249
local and global rotation in Gödel model universe of "first type," 242, 243
magnetic field induced by, 348–349
of Earth, 361–362
of Earth exosphere, 345
of universe and inertia, 251
of universe and measurement of de Sitter effect, 136, 255
of universe, experimental limits on, 251, 254–255, from comparison between Lunar Laser Ranging and Very Long Baseline Interferometry, 255, from isotropy of cosmic microwave background radiation, 251, 254–255, from proposed experiment with a lander on Phobos, 255
of a vector induced by curvature and boundary of the boundary principle, 61
Rotation, or vorticity vector, 238–239, 245, 247. See also rotation and rotation tensor
Rotational orientation of Earth, UT1, 134–136, 345, 352
Rotational symmetry, 242. See also "rotations" and Killing vector
Rotations and isometries of Friedmann-Robertson-Walker metric, 195–199, 223
Rotation sensors, 347–350. See also gyroscopes
optical, 361, 365–367, 371, 373–374
ring laser, 349–350, 361, 365–367, 371, 373–374

superconducting, 332–333, 347–349
"Rotations" on a manifold, 36–37, 195, 196–199, 223
Rotation tensor, 234–240, 235–236, 238
Rotation vector, 238–239, 245, 247. See also rotation and rotation tensor
null in Friedmann models, 202
Rubincam, 342
Rubincam effect, 340–343
Runge-Lenz vector, 322

SAGITTARIUS, 160–162. See also LINE, Laser Interferometer Near Earth
Sagnac, 373
"Sagnac effect," and general relativistic "Sagnac effect," 106–108, 350, 366–367, 371, 373–374. See also Harress
Satellite Laser Ranging, SLR, 335, 345, 372. See also LAGEOS and Starlette
Saturn, 103
Scalar curvature, 22, 201, 202, 280, 355. See also Ricci curvature scalar
Scalar, expansion scalar, 26, 234–240, 238
Scalar field in Brans-Dicke theory, 109–110
Scalar, initial-value, equation, 292, 294, 305
Scalar polynomial curvature singularities, 69
Scale factor
conformal, 288
exponential expansion in inflationary models, 218
in Friedmann cosmological models, 203, 208, 210, 212
Schiff, 332
Schiff conjecture, 163
Schoen and Yau positive energy theorem, 48–49
Schoen and Yau theorem on global topology of a three-manifold, 227–228
Schur theorem, 201
Schwarzschild
geometry, 36–41, 253, 316, 351, 358
geometry and foliations, 277
horizon, 65, 354. See also Schwarzschild radius
metric, 40
metric, maximally extended, 66
radius, 62
Sciama, 274, 388
Sciama sum for inertia, 299, 392–393
Science of Mechanics, 386. See also Mach
Second fundamental form, 24, 226, 227, 276–277, 286, 293, 306. See also extrinsic curvature and mathematical appendix
Sectional curvature, 199–200. See also mathematical appendix
"Selectrons" and cosmology, 193, 213
Self-consistent solutions with "timelike wormholes," 243
Semimajor axis, 322
Shapiro, 127
Shapiro time delay of electromagnetic waves, 122
Shear, 234–240, 235, 238, 327
coefficient of shear viscosity, 26
in Bianchi IX cosmological models, 247
null in Friedmann models, 202
null in Gödel model universe of "first type," 242
tensor, 26, 234, 235, 238
Shell
internal solution for a nonrotating, empty, spherically shell (Minkowski metric), 41
Shell, rotating, 322–324
"acceleration" of a test particle inside a, 323–324, 324–326
"Coriolis and centrifugal forces" inside a, 323–324, 324–326
dragging of gyroscopes inside a, 323
field inside a, 323–324, 324–326
gravitomagnetism inside a, 323–324, 324–326
Shepley, 185, 245
Shift
of frequency of counterrotating electromagnetic waves, 365–367, 371, 373–374
of pericenter, 141–147, due to gravitomagnetic field, 144, 322, 331. See also pericenter advance
Shift function, 294
Simplex, four-simplex, 52
Simultaneity
between nearby events, 105
between two events, condition of, 104
in general relativity, 100–102, 104–108
in a static spacetime, 100, 102
in a stationary spacetime, 101
Simultaneous events in a static spacetime, 100, 102
Singularities, 69–71

Singularities (*cont.*)
absence of cosmological singularities in Gödel model universe of "first type," 243
and initial-value formulation, 397
and quantum gravity, 71
coordinate, 65, 354
cosmological, 212, 230, 231, 240
curvature, 354
Hawking-Penrose theorem on, 71
parallelly propagated curvature singularities on, 69
Penrose theorem, 70
scalar polynomial curvature singularities, 69
true geometrical, 65, 69
Singularity
initial, in cosmology, 212, 240
of Schwarzschild geometry and foliations, 277
Sirius B, 103
Slipher, 203
Slow motion and weak gravitational field, 164. See also post-Newtonian
"Small universe" hypothesis, 219–220
Solar oscillations and Sun quadrupole moment, 145–146
Solar Probe, or STARPROBE, or VULCAN, 145
Solar radiation pressure, 339–343
Solution of Einstein field equation with mass and charge (Reissner-Nordstrøm), 41
Solution of Einstein field equation with mass, charge and angular momentum (Kerr-Newman), 41
Solution of initial-value equations, 289–292
Solution of linearized Einstein field equation and gravitational waves, 73
Source, confinable and unconfinable, 290
Southern sky redshift survey, 186
Southern-Wall, 186
Space
absolute, 249, 250, 384–387, 394
absolute, in Galilei-Newton mechanics, 249, 250
asymptotically flat space and inertia, 249–253, 271–274, 300, 304, 306
compact, 220–231, 270, 273–275, 396–399
Space curvature. See curvature
tests of, 116–140, 351–352
Space interferometers. See Gravitational waves, proposed measurements of
Spacelike vector, 20
Space-projection tensor, 234–235
Spacetime
homogeneous, in space and time, model universes, 240–244, 242
in general relativity and other metric theories, 87–88, 139. See also Lorentzian manifold
Lorentzian manifold, 20
Spacetime curvature, 4. See also curvature
Spacetime geometry, 19–21, 116–117. See also curvature
and mass, 296–297
and other fields, 296
Spacetime location invariance and equivalence principle, 109–111
Spacetime manifold and singularities, 69–70
Spacetime metric and local inertial frames, 13–18, 301, 302, 395
Spatial curvature and closure in time in Friedmann models, 209–214
Spatially compact cosmological models, 220–231, 250, 252–254, 300, 303, 304
and closure in time, 220–231
and inertia, 300, 303, 304, 250, 252–254
spatially noncompact model universe, 303, 304
Spatially homogeneous and isotropic model universes, 193–214. See also Friedmann models
Spatially homogeneous cosmological models, 193–194, 231–234, 240–249
Speed of light, 126, 365, 393, 397
Sphere. See also mathematical appendix
model universes with spatial topology of three-sphere, or two-sphere times a circle, 226–231
n-sphere, 224
one-sphere, 224
quotient a subgroup, 227
three-sphere, 222, 278
three-sphere in Friedmann cosmology, 220, 222
three-sphere with wormholes, or handles, 278
two-sphere, 221
with *n*-handles, 227–229
Spherically symmetric
and static, post-Newtonian parametrized, metric, 119
geometrical quantities, 36
manifold, 36

metric, general expression of, 38
metric, in a particular coordinate system, 39
vacuum solution of Einstein field equation, 40. See also Schwarzschild
Spinning black hole, 62, 65, 204, 327–328
Spinning particle, equation of, 352
Spin tensor of spinning particle, 352. See also angular momentum in relativity
Spin four-vector, intrinsic, of spinning particle, 352. See also angular momentum four-vector
Sponginess, 185
"Square" of conformal distortion tensor, 292, 305
SQUID, 333, 349
Stanford gyroscope, 332–333
Starlette, 97
Star operator
on forms, 59
on polyvectors, 59
STARPROBE, or Solar Probe, or VULCAN, 145
Stars
distant, 134–136, 251, 255, 352. See also quasars
"fixed," 329
Static metric, 40, 100, 102
Static spacetime, 40, 100, 102
and coordinate time, 102
and simultaneity, 102
Stationary metric, 40, 100, 101
Stationary model universe, 241
Stationary spacetime, 40, 100
and coordinate time, 101
and simultaneity, 101
"Steady-state" model universe, 243
Stefan-Boltzmann law, 341
Stefan constant, 341
STENSOR, 356
STEP, Satellite Test of Equivalence Principle, 93
Stokes' theorem, 54–55. See also mathematical appendix
Straight lines postulate, 19
Strings, cosmic, 186, 219
String theories, 10
Strong energy condition, 71, 230, 231
Structure constants, 232. See also mathematical appendix
Sum, connected, of manifolds, 227, 230
Sum for inertia, 5, 299, 329–330
Sciama sum for inertia, 392–393
Sum over all histories, Feynman, 279–280
Sun, 103, 120, 125, 126
eclipses, 110
gravitational redshift of infrared lines, 103
quadrupole moment and perihelion advance, 144–146
quadrupole moment, measurements from solar oscillations, 145–146
quadrupole moment, measurements of, 145–146
Sun-Earth-Moon system, 110, 113, 114
Sunlight, 390. See also solar radiation pressure
Superclusters, 185–186
Superconducting
gradiometer, 350–351
gyroscopes, 332–333, 347–349 GP-B, 332–333
Superconducting Quantum Interference Device, SQUID, 333, 349
Superconductor, 348–349
magnetization by rotation of, 348–349
Superfluid gyroscopes, 347–349
Supermassive, rotating, black hole, 327–328, observations of, in M87, 62, 204. See also jets
Supernovae, 207
Superspace, 283
Supplementary inclination satellites to measure the gravitomagnetic field, 337–339, 338. See also LAGEOS III
Surface, Cauchy. See Cauchy surface
Surface element two-dimensional, 54
Surfaces with different curvature, 3, 221, 225
Survey
deep pencil-beam, 186
southern sky redshift, 186
Symmetries of a tensor field, 36, 37, 193, 194, 231–232
Symmetry, rotational, 242
Synchronization. See also light pulses
along a closed path, 104–108
and gravitomagnetic field, 105, 107
of clocks, 104–108
of clocks on spacelike hypersurfaces of a cosmological model, 245
Synchronous coordinate system, 245
Synge, 35

Taub solution, 231, 297
Taylor, 295

Taylor and Hulse, 146
Telescope, Hubble Space Telescope, 123, 124, 204, 207
Telescopes and Gyroscopes, 134, 135, 332, 333, 361. See also gyroscopes
  a mathematical representation of, 130–133
Temperature of a black hole, 69
Temperature, effective, of a resonant detector, 153
TENKO, gravitational-wave laser interferometer, 156
Tensor, 20. See also mathematical appendix
  decomposition of a symmetric two-covariant tensor, 286
Tensor calculus, 19–21, 403–436 (mathematical appendix)
Test particles, 13–14, 16, 29–30, 88. See also geodesic
  gravitomagnetic field effect on, 321–323
  to determine spacetime curvature, 34–36, 358–360, minimum number of, 34–36
Tests of Einstein geometrodynamics, chapter 3 and chapter 6: 87–184, 315–383
  table of main solar system tests, 89
Tests of weak equivalence principle, 91–97, 92
Tetrad
  components of local orthonormal comoving, 132
  local orthonormal comoving, 130, 237
  orthonormal, in Bianchi IX model universes, 246
Texas Mauritanian Eclipse Team, 120
Texture, 185
Thermal inertia, 342, 343
Thermal thrust, 341
Thirring, 7, 322, 326, 395
Thirring-Lense, or Lense-Thirring effect, 128–134, 133, 134, 299, 316, 321–323, 322
Thomas precession, 133, 348
Thomson, 388
Thorne, 243
Three-axes gradiometer, 350–351
Three-cube and its boundary, 51
Three-geometry, 279, 283
  conformal, 279, 283–284, 285, 287, 289, 293, 304
Three-sphere, topology, and cosmology, 226–231
Thrust, particle, on Earth satellite, 339, 342, 345
Tides, solid and ocean Earth tides, and LAGEOS III experiment, 339, 343–344
Time delay of electromagnetic waves, 122–128
  tests of, 127–128. See also delay in propagation time of electromagnetic waves
Time dilation of clocks, gravitational, 97, 99–108, 138, 139, 140
  around a black hole, 64
  general formula of, 108
  in the solar system, 103
  tests of, 87, 103
Time-independent model universe, 240
Timelike closed curves
  absence of, in Gödel rotating model universes of "second type," 244, 249
  and causality, 243–244
  and Gödel model universe of "first type," 243
Timelike convergence condition, 71, 230
Timelike Killing vector, 40, 241
Timelike vector, 20
"Timelike wormhole," 243
Time, many-fingered, 276
Time-projection tensor, 234–235
"Top-down" cosmological scenario, 192
Topology. See also mathematical appendix
  of space and spacetime, 270
  spatial, of a recollapsing model universe (closed in time), 185, 220–231, 277, 278
  spatial, of three-sphere with wormholes and magnetic field, 274
  spatial, of a model universe, 220–231, 250, 252–254
Torque
  "torque" on a gyroscope, in general relativity, 320–321
  on a magnetic dipole, in electromagnetism, 318
Torsion. See also mathematical appendix
  tensor, 137
  2-forms, 57–58
Torsion balance to test weak equivalence principle, 91–93, 92
Torsion, condition of null torsion, 57–58
Torus
  flat three-torus, 224–226
  n-holed, 227–229
  n-torus, 224
  three-torus, 278

"three-torus, flat, universes," expanding forever, 224–226
two-torus, 221, 224–226, 225, flat, 224–226, with positive, negative and zero curvature, 221, 224–225
Total conformal distortion tensor, 291
Traceless tensor, 286–287
Trace of extrinsic curvature, 276
Transducer of resonant detectors, 153, 154
Transformation from a local tetrad to a local orthonormal tetrad comoving with a particle, 130–132
Transitive, simply, group of isometries, 240–241
Translations
and isometries of Friedmann-Robertson-Walker metric, 195, 196–199, 223. See also Killing vector
simple, in time and space, 240–241
on a manifold, 195, 196–198
Transport, Fermi-Walker, 128, 134, 137, 149, 237, 238, 242, 300, 302, 361. See also Fermi-Walker transport and mathematical appendix
Transverse tensor, 287
Transverse-traceless gauge, 73
TRIAD, 363
True anomaly, 322
Twisted torus, or Klein bottle, 50
Twistors and quasilocal definition of energy, 48
Two-cube and its boundary, 51

UHURU, 62
ULYSSES, 163
Unconfinable source, 290
Uniqueness and existence of initial-value equations, 291, 292
Uniqueness of free-fall, tests of, 91–97. See also weak equivalence principle, tests of
Unit tensor, Cartan's, 59
Universe, 185–193, 193–255
"age" of, in Friedmann cosmological models, 212
"small universe" hypothesis, 219–220
Unruh, 274
UT1, rotational orientation of Earth, 134–136, 345, 352

"Vacuum, false" in inflationary models, 218
Vacuum solutions of Einstein field equation, 40–42
Vacuum, spherically symmetric solutions of Einstein field equation, 40–42
Van Patten and Everitt, 331–332, 336. See also Yilmaz
Variation
first, of integral of squared spacetime interval, 29–30. See also geodesic equation
second, of integral of squared spacetime interval, 32. See also geodesic deviation equation
Variational principle
for field equation in geometrodynamics, 21–25
for field equation, quantities to be fixed at the boundary, 23–24, 279–282
in mechanics, 281–282
in quantum gravity, 281
in quantum mechanics, 279–281
Palatini method, 25
Variation of metric tensor, 22, 26–27
Variation of Ricci tensor, 23, 27
Variations in time and space of gravitational constant, limits on, 110–111, 335. See also deviations from the inverse square law
Variations in time and space of nongravitational constants, hypotheses on, 109
Vector initial-value equation, 289, 294, 305
"Velocity-coded" memory, gravitational waves with, 76
Venus, 110, 127
Velocity versus distance relation, 186, 203
Very Long Baseline Array, VLBA, 121
Very Long Baseline Interferometry, VLBI, 7, 121–122, 134–136, 212, 255, 345, 346, 348, 352
and deflection of electromagnetic waves, 121–122
Very strong equivalence principle, 13–18
and gravitational constant, 109–111
tests of, 112–115
Vessel, Newton's, 328, 385
critique by Mach, 386–387
Einstein ideas on, 269–270
Vessot and Levine test of gravitational time dilation, 103, 140
V 404 Cygni, black hole candidate, 62

Vierbein, local orthonormal, 130, 237. See also tetrad
Vikings, 110, 127
VIRGO, gravitational-wave laser interferometer, 156
Virgo cluster and spiral galaxy M100, 207
Viscosity, 327
  coefficient of bulk viscosity, 26
  coefficient of shear viscosity, 26
VLBI frame, 135, 136, 352. See also Very Long Baseline Interferometry
Voids, cosmological, 185–186
Volume element, four-dimensional, 54
Volume expansion, 234–240, 238. See also expansion
Vorticity, or rotation vector, 238–239, 245, 247
Vorticity, or rotation tensor, 234–240, 235–236, 238
Voyager 1, 103
Voyager 2, 128
VULCAN, or Solar Probe, or STARPROBE, 145

Wall, Great, 186
Walls, cosmological, 185, 186
Wall, Southern-Wall, 186
Wave equation
  for metric perturbations, 73
  for weak-field gravitational waves, 73, 317
Wave solution, weak field, of field equation, 72
Weak energy condition, 49, 71, 228
Weak equivalence principle, 13–14
  tests of, 90, 91–97
Weak gravitational field and slow motion, 164. See also post-Newtonian
Weak Interacting Massive Particles, WIMPs, 213
Weber, 154
Wedge product, or exterior product, 51–52. See also mathematical appendix
Weinberg, 328
Weyl tensor, 34–35, 356. See also mathematical appendix
Wheeler, 49, 245, 248
  and Einstein ideas on spatially compact universe and inertia, 224, 231, 244, 253
  and geons, 77
  and the name "black hole," 62
Wheeler's view of physics, 49. See also "Law without law"
Whittaker, 316
Will and Nordtvedt, 163
WIMPs, Weak Interacting Massive Particles, 213
Wormholes, 66
  and magnetic field, 274
  timelike, and causality, 243

X-ray cosmological background, 186
X-rays, 390

Yarkovsky effect, 340–343
Yilmaz space proposal to detect the gravitomagnetic field, 331, 336
York, 277–292
York time, 276–277, 281, 283, 285, 292, 293, 304
Yukawa-type gravitational term, hypotheses of, and limits on, 93–97, 98

Zel'dovich, 192
Zonal, even, harmonic coefficients, 336
z redshift parameter of a galaxy, 203

Fundamental and Astronomical Constants and Units

Speed of light	$c = 2.9979250(10) \times 10^{10}\text{cm/s} = 1$
Gravitational constant	$G = 6.670(4) \times 10^{-8}\text{cm}^3\text{s}^{-2}/\text{g}^{-1} = 1$
Combinations of G and c	$G/c^2 = 7.421 \times 10^{-29}\text{cm/g} = 1$
	$c^5/G = 3.631 \times 10^{59}\ \text{erg/s} = 1$
	$G/c = 2.225 \times 10^{-18}\ \text{cm}^2\ \text{Hz/g} = 1$
	$c^2/G^{\frac{1}{2}} = 3.480 \times 10^{24}\ \text{gauss cm} = 1$
Planck constant	$2\pi\hbar = h = 6.62620(5) \times 10^{-27}\text{erg s}$
	$\hbar = 1.054592(8) \times 10^{-27}\text{erg s}$
Planck distance	$\left(\frac{\hbar G}{c^3}\right)^{1/2} = 1.616 \times 10^{-33}\text{cm}$
Planck time	$\left(\frac{\hbar G}{c^5}\right)^{1/2} = 5.391 \times 10^{-44}\text{s}$
Planck mass	$\left(\frac{\hbar c}{G}\right)^{1/2} = 2.177 \times 10^{-5}\text{g}$
Planck density	$\frac{c^5}{\hbar G^2} = 5.157 \times 10^{93}\ \text{g/cm}^3$
Electron charge	$e = 4.80325(2) \times 10^{-10}\left(\text{g cm}^3/\text{s}^2\right)^{\frac{1}{2}}$
	$= 1.381 \times 10^{-34}\text{cm}$
Fine-structure constant	$\alpha = \frac{2\pi e^2}{hc} = 7.297351(11) \times 10^{-3} = \frac{1}{137.0360(2)}$
Mass of electron	$m_e = 9.1093897(54) \times 10^{-28}\text{g}$
	$= 8.1873 \times 10^{-7}\text{erg} = 0.51099906(15)\text{MeV}/c^2$
	$m_e c^2 = 6.764 \times 10^{-56}\text{cm}$
Proton mass	$m_p = 1.6726231(10) \times 10^{-24}\text{g}$
	$= 1.50327 \times 10^{-3}\text{erg} = 938.27231(28)\text{MeV}/c^2$
	$m_p c^2 = 1.2419 \times 10^{-52}\text{cm}$
Bohr radius	$a_0 = \frac{\hbar^2}{m_e e^2} = 0.5291775(8) \times 10^{-8}\text{cm}$
Compton wavelength	$\frac{h}{m_e c} = 2.426310 \times 10^{-10}\text{cm}$
	$\frac{\hbar}{m_e c} = 3.861592 \times 10^{-11}\text{cm}$

Classical electron radius	$r_0 = \frac{e^2}{m_e c^2} = 2.81794 \times 10^{-13}\text{cm}$
Atomic mass unit	$931.49432(28)\text{MeV}/c^2$ $= 1.6605402(10) \times 10^{-27}\text{kg}$
Atomic unit of energy	$\frac{e^2}{a_0} = 4.35983 \times 10^{-11}\text{erg}$ $= 27.21165\ \text{eV}$ $= 3.602 \times 10^{-60}\text{cm}$
Atomic unit of angular momentum	$\hbar = \frac{h}{2\pi} = 1.054592(8) \times 10^{-27}\text{g cm}^2/\text{s}$
Boltzmann constant	$k = 1.38062(6) \times 10^{-16}\text{erg K}^{-1}$ $= 8.6171 \times 10^5\text{eV K}^{-1}$
Radiation density constant	$a = \frac{8\pi^5 k^4}{15 c^3 h^3}$ $= 7.56464 \times 10^{-15}\text{erg cm}^{-3}\ \text{K}^{-4}$ (blackbody radiation energy density $= a\,T^4$)
Stefan-Boltzmann constant	$\sigma = \frac{ac}{4} = 5.66956 \times 10^{-5}\text{erg cm}^{-2}\ \text{K}^{-4}\ \text{s}^{-1}$ (blackbody radiation emittance $= \sigma\,T^4$)
Energy, mass, temperature	$1\ \text{eV} = 1.602192 \times 10^{-12}\ \text{ergs}$ $= 1.16048 \times 10^4\ \text{K}$ $= 1.78268 \times 10^{-33}\ \text{g} = 1.324 \times 10^{-61}\text{cm}$
Atomic second	$s_A = 9192631770$ caesium cycles
Light year	$1\ \text{lt-yr} = 9.460530 \times 10^{17}\text{cm}$
Parsec	$1\text{pc} = 3.085678 \times 10^{18}\text{cm}$ $= 3.261633$ lt-yr $= 206264.806$ AU
Astronomical Unit of distance = mean Sun-Earth distance = semimajor axis of Earth orbit	$1\ \text{AU} = 1.495979(1) \times 10^{13}\text{cm}$
Tropical year (1900.0)	$= 31556925.9747\text{s}_\text{E}$ (ephemeris seconds)
Period of rotation of Earth (referred to fixed stars)	$= (86164.09892 + 0.0015T)\ \text{s}_\text{E}$ (T is epoch from 1900.0 in centuries)
Ephemeris day	$d_\text{E} = 86400\ \text{s}_\text{E}$
Tropical year (equinox to equinox)	$= (365.24219878 - 0.00000616\ T)\text{d}_\text{E}$ $= (31556925.9747 - 0.530\ T)\text{s}_\text{E}$

Sidereal year (fixed stars)	$= (365.25636556 + 0.00000011\ T)\mathrm{d_E}$ $= (31558149.984 + 0.010\ T)\mathrm{s_E}$
Moon synodical month (new moon to new moon)	$= (29.5305882 - 0.0000002\ T)$days
Moon sidereal month (fixed stars)	$= (27.3216610 - 0.0000002T)$days
Period of Moon's node, nutation period	= 18.61 tropical yr
Period of Earth satellite, LAGEOS	= 3.758 hr
Period of LAGEOS node	= 1046 days
Sun mass	$M_\odot = 1.989(1) \times 10^{33}$ g $\frac{GM_\odot}{c^2} = 1.47664(2) \times 10^5$ cm
Sun radius	$R_\odot = 6.9599(7) \times 10^{10}$ cm $\frac{M_\odot}{R_\odot} = 2.122 \times 10^{-6}$
Sun radiation	$\mathcal{L}_\odot = 3.826(8) \times 10^{33}$ erg/s
Sun sidereal rotation in sunspot zone	$14^\circ.44 - 3^\circ.0\ \sin^2\phi$ per day (varies with latitude ϕ)
Sun moment of inertia	5.7×10^{53}g cm^2
Sun angular momentum, *estimated value based on surface rotation*	1.63×10^{48}g cm^2 s^{-1}
Sun oblateness: semidiameter difference, equator-pole	$= 0''.05$
Sun quadrupole moment coefficient	$J_{2\odot} = (I_{\odot R} - I_{\odot \mathrm{E}})/M_\odot R^2_{\odot \mathrm{E}}$ ($I_{\odot R}$ and $\mathrm{I}_{\odot \mathrm{E}}$ are Sun moments of inertia about rotation and equatorial axis). Various measurements of $J_{2\odot}$ range from about 5.5×10^{-6} to about 1.7×10^{-7}
Earth mass	$M_\oplus = 5.976(4) \times 10^{27}$ g $\frac{GM_\oplus}{c^2} = 0.4438$ cm
Earth equatorial radius	$R_{\oplus \mathrm{E}} = 6378.164(3) \times 10^5$ cm $\frac{M_\oplus}{R_\oplus} = 6.958 \times 10^{-10}$

Earth angular velocity (1900)	$7.29211515 \times 10^{-5}$ rad s^{-1}
Earth moment of inertia about rotation axis	$I_{\oplus R} = 0.3306\, M_\oplus\, R^2_{\oplus \mathrm{E}} = 8.04 \times 10^{44}$ g cm^2
Earth angular momentum	5.861×10^{40} cm^{2}g s^{-1} $= 145$ cm^2
Earth quadrupole moment coefficient	$J_{2\oplus} \equiv J_2 = (I_{\oplus R} - I_{\oplus \mathrm{E}})/M_\oplus R^2_{\oplus \mathrm{E}}$ $= 1082.64 \times 10^{-6}$ (see § 6.7)
Other Earth even and odd zonal harmonics coefficients	$J_4 = -1.58 \times 10^{-6}$, $J_3 = -2.54 \times 10^{-6}$ $J_6 = +0.59 \times 10^{-6}$, $J_5 = -0.22 \times 10^{-6}$ $J_8 = -0.2 \times 10^{-6}$, $J_7 = -0.40 \times 10^{-6}$ $J_{10} = -0.4 \times 10^{-6}$, $J_9 = +0.05 \times 10^{-6}$
Obliquity of ecliptic (instantaneous ecliptic)	$\epsilon = 23^\circ\, 27'\, 8''.26 - 46''.84\, T$ (T is in centuries from 1990)
Moon mass	$M_{☽} = 7.350 \times 10^{25}$ g $\frac{GM_{☽}}{c^2} = 5.457 \times 10^{-3}$ cm
Mean Moon radius	$\overline{R}_{☽} = 1738.2$ km
Moon mean distance from Earth	= 384,401 km (range from 356,400 to 406,700 km)
Inclination of Moon orbit to ecliptic	$= 5^\circ\, 8'\, 43''$ oscillating of $\pm\, 9'$ with period of 173 days
Milky Way Galaxy diameter:	$\sim 10^5$ lt-yr
Milky Way thickness	$\sim 6 \times 10^3$ lt-yr
Milky Way Galaxy mass	$1.4 \times 10^{11}\, M_\odot$
Hubble constant	$H_\circ = 100 h_\circ \frac{\mathrm{km}}{\mathrm{s\,Mpc}} \cong \left[\frac{3\times 10^{17}\mathrm{s}}{h_\circ}\right]^{-1}$ $\cong \left[\frac{10^{10}\ \mathrm{yr}}{h_\circ}\right]^{-1}$, where $h_\circ = \{0.4,\ 1\}$, (see §4.2)
Cosmological constant, Λ	$\vert\Lambda\vert < 3 \times 10^{-52}$ m^{-2}

Deceleration parameter		$q_\circ = 1.0 \pm 0.8$
Cosmological critical density	$\varepsilon_c \equiv \frac{3H_\circ^2}{8\pi G}$	$= 1.88 \times 10^{-29}\ h_\circ^2$ g cm^{-3}, (see §4.2)
Cosmological luminous matter density		$< \varepsilon_L > \sim\ 2 \times 10^{-31}$ g cm^{-3}
		$= 1 \times 10^{-7}$ atoms cm^{-3}
		$= 3 \times 10^9\ M_\odot/\text{Mpc}^3$

Planets

		Orbital Semimajor Axis (km)	Sidereal Period (Tropical years: yr)	Eccentricity	Inclination to Ecliptic	Equatorial Radius (km)	Mass ($M_\oplus$)	Sidereal Rotation Period Equatorial
Mercury	☿	57.9×10^6	0.24085	0.205628	7°0′15″	2425	0.0554	59 days
Venus	♀	108.2×10^6	0.61521	0.006787	3°23′40″	6070	0.815	244.3 days
Earth	⊕	149.6×10^6	1.00004	0.016722	. . .	6378	1.000	$23^h56^m4^s.1$
							$(M_\oplus + M_☽ = 1.0123\ M_\oplus)$	
Mars	♂	227.9×10^6	1.88089	0.093377	1°51′0″	3395	0.1075	$24^h37^m22^s.6$
Jupiter	♃	778.3×10^6	11.86223	0.04845	1°18′17″	71,300	317.83	$9^h50^m30^s$
Saturn	♄	1427.0×10^6	29.4577	0.05565	2°29′22″	60,100	95.147	10^h14^m
Uranus	♅	2869.6×10^6	84.0139	0.04724	0°46′23″	24,500	14.54	10^h49^m
Neptune	♆	4496.6×10^6	164.793	0.00858	1°46′22″	25,100	17.23	15^h48^m
Pluto	♇	5900×10^6	247.7	0.250	17°10′	3200	0.17	6 days 9^h